VOLUME FIVE HUNDRED AND NINETY ONE

METHODS IN ENZYMOLOGY

DNA Repair Enzymes: Cell, Molecular, and Chemical Biology

METHODS IN ENZYMOLOGY

Editors-in-Chief

ANNA MARIE PYLE
Departments of Molecular, Cellular and Developmental Biology and Department of Chemistry
Investigator, Howard Hughes Medical Institute
Yale University

DAVID W. CHRISTIANSON
Roy and Diana Vagelos Laboratories
Department of Chemistry
University of Pennsylvania
Philadelphia, PA

Founding Editors

SIDNEY P. COLOWICK and NATHAN O. KAPLAN

VOLUME FIVE HUNDRED AND NINETY ONE

METHODS IN ENZYMOLOGY

DNA Repair Enzymes: Cell, Molecular, and Chemical Biology

Edited by

BRANDT F. EICHMAN
Vanderbilt University
Nashville, TN, United States

ELSEVIER

ACADEMIC PRESS
An imprint of Elsevier

Academic Press is an imprint of Elsevier
50 Hampshire Street, 5th Floor, Cambridge, MA 02139, United States
525 B Street, Suite 1800, San Diego, CA 92101–4495, United States
The Boulevard, Langford Lane, Kidlington, Oxford OX5 1GB, United Kingdom
125 London Wall, London, EC2Y 5AS, United Kingdom

First edition 2017

Notices
Knowledge and best practice in this field are constantly changing. As new research and experience broaden our understanding, changes in research methods, professional practices, or medical treatment may become necessary.

Practitioners and researchers must always rely on their own experience and knowledge in evaluating and using any information, methods, compounds, or experiments described herein. In using such information or methods they should be mindful of their own safety and the safety of others, including parties for whom they have a professional responsibility.

To the fullest extent of the law, neither the Publisher nor the authors, contributors, or editors, assume any liability for any injury and/or damage to persons or property as a matter of products liability, negligence or otherwise, or from any use or operation of any methods, products, instructions, or ideas contained in the material herein.

ISBN: 978-0-12-811846-7
ISSN: 0076-6879

Publisher: Zoe Kruze
Acquisition Editor: Zoe Kruze
Editorial Project Manager: Helene Kabes
Production Project Manager: Magesh Kumar Mahalingam
Cover Designer: Greg Harris

Typeset by SPi Global, India

CONTENTS

CONTRIBUTORS

Jennifer Allen
University of California—Santa Cruz, Santa Cruz, CA, United States

Phillip L. Bartels
California Institute of Technology, Pasadena, CA, United States

Jacqueline K. Barton
California Institute of Technology, Pasadena, CA, United States

Stefania Bertora
DNA Metabolism Laboratory, IFOM—The FIRC Institute of Molecular Oncology, Milan, Italy

Cynthia J. Burrows
University of Utah, Salt Lake City, UT, United States

Manel Camps
University of California—Santa Cruz, Santa Cruz, CA, United States

Denisse Carvajal-Maldonado
Saint Louis University School of Medicine, St. Louis, MO, United States

Alejandra Castaño
Stony Brook University, Stony Brook, NY, United States

Layla Cervantes
University of California—Santa Cruz, Santa Cruz, CA, United States

David Cortez
Vanderbilt University School of Medicine, Nashville, TN, United States

Vincenzo Costanzo
DNA Metabolism Laboratory, IFOM—The FIRC Institute of Molecular Oncology, Milan, Italy

James M. Daley
Yale University School of Medicine, New Haven, CT, United States

Anthony J. Davis
University of Texas Southwestern Medical Center, Dallas, TX, United States

Yingxin Deng
California Institute of Technology, Pasadena, CA, United States

Yun Ding
University of Utah, Salt Lake City, UT, United States

Aidan J. Doherty
Genome Damage and Stability Centre, School of Life Sciences, University of Sussex, Brighton, United Kingdom

Julie A. Dragon
The Markey Center for Molecular Genetics, University of Vermont, Burlington, VT, United States

Vitaly Epshtein
New York University School of Medicine, New York, NY, United States

Aaron M. Fleming
University of Utah, Salt Lake City, UT, United States

Thomas G.W. Graham
Harvard Medical School, Boston, MA, United States

Thomas A. Guilliam
Genome Damage and Stability Centre, School of Life Sciences, University of Sussex, Brighton, United Kingdom

Jia Guo
Biodesign Institute & School of Molecular Sciences, Arizona State University, Tempe, AZ, United States

Venu Kamarthapu
New York University School of Medicine; Howard Hughes Medical Institute, New York University School of Medicine, New York, NY, United States

Youngho Kwon
Yale University School of Medicine, New Haven, CT, United States

Delphine Lemacon
Saint Louis University School of Medicine, St. Louis, MO, United States

Joshua Lilly
University of California—Santa Cruz, Santa Cruz, CA, United States

Joseph J. Loparo
Harvard Medical School, Boston, MA, United States

Carolyn G. Marsden
The Markey Center for Molecular Genetics, University of Vermont, Burlington, VT, United States

Joao Matos
Institute of Biochemistry, ETH Zürich, Zürich, Switzerland

Manas Mondal
Biodesign Institute & School of Molecular Sciences, Arizona State University, Tempe, AZ, United States

Evgeny Nudler
New York University School of Medicine; Howard Hughes Medical Institute, New York University School of Medicine, New York, NY, United States

Elizabeth O'Brien
California Institute of Technology, Pasadena, CA, United States

Federica Pezzimenti
DNA Metabolism Laboratory, IFOM—The FIRC Institute of Molecular Oncology, Milan, Italy

Annabel Quinet
Saint Louis University School of Medicine, St. Louis, MO, United States

Upasana Roy
Stony Brook University, Stony Brook, NY, United States

Janapriya Saha
University of Texas Southwestern Medical Center, Dallas, TX, United States

Vincenzo Sannino
DNA Metabolism Laboratory, IFOM—The FIRC Institute of Molecular Oncology, Milan, Italy

Orlando D. Schärer
Stony Brook University, Stony Brook, NY, United States; Center for Genomic Integrity, Institute for Basic Science; Ulsan National Institute of Science and Technology, Ulsan, Korea

Melissa Standley
University of California—Santa Cruz, Santa Cruz, CA, United States

Patrick Sung
Yale University School of Medicine, New Haven, CT, United States

Joann B. Sweasy
The Markey Center for Molecular Genetics, University of Vermont, Burlington, VT; Yale University School of Medicine, New Haven, CT, United States

Alan E. Tomkinson
University of New Mexico School of Medicine, Albuquerque, NM, United States

Alessandro Vindigni
Saint Louis University School of Medicine, St. Louis, MO, United States

Susan S. Wallace
The Markey Center for Molecular Genetics, University of Vermont, Burlington, VT, United States

Johannes C. Walter
Harvard Medical School; Howard Hughes Medical Institute, Boston, MA, United States

Shih-Ya Wang
University of Texas Southwestern Medical Center, Dallas, TX, United States

Stephen C. West
The Francis Crick Institute, London, United Kingdom

Nathaniel E. Wiest
University of New Mexico School of Medicine, Albuquerque, NM, United States

PREFACE

Studies on the effects of radiation on biological organisms in the 1930s and subsequent work on mutagenesis and the discovery of DNA as the genetic material in the 1940s spawned an entire field of science focused on DNA repair. Since that time, the field has discovered a myriad of sources, types, and biological consequences of chemical transformation to DNA (i.e., DNA damage). This led to elucidation of a number of distinct DNA repair pathways that preserve the integrity of the genetic information, some of which function to restore aberrant DNA to an undamaged state, while others help the cell tolerate, or function in spite of, DNA damage. Genetic defects in these repair pathways are often linked to chromosome instability and predisposition of individuals to cancer and other diseases. Importantly, our understanding of the fundamental science behind DNA repair and the mechanisms of the enzymes involved has led to a number of therapeutic strategies to treat these diseases.

In these two volumes of *Methods in Enzymology*, leading investigators in the field of DNA repair present some of the most important, cutting-edge techniques used to probe DNA repair mechanisms across multiple scales—from cells to atoms. Each chapter focuses on a specific type of DNA damage or repair pathway, including base excision repair (BER), nucleotide excision repair (NER), mismatch repair (MMR), double-strand break repair (DSBR), replication-coupled DNA damage response (DDR), homologous recombination (HR), and DNA synthesis by specialized polymerases. Many of the methods described here are not specific to a particular enzyme or pathway, and thus the chapters are organized by a technique rather than by a specific type of DNA repair. Some authors provide detailed protocols and considerations when carrying out a particular experiment, while others present their philosophies for integrating multiple fields of investigation. The first volume (591) focuses largely on cellular, molecular, and chemical biology methods to investigate DNA damage and repair functions in a genomic or cellular context. The second volume (592) is focused on structural, single-molecule, and kinetic methods aimed at elucidating detailed enzymatic and mechanistic information.

Volume 591 ("DNA Repair Enzymes: Cell, Molecular, and Chemical Biology") begins at the DNA replication fork. The first two chapters (Wiest and Tomkinson; Cortez) describe benefits, limitations, and improvements to the use of iPOND (isolation of proteins on nascent DNA) to identify DDR activities at sites of DNA synthesis. This is followed by the uses of DNA fiber analysis to monitor replication fork progression and stalling in the presence of DNA damage, as presented by Quinet et al. In the fourth chapter, Mondal and Guo describe a powerful method to detect strand-specific repair of low levels of DNA damage in cells, followed by a chapter by Saha and colleagues detailing how to monitor DSBR at various stages of the cell cycle. In the sixth chapter, Marsden and coworkers outline an integrated computational and cell biology pipeline to identify and characterize functionally significant mutations in BER genes, followed by a chapter from Standley et al. describing methods to detect bacterial mutations. In the eighth chapter, Fleming and colleagues describe innovative sequencing methods to detect 8-oxoguanine, a principal product of DNA oxidation in the genome. Chapters 9 and 10 provide protocols for the use of cell-free extracts of *Xenopus* oocytes; the first by Sannino et al. describes general aspects of replication-coupled DDR, and the second by Graham et al. outlines ensemble and single-molecule approaches to study DSBR mechanisms. In Chapter 11, Matos and West provide protocols to measure enzymatic activity of structure-selective endonucleases from yeast and human extracts. In vitro and cellular methods to monitor transcription-coupled NER are outlined in Chapter 12 by Epshtein et al., followed in Chapter 13 by an in vitro system by Kwon and colleagues to reconstitute repair DNA synthesis activities at D-loops during HR. In Chapter 14, Guilliam and Doherty provide a historical and practical description of measuring DNA primase activities in vitro. In Chapter 15, Barton and colleagues provide a comprehensive review of their innovative DNA electrochemistry platform that takes advantage of DNA change transport properties to probe binding and catalysis of redox-active proteins, including iron–sulfur cluster-containing DNA repair systems. Volume 591 concludes with useful instructions by Castaño and colleagues for preparation of stable DNA interstrand cross-links that can be used in vitro and in cells to study all aspects of DNA metabolism.

Volume 592 ("DNA Repair Enzymes: Structure, Biophysics, and Mechanism") includes a series of papers on structural techniques, including X-ray crystallography, small-angle X-ray scattering (SAXS), nuclear magnetic resonance (NMR), and electron microscopy, as well as single-molecule and enzyme kinetic approaches to probe mechanistic details of enzymes involved in all types of excision and break repair. The first two chapters (Gradia et al. and Rees et al.) provide state-of-the-art methods for efficient cloning to produce multisubunit complexes in large quantities for structural and biochemical studies. Thompson and colleagues in the third chapter describe SAXS and

NMR methodologies to study conformational states and dynamic properties of multidomain proteins, using DNA primase and replication protein A as examples. In the fourth chapter, Friedhoff and coworkers share their cross-linking approach to trap transient conformational states of proteins involved in MMR. Chapters 5–7 focus on several contemporary challenges to X-ray crystallographic studies. Malaby et al. recount how they overcame challenges in expression and purification of full-length and active deletion constructs of human DNA pol θ, a specialized DNA polymerase involved in DSBR alternative end-joining. Next, Figiel and Nowotny use RNase H2 to summarize crystallographic approaches to protein–nucleic acid complexes, followed by strategies from Chirgadze and colleagues used to push the resolution of the crystal structure of the 4128-residue DNA-dependent protein kinase catalytic subunit important for regulation of nonhomologous end-joining DSBR. The next four chapters (Sawicka et al., LeBlanc et al., Kong et al., Soniat et al.) focus on the examination of physical behaviors of DNA repair machines using single-molecule methods, including electron, atomic force, and fluorescence microscopies coupled to innovative DNA visualization techniques such as DNA tightropes (Kong et al.) and DNA curtains (Soniat et al.). In the 12th chapter, Samara and coworkers illustrate their powerful time-resolved X-ray crystallographic methods that have pushed the boundaries of our understanding of catalysis of nucleic acid synthesis and degradation. Chapters 13–15 beautifully describe different methods and uses of enzyme kinetics—Powers and Washington include a variety of approaches to monitor catalysis and binding activities of translesion DNA polymerases; Coey and Drohat detail the design, execution, and interpretation of single- and multiple-turnover kinetics experiments using DNA glycosylases as an example; Hendershot and O'Brien follow with kinetic strategies to characterize DNA-binding and nucleotide-flipping mechanisms used by DNA glycosylases and many other DNA processing enzymes. The two-volume series concludes with a discussion from Brosey and colleagues, who tie together elements of both volumes by discussing examples of studies that integrate information from structural biochemistry and cell biology to develop a comprehensive understanding of these multifaceted DNA damage responses.

This collection is the culmination of 80 years of ingenuity and discovery in nucleic acid biology and is timely addition to the field, following the 2015 Nobel Prize in Chemistry to Modrich, Lindahl, and Sancar for their mechanistic studies of DNA repair. My hope is that these chapters will help inspire new innovations at the frontier of DNA repair research, while also serving as a guide to scientists engaged in all aspects of molecular biology. It is an exciting time as we address the next challenges focused on how mechanisms, pathways, and regulation of DNA repair intersect with those of human disease, mutagenesis, and evolution. As in other volumes of *Methods in Enzymology*, the approaches described here are applicable across disciplines and therefore have the potential to cross-pollinate and inspire new ideas in other areas of investigation.

BRANDT F. EICHMAN
Vanderbilt University

CHAPTER ONE

Optimization of Native and Formaldehyde iPOND Techniques for Use in Suspension Cells

Nathaniel E. Wiest, Alan E. Tomkinson[1]
University of New Mexico School of Medicine, Albuquerque, NM, United States
[1]Corresponding author: e-mail address: atomkinson@salud.unm.edu

Contents

Abstract

The isolation of proteins on nascent DNA (iPOND) technique developed by the Cortez laboratory allows a previously unparalleled ability to examine proteins associated with replicating and newly synthesized DNA in mammalian cells. Both the original, formaldehyde-based iPOND technique and a more recent derivative, accelerated native iPOND (aniPOND), have mostly been performed in adherent cell lines. Here, we describe modifications to both protocols for use with suspension cell lines. These include cell culture, pulse, and chase conditions that optimize sample recovery in both protocols using suspension cells and several key improvements to the published aniPOND technique that reduce sample loss, increase signal to noise, and maximize sample recovery.

Methods in Enzymology, Volume 591
ISSN 0076-6879
http://dx.doi.org/10.1016/bs.mie.2017.03.001

Additionally, we directly and quantitatively compare the iPOND and aniPOND protocols to test the strengths and limitations of both. Finally, we present a detailed protocol to perform the optimized aniPOND protocol in suspension cell lines.

1. INTRODUCTION

Techniques to examine the dynamics of protein association and dissociation at replication forks and with newly synthesized DNA in mammalian cells have until recently lagged behind their counterparts in lower eukaryotes. In *Saccharomyces* sp., for example, replication origins are generally predictable due to their dependence on defined and validated sequence elements (Hyrien, 2015; Nieduszynski, 2006), thereby allowing for the proteins associated with replicating DNA to be monitored by employing chromatin immunoprecipitation to capture proteins that are bound adjacent to recently fired origins (Kanemaki & Labib, 2006; Trujillo & Osley, 2012). In contrast, metazoan replication origins are less predictable and sequence independent (Hyrien, 2015), preventing the same approach from being applied to study replication of the mammalian genome and epigenome. While techniques such as BrdU coimmunofluorescence and singe-molecule fluorescence resonance energy transfer imaging can provide valuable information about protein occupancy and relative positioning on replicating DNA (Duderstadt, Reyes-Lamothe, van Oijen, & Sherratt, 2014), techniques in mammalian cells to spatiotemporally monitor the dynamics of protein association and dissociation at replication forks and with newly replicated DNA as the fork moves away have been lacking. The development of the isolation of proteins on nascent DNA (iPOND) technique, first described in 2011 by the Cortez laboratory (Sirbu, Couch, & Cortez, 2012; Sirbu et al., 2011), has led to novel insights into the repertoire of proteins present at active and stalled replication forks and the temporal links between replicative DNA synthesis, nucleosome assembly, and chromatin maturation (Fig. 1A).

The iPOND technique is based upon the altered chemical properties of 5-ethynyl-2′-deoxyuridine (EdU), a thymidine analog containing a reactive alkyne group that is readily incorporated into living cells both in vitro and in vivo (Chehrehasa, Meedeniya, Dwyer, Abrahamsen, & Mackay-Sim, 2009). Addition of EdU to the media results in the incorporation of EdU into the newly synthesized DNA in place of thymidine. The extent of EdU incorporation is determined by the length of incubation in the EdU-containing

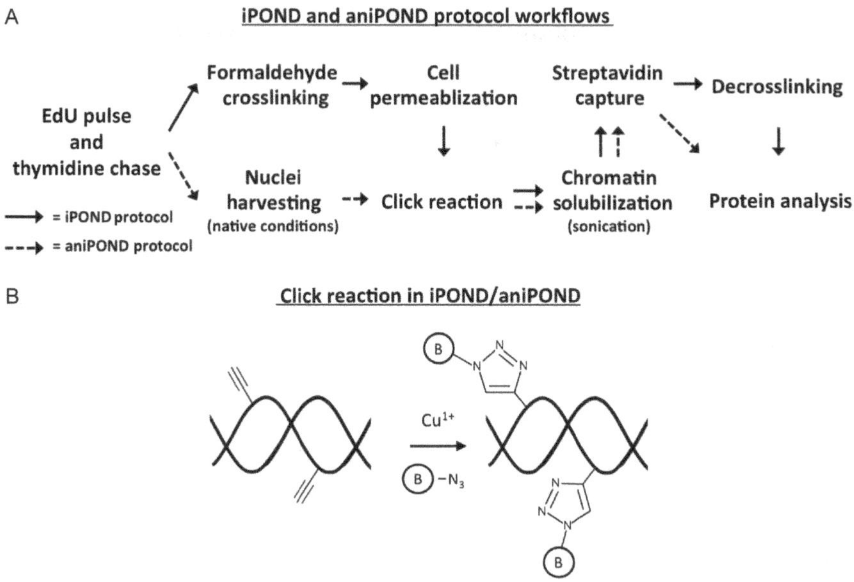

Fig. 1 The iPOND and aniPOND protocol workflows and click reaction. (A) The major steps for both iPOND and aniPOND are illustrated. (B) An illustration of the iPOND/aniPOND click reaction, in which the alkyne moieties from the incorporated EdU are covalently linked to the azide moieties of biotin-azide in the presence of reduced copper.

media. Furthermore, incorporation can be effectively terminated by the replacement of the EdU-containing media with thymidine-containing media. By incubating with thymidine for different times following the EdU pulse, chromatin can be isolated at different stages postreplicative synthesis to monitor events such as histone deposition and chromatin maturation. After crosslinking with formaldehyde, EdU-containing genomic DNA is covalently conjugated to biotin via a copper-catalyzed azide–alkyne cycloaddition reaction (Fig. 1B), known as a click chemistry reaction (Presolski, Hong, & Finn, 2009). The click chemistry reaction in the iPOND protocol involves the conjugation of biotin-azide to EdU in the presence of Cu^{1+} that is generated by the reduction of copper sulfate in the presence of sodium ascorbate (Sirbu et al., 2012). A consequence of Cu^{1+} generated in the click reaction is the fragmentation of DNA (Meneghini, 1997), which in the iPOND and aniPOND protocols leads to DNA fragments with a mean distribution of ~150 bp.

After the click reaction, the biotin-labeled DNA with accompanying covalently linked proteins is affinity purified using streptavidin beads.

Proteins associated with the newly synthesized DNA are eluted from the streptavidin beads by thermal decrosslinking in the presence of sodium dodecyl sulfate (SDS) and then identified by either immunoblotting or mass spectrometry (see "Proteomic analyses of the eukaryotic replication machinery" by David Cortez). To date, the iPOND protocol has been utilized to catalog the proteins present at replication forks (including the identification of the new replication protein Znf24) (Lopez-Contreras et al., 2013; Sirbu et al., 2013), probe the changes in replication fork protein composition under stress conditions including fork stalling, fork collapse, and hypoxia (Dungrawala et al., 2015; Min et al., 2013; Olcina, Giaccia, & Hammond, 2016; Sirbu et al., 2011; Wang et al., 2015), identify the epigenetic regulators present on replicating DNA in embryonic stem cells (Aranda, Rutishauser, & Ernfors, 2014), monitor the association of clinically relevant target proteins with replicating DNA (Wells et al., 2013), and examine the factors recruited to replicating viral genomes (Dembowski & DeLuca, 2015).

A limitation of the iPOND technique is the need for decrosslinking of the formaldehyde-fixed chromatin before proteins can be analyzed. Thermal decrosslinking, especially of large proteins and protein complexes, is an inefficient process that can limit the recovery of many proteins. To address this, a modified protocol named accelerated native iPOND (aniPOND) was reported in 2013 (Leung, Abou El Hassan, & Bremner, 2013). While based on the same click reaction chemistry as iPOND, the chromatin fraction containing EdU-labeled DNA is isolated under native, nondenaturing conditions in the aniPOND protocol, thereby eliminating the need for decrosslinking (Fig. 1A). The aniPOND technique was reported to have an increased overall protein yield compared to iPOND, a finding that was borne out in a study examining replication factors recruited to Herpes simplex virus 1 genomes (Dembowski & DeLuca, 2015), and to increase recovery of large chromatin remodeling complexes (Leung et al., 2013). Given the different approaches to capturing EdU-associated proteins, it is likely that iPOND and aniPOND are complementary techniques that may provide information on different but overlapping sets of proteins associated with replicating DNA and newly deposited chromatin.

Among publications utilizing either the iPOND or aniPOND techniques, only two studies have utilized suspension cells (Sirbu et al., 2013; Wells et al., 2013). Suspension cell lines such as lymphocytes may present an attractive alternative to adherent cell lines for certain studies, such as when specialized genetic models are present in suspension cells or in the study of replication abnormalities in leukemia and lymphoma cell lines. Here, we describe optimizations for performing both iPOND and aniPOND experiments with suspension cells. In the following sections, we will detail: (i) growth and cell handling conditions for suspension cells to avoid sample loss during the pulse and chase steps of both iPOND and aniPOND; (ii) critical modifications to the published aniPOND protocol that reduce sample loss, increase chromatin recovery, and reduce nonspecific background; (iii) a direct, quantitative comparison of the iPOND and optimized aniPOND protocols using suspension cells; and (iv) a detailed protocol for optimized aniPOND utilizing quantitative near-infrared (NIR) fluorescence immunoblotting.

2. SUSPENSION CELL GROWTH AND HANDLING FOR IPOND AND ANIPOND

The aniPOND and iPOND techniques require from 60 to 100 million cells per standard sample, respectively (Leung et al., 2013; Sirbu et al., 2012). It is important that the cells cultured for these experiments are growing optimally to ensure maximum EdU incorporation and reproducibility between experiments. Later, we describe variables that are important for the growth of B-lymphocytes and strategies to obtain maximal growth that can be applied to other suspension cell lines. In addition, the considerations for cell handling during the initial pulse and chase phases of iPOND/aniPOND are different for suspension cells compared with adherent cell lines. We describe how to avoid sample loss during these steps with suspension cells to ensure maximum downstream signal recovery.

2.1 Optimizing Growth Conditions to Obtain iPOND/aniPOND Cell Numbers

The growth of both adherent and suspension cells is sensitive to multiple environmental variables including temperature, pH, O_2 and CO_2 content, nutrient and metabolite concentration, and the presence of growth factors from serum, nearby cells, and exogenous stimulating factors. Before attempting iPOND or aniPOND, it is critical to optimize growth conditions in order to reproducibly obtain large cultures of rapidly dividing cells. Utilizing poorly and inconsistently growing cell populations for iPOND and aniPOND may lead to problems with EdU incorporation and reproducibility between experiments. Later, we describe key variables that impacted the growth of a mouse B-cell line in suspension.

First, we found that the surface area for gas exchange was a critical factor in determining the cell density that can be reached before cell proliferation starts to plateau. While some laboratories grow suspension cells in flasks that are upright, simply laying flasks on their side to increase surface area for gas exchange significantly increased the proliferative capacity of mouse B-cells, especially at high cell densities (Fig. 2A). Furthermore, in addition to increasing that maximum number of cells obtained, cells grown in flasks on their side doubled ~40% faster than those grown in flasks upright (Fig. 2A). As an alternative, spinner flasks may be used to constantly mix media and increase aeration during growth. Second, certain suspension cell lines such as lymphocytes are dependent upon β-mercaptoethanol (βME) for proliferation and survival (Metcalf et al., 1975). For example, when the βME is omitted from the media, mouse B-cell growth stops completely (Fig. 2B). Since βME is volatile and loses its reducing potential over time while in storage, it must be added freshly to media in flasks during cell dilution rather than to the stock media. Third, it is very important to avoid cell overgrowth. For many suspension cell lines, this occurs at approximately 1.5×10^6 cells/mL and should be determined experimentally under the optimized growth conditions (Fig. 2A). If EdU labeling is carried out in overgrown cell populations, a significant but variable fraction of the cells will not be replicating. Furthermore, overgrowth may result in the introduction of confounding factors such as epigenetic changes into the cell

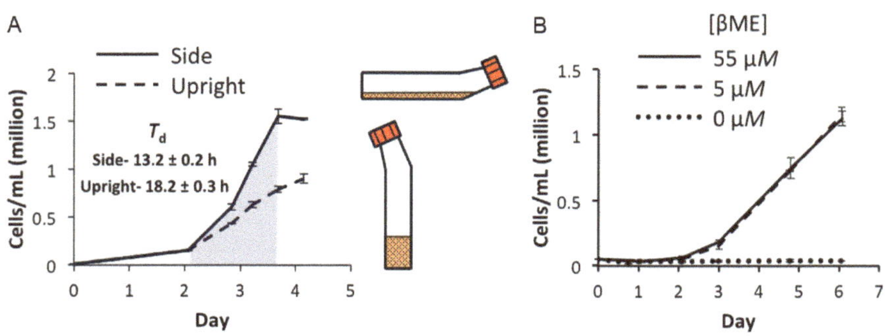

Fig. 2 Important considerations for suspension cell growth. (A) Growth curves for CH12F3 mouse B-cell lymphoma cells incubated in 100 mL of media in T175 flasks either upright or on their sides to increase surface area for gas exchange. Doubling times (T_d) were calculated using the least squares fitting method over the exponential phase (*shaded*). (B) Growth curves for CH12F3 cells grown with the indicated concentrations of β-mercaptoethanol (βME). For all conditions, $n = 2$ independent biological replicates. Error bars ± S.D.

population. Thus, we recommend restarting the population from early passage frozen cells if overgrowth occurs to increase consistency between experiments.

2.2 Handling of Suspension Cells During Pulse and Chase

The EdU pulse-labeling and chase phases of the iPOND and aniPOND protocols are essentially identical (Fig. 1A). Cells that have been grown to large numbers in a manner that maximizes proliferative capacity (see earlier) are labeled with EdU. After incubation in the EdU-containing media for a defined time, cells are either processed immediately, or resuspended in thymidine-containing medium and then incubated for different times prior to processing. A disadvantage of utilizing suspension cells in the iPOND and aniPOND protocols compared to adherent cells is the need for centrifugation to pellet cells in order to resuspend them in the chase medium. Our observations indicate that mouse B-cells undergo morphological changes after centrifugation (Fig. 3A–C) and became sticky, adhering to the surfaces of the tissue culture flasks even in flasks that are hydrophobically coated (not shown). As a result of this centrifugation-induced adherence with flask surfaces, approximately half of the downstream sample was lost when cells were resuspended in chase medium and incubated in flasks on their side (Fig. 3D). By instead setting flasks upright during the chase step, the B-cells had much less surface area to adhere to and settled more slowly, leading to an almost complete elimination of the sample loss observed in chase samples (Fig. 3E). Thus, while a greater surface area for gas exchange facilitates maximal growth to the cell numbers required for iPOND and aniPOND, this surface area becomes a liability when performing the chase step because centrifugation induces cellular stress and morphological changes that promote adherence to flask surfaces.

3. OPTIMIZATIONS TO THE ANIPOND PROTOCOL TO INCREASE FUNCTIONALITY

While the aniPOND protocol was reported to offer multiple advantages over the traditional formaldehyde-based iPOND technique, including faster processing time and greater sensitivity (Leung et al., 2013), we experienced technical difficulties when initially implementing the protocol in both SV40-immortalized human fibroblasts and mouse B-cells. Specifically, we observed a high degree of sample loss and nonspecific binding to the

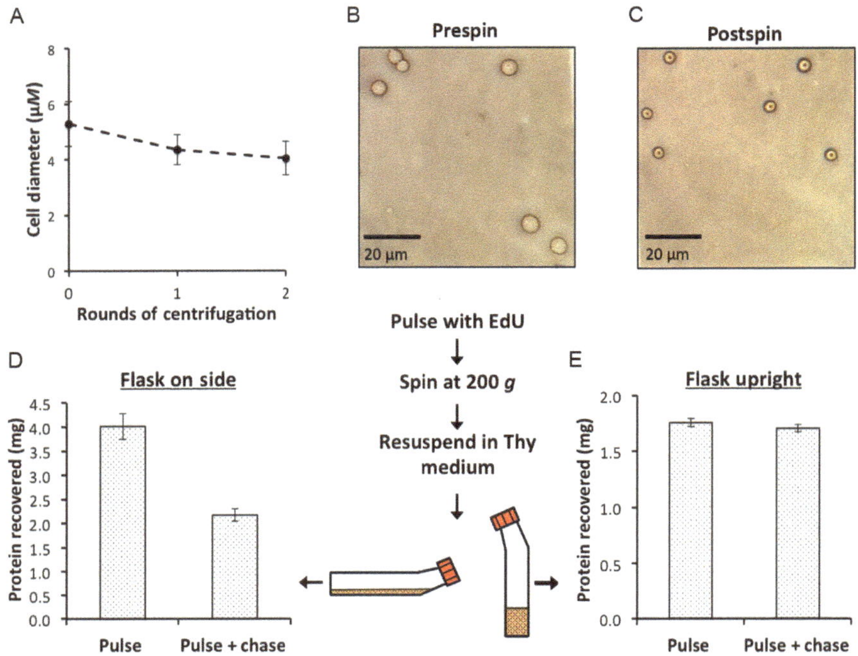

Fig. 3 Centrifugation of mouse B-cells during the iPOND/aniPOND pulse and chase protocol leads to morphological changes and preventable sample loss. CH12F3 mouse B-cell lymphoma cells were subjected to rounds of standard cell centrifugation (200 × *g*, 5 min). The diameters of cells (A) were measured by Image J software analysis of photos taken of cells both before (B) and after (C) centrifugation. A minimum of 15 cell diameters was measured per round, and error bars represent the S.D. of measurements from two independent samples. Note the biconcave morphology of postspin cells. After resuspending cells in thymidine medium, the chase sample flasks were either incubated on the side or upright before proceeding. To test the downstream sample recovery, the protein content of sonicated chromatin was measured in cells that were either incubated in thymidine medium in flasks on their sides and then processed for iPOND (D), or cells that were incubated in thymidine medium in flasks that were positioned upright and then processed for aniPOND (E).

streptavidin beads. Through systematically troubleshooting the steps of the original protocol, we have made multiple modifications to the original aniPOND protocol that reproducibly reduce sample loss, increase signal to noise, and maximize sample recovery in aniPOND experiments with mouse B-cells that we present later. In describing the revised protocol later, we discuss the modifications in the context of chromatin biology to facilitate adaptation of the aniPOND technique to other cell lines.

3.1 Optimized Sonication Regimen

After isolating nuclei containing the EdU-labeled DNA as described in the published aniPOND protocol (Leung et al., 2013), the EdU-containing genomic DNA is conjugated to biotin-azide by the click reaction within the nuclei (Fig. 1B). The next major step in both the iPOND and aniPOND protocols is to solubilize the chromatin by sonication to generate fragments of chromatin that are amenable to pulldown with streptavidin beads (Fig. 1A). The original iPOND protocol (Sirbu et al., 2012) calls for sonicating on ice with 20-s pulses followed by 40 s rests between pulses using a microtip sonicator on 13–16 W output, with the number of rounds depending on the sample volume. Successful solubilization is immediately observable by clarification of the lysate. The published aniPOND protocol (Leung et al., 2013) uses a more stringent sonication regimen that incorporates 12 × 10-s pulses on ice at output setting 3 to 4 (~10 W of output) with 10 s rests in-between pulses. Additionally, the aniPOND protocol incorporates two prior wash steps (sonication washes), in which nuclei are resuspended in buffer, sonicated for 10 s, spun down, and then resuspended in buffer again (Fig. 4A). In our initial attempts to perform that aniPOND protocol, we observed that the sonication washes appeared to solubilize significant portions of the nuclei since the nuclei pellets were much smaller after each of the sonication wash steps. Indeed, when we quantitated the amount of both protein and DNA (therefore chromatin) solubilized by the two wash steps, we found that these two initial wash sonications released nearly as much chromatin as the final solubilization step involving 12 sonication pulses (Fig. 4B). Thus, as much as 50% of the chromatin is lost during the sonication wash steps. We also observed that the final aniPOND chromatin solubilization protocol—12 rounds of 10 s on, 10 s off, on ice—was leading to excessive foaming and splashing of sample in the later rounds, even on the lowest recommended setting (output 3). Since oversonication can cause protein aggregation (Stathopulos et al., 2004), in addition to potentially damaging proteins due to sample overheating, we tested if the giving the samples more rest on ice between sonication assisted in chromatin recovery. Indeed, we found that performing the bursts in rounds of three times 10 s on, 10 s off, with at least a minute rest on ice before the next round of three, markedly increased sample recovery (Fig. 4C), indicating that care needs to be taken in the final solubilization step to avoid overheating.

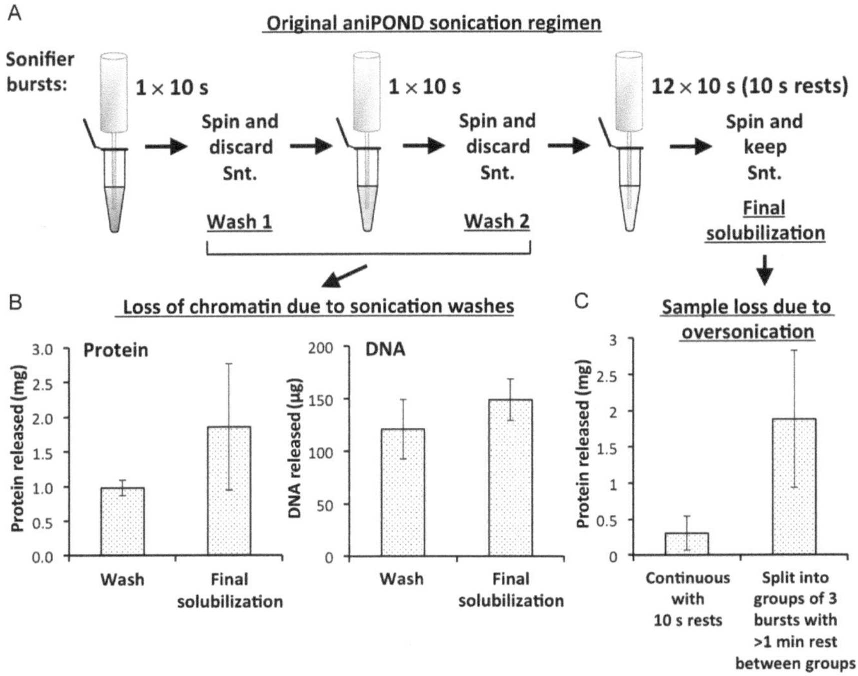

Fig. 4 The original aniPOND sonication regimen leads to sample loss at two steps. (A) Schematic of the original aniPOND sonication regimen. (B) Supernatant was recovered after spinning down nuclei following the sonication wash steps (Wash) and assayed for protein via Bradford assay and DNA by nanodrop spectrophotometer (Thermo Fisher Sci., Waltham, MA) following column DNA purification. Results were juxtaposed to protein and DNA measurements obtained following the final solubilization of the nuclei (Final Solubilization). (C) The original final solubilization protocol of 12 rounds of 10″ on, 10″ off, on ice (Continuous with 10 s rests) was compared to a modified protocol consisting of four groups of three rounds of 10″ on, 10″ off, on ice with at least 1 min of rest between groups on ice (Split into groups of 3 bursts with >1 min rest between groups). For all samples, signal represents the mean of two independent biological replicates. Error bars ± S.D.

As solubilization of chromatin by sonication is critical for maximum recovery of biotin-labeled DNA, we investigated the factors that influence sonication efficiency. A previous study on chromatin compaction demonstrated that increasing levels of monovalent cations lead to greater degrees of chromatin compaction in vitro, with peak in vitro compaction occurring at greater than approximately 60 m*M* NaCl (Thoma, Koller, & Klug, 1979). In addition, the same study demonstrated that relatively low concentrations of divalent cation, for example, 0.5 m*M* Mg^{2+}, also cause maximum in vitro compaction. In the published aniPOND protocol (Leung et al., 2013), cells are first harvested in a nucleus extraction buffer containing 3 m*M* Mg^{2+}, followed by washing and the click reaction that occur in phosphate-buffered saline (PBS)-based buffers containing ~130 m*M* NaCl. These buffers have cation concentrations higher than that shown to maximally compact chromatin in vitro. Notably, collapsed chromatin morphology was observed by electron microscopy in nuclei prepared using the same nonionic detergent and concentration as the published aniPOND protocol (Stuart, Clawson, Rottman, & Patterson, 1977). Thus, we expect that under these conditions the chromatin inside the isolated nuclei will be collapsed into more compacted structures that may be initially resistant to sonication. To create conditions in which compacted, potentially sonication-resistant chromatin relaxes and becomes vulnerable to sonication, we removed the two sonication wash steps in the original aniPOND sonication regimen (Fig. 4A) and replaced them with two 30 min rotations in the aniPOND protocol buffer B1, a low salt, nonionic detergent sonication buffer containing EDTA to chelate divalent cations (25 m*M* NaCl, 2 m*M* EDTA, 50 m*M* Tris–HCl pH 8.0, and 1% IGEPAL CA630). This alteration prior to sonication increased the fraction of chromatin solubilized from less than 50% to approximately 80% (Fig. 5A), presumably by creating conditions that promote in vitro chromatin relaxation. Comparing the fractions of histone H4 solubilized by sonication to the remaining insoluble histone H4 after sonication (pellet) by immunoblotting revealed that, in line with the protein measurements, preincubation in buffer B1 resulted in the majority of the histone H4 being solubilized (Fig. 5B). Based on these results, we designed the optimized sonication regimen displayed in Fig. 5C that both reduces sample loss during sonication and increases the amount of chromatin solubilized.

3.2 Preventing Chromatin Precipitation in Sonicated Chromatin

Following chromatin solubilization by sonication as discussed earlier, the next step in the published aniPOND protocol (Leung et al., 2013) is to dilute the chromatin solubilized in low salt buffer B1 equally with physiologic salt buffer B2 (150 m*M* NaCl, 2 m*M* EDTA, 50 m*M* Tris–HCl pH 8.0, and 0.5% IGEPAL CA630) to bring the NaCl concentration closer to physiologic levels before incubation with streptavidin beads. Under these conditions, the solubilized chromatin isolated as described earlier turned opaque upon addition of buffer B2 (Fig. 6A). This is consistent with early observations from the chromatin literature reporting that increasing the salt concentrations of native chromatin preparations toward physiologic salt concentrations led to decreased solubility of histones and precipitation of a

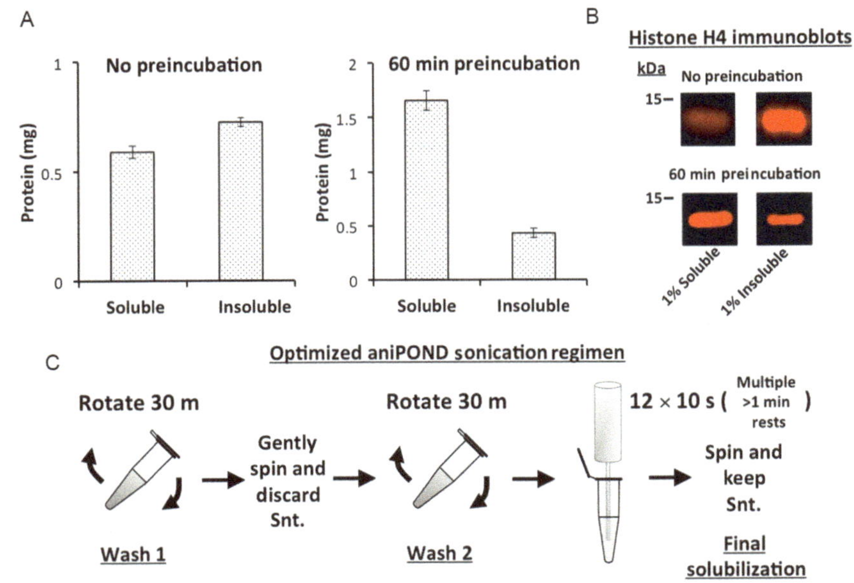

Fig. 5 Preincubation of nuclei in low salt, EDTA-containing buffer B1 leads to more efficient solubilization of chromatin. (A) Comparisons of protein released by sonication to the remaining insoluble protein levels (pellet) in aniPOND samples either resuspended in B1 and sonicated immediately (No Preincubation) or in samples rotated twice for 30 min in B1 before sonication (60 min Preincubation). Signal represents the mean of two independent biological replicates. Error bars ± S.D. (B) Representative immunoblots with histone H4 antibody (Abcam ab17036, Cambridge, United Kingdom) comparing 1% of the sonication-solubilized protein fraction with 1% of the insoluble (pellet) fraction. 1% soluble and 1% insoluble bands for "No Preincubation" and "60 min Preincubation" blots were run on the same gels testing multiple conditions. The indicated bands were cropped from the image captures obtained by acquisition of near-infrared Western blots using a Li-Cor Odyssey Fc instrument. Uncropped images are readily available upon request. (C) Schematic of the optimized aniPOND sonication regimen. Two gentle rotation washes in B1 buffer designed to encourage chromatin decompaction followed by a final solubilization step with 12 rounds of sonication and multiple rounds of >1 min rests (we recommend six groups of 2× 10″ on, 10″ off, on ice with at least 1 min of additional rest on ice between groups).

fraction of the chromatin containing, in addition to other proteins, essentially all histone H1 (Tatchell, 1978). Notably, maximum nucleosome insolubility occurred in that study at about ~0.1 *M* NaCl, a concentration close to that obtained by diluting the chromatin solubilized in buffer B1 (25 m*M* NaCl) with buffer B2 (150 m*M* NaCl) in the published aniPOND protocol (Leung et al., 2013). Indeed, upon microscopic examination, we observed that diluting the low salt B1 buffer-sonicated chromatin with physiologic salt buffer B2 resulted in the formation of visible aggregates (Fig. 6B). After centrifugation, we determined that these aggregates contained about half the

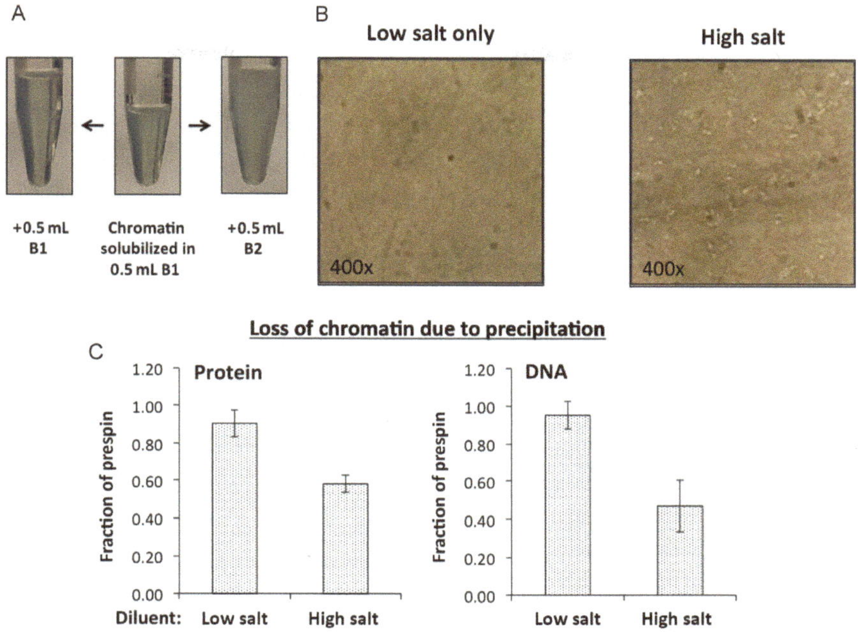

Fig. 6 Dilution of chromatin solubilized by sonication in low salt buffer B1 with physiologic salt buffer B2 leads to chromatin aggregation and sample loss after clarification centrifugation. (A) Photographs of 1.5 mL tubes containing chromatin solubilized by sonication in 0.5 mL low salt buffer B1 before and after addition of 0.5 mL of physiologic salt buffer B2 or more buffer B1. (B) Photographs of preparations of chromatin solubilized by sonication in low salt buffer B1 that has been diluted either with more buffer B1 (Low Salt Only) or with physiologic salt buffer B2 (High Salt) at 400 × magnification with a light microscope. Note the multitudinous small white aggregates in the high salt sample. (C) The contents of the aggregates formed upon adding either low salt buffer B1 (Low Salt) or physiologic salt buffer B2 (High Salt) to chromatin solubilized in buffer B1 were evaluated by measuring the protein and DNA content of the chromatin fraction before and after a clarification centrifugation spin. Signal represents the mean of two independent biological replicates. Error bars ± S.D.

total chromatin, whereas almost no chromatin was lost if it was maintained at a low salt concentration (Fig. 6C). Thus aniPOND samples, which are essentially concentrated low-salt native chromatin preparations, are highly sensitive to increasing salt concentrations and so we recommend maintaining the solubilized chromatin in low salt B1 buffer.

3.3 Eliminating Sources of Background

After the biotin-labeled fraction of the solubilized chromatin has been pulled down using streptavidin beads, bound proteins are eluted by boiling in SDS-containing sample buffer. The eluted fraction contains proteins that were pulled down both by the streptavidin–biotin interaction, as well as any proteins that nonspecifically associated with the beads. Both the iPOND and aniPOND protocols utilize either a no-EdU or a no-click control (NCC) sample, in which either the DNA is not labeled with EdU or the click reaction does not contain biotin-azide (in both cases, the newly synthesized DNA is not conjugated to biotin), in order to account for the nonspecific binding of proteins to the beads.

We observed on multiple occasions when using the published aniPOND protocol of diluting chromatin solubilized in low salt buffer B1 with physiologic salt buffer B2 that there was a high degree background binding in NCC samples that included both histone (histone H4) and nonhistone (PCNA) chromatin proteins (Fig. 7A). Since histones comprise approximately half of the protein content of chromatin, some level of histone background may be anticipated (van Holde, 1989), whereas PCNA is present at much lower levels than histone proteins. The problem of nonspecific PCNA binding in NCC samples was resolved by the changes in the chromatin preparation described earlier that reduced chromatin aggregation. In samples that were maintained in low salt buffer B1 after sonication, no PCNA was observed in the NCC pulldown lanes and PCNA unloading from newly replicated DNA was clearly observable (Fig. 7B).

While the nonspecific PCNA binding was caused by chromatin aggregation, the nonspecific binding of histones was not reduced by preventing chromatin aggregation. This prompted us to consider the characteristics of the beads themselves. Based on recently published iPOND experiments (Dungrawala et al., 2015), we were using Dynabeads MyOne C1 Streptavidin beads (Thermo Fisher Sci., Waltham, MA). These beads have a hydrophilic, negatively charged surface that can potentially form charge-based interactions with positively charged proteins. To determine if charge-based interactions may contribute to nonspecific protein binding, we compared negatively charged Dynabeads with neutral streptavidin agarose beads (Fig. 7C). In this experiment, both sets of beads yielded signal above noise for PCNA, with larger specific signal obtained with the streptavidin agarose. While similar high levels histone H4 were retained on the negatively charged Dynabeads in both the NCC and EdU pulse samples, very little H4 binding to the associated with the neutral streptavidin agarose beads was detected in either sample. We estimated that negatively charged Dynabeads had ~60× more nonspecifically bound histone H4 background per μm^2 of surface area than streptavidin agarose (Fig. 7D). The absence of specific histone H4 binding in a 10min EdU pulse sample is consistent with subsequent studies indicating that H4 deposition occurs rapidly after ~15min. Thus, the neutrally charged streptavidin agarose beads

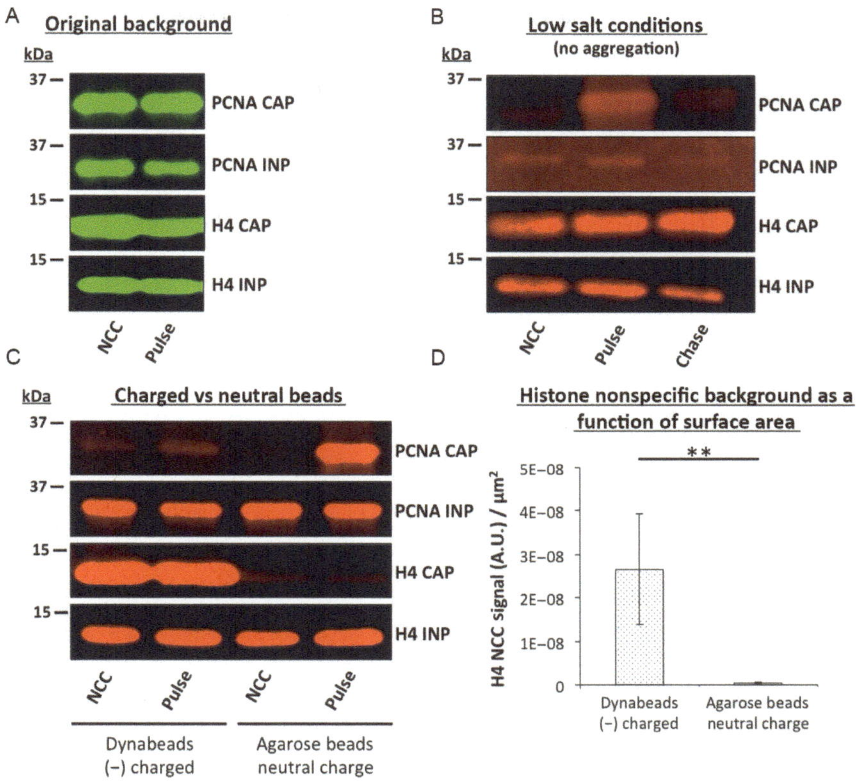

Fig. 7 Sources of background in the aniPOND protocol. (A) CH12F3 mouse B-cells were processed for aniPOND using the published method of diluting chromatin solubilized in low salt buffer B1 with physiologic salt buffer B2 before pulldown with Dynabeads MyOne Streptavidin C1 beads (Thermo Fisher Sci., Waltham, MA). CAP, 50% of proteins captured by beads; INP, 1% of prepulldown chromatin; NCC, no-click control; Pulse, 10min EdU pulse. (B) As in (A), but diluting chromatin solubilized in low salt buffer B1 with additional low salt buffer B1 to prevent chromatin aggregation. Chase = 10min EdU pulse followed by 60min thymidine chase. (C) As in (B), but comparing pulldown with either Dynabeads or Pierce High Capacity Streptavidin Agarose (Thermo Fisher Sci., Waltham, MA) as indicated. (D) Histone H4 NCC signal from 4 to 5 independent experiments as quantified by Li-Cor near-infrared Western blotting was normalized to the total surface area of the beads added for pulldown. Statistical significance between average H4 NCC signal per μm^2 was assessed by one-way ANOVA. ** = $P < 0.01$. Beads surface area was calculated using publically available product information.

accurately recapitulated the presence of PCNA at the replication fork while avoiding nonspecific histone binding.

3.4 Optimization Limitations

A limitation to the optimizations that we have presented earlier is that we have only studied the aniPOND protocol in suspension cells with mouse B-lymphocytes. While we believe that many of our observations reflect fundamental properties of chromatin biology and should thus be broadly applicable, it is nonetheless possible that other types of suspension cell lines will behave differently in both the growth conditions necessary to generate aniPOND cell numbers as well as during the protocol itself.

4. COMPARISON OF IPOND AND ANIPOND IN SUSPENSION CELLS

Previous comparisons of the iPOND and aniPOND techniques have suggested that the protein yield is higher in aniPOND compared with iPOND, and that the two techniques may isolate different but overlapping sets of proteins (Dembowski & DeLuca, 2015; Leung et al., 2013). We reasoned that, while certain large proteins and protein complexes may decrosslink poorly in the iPOND protocol, proteins that interact transiently with replicating DNA may require crosslinking for detection. To quantitatively address the strengths and limitations of both protocols, we directly compared the iPOND and optimized aniPOND techniques using identical quantities of mouse B-cells over a 60-min time course (Fig. 8A) and calculated the beads capture signals as a percent of input using NIR immunoblotting. Consistent with the hypothesis that formaldehyde crosslinking assists in the capture of transient interactions, Lig1 was only detectable in both capture and input samples using iPOND (Fig. 8B). In contrast, PCNA, which is topologically linked to DNA at the replication fork, was detected efficiently by both iPOND and aniPOND, with ~50% more efficient capture in the aniPOND protocol (Fig. 8C). There was a large difference in the efficiency of histone H4 capture between the iPOND and aniPOND protocols, with ~5 × more histone H4 captured by aniPOND (Fig. 8D). Nonetheless, both techniques demonstrated the expected turnover of the H4K5ac mark (Fig. 8E). A comparison of the prepulldown chromatin content (input) of the iPOND and aniPOND samples demonstrated that almost no Lig1 is present in aniPOND chromatin preparations (Fig. 8F), while PCNA, H4, and H4K5ac are present. Thus, iPOND detects transiently interacting proteins that are lost from chromatin during aniPOND sample preparation, whereas the aniPOND protocol very efficiently captures chromatin proteins that remain associated under native conditions.

One of the rationales for the development of the aniPOND protocol was to avoid thermal decrosslinking (Leung et al., 2013). To test the requirement for thermal decrosslinking, we boiled iPOND samples for different amounts of time and compared to aniPOND samples (Fig. 9). We found that 25 min of boiling, as suggested in the iPOND protocol (Sirbu et al., 2012), was sufficient to decrosslink both Lig1 and histone H4 (Fig. 9). However, in line

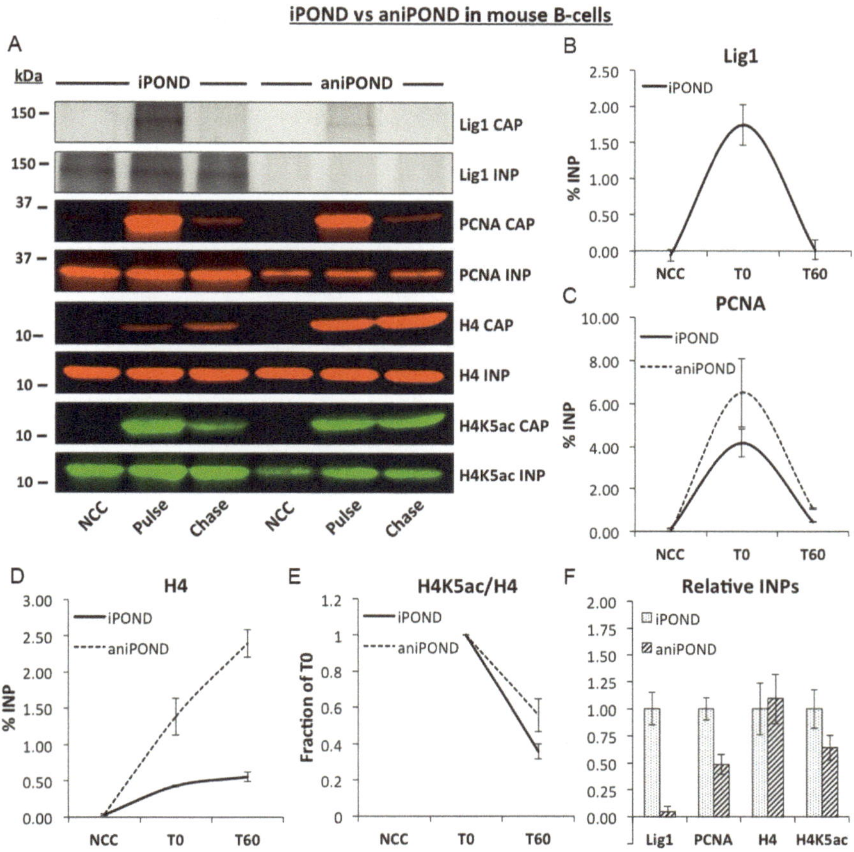

Fig. 8 Direct comparison of the iPOND and optimized aniPOND techniques. (A) 90 million CH12F3 mouse B-cell lymphoma cells were processed for either iPOND according to the published protocol (Sirbu et al., 2012), or optimized aniPOND as described in this chapter. No-click control (NCC), 15 min EdU pulse (Pulse), and 15 min EdU pulse followed by 60 min thymidine chase (Chase) samples were performed for each protocol simultaneously. 50% of proteins captured by streptavidin beads (CAP) and 1% of prepulldown input (INP) were analyzed on the same SDS-PAGE gels for the indicated proteins. Antibodies used were Rabbit anti-Lig1 (in house) 1:2500, Mouse anti-PCNA (Santa Cruz sc-56, Dallas, TX) 1:200, Mouse anti-H4 (Abcam ab17036, Cambridge, United Kingdom) 1:1000, and Rabbit anti-H4K5ac (Abcam 51997, Cambridge, United Kingdom) 1:10,000, followed by incubation with Goat anti-Rabbit 800 nm 1:5000 or Goat anti-Mouse 680 nm 1:5000 NIR fluorescent secondary antibodies and detection with an Odyssey NIR imaging system (Li-Cor Biosciences, Lincoln, NE). The capture of Lig1 (B), PCNA (C), and H4 (D) was assessed as a percent of input, with the pulse samples indicated as T0 and the chase samples indicated as T60, and the removal of the H4K5ac mark was monitored (E). Prepulldown input signals for the indicated proteins were compared as a fraction of iPOND INP (F). For all quantitations, signal represents the mean of two independent biological replicates. Error bars ± S.D.

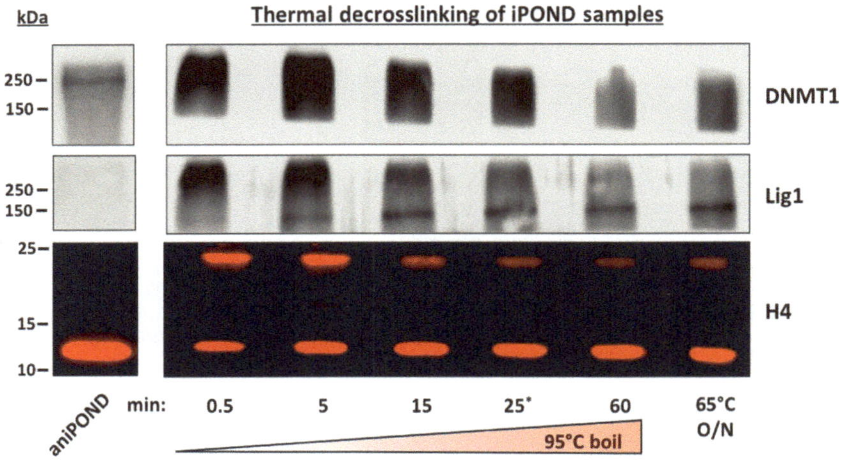

Fig. 9 Assessing the efficiency of thermal decrosslinking of iPOND samples. CH12F3 mouse B-cell lymphoma cells were processed for either iPOND or aniPOND. iPOND input samples were boiled for varying amounts of time as indicated or placed at 65°C overnight. * indicates the recommended boiling time in the published protocol (Sirbu et al., 2012). 1% input samples were resolved on SDS-PAGE gels and probed for DNMT1 (Mouse anti-DNMT1 1:200, Santa Cruz sc-271729, Dallas, TX), Lig1 (Rabbit anti-Lig1 1:2500, in house), and histone H4 (Mouse anti-H4 1:1000, Abcam 17036, Cambridge, United Kingdom), followed by detection of DNMT1 and Lig1 with HRP-conjugated secondary antibodies (Bio-Rad, Hercules, CA) or histone H4 with Goat anti-Mouse 680 nm 1:5000 NIR secondary antibody (Li-Cor Biosciences, Lincoln, NE).

with the hypothesis that large chromatin modifying enzymes inefficiently decrosslink, DNMT1 never resolved out of a high molecular weight smear into a clearly identifiable band, even after 1 h of boiling (Fig. 9). In contrast, DNMT1 was clearly detectable in the aniPOND sample. Based on these data, we recommend utilizing iPOND to test the association of transiently interacting proteins with newly synthesized DNA, and we recommend utilizing the optimized aniPOND protocol described in this chapter to examine histone proteins and large chromatin modifying complexes that inefficiently decrosslink. Together, these results provide evidence for the strengths and limitations of both iPOND and aniPOND, and highlight the complementary nature of these techniques.

5. PROTOCOL FOR OPTIMIZED ANIPOND IN SUSPENSION CELLS

Below is the optimized protocol for aniPOND using suspension cells that is based upon the published aniPOND protocol (Leung et al., 2013) and incorporates the modifications described earlier. The protocol is designed for 8×10^7 suspension cells.

5.1 Suspension Cell Growth to aniPOND Experimental Numbers

5.1.1 Equipment

- Cell culture incubator
- Biological safety cabinet
- 250-mL canted neck suspension culture flasks with ventilation cap (USA Sci., cat. no. 5665-8190, Ocala, FL)
- T175 canted neck flasks with ventilation cap (Sarstedt, cat. no. 83.3912.002, Nümbrecht, Germany)
- T225 ventilation cap flasks (USA Sci., cat. no. CC7682-4822, Ocala, FL)
- Hemocytometer (Fisher Sci., cat. no. 02-671-10, Waltham, MA)

5.1.2 Buffers and Reagents

- Cell culture medium, e.g., RPMI 1640 (Thermo Fisher Sci., cat. no. 11875135, Waltham, MA) and additives including 10% fetal bovine serum (Sigma-Aldrich, cat. no. F2442, St. Louis, MO) and 55 m*M* (100 ×) βME for tissue culture (Thermo Fisher Sci., cat. no. 21985023, Waltham, MA).

5.1.3 Procedure

1. Rapidly thaw a vial of early passage suspension cells, such as CH12F3 mouse B-cell lymphoma cells, in a 37°C water bath and immediately dilute into the appropriate prewarmed media supplemented with necessary additives such as 10% fetal bovine serum and 55 μ*M* βME that are required for optimal growth.
2. Perform routine subculturing in 20 mL of appropriate media in a hydrophobic-coated ventilation cap flask laid flat to increase surface area for gas exchange in a cell culture incubator set to the appropriate temperature and CO_2 (standard settings are 37°C and 5% CO_2 content).
3. Establish a growth curve for your cell line(s), such as that in Fig. 2A, by periodic counting with a hemocytometer. Calculate the doubling time (T_d) of the cells in exponential-phase growth and ensure that the calculated T_d matches the reported T_d in the literature.
4. Passage cells by diluting into fresh medium before overgrowth is reached, preferably in mid exponential phase. Keep careful track of the passage number of the cells and do not over passage. For many cell lines, this means using before ~20 passages after thawing.
5. Once optimal growth conditions are established, dilute cells for aniPOND using the T_d of the cell line(s) to estimate the number of cells necessary to obtain 8×10^7 cells in ~48 h. For example, if the $T_d = 12$ h

and the cells are desired to be ready in 48 h, 5×10^6 cells would be diluted into 80 mL of warm media in a T175 canted neck flask. In this manner, 8×10^7 cells will be ready in 48 h at a density of 1×10^6/mL, which is mid exponential phase for many cell lines.

6. Count the cells by hemocytometer during the growth phase up to the starting point of the aniPOND protocol to ensure that doubling time is optimal.
7. One day prior to starting the aniPOND protocol, place media for EdU pulse and thymidine chase in the incubator to equilibrate temperature and CO_2 content. For the EdU pulse media, calculate 2 mL/80 mL culture (make up 25% extra to account for losses due to evaporation, etc.). For the thymidine chase media, calculate 80 mL for each chase sample (make up 10% extra to account for losses due to evaporation, etc.).

5.1.4 Notes

1. It is essential that the cells utilized have an established growth curve in your laboratory and are growing in an optimal manner. Frequent cell counting will allow determination of doubling-times to determine if proliferation is optimal.
2. While the addition of antibiotics to the media may help prevent bacterial contamination, this may also mask underlying contamination. We suggest performing routine subculturing without antibiotics and performing monthly PCR-based *Mycoplasma* testing.
3. Multiple routine passages may be performed in the same flask to reduce costs. This does, however, increase the risk of contamination over time. We suggest changing routine subculture flasks on a weekly basis.

5.2 EdU Pulse, Thymidine Chase, and Click Reaction

5.2.1 Equipment

- Cell culture incubator
- Biological safety cabinet
- Swinging-bucket tabletop centrifuge for 15- and 50-mL tubes
- Rotating platform for 1.5- and 15-mL tubes at 4°C (in a cold room or refrigerator)

5.2.2 Buffers and Reagents

- Flasks with 8×10^7 suspension cells in mid exponential growth phase in 80 mL of media
- Flask with preequilibrated media for EdU pulse (2 mL/sample, plus some extra)

- Flask(s) with preequilibrated media for thymidine chase (80 mL/chase sample, plus some extra)
- Dimethyl sulfoxide (DMSO, EMDMillipore cat. no. MX1458-6, Billerica, MA)
- EdU (Sigma-Aldrich, cat. no. T511285, St. Louis, MO). Dissolve in DMSO to a final concentration of 10 m*M*. Aliquot and store at −20°C protected from light for up to 1 year. Thaw at 37°C immediately before use.
- Thymidine (Sigma-Aldrich, cat. no. T9250, St. Louis, MO). Dissolve in PBS to a final concentration of 100 m*M*, filter sterilize, aliquot, and store at −20°C for up to 3 years. Thaw at 37°C immediately before use.
- Biotin-PEG3-azide (Sigma-Aldrich, cat. no. 762024, St. Louis, MO). Dissolve in DMSO to a final concentration of 50 m*M*. Aliquot and store at −20°C protected from light for up to 1 year. Thaw at 37°C immediately before use.
- NaCl. A 5 *M* stock in double-distilled, filter sterilized, and stored at 4°C.
- HEPES. A 500 m*M* stock in double-distilled water, adjusted to pH 7.2 with KOH, filter sterilized, and stored at 4°C.
- $MgCl_2$. A 300 m*M* stock in double-distilled water, filter sterilized, and stored at 4°C.
- Sucrose. A 1.2 *M* stock in double-distilled water, filter sterilized, and stored at 4°C.
- IGEPAL CA630 (Sigma-Aldrich, cat. no. I8896, St. Louis, MO). A 10% v/v stock in double-distilled water, filter sterilized, and stored at 4°C for a few months.
- Prechilled NEB (20 m*M* HEPES-KOH pH 7.2, 50 m*M* NaCl, 3 m*M* $MgCl_2$, 300 m*M* sucrose, 0.5% IGEPAL CA630) freshly prepared.
- Prechilled PBS (137 m*M* NaCl, 2.68 m*M* KCl, 1.47 m*M* KH_2PO_4, 9.55 m*M* Na_2HPO_4, pH 7.45). A 10× autoclaved stock solution can be prepared and stored at room temperature (RT).
- Copper (II) sulfate pentahydrate (Sigma-Aldrich, cat. no. 203165, St. Louis, MO). Dissolve to a final concentration of 100 m*M* in double-distilled H_2O and store at RT for up to 3 months.
- (+)-Sodium L-ascorbate (Sigma-Aldrich, cat. no. A4034, St. Louis, MO). Dissolve to a final concentration of 100 m*M* in double-distilled H_2O and place on ice immediately before use in click reaction setup.
- Click/NCC reaction mixture. To prepare 10 mL for one sample, add *in order*: 8.8 mL ice-cold PBS, 5 μL of 50 m*M* biotin-PEG3-aizde (or 5 μL of DMSO for NCC), 1 mL of 100 m*M* (+)-sodium L-ascorbate, and 200 μL of 100 m*M* copper (II) sulfate. Scale up the quantity of reaction mixture

for the number of samples, and always prepare fresh click reaction/NCC mixture.

5.2.3 Procedure

1. Prepare the overnight preequilibrated media for EdU pulse by adding EdU to a final concentration of 410 μ*M*.
2. Prepare the overnight preequilibrated media for thymidine chase by adding thymidine to a final concentration of 10 μ*M*.
3. To begin pulse, add 2 mL of the 410 μ*M* EdU in media to the 80 mL of cells in the first flask (final concentration, 10 μ*M* EdU), gently mix with 5–10 rotations, start a timer counting up for time of pulse, and set the flask in the 37°C incubator standing upright to reduce surface area for cells to stick. It is imperative that every sample be treated in the same manner (see *Notes* later).
4. After 7.5 min has expired on timer for pulse, remove pulse flask(s) from the incubator to the biological safety cabinet.
5. Pour cells from pulse flask(s) into 2 × 50 mL labeled tubes each and set in swinging-bucket tabletop centrifuge.
6. After 9 min has expired on timer for pulse, centrifuge at 200 × *g* for 4 min at RT.

   For chase samples only during centrifugation:
   - Pour ~35 mL of RT PBS into the flask(s). Rotate the PBS around and aspirate off to wash out EdU-containing pulse media.
   - Add 55 mL of warm thymidine media to the flask(s) (see *Notes* later).
7. After centrifugation has finished (~13 min on pulse timer), gently remove tubes to biological safety cabinet and aspirate pulse medium without disturbing pellets.
8. At exactly 15 min on the pulse timer, either completely resuspend the cell pellet in 25 mL of thymidine chase media and proceed to step 9 for chase samples, or resuspend in 10 mL of ice-cold nucleus extraction buffer (NEB) for pulse or NCC samples and proceed to step 15. Combine the two ½ pellets from the two 50 mL tubes for each sample in the same 25 mL of chase media or 10 mL NEB.
9. CHASE SAMPLES: Pipette the cells in 25 mL of thymidine media into the PBS-washed flasks already containing 55 mL of warm chase media for a final chase volume of 80 mL. Start a timer counting up for chase and remove to 37°C incubator. Set the flasks standing upright in the incubator to reduce surface area for cells to adhere to.

10. 7.5 min before the desired chase time (for example, at min 22.5 of a desired 30-min chase sample), remove chase flask(s) from the incubator to the biological safety cabinet.
11. Pour cells from pulse flask(s) into 2 × 50 mL labeled tubes and set in swinging-bucket tabletop centrifuge.
12. 6 min before the desired chase time, centrifuge at 200 × *g* for 4 min at RT.
13. After spin has finished (~2 min before desired chase time), gently remove tubes to biological safety cabinet and aspirate chase medium without disturbing pellets.
14. When the desired chase time has arrived, immediately resuspend the pellet in 10 mL of ice-cold NEB. Combine the two ½ pellets from the two 50 mL tubes for each sample in the same 10 mL of ice-cold NEB.
15. ALL SAMPLES: Rotate in NEB at 4°C for 15 min to obtain nuclei.
16. Centrifuge the nuclei for 10 min at 500 × *g* at 4°C to pellet.
    - Optional: After centrifugation, remove 1 mL of solubilized proteins above the nuclei to a 1.5 mL tube for protein measurement later. The amount of soluble protein is proportional to the total starting amount of cells.
17. Aspirate the NEB and resuspend the nuclei in 10 mL ice-cold PBS with a serological pipette to wash.
18. Centrifuge the nuclei for 10 min at 500 × *g* at 4°C to pellet.
19. During the centrifugation, prepare the click reaction/NCC mixture as described in Section 5.2.2.
20. Aspirate the PBS wash and completely resuspend nuclei in 10 mL of click/NCC reaction mixture.
21. Rotate samples at 4°C for 60 min.
22. Centrifuge the nuclei for 10 min at 500 × *g* at 4°C to pellet.
23. Aspirate the click/NCC reaction mixture and resuspend the nuclei in 10 mL ice-cold PBS with a serological pipette to wash.
24. Centrifuge the nuclei for 10 min at 500 × *g* at 4°C to pellet.
25. Either proceed to Section 5.3 immediately or freeze nuclei on dry ice and store at −80°C for up to 2 weeks before proceeding.

5.2.4 Notes

1. Consistency between aniPOND samples is critical. Large variations in the time of EdU pulse or thymidine chase between samples will confound results. We have established a set of conventions that we adhere to for every aniPOND sample:

I. Pulses are designed so that the *total time* in EdU is 15 min. This includes time during spins and after pelleting before resuspension in thymidine chase media or NEB.

II. Centrifugations are performed at 200 × *g* for 4 min at RT with moderate acceleration and deceleration.

III. If performing two samples simultaneously (for example, a treated and untreated), leave exactly 1 min between samples and maintain this separation and the order of sample processing until nuclei harvesting. This allows time for individual sample processing so that total time in EdU or thymidine media for each individual sample remains constant.

IV. If staggering multiple time points (for example, a 30 min chase with a 60 min chase), allot 1 min of extra time to pour samples into 50 mL tubes before starting centrifugation (step 10).

V. Record the total time of each sample in EdU and thymidine media (for chase samples).

2. Nuclei should largely remain individual and not form clumps during the nucleus extraction, PBS washing, and click reaction steps. If macroscopically visible clumps of nuclei form, ensure that the $MgCl_2$ concentration of the NEB is correct, the PBS formulation contains at least 130 m*M* NaCl, and that nuclei spins are being performed at 500 × *g*.
3. The nuclei may change color slightly after click reaction (from white to slightly yellow/green). Occasionally, a small amount of Cu^{1+} will form a brown precipitate that spins down with the nuclei. This precipitate does not interfere with downstream processing.
4. It is imperative that flasks are completely washed before addition of thymidine chase medium. Residual EdU-containing medium will lead to paradoxical EdU incorporation during chase steps, confounding the results.

5.3 Solubilization and Pulldown of Biotin-Labeled Chromatin

5.3.1 Equipment

- Rotating platform for 1.5 mL tubes at 4°C
- Refrigerated microcentrifuge for 1.5 mL tubes at 4°C
- Microtip sonicator such as a Branson 250 cell disruptor with double step microtip (Emerson, cat. no. 101-063-196 and VWR 33996-243, Eden Prairie, MN)
- Spectrophotometer and cuvettes for protein measurements
- Nanodrop spectrophotometer (Thermo Fisher Sci., Waltham, MA) for DNA measurements
- DNA purification columns (such as Qiagen, cat. no. 28104, Hilden, Germany)
- Agarose gel electrophoresis system

- UV lamp gel imaging system (such as Bio-Rad, cat. no. 1708195, Hercules, CA)

5.3.2 Buffers and Reagents

- Same as in Section 5.2, plus:
- EDTA. A 500 m*M* stock in double-distilled water, adjusted to pH 8.0, filter sterilized, and stored at 4°C.
- Tris–HCl pH 6.8 and pH 8.0. A 1 *M* stock in double-distilled water, adjusted to correct pH with HCl, autoclaved, and stored at RT.
- SDS. A 10% solution prepared in double-distilled water and stored at RT.
- $CaCl_2$. A 1 *M* stock in double-distilled water prepared and stored at RT.
- Bromophenol blue.
- Bovine serum albumin (BSA).
- Agarose for DNA analysis (such as BioExpress, cat. no. E-3120, Kaysville, UT)
- Ethidium bromide.
- RNAse A (such as Thermo Fisher Sci., cat. no. 12091021, Waltham, MA).
- Pronase (such as Sigma, cat. no. P6911, St. Louis, MO). A 20 mg/mL stock in double-distilled water prepared and stored at −20°C.
- Buffer B1 (25 m*M* NaCl, 2 m*M* EDTA, 50 m*M* Tris–HCl pH 8.0, 1% IGEPAL CA630). Prepare fresh before use and add protease inhibitors (such as Sigma-Aldrich, cat. no. P8340, St. Louis, MO).
- High capacity streptavidin agarose (Pierce, cat. no. 20359, Waltham, MA).
- 2 × Laemli buffer (4% SDS, 20% glycerol, 125 m*M* Tris–HCl pH 6.8, 0.01% bromophenol blue). Before use, add 50 μL of concentrated (14.3 *M*) βME to 950 μL 2 × Laemli buffer (final concentration, 5% v/v βME).
- TE buffer (10 m*M* Tris–HCl pH 8, 1 m*M* EDTA).
- Protein assay dye reagent (Bio-Rad, cat. no. 500-0006, Hercules, CA).

5.3.3 Procedure

1. Resuspend nuclei in 0.5 mL of ice-cold buffer B1 supplemented with protease inhibitors and transfer to a 1.5 mL tube.
2. Rotate at 4°C for 30 min.
3. Centrifuge nuclei for 10 min at 500 × *g* at 4°C to pellet.
4. Aspirate supernatant.
5. Resuspend nuclei in 0.5 mL of ice-cold buffer B1 supplemented with protease inhibitors.
6. Rotate at 4°C for 30 min.
7. Solubilize chromatin with a microtip sonicator. Keep tubes on ice at all times and perform 6 rounds of 2 × 10 s on, 10 s off on output 3 to 4

(~10 W output) for a total of 12 × 10-s bursts. Between rounds, rest tubes on ice for at least 1 min.

8. Centrifuge samples for 10 min at 16.1k × *g* at 4°C.
9. Carefully remove the clarified, sonicated chromatin fraction to a new 1.5 mL tube.
10. Measure the volume of sonicated chromatin and bring the volume to 1 mL by adding ice-cold buffer B1 supplemented with protease inhibitors.
11. Gently mix the sonicated chromatin samples and remove 20 μL (2%) to a new 1.5 mL tube as input (INP). Add 20 μL 2 × Laemli buffer (with βME) to the input sample and store at −20°C until the next day.
12. OPTIONAL: Remove 10 μL (1%) to a new 1.5 mL tube for DNA analysis.

    DNA Analysis Steps:

    I. Add 79 μL of TE buffer and 1 μL of 10–20 mg/mL RNAse A and incubate at 37°C for 30 min to degrade RNA.
    II. Add 10 μL of 20 mg/mL pronase and 1 μL of 1 *M* $CaCl_2$ and incubate at 42°C for 2 h to digest proteins.
    III. Column purify DNA using a commercial kit.
    IV. Measure DNA quantity with a nanodrop spectrophotometer.
    V. Visualize DNA fragment size distribution by running >500 ng on a 1.3% agarose/TAE gel, staining with 0.2 μg/mL ethidium bromide, and visualizing with a UV lamp gel imaging system. Compare fragments to an appropriate DNA ladder.
13. Measure the protein content of sonicated chromatin and the optional soluble fraction from Section 5.2 step 16 by Bradford assay with BSA standards.
14. Prepare streptavidin beads to be added to chromatin samples for pulldown. Each sample requires 100 μL of bead slurry (50% slurry). Remove enough volume of well-mixed beads (plus enough for two extra samples) to a 1.5 mL tube.
15. Mark the level of the beads slurry on the side of the tube.
16. Centrifuge the beads slurry for 2 min at 500 × *g* at 4°C.
17. Let the slurry sit on ice for 2 min to completely settle.
18. Gently remove the supernatant and add at least the same volume as the original beads slurry volume of ice cold buffer B1 to wash (no protease inhibitors necessary). For example, if 600 μL of beads slurry was removed, at least 600 μL of buffer B1 would be added to wash the beads.
19. Repeat steps 16–18 two more times for a total of three beads washes.
20. After the third wash, bring the volume of buffer B1 to the level of the original bead slurry volume as marked.

21. Pipette 100 μL of buffer B1-washed beads (50% slurry) into the sonicated chromatin samples, using cut-tip P200 tips, and mixing the beads well between additions to ensure equal addition to the different samples.
22. Rotate samples with beads overnight at 4°C.

5.3.4 Notes

1. Sonicator microtips can corrode and pit over time, leading to reduced output. Periodic polishing with metal sandpaper (such as 3M, cat. no. 11694, Maplewood, MN) when corrosion is observed prolongs the life of the microtip before replacement is required.
2. It is recommended to check both protein and DNA recovery in each sample as a control in order to provide a measure of the efficiency of sonication.
3. Solubilized chromatin released by sonication should have a mean distribution of ~150 bp upon DNA analysis. Most of the DNA fragmentation occurs during the click reaction by the actions of Cu^{1+} (see Fig. 10).

5.4 Protein Elution and Quantitative Western Analysis

5.4.1 Equipment

- Same as in Section 5.3, plus:
- 200 μL tapered gel loading tips (such as USA Sci., cat. no. 1252-0600, Ocala, FL)
- Low fluorescence background PVDF membrane (such as Bio-Rad, cat. no. 1620261, Hercules, CA) or nitrocellulose
- Blocking solution for NIR Western blotting (such as Li-Cor, cat. no. 927-50000, Lincoln, NE)
- Odyssey NIR imaging system (Li-Cor Biosciences, Lincoln, NE)

5.4.2 Buffers and Reagents

- Same as in Section 5.3, plus:
- Tris-buffered saline (TBS; 150 m*M* NaCl, 50 m*M* Tris–HCl, pH 7.5)
- TBST (TBS supplemented with 0.1% Tween-20)

5.4.3 Procedure

1. Centrifuge the bead capture samples for 2 min at 500 × *g* at 4°C.
2. Incubate on ice for 2 min to completely settle.
3. Gently aspirate the supernatant to ~100 μL above the beads and add 1 mL of buffer B1 (no protease inhibitors necessary).
4. Rotate samples for 5 min at 4°C to wash.
5. Repeat steps 1–4 three more times for a total of four washes.

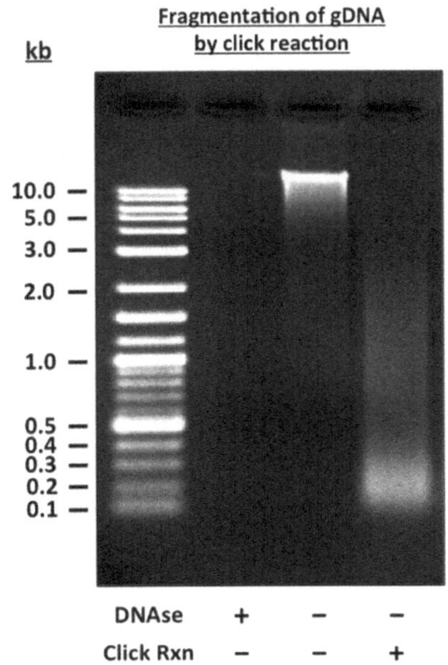

Fig. 10 Fragmentation of genomic DNA (gDNA) by Cu^{1+} produced in the click reaction.

6. After the final wash, centrifuge the beads capture samples for 2 min at 500 × *g* at 4°C.
7. Incubate on ice for 2 min to completely settle.
8. Gently aspirate the supernatant to ~100 μL above the beads, then pull the remainder of the wash from below the beads using 200 μL tapered gel loading tips. Press the tip against the bottom of the tube before pulling up to avoid taking up any beads.
9. Resuspend the washed beads from the capture samples in 100 μL of 2 × Laemli buffer (with βME) with a cut-tip P200 tip.
10. Thaw INPs on ice.
11. Boil the beads capture (CAP) and INP samples for 15 min. Place safety caps on the tubes to prevent the caps from popping open during boiling.
12. Vortex CAP samples for 3 s, and set both CAP and INP samples on ice for ~1 min to cool to RT.
13. Centrifuge the CAP and INP samples for at least 1 min at 1500 × *g* at RT.
14. Remove the eluted CAP samples in Laemli buffer from below the beads using 200 μL tapered gel loading tips and transfer to new 1.5 mL tubes.
15. The CAP and INP samples can now either be stored at −20°C or processed for quantitative Western blot analysis immediately. If stored

before use, thaw on ice and then briefly boil (2 min), cool on ice, centrifuge, and then proceed.

16. For quantitative Western analysis of the control proteins PCNA and histone H4, freshly prepare a 12% SDS-PAGE gel and perform standard gel electrophoresis. For NIR detection of proteins, which we have found to be less sensitive than HRP detection, it is useful to load 20 μL of INP sample (1%) and 50 μL of CAP sample (50%).
17. Transfer proteins onto either nitrocellulose or low infrared-background PVDF membrane. Standard PVDF is unsuitable due to variable and potentially high fluorescent background.
18. Block membranes in NIR blocking solution. It is useful to dilute NIR blocking solution 1:3 in TBS, as full-strength NIR blocking solutions can reduce antibody binding. Traditional milk-blocking solutions should be avoided as they may increase fluorescent background.
19. After blocking, cut the membrane at the 20 kDa marker. Incubate the >20 kDa membrane segment in Mouse anti-PCNA (Santa Cruz, cat. no. sc-56, Dallas, TX) diluted 1:200 in TBST and the <20 kDa segment in Mouse anti-H4 (Abcam, cat. no. ab17036, Cambridge, United Kingdom) diluted 1:1000 in TBST. Incubate in primary antibody for 4 h at RT or overnight at 4°C.
20. Wash the membranes four times for 5 min with TBST at RT.
21. Incubate in secondary antibody for 1 h at RT protected from light. For both primary antibodies, use Goat anti-Mouse 680 nm (Li-Cor Biosciences, cat. no. 925-68070, Lincoln, NE) diluted 1:5000 in TBS.
22. Wash the membranes four times for 5 min with TBST at RT.
23. Wash the membranes one time for 5 min with TBS.
24. Image blots with an Odyssey NIR imaging system using the 680 nm channel. Expected results (see Fig. 8): PCNA should be present on DNA with a 15 min EdU pulse but should offload after thymidine chase. Histone H4 may be begin to be present with a 15 min EdU pulse and will increase in signal with thymidine chase as more histone is deposited postreplication. NCC CAP samples should have little to no signal.

5.4.4 Notes

1. For NIR Western blotting, great care must be taken when loading molecular weight marker ladders, as the dyes that stain marker proteins can strongly fluoresce in the NIR channels. A useful ladder is all blue prestained ladder (Bio-Rad, cat. no. 1610373, Hercules, CA), which fluoresces strongly in the 680 nm channel. Load no more than 2.5 μL.

5.5 Troubleshooting

Problem	Reason(s) and Solution
Cells are growing poorly (T_d does not match published values)	Suboptimal proliferation may indicate the absence of a required growth factor, incorrect media conditions, suboptimal oxygenation, epigenetic changes from cell mistreatment, or *Mycoplasmsa* or other contamination. Ensure that the correct media and supplements are being used. Thaw a fresh stock from a vial of early passage cells. Grow cells in flasks on their side to increase surface area for gas exchange and test for *Mycoplasma* contamination. Culturing cells without antibiotics will allow any low-level contamination to become evident.
Inefficient pulldown of control proteins (PCNA, H4) in click reaction samples	Failure at various steps can lead to low signal in click reaction samples. We recommend the following steps to resolve this issue: 1. Failure of click reaction. Any of the components of the click reaction may go off. We suggest ordering new EdU, biotin-PEG3-azide, and sodium ascorbate. To test if Cu^{1+} is being generated in the click reaction, a mock click reaction with purified genomic DNA (use a kit such as Thermo Fisher Sci., cat. no. K1820-01, Waltham, MA) can be performed and DNA fragmentation can be assessed by agarose gel electrophoresis with ethidium bromide staining (see Fig. 10). Intact fragmentation excludes the (+)-sodium L-ascorbate and $CuSO_4$ as the source of the problem. 2. Antibody or Western blotting issues. A robust INP signal should be present for PCNA and H4. A weak INP signal suggests that either the antibody is expired or a step in the Western blotting process has failed, such as the transfer. 3. Failure of chromatin solubilization. To test if chromatin is being solubilized by sonication, compare the PCNA and H4 content of the insoluble pellet to the solubilized chromatin postcentrifugation (see Fig. 5B). If a majority of chromatin is being retained in the pellet, consider maintenance of the sonicator by polishing/replacing the microtip. Additionally, ensure that buffer B1 has the proper composition. 4. Failure of beads pulldown or elution. Order new high capacity streptavidin beads and ensure that the SDS elution buffer (Laemli buffer) has correct composition and is supplemented with βME before use. 5. Reagent supplier. We have experienced multiple bad batches of click reagents from a single supplier. Consider switching suppliers if pulldown is not restored after performing solution steps 1–4 above.
Weak signal for protein of interest	If performing quantitative near-infrared (NIR) fluorescence Western blotting, reduce the stringency of the wash steps and increase the concentration of NIR 2° antibody. Some proteins may require visualization with HRP-conjugated 2° antibody. High sensitivity chemiluminescence substrate (such as Thermo Fisher Sci., cat. no. 34095, Waltham, MA) development of blots probed with HRP-conjugated 2° antibody will allow detection of low abundance proteins that fall below the threshold of NIR detection. If signal is still not observed, increase the amount of starting material and verify that published iPOND or aniPOND mass spectrometry experiments detect your protein of interest on replicating DNA.

ACKNOWLEDGMENTS

The authors thank members of the Tomkinson laboratory (with special mention to Dr. Annahita Sallmyr), as well as members of the laboratories of Dr. David Cortez and Dr. Rod Bremner for helpful discussions and observations regarding both the iPOND and aniPOND protocols. This work was supported by the University of New Mexico Comprehensive Cancer Center (P30 CA118100) and National Institute of Health Grants R01 GM57479 and CA92584 (to A.E.T.).

REFERENCES

Aranda, S., Rutishauser, D., & Ernfors, P. (2014). Identification of a large protein network involved in epigenetic transmission in replicating DNA of embryonic stem cells. *Nucleic Acids Research*, *42*(11), 6972–6986. http://dx.doi.org/10.1093/nar/gku374.

Chehrehasa, F., Meedeniya, A. C. B., Dwyer, P., Abrahamsen, G., & Mackay-Sim, A. (2009). EdU, a new thymidine analogue for labelling proliferating cells in the nervous system. *Journal of Neuroscience Methods*, *177*(1), 122–130. http://dx.doi.org/10.1016/j.jneumeth.2008.10.006.

Dembowski, J. A., & DeLuca, N. A. (2015). Selective recruitment of nuclear factors to productively replicating herpes simplex virus genomes. *PLoS Pathogens*, *11*(5). e1004939. http://dx.doi.org/10.1371/journal.ppat.1004939.s001

Duderstadt, K. E., Reyes-Lamothe, R., van Oijen, A. M., & Sherratt, D. J. (2014). Replication-fork dynamics. *Cold Spring Harbor Perspectives in Biology*, *6*(1), a010157. http://dx.doi.org/10.1101/cshperspect.a010157.

Dungrawala, H., Rose, K. L., Bhat, K. P., Mohni, K. N., Glick, G. G., Couch, F. B., et al. (2015). The replication checkpoint prevents two types of fork collapse without regulating replisome stability. *Molecular Cell*, *59*(6), 998–1010. http://dx.doi.org/10.1016/j.molcel.2015.07.030.

Hyrien, O. (2015). Peaks cloaked in the mist: The landscape of mammalian replication origins. *The Journal of Cell Biology*, *208*(2), 147–160. http://dx.doi.org/10.1242/jcs.086561.

Kanemaki, M., & Labib, K. (2006). Distinct roles for Sld3 and GINS during establishment and progression of eukaryotic DNA replication forks. *The EMBO Journal*, *25*(8), 1753–1763. http://dx.doi.org/10.1038/sj.emboj.7601063.

Leung, K. H. T., Abou El Hassan, M., & Bremner, R. (2013). A rapid and efficient method to purify proteins at replication forks under native conditions. *BioTechniques*, *55*(4), 204–206. http://dx.doi.org/10.2144/000114089.

Lopez-Contreras, A. J., Ruppen, I., Nieto-Soler, M., Murga, M., Rodriguez-Acebes, S., Remeseiro, S., et al. (2013). A proteomic characterization of factors enriched at nascent DNA molecules. *Cell Reports*, *3*(4), 1105–1116. http://dx.doi.org/10.1016/j.celrep.2013.03.009.

Meneghini, R. (1997). Iron homeostasis, oxidative stress, and DNA damage. *Free Radical Biology and Medicine*, *23*(5), 783–792.

Metcalf, D., Nossal, G. J., Warner, N. L., Miller, J. F., Mandel, T. E., Layton, J. E., et al. (1975). Growth of B-lymphocyte colonies in vitro. *The Journal of Experimental Medicine*, *142*(6), 1534–1549.

Min, W., Bruhn, C., Grigaravicius, P., Zhou, Z.-W., Li, F., Krüger, A., et al. (2013). Poly (ADP-ribose) binding to Chk1 at stalled replication forks is required for S-phase checkpoint activation. *Nature Communications*, *4*, 2993. http://dx.doi.org/10.1038/ncomms3993.

Nieduszynski, C. A. (2006). Genome-wide identification of replication origins in yeast by comparative genomics. *Genes & Development*, *20*(14), 1874–1879. http://dx.doi.org/10.1101/gad.385306.

Olcina, M. M., Giaccia, A. J., & Hammond, E. M. (2016). Isolation of Proteins on Nascent DNA in Hypoxia and Reoxygenation Conditions. Dans In C. Koumenis, L. Coussens, A. Giaccia, & E. Hammond (Eds.), *Advances in Experimental Medicine and Biology: 899. Tumor Microenvironment* (pp. 27–40). Cham: Springer International Publishing. http://dx.doi.org/10.1007/978-3-319-26666-4_3.

Presolski, S. I., Hong, V. P., & Finn, M. G. (2009). *Copper-catalyzed azide-alkyne click chemistry for Bioconjugation*. Hoboken, NJ, USA: John Wiley & Sons, Inc. http://dx.doi.org/10.1002/9780470559277.ch110148

Sirbu, B. M., Couch, F. B., & Cortez, D. (2012). Monitoring the spatiotemporal dynamics of proteins at replication forks and in assembled chromatin using isolation of proteins on nascent DNA. *Nature Protocols*, 7(3), 594–605. http://dx.doi.org/10.1038/nprot.2012.010.

Sirbu, B. M., Couch, F. B., Feigerle, J. T., Bhaskara, S., Hiebert, S. W., & Cortez, D. (2011). Analysis of protein dynamics at active, stalled, and collapsed replication forks. *Genes & Development*, *25*(12), 1320–1327. http://dx.doi.org/10.1101/gad.2053211.

Sirbu, B. M., McDonald, W. H., Dungrawala, H., Badu-Nkansah, A., Kavanaugh, G. M., Chen, Y., et al. (2013). Identification of proteins at active, stalled, and collapsed replication forks using isolation of proteins on nascent DNA (iPOND) coupled with mass spectrometry. *Journal of Biological Chemistry*, *288*(44), 31458–31467. http://dx.doi.org/10.1074/jbc.M113.511337.

Stathopulos, P. B., Scholz, G. A., Hwang, Y.-M., Rumfeldt, J. A. O., Lepock, J. R., & Meiering, E. M. (2004). Sonication of proteins causes formation of aggregates that resemble amyloid. *Protein Science: A Publication of the Protein Society*, *13*(11), 3017–3027. http://dx.doi.org/10.1110/ps.04831804.

Stuart, S. E., Clawson, G. A., Rottman, F. M., & Patterson, R. J. (1977). RNA transport in isolated myeloma nuclei. Transport from membrane-denuded nuclei. *The Journal of Cell Biology*, *72*(1), 57–66.

Tatchell, G. K. (1978). *Physical structure and reconstitution of chromatin core particles. PhD Thesis*. Oregon State University. http://hdl.handle.net/1957/43021.

Thoma, F., Koller, T., & Klug, A. (1979). Involvement of histone H1 in the organization of the nucleosome and of the salt-dependent superstructures of chromatin. *The Journal of Cell Biology*, *83*(2 Pt. 1), 403–427.

Trujillo, K. M., & Osley, M. A. (2012). A role for H2B ubiquitylation in DNA replication. *Molecular Cell*, *48*(5), 734–746. http://dx.doi.org/10.1016/j.molcel.2012.09.019.

van Holde, K. E. (1989). *Springer series in molecular biology: Chromatin*. (S.l.) New York, NY: Springer-Verlag New York Inc.

Wang, G., Li, Y., Wang, P., Liang, H., Cui, M., Zhu, M., et al. (2015). PTEN regulates RPA1 and protects DNA replication forks. *Nature Publishing Group*, *25*(11), 1189–1204. http://dx.doi.org/10.1038/cr.2015.115.

Wells, C. E., Bhaskara, S., Stengel, K. R., Zhao, Y., Sirbu, B., Chagot, B., et al. (2013). Inhibition of histone deacetylase 3 causes replication stress in cutaneous T cell lymphoma. *PloS One*, *8*(7). e68915. http://dx.doi.org/10.1371/journal.pone.0068915.s005.

CHAPTER TWO

Proteomic Analyses of the Eukaryotic Replication Machinery

David Cortez[1]
Vanderbilt University School of Medicine, Nashville, TN, United States
[1]Corresponding author: e-mail address: david.cortez@vanderbilt.edu

Contents

Abstract

DNA replication in a human cell involves hundreds of proteins that copy the DNA accurately and completely each cell division cycle. In addition to the core DNA copying machine (the replisome), accessory proteins work to respond to replication stress, correct errors, and repackage the DNA into appropriate chromatin structures. New proteomic tools have been invented in the past few years to facilitate the purification, identification, and quantification of the replication, chromatin maturation, and replication stress response machineries. These tools, including iPOND (isolation of proteins on nascent DNA) and NCC (nascent chromatin capture), have yielded discoveries of new proteins involved in these processes and insights into the dynamic regulatory processes ensuring genome and chromatin integrity. In this review, I will introduce these experimental approaches and examine how they have been utilized to define the replication fork proteome.

Methods in Enzymology, Volume 591
ISSN 0076-6879
http://dx.doi.org/10.1016/bs.mie.2017.03.002

1. INTRODUCTION

The genome must be copied completely and accurately each cell division cycle, and the newly synthesized DNA must be repackaged into chromatin in ways that control epigenetic regulation of gene expression and other aspects of chromosome function. Numerous challenges impede these processes, including DNA damage, difficult to replicate DNA sequences, and conflicts with transcription. Large and dynamic protein machines overcome these challenges and maintain genome stability.

The replisome is the DNA copy machine that contains helicases to unwind the DNA duplex, polymerases to copy the DNA, as well as nucleases and ligases to process the discontinuous stretches of DNA on the lagging strand (Bell & Dutta, 2002). The replisome also contains accessory factors that increase efficiency and serve to couple leading and lagging strand synthesis. Proteins involved in chromatin disassembly, reassembly, and modification are tethered to the replisome ensuring a tight coordination between chromatin and DNA replication (Probst, Dunleavy, & Almouzni, 2009). Error correction mechanisms such as mismatch repair proteins, and replication stress response proteins that deal with replication challenges are also tethered via direct interactions with replisome subunits or are recruited to replication forks following changes in DNA structure (Cimprich & Cortez, 2008; Moldovan, Pfander, & Jentsch, 2007; Nam & Cortez, 2011).

Obtaining an inventory of all of the proteins involved in DNA and chromatin replication is a requirement to fully understand the process. Certainly, most of the core essential components of the eukaryotic replisome are known. However, there continues to be a steady stream of discoveries of new accessory factors that influence the fidelity and efficiency of DNA replication. Furthermore, we have a long way to go to understand how these proteins are regulated and coordinated within these dynamic machines especially in circumstances when a replication fork encounters an obstacle.

In this review, I will discuss recent advances in proteomic methods to purify, identify, and characterize replication fork-associated proteins, concentrating on approaches applicable to mammalian cells. These methods have the potential to complete the inventory of the DNA and chromatin replication proteome and will be useful in interrogating the regulation of this machinery. I will summarize the findings from these approaches, highlighting their strengths and weaknesses.

2. REPLISOME PURIFICATION APPROACHES

The most obvious way to identify proteins that replicate DNA and chromatin is to purify these protein machines and use mass spectrometry to identify them. One approach to purification is to use a known replisome subunit as the bait to fish out the interacting complex. However, this approach does not necessarily distinguish between active replisomes and complexes that are either not bound to DNA or not engaged in active synthesis. In addition, multiple protein baits would need to be utilized to try to ensure most of the machinery is purified.

A second approach is to compare chromatin-associated proteins in replicating vs quiescent cells with the assumption that replisome subunits will only be associated with chromatin during DNA replication. The vast numbers and abundance of chromatin proteins compared to replication-specific proteins in cells makes this approach difficult, and again it does not necessarily distinguish between active and inactive complexes. Nonetheless, the robustness of replication in *Xenopus* egg extracts provides an opportunity to use this approach in an in vitro replication system. Chromatin mass spectrometry (CHROMASS) takes advantage of this system and was used to identify proteins that are recruited to chromatin that has been damaged with a DNA crosslinking agent (Raschle et al., 2015). Thus, CHROMASS is especially useful in situations where an investigator wants to interrogate chromatin composition in the context of defined DNA damage structures. However, this is an in vitro replication system, it does not directly answer whether a protein is enriched at a replication fork, and is not easily adapted to mammalian cells.

Two other purification strategies, isolation of proteins on nascent DNA (iPOND) and nascent chromatin capture (NCC) were recently developed that overcome these limitations (Alabert et al., 2014; Sirbu et al., 2011). Instead of targeting the machinery itself for purification, these methods purify the newly synthesized DNA and then interrogate the associated proteins. Proteins involved in DNA and chromatin replication are identified by their enrichment with nascent DNA compared to bulk chromatin.

2.1 Isolation of Proteins on Nascent DNA

iPOND takes advantage of the nucleoside analog 5-ethynyl-2′-deoxyuridine (EdU). When added to cell culture, EdU is quickly imported into the cell, phosphorylated, and incorporated into newly synthesized

DNA by the replisome polymerases (Salic & Mitchison, 2008). EdU contains an alkyne group that can be covalently linked to an azide-containing molecule using click chemistry. Thus, biotin can be conjugated to the EdU to facilitate a single-step affinity purification of the newly synthesized DNA and bound proteins using streptavidin (Fig. 1A). A sonication procedure combined with copper-catalyzed DNA hydrolysis during the click chemistry step generates protein-bound DNA fragments ranging from 100 to 200 base pairs. Thus, the method can provide high spatial resolution largely determined by the rate of DNA polymerization and EdU labeling time (Dungrawala & Cortez, 2015; Sirbu, Couch, & Cortez, 2012; Sirbu et al., 2011). The eukaryotic replication fork elongates at rates between 0.5 and 2 kb/min with considerable variation in speed depending on factors such as early or late S-phase, DNA sequence, and chromatin composition. Typically, 5–10 min incorporation times are utilized in iPOND experiments although labeling times of as short as 2.5 min have been reported (Sirbu et al., 2011). Removing the EdU from the growth media and incubating in the

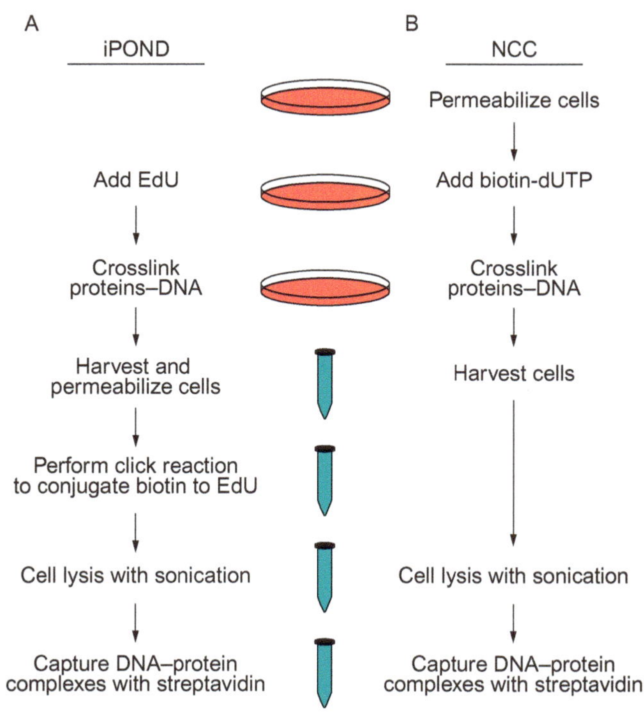

Fig. 1 Schematic of the (A) iPOND and (B) NCC approaches to purifying proteins associated with nascent DNA.

absence of EdU can generate "chase" samples that provide spatial information of where proteins are with respect to the fork. This chase sample is essential to distinguish the proteins that participate in DNA and chromatin replication vs those that are components of bulk chromatin (Fig. 2).

Six studies utilized iPOND combined with different types of mass spectrometry to identify the replication fork proteome (Table 1). All compared an EdU-labeled sample with a chase sample to find proteins enriched on nascent DNA compared to mature chromatin. The primary differences in the protocols were in how the quantitative mass spectrometry was performed. Later, I summarize the results from each of these studies emphasizing their strengths and weaknesses.

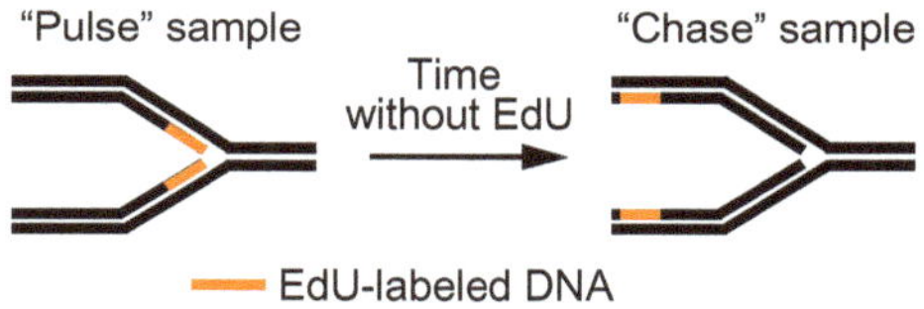

Fig. 2 A pulse–chase protocol is essential to identify proteins that are specifically enriched near the replication fork vs proteins that are associated more generally with chromatin. This methodology also facilitates analysis of chromatin maturation.

Table 1 Summary of Manuscripts Using iPOND or NCC to Identify Proteins Associated With Replication Forks

Study	Method	Mass Spectrometry Method	Number of Replicates	Number of Proteins
Lopez-Contreras et al. (2013)	iPOND	Label-free	6	48
Sirbu et al. (2013)	iPOND	Label-free	5	84
Lossaint et al. (2013)	iPOND	Label-free	3/2[a]	n/a[b]
Aranda, Rutishauser, and Ernfors (2014)	iPOND	Label-free	4	207
Alabert et al. (2014)	NCC	SILAC	3	462
Dungrawala et al. (2015)	iPOND	SILAC	3	218
Lecona et al. (2016)	iPOND	iTRAQ	1	n/a[b]

[a]3 "pulse" samples and 2 "chase" samples.
[b]These investigators did not attempt to designate which proteins should be considered significantly enriched at forks.

2.1.1 iPOND-Label-Free MS

Four groups utilized label-free mass spectrometry quantitation procedures combined with iPOND (Aranda et al, 2014; Lopez-Contreras et al., 2013; Lossaint et al., 2013; Sirbu et al., 2013). Label-free mass spectrometry utilizes the information derived from tandem mass spectrometry such as number of peptide spectra for a given protein or the abundance of the precursor ion to assign an abundance value to that protein within the dataset. The EdU pulse and chase samples are collected and processed independently, and then the data compared after collection (Fig. 3A). Label-free mass spectrometry methods are compatible with any source of proteins and are relatively inexpensive. However, label-free methods are

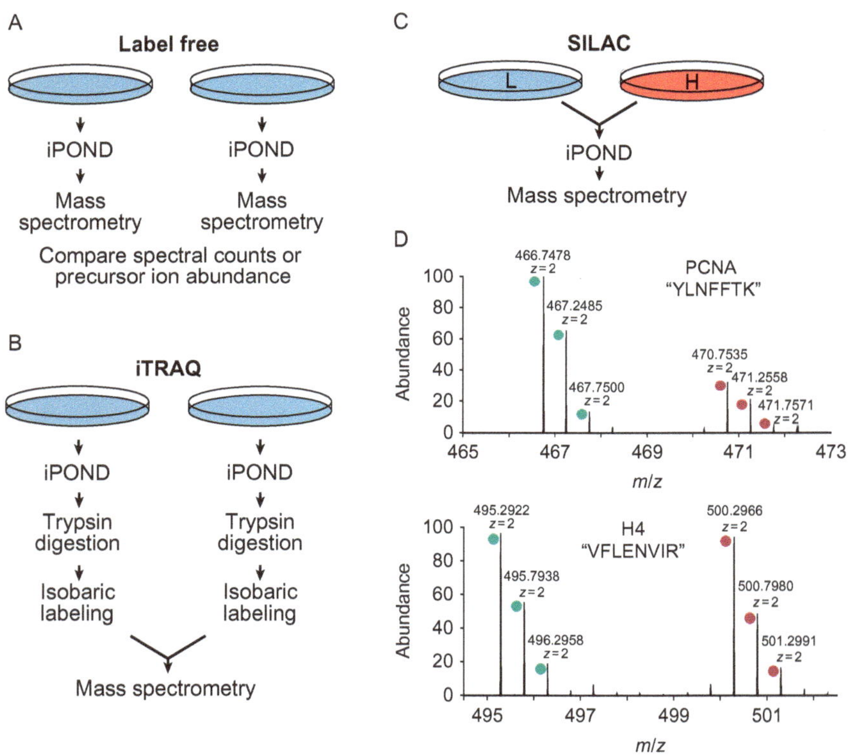

Fig. 3 Comparison of mass spectrometry approaches that have been combined with iPOND. (A) Label-free methods depend on comparing datasets that are generated independently. (B) iTRAQ improves the quantitation precision but does not eliminate variability associated with purification. (C) SILAC improves precision and reduces variation since heavy and light samples are combined prior to purification. (D) Example of data obtained from a SILAC experiment in which the light sample was labeled with EdU and the heavy sample was the chase.

much less quantitatively precise and can suffer from reproducibility problems.

Three different label-free methods have been combined with iPOND. Two studies utilized spectral counting (Lossaint et al., 2013; Sirbu et al., 2013). By counting the number of spectra corresponding to peptides from a given protein in either the pulse or chase sample, a ratio was derived to describe the relative amounts of a protein at the replication fork vs in bulk chromatin. Spectral counting is simple; however, it is also the least accurate quantification method. One of these studies performed by our research group identified 84 proteins as enriched on nascent DNA after completing five replicate samples (Sirbu et al., 2013). The second study performed smaller number of replicates and did not attempt to provide a cutoff for what should be considered significant or provide a measure of reproducibility in the measurements (Lossaint et al., 2013).

The 84 proteins identified in our study had highest GO enrichment values for DNA metabolic process. However, the data suffer from a high false-negative rate, which is illustrated by the failure to identify many subunits of the replicative helicase (Sirbu et al., 2013). Examining the 100 proteins with the largest pulse/chase ratios in the Lossaint et al. dataset reveals that these proteins are most highly enriched in carboxylic acid metabolic processes (37 proteins) and DNA replication is only the seventh most enriched biological process (16 proteins) (Lossaint et al., 2013). Thus, the limitations of spectral counting as a quantitative measurement can generate high false-positive and false-negative discovery rates.

A second label-free mass spectrometry method employed in conjunction with iPOND is a variation of spectral counting which utilizes an algorithm that assigns a probability-based score (Mascot score) to each protein identified in a sample. Mascot scores were initially designed to indicate a probability of correct protein identification so it is unclear how effectively they quantitate protein abundance (Perkins, Pappin, Creasy, & Cottrell, 1999). Nonetheless, the investigators using this method identified 207 proteins enriched on nascent DNA based on the ratio of Mascot scores in pulse vs chase samples (Aranda et al., 2014). The list is highly enriched in DNA metabolism and replication proteins, but it does not contain a quantitative measurement of reproducibility. Furthermore, for many proteins, quantitative ratios were not generated. Despite these limitations, the dataset generated seems superior to either of the spectral counting datasets based on the number of known replication proteins identified and their enrichment values in gene ontology analyses.

The third label-free method coupled to iPOND quantitated the precursor ion abundance in the MS–MS data (Lopez-Contreras et al., 2013). This method is more accurate than spectral counting but still suffers from relatively high variability. The lack of precision is illustrated by the large variations in the reported abundance of the six subunits of the MCM2–7 complex in this dataset. For example, among the biological replicates, the log2 of the abundance ratio of MCM2 varied from −5.75 to +11.46 (where a positive value indicates enrichment at the fork, zero indicates no enrichment, and a negative value indicates exclusion from the fork). To try to compensate for the lack of precision, six biological replicates were performed and only proteins that were observed to be enriched at least eightfold in at least all experiments but one in which they were observed were reported to be enriched on the nascent DNA. This list included 48 proteins and the authors acknowledge the method suffers from a high false-negative rate. Only 3 of the 6 MCM2–7 subunits met the criteria for a fork-associated protein. Despite the stringent criteria applied to the dataset, it also seems likely that it contains false-positives since predominantly cytoplasmic proteins like Tubulin and a ribosomal protein were scored as hits.

Overall, the value of combining iPOND with the label-free mass spectrometry methods is primarily limited to the generation of a starting list for a candidate approach to finding new replication machinery proteins. The high false-positive and false-negative rates preclude a description of the replication fork proteome from these datasets. Two more quantitative mass spectrometry methods, iTRAQ and SILAC provide superior data to overcome these limitations.

2.1.2 iPOND-iTRAQ-MS

iTRAQ (isobaric tags for relative and absolute quantification) involves labeling peptides from different experimental conditions with different mass tags, combining the labeled peptides, and then performing mass spectrometry (Fig. 3B). iTRAQ yields highly quantitative results and has some other advantages. Since it involves labeling peptides after cell lysis, it can be used in systems like samples from intact tissues where other methods such as SILAC (stable isotope labeling with amino acids in cell culture) are difficult or impossible. Second, iTRAQ can be used to compare four or more samples in a single experiment. Unfortunately, the method tends to minimize abundance differences—compressing the relative abundance ratios (Rauniyar & Yates, 2014). It also is less precise and reproducible than SILAC due to differences in peptide labeling efficiencies and the need to process individual samples through the purification procedure separately prior to combining the labeled peptides for mass spectrometry.

The dataset reported by Lecona et al. illustrates the abundance ratio compression problem (Lecona et al., 2016). Only 31 proteins were found to be at least twofold enriched on the nascent DNA with the highest log2 abundance ratio being 2.1 for PCNA. Most known replication fork proteins including all the subunits of the replicative helicase were enriched at ratios that would be difficult to know are significant. For example, the average log2 ratio for the MCM2–7 complex subunits was only 0.26 and 570 proteins had higher or equal enrichment than the lowest MCM subunit (MCM4, log2 = 0.15). This dataset also did not include biological replicates. Perhaps for these reasons the authors did not attempt to define which proteins were actually enriched on the nascent DNA and instead used the dataset to identify candidates for further analyses.

2.1.3 iPOND-SILAC-MS

Our group combined iPOND with SILAC mass spectrometry (Dungrawala et al., 2015). SILAC relies on the incorporation of isotopically distinct amino acids during protein synthesis in cells (Ong & Mann, 2005). Two cell populations are prepared by growing in "light" or "heavy" media for several generations to isotopically label nearly 100% of the proteome. These cell populations are equivalent to one another with the exception of the small mass differences in each protein. Thus, EdU-pulse and chase samples can be compared directly by examining the abundance of the heavy vs light version of a peptide. SILAC minimizes experimental variability since the samples to be compared are mixed prior to performing the iPOND procedure (Fig. 3C and D). It can also provide highly quantitative results with measured variability of less than 20% in many cases (Dungrawala et al., 2015). Disadvantages include the relatively high cost and limitation to systems that are amenable to metabolic labeling like cell culture. Generally, only two samples can be compared although this can be stretched to additional comparison by utilizing high precision mass spectrometers.

The advantages of iPOND-SILAC-MS are readily apparent from the datasets that are generated. In the studies from our lab, three iPOND biological replicates generated a list of 218 proteins with average enrichment ratios at replication forks compared to mature chromatin of greater than two (Dungrawala et al., 2015). These 218 proteins included all of the known replication elongation proteins except GINS2, POLD4, POLE3, DNA2, and RPA3 (Fig. 4A). Four of these proteins are small with few

A

UniProt accession	Gene symbol	Reason
Dungrawala et al., 2015		
P35244	RPA3	Not observed; 14 kDa
Q9Y248	GINS2	Not observed; 21 kDa
Q9HCU8	POLD4	Not observed; 12 kDa
P51530	DNA2	Log2 enrichment 0.87
Q9NRF9	POLE3	Not observed; 17 kDa
Alabert et al., 2015		
P35244	RPA3	Filtered out improperly
Q9HCU8	POLD4	Not observed; 12 kDa
P51530	DNA2	Not observed; 120 kDa
P39748	FEN1	Log2 enrichment 0.4
Q9BRT9	GINS4	Filtered out improperly
Q14691	GINS1	Filtered out improperly
Q9NRF9	POLE3	Log2 enrichment –0.15

B

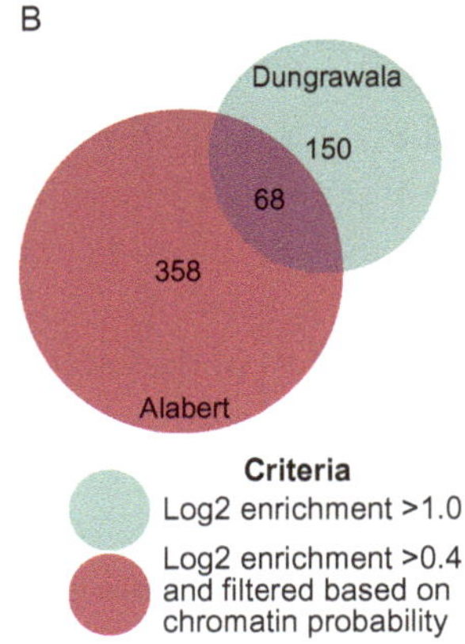

C

ToppGene pathway	*P* value	No. of proteins	Total in pathway
Dungrawala et al., 2015			
DNA repair	2.70E–08	10	113
Fanconi anemia pathway	2.32E–07	7	53
Base excision repair	2.46E–07	6	33
DNA strand elongation	4.89E–06	5	31
Base excision repair	1.52E–05	4	19
Resolution of abasic sites (AP sites)	1.52E–05	4	19
Unwinding of DNA	8.53E–05	3	11
Resolution of AP sites via the single-nucleotide replacement pathway	1.13E–04	3	12
DNA replication	2.08E–04	4	36
E2F transcription factor network	3.29E–04	5	73
Alabert et al., 2014			
Cell cycle	3.41E–17	42	514
Ribosome biogenesis in eukaryotes	3.50E–16	19	85
M Phase	4.58E–14	26	232
Cell cycle, mitotic	3.11E–13	33	416
Nuclear envelope breakdown	6.69E–12	12	43
Mitotic anaphase	3.74E–10	19	179
Mitotic metaphase and anaphase	4.12E–10	19	180
Mitotic prophase	7.02E–10	12	62
Spliceosome	1.23E–09	16	131
Resolution of sister chromatid cohesion	3.38E–09	14	103

Fig. 4 Comparison of the iPOND-SILAC and NCC-SILAC datasets. (A) Known replication fork proteins that were not identified or filtered out by the bioinformatics approach used in the NCC study. (B) Comparison of the total numbers of proteins identified in each dataset and the criteria utilized. (C) Gene ontology analysis of the 150 and 358 proteins uniquely identified in the Dungrawala et al. or Alabert et al. datasets.

tryptic peptides likely explaining why their abundance was difficult to quantitate. The fourth, DNA2, was observed but did not quite meet the stringent cutoff for significance. Thus, the method has a lower false-negative rate than any of the other methods. It also likely has a low false-positive rate (see later).

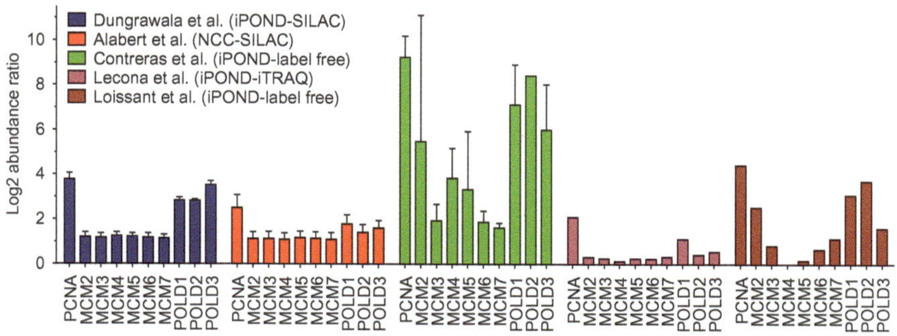

Fig. 5 Comparison of the enrichment of selected replisome proteins at replication forks calculated in five proteomic datasets. A log2 transformation of the mean enrichment comparing fork/chromatin (pulse/chase) is depicted. Larger positive values indicate increased enrichment at forks compared to bulk chromatin. Error bars were calculated as SEM where possible. The data illustrate the reproducibility and precision of SILAC quantitation compared to other methods.

The robustness of the method is evident from the highly similar quantitative data obtained for the subunits of the MCM2–7 complex, which differed in mean enrichment values by less than 10% (Fig. 5). In fact, the quantitation is so robust that it can predict protein complexes (Dungrawala et al., 2015). Proteins that function as part of a stoichiometric protein complex should be captured equally in any given iPOND experiment. Unsupervised hierarchical clustering of the approximately 1500 proteins quantitated in multiple experiments was capable of predicting both known and new protein complexes based on similarities in their subunit abundance ratios across different experiments. This property of the iPOND-SILAC-MS data was used to identify ZNF644 as a new subunit of the G9a/GLP methyltransferase complex that travels just behind the replication fork to modify newly deposited histones and illustrates the accuracy and reproducibility of iPOND-SILAC-MS (Dungrawala et al., 2015).

2.1.4 Variations on iPOND

Variations to the iPOND procedure that may be useful in specific circumstances have been reported. A similar methodology was developed independently and named Dm-ChP for DNA-mediated chromatin pull-down (Kliszczak, Rainey, Harhen, Boisvert, & Santaocanale, 2011). This method was combined with mass spectrometry, but the lack of a chase sample precludes analysis of which proteins are specifically enriched at replication forks vs simply being chromatin associated. In another iPOND variation, we omitted the formaldehyde crosslinking step to examine histone modifications in a procedure we termed native iPOND corresponding to native chromatin immunoprecipitation methods (Sirbu et al., 2012). A similar procedure was also reported by another group and called aniPOND (Leung, Abou El Hassan, & Bremner, 2013). An advantage of native iPOND is that the formaldehyde crosslinking step in regular iPOND can interfere with detection of proteins by Western blotting due to destruction of epitopes or incomplete crosslink reversal yielding aberrant migration on gels. This issue is especially problematic for large proteins, but does not interfere with detection by mass spectrometry.

2.2 Nascent Chromatin Capture

NCC is similar to iPOND except that biotin-dUTP is utilized instead of EdU (Fig. 1B and Alabert et al., 2014). Thus, no biotin-conjugation step is required in NCC. However, unlike EdU, biotin-dUTP cannot be imported into intact cells. Thus, a cell permeabilization step such as incubating cells in hypotonic conditions must be utilized to allow the biotin-dUTP to cross the cell membrane. Similar to iPOND, a chase step after the biotin-dUTP labeling allows the identification of proteins that are enriched near the fork compared to those that are simply bound to chromatin.

2.2.1 Comparison of iPOND and NCC

Both iPOND and NCC depend on the polymerase to be capable of incorporating and extending from the modified base. Furthermore, both depend on the cell being unable to recognize the modified base as DNA damage and assume the modification does not interfere with the binding of proteins to the DNA. In short-term assays these assumptions appear to be justified; however, there is a long literature indicating that these types of base modifications are not completely neutral. For example, incorporation of halogenated nucleoside analogs like BrdU can activate the DNA damage response (Masterson & O'Dea, 2007). Less is known about EdU and biotin-dUTP. Assays of DNA replication did not observe any consequences of EdU in short-term assays, although long-term assays indicated decreased cell proliferation and increased DNA damage signaling (Kohlmeier, Maya-Mendoza, & Jackson, 2013). It is likely that the much larger biotin group on biotin-dUTP could cause similar or worse problems.

iPOND has the advantage over NCC in that it is quite simple to add or remove the EdU from the culture media to label for defined times. Thus, it is possible to incorporate the EdU at stressed forks such as those challenged by hydroxyurea or DNA-damaging agents, which may incorporate very slowly and require labeling periods of several hours. Since the biotin-nucleotide used in NCC is not cell permeable, long-term labeling is not possible.

An additional concern with NCC is the need to alter the cell growth conditions to permeabilize cells to the biotin-dUTP. The osmotic stress used to achieve permeabilization can affect multiple cellular processes. For example, osmotic stress activates the ATM DNA damage response kinase which can modify replisome proteins and alter DNA replication kinetics (Bakkenist & Kastan, 2003). Thus, permeabilization conditions need to be optimized for each cell type to minimize unwanted effects.

The Groth lab has combined NCC with SILAC-MS and reported the identification of 426 proteins enriched on nascent DNA (Alabert et al., 2014). This list was selected by a combination of enrichment criteria from the mass spectrometry data (with a log2 enrichment cutoff of >0.4 that yielded a list of 1296 proteins) and filtering by bioinformatics comparison to a previous dataset of chromatin-associated proteins. This list of 426 proteins has relatively few false-negatives with only seven of the well-established replication fork elongation proteins missing the cutoff because of filtering, lack of sufficient enrichment, or nondetection (Fig. 4A). As might be expected, the use of SILAC provides highly quantitative data. Examining the reproducibility of the MCM2–7 complex subunit quantifications illustrates its quantitation accuracy is second only to the iPOND-SILAC-MS dataset and both are far more reproducible than label-free mass spectrometry (Fig. 5).

Comparing the 426 proteins from the NCC-SILAC approach to the 218 in the iPOND-SILAC dataset indicates that the NCC methodology likely yielded a much higher false-positive detection rate. It is possible to decrease the false-positive rate by requiring a larger enrichment value such as the two-fold enrichment criteria in the iPOND-SILAC-MS dataset, but that would also increase the false-negative frequency. There are 68 proteins in common between the two datasets (Fig. 4B and Table 2). These are mostly core replisome proteins, proteins tethered to the replisome to facilitate chromatin deposition and remodeling, or replication stress response proteins. Subtracting the common proteins from the iPOND data still yields a list of 150 proteins highly enriched for processes related to DNA replication including DNA repair (Fig. 4C). On the other hand, subtracting these proteins from the NCC dataset leaves a list of 358 proteins enriched in nuclear functions, but not necessarily DNA metabolism (Fig. 4C).

Table 2 Replication Fork Proteins Identified in Both Dungrawala et al. (2015) and Alabert et al. (2014)

Uniprot	Symbol	Gene ID
O75419	CDC45	8318
Q92674	CENPI	2491
Q13111	CHAF1A	10,036
Q13112	CHAF1B	8208
Q8WVB6	CHTF18	63,922
Q9HAW4	CLSPN	63,967
Q6PJP8	DCLRE1A	9937
P26358	DNMT1	1786
Q9BVC3	DSCC1	79,075
Q9NZJ0	DTL	51,514
Q56NI9	ESCO2	157,570
Q9UQ84	EXO1	9156
Q9BXW9	FANCD2	2177
Q9NVI1	FANCI	55,215
P16383	GCFC2	6936
Q9BRX5	GINS3	64,785
Q15004	PCLAF	9768
P18858	LIG1	3978
Q7L590	MCM10	55,388
P49736	MCM2	4171
P25205	MCM3	4172
P33991	MCM4	4173
P33992	MCM5	4174
Q14566	MCM6	4175
P33993	MCM7	4176
Q6ZRQ5	MMS22L	253,714
P49959	MRE11A	4361

Table 2 Replication Fork Proteins Identified in Both Dungrawala et al. (2015) and Alabert et al. (2014)—cont'd

Uniprot	Symbol	Gene ID
P43246	MSH2	4436
P20585	MSH3	4437
P52701	MSH6	2956
O60934	NBN	4683
Q86W56	PARG	8505
P12004	PCNA	5111
Q14181	POLA2	23,649
P28340	POLD1	5424
P49005	POLD2	5425
Q15054	POLD3	10,714
Q07864	POLE	5426
P56282	POLE2	5427
P49642	PRIM1	5557
P49643	PRIM2	5558
Q9Y606	PUS1	80,324
Q9NS91	RAD18	56,852
Q92878	RAD50	10,111
Q9Y4B4	RAD54L2	23,132
Q99638	RAD9A	5883
O94762	RECQL5	9400
P35251	RFC1	5981
P35250	RFC2	5982
P40938	RFC3	5983
P35249	RFC4	5984
P40937	RFC5	5985
Q5TBB1	RNASEH2B	79,621
P27694	RPA1	6117

Continued

Table 2 Replication Fork Proteins Identified in Both Dungrawala et al. (2015) and Alabert et al. (2014)—cont'd

Uniprot	Symbol	Gene ID
P15927	RPA2	6118
Q9BQI6	SLF1	84,250
Q8IX21	SLF2	55,719
Q9NZC9	SMARCAL1	50,485
Q08945	SSRP1	6749
Q96FV9	THOC1	9984
Q9UNS1	TIMELESS	8914
Q9BVW5	TIPIN	54,962
Q92547	TOPBP1	11,073
O94842	TOX4	9878
Q9BSV6	TSEN34	79,042
O94782	USP1	7398
O75717	WDHD1	11,169
Q6PJT7	ZC3H14	79,882

3. ADDITIONAL APPLICATIONS

3.1 Purifying Replication Stress Response Proteins

In addition to identifying proteins at unperturbed DNA replication forks, iPOND has been utilized to interrogate the replication fork proteome after a challenge that perturbs replication (Table 3 and Dungrawala et al., 2015; Lossaint et al., 2013; Olcina et al., 2016; Ribeyre et al., 2016; Sirbu et al., 2013). For example, combining the EdU label with drugs like hydroxyurea or aphidicolin allowed the identification of proteins at stalled replication forks. Combining EdU with camptothecin or alternatively with hydroxyurea and a selective inhibitor of the ATR checkpoint kinase identifies proteins recruited to collapsed replication forks (Dungrawala et al., 2015; Ribeyre et al., 2016; Sirbu et al., 2013). iPOND is particularly good for these types of experiments since the EdU labeling time can be varied easily. Thus, the amount of DNA labeling can be optimized to ensure equal

Table 3 Summary of Studies That Examined Replication Stress Proteomes

Study	Perturbation	Method	Number of Replicates	Number of Proteins
Sirbu et al. (2013)	Hydroxyurea	iPOND Label-free	5	139[a]
Sirbu et al. (2013)	Hydroxyurea + ATR inhibitor	iPOND Label-free	5	137[a]
Lossaint et al. (2013)	Hydroxyurea	iPOND Label-free	3	n/a
Dungrawala et al. (2015)	Hydroxyurea	iPOND-SILAC	18	192
Dungrawala et al. (2015)	Aphidicollin	iPOND-SILAC	2	n/a
Dungrawala et al. (2015)	Hydroxyurea + ATR inhibitor	iPOND-SILAC	11	151
Dungrawala et al. (2015)	Aphidicolin + ATR inhibitor	iPOND-SILAC	2	n/a
Ribeyre et al. (2016)	Camptothecin	iPOND-Label-free	6	21
Olcina, Giaccia, and Hammond (2016)	Hypoxia	iPOND	n/a	n/a
Raschle et al. (2015)	Psoralen	Chromass-Label-free	42[b]	198[c]

[a]These numbers included proteins that are at undamaged forks since the comparison was to a chase sample.
[b]Additional samples were analyzed containing other inhibitors in addition to psoralen.
[c]112 of these proteins were enriched in a replication-dependent manner.

amounts of capture. Often these experiments are done with a short pre-incubation with EdU followed by addition of the replication stress agent. Since even high doses of hydroxyurea may not completely stop replication fork movement (Dungrawala et al., 2015), it is important to maintain EdU in the growth media during long HU time courses to ensure the DNA adjacent to the fork continues to be labeled. Alternatively, the drug can be added prior to the EdU, but it is important to normalize the amount of EdU incorporation to facilitate sample comparisons.

Another consideration in using iPOND with genotoxic drugs or genetic perturbations is that some conditions could alter replication initiation. For example, comparison of the stalled replication fork proteome after hydroxyurea treatment with the proteome after combining hydroxyurea with an ATR inhibitor initially suggested that most replisome proteins actually became more abundant in the ATR-inhibited sample (Dungrawala et al., 2015). However, the abundance differences are entirely explained by increased replication initiation in the ATR-inhibited samples due to the inhibition of the checkpoint response. Thus, comparisons like these need to consider whether equal numbers of forks are being purified in addition to considerations of whether equal amounts of nascent DNA are captured.

As with experiments examining the unperturbed replisome, the mass spectrometry methodology makes a substantial difference in the results. The most complete datasets completed with iPOND-SILAC mass spectrometry have identified 192 proteins recruited to stalled forks and 151 recruited to collapsed forks (Dungrawala et al., 2015). Among these proteins are the known DNA damage response proteins, but a significant number of potential new replication stress response proteins are also identified. For example, a new ATR-activating protein, ETAA1, was identified from iPOND-derived stalled replication fork proteomes (Bass et al., 2016).

3.2 Analysis of Chromatin Deposition and Maturation

Both iPOND and NCC are particularly useful for studying the process of chromatin deposition and maturation following DNA replication. By examining multiple chase time points, it is possible to follow the assembly of the histones on the nascent DNA and changes in their posttranslational modifications. For example, the timing of histone H1 deposition in relation to DNA replication, the changes in histone acetylation and their genetic dependencies, and H2AX phosphorylation spreading from a stalled fork were followed by iPOND (Nagarajan et al., 2013; Sirbu et al., 2011). The Groth lab combined NCC with pulsed SILAC-MS to monitor changes in the modifications of new histones deposited on the nascent DNA as a function of time after DNA replication. This procedure allowed them to determine how rapidly histone marks are copied to the newly synthesized histones after DNA replication (Alabert et al., 2015).

3.3 Analysis of Viral Replication and Other Opportunities

Hundreds of studies have utilized iPOND and NCC since their development primarily to study DNA replication, chromatin deposition/maturation, and replication stress responses. There are also opportunities beyond studying nuclear genome replication. In principle, any process that involves the synthesis of new DNA could be analyzed. For example, iPOND was recently utilized to identify proteins that function in Herpes Simplex Virus replication, genome processing, and packaging (Dembowski & DeLuca, 2015). Other opportunities include the study mitochondrial DNA metabolism, DNA repair, chromatin reestablishment after DNA repair synthesis, and break-induced replication. Additionally, the DNA that is captured in the iPOND protocol can be analyzed instead of the proteins. This served as the basis for the analysis of the nucleosome landscape following DNA replication in a methodology called MINCE-Seq (Ramachandran & Henikoff, 2016).

4. SUMMARY AND CONCLUSIONS

The development of procedures to purify newly synthesized DNA and associated proteins has provided an opportunity to define the replication fork proteome and identify proteins that contribute to DNA and chromatin replication. The 68 proteins in common between the iPOND-SILAC-MS and NCC-SILAC-MS nascent DNA datasets are the highest confidence list of replication fork-associated proteins in mammalian cells (Table 2). Most of these proteins are well-known parts of the DNA copying, replication stress response, or chromatin modification machinery. However, approximately 10% of these proteins have never been experimentally linked to DNA replication previously. Furthermore, the proteins unique to each of the iPOND-SILAC-MS and NCC-SILAC-MS datasets provide high probability candidates for further study. Since each of these datasets was derived from a single cell line with only three biological replicates, further interrogation of the fork proteomes using these methods promises to be a high value approach. Furthermore, combining these techniques with drug and genetic perturbations, methods to label new and recycled histones, and mass spectrometry approaches to study posttranslational modifications provide opportunities to better understand the regulation of DNA and chromatin replication and responses to replication challenges.

ACKNOWLEDGMENTS

Research using iPOND in the Cortez lab is supported by NIH Grant R01GM116616 and the Breast Cancer Research Foundation. I thank Dr. Huzefa Dungrawala for his input.

REFERENCES

Alabert, C., Barth, T. K., Reveron-Gomez, N., Sidoli, S., Schmidt, A., Jensen, O. N., et al. (2015). Two distinct modes for propagation of histone PTMs across the cell cycle. *Genes & Development*, *29*, 585–590.

Alabert, C., Bukowski-Wills, J. C., Lee, S. B., Kustatscher, G., Nakamura, K., de Lima Alves, F., et al. (2014). Nascent chromatin capture proteomics determines chromatin dynamics during DNA replication and identifies unknown fork components. *Nature Cell Biology*, *16*, 281–293.

Aranda, S., Rutishauser, D., & Ernfors, P. (2014). Identification of a large protein network involved in epigenetic transmission in replicating DNA of embryonic stem cells. *Nucleic Acids Research*, *42*, 6972–6986.

Bakkenist, C. J., & Kastan, M. B. (2003). DNA damage activates ATM through intermolecular autophosphorylation and dimer dissociation. *Nature*, *421*, 499–506.

Bass, T. E., Luzwick, J. W., Kavanaugh, G., Carroll, C., Dungrawala, H., Glick, G. G., et al. (2016). ETAA1 acts at stalled replication forks to maintain genome integrity. *Nature Cell Biology*, *18*, 1185–1195.

Bell, S. P., & Dutta, A. (2002). DNA replication in eukaryotic cells. *Annual Review of Biochemistry*, *71*, 333–374.

Cimprich, K. A., & Cortez, D. (2008). ATR: An essential regulator of genome integrity. *Nature Reviews. Molecular Cell Biology*, *9*, 616–627.

Dembowski, J. A., & DeLuca, N. A. (2015). Selective recruitment of nuclear factors to productively replicating herpes simplex virus genomes. *PLoS Pathogens*, *11*. e1004939.

Dungrawala, H., & Cortez, D. (2015). Purification of proteins on newly synthesized DNA using iPOND. *Methods in Molecular Biology*, *1228*, 123–131.

Dungrawala, H., Rose, K. L., Bhat, K. P., Mohni, K. N., Glick, G. G., Couch, F. B., et al. (2015). The replication checkpoint prevents two types of fork collapse without regulating replisome stability. *Molecular Cell*, *59*, 998–1010.

Kliszczak, A., Rainey, M., Harhen, B., Boisvert, F., & Santaocanale, C. (2011). DNA mediated chromatin pull-down for the study of chromatin replication. *Scientific Reports*, *1*, 1–7.

Kohlmeier, F., Maya-Mendoza, A., & Jackson, D. A. (2013). EdU induces DNA damage response and cell death in mESC in culture. *Chromosome Research: An International Journal on the Molecular, Supramolecular and Evolutionary Aspects of Chromosome Biology*, *21*, 87–100.

Lecona, E., Rodriguez-Acebes, S., Specks, J., Lopez-Contreras, A. J., Ruppen, I., Murga, M., et al. (2016). USP7 is a SUMO deubiquitinase essential for DNA replication. *Nature Structural & Molecular Biology*, *23*, 270–277.

Leung, K. H., Abou El Hassan, M., & Bremner, R. (2013). A rapid and efficient method to purify proteins at replication forks under native conditions. *BioTechniques*, *55*, 204–206.

Lopez-Contreras, A. J., Ruppen, I., Nieto-Soler, M., Murga, M., Rodriguez-Acebes, S., Remeseiro, S., et al. (2013). A proteomic characterization of factors enriched at nascent DNA molecules. *Cell Reports*, *3*, 1105–1116.

Lossaint, G., Larroque, M., Ribeyre, C., Bec, N., Larroque, C., Decaillet, C., et al. (2013). FANCD2 binds MCM proteins and controls replisome function upon activation of s phase checkpoint signaling. *Molecular Cell*, *51*, 678–690.

Masterson, J. C., & O'Dea, S. (2007). 5-Bromo-2-deoxyuridine activates DNA damage signalling responses and induces a senescence-like phenotype in p16-null lung cancer cells. *Anti-Cancer Drugs*, *18*, 1053–1068.

Moldovan, G. L., Pfander, B., & Jentsch, S. (2007). PCNA, the maestro of the replication fork. *Cell*, *129*, 665–679.

Nagarajan, P., Ge, Z., Sirbu, B., Doughty, C., Agudelo Garcia, P. A., Schlederer, M., et al. (2013). Histone acetyl transferase 1 is essential for mammalian development, genome stability, and the processing of newly synthesized histones H3 and H4. *PLoS Genetics*, *9*. e1003518.

Nam, E. A., & Cortez, D. (2011). ATR signalling: More than meeting at the fork. *The Biochemical Journal*, *436*, 527–536.

Olcina, M. M., Giaccia, A. J., & Hammond, E. M. (2016). Isolation of proteins on nascent DNA in hypoxia and reoxygenation conditions. *Advances in Experimental Medicine and Biology*, *899*, 27–40.

Ong, S. E., & Mann, M. (2005). Mass spectrometry-based proteomics turns quantitative. *Nature Chemical Biology*, *1*, 252–262.

Perkins, D. N., Pappin, D. J., Creasy, D. M., & Cottrell, J. S. (1999). Probability-based protein identification by searching sequence databases using mass spectrometry data. *Electrophoresis*, *20*, 3551–3567.

Probst, A. V., Dunleavy, E., & Almouzni, G. (2009). Epigenetic inheritance during the cell cycle. *Nature Reviews. Molecular Cell Biology*, *10*, 192–206.

Ramachandran, S., & Henikoff, S. (2016). Transcriptional regulators compete with nucleosomes post-replication. *Cell*, *165*, 580–592.

Raschle, M., Smeenk, G., Hansen, R. K., Temu, T., Oka, Y., Hein, M. Y., et al. (2015). DNA repair. Proteomics reveals dynamic assembly of repair complexes during bypass of DNA cross-links. *Science*, *348*, 1253671.

Rauniyar, N., & Yates, J. R., 3rd. (2014). Isobaric labeling-based relative quantification in shotgun proteomics. *Journal of Proteome Research*, *13*, 5293–5309.

Ribeyre, C., Zellweger, R., Chauvin, M., Bec, N., Larroque, C., Lopes, M., et al. (2016). Nascent DNA proteomics reveals a chromatin remodeler required for topoisomerase I loading at replication forks. *Cell Reports*, *15*, 300–309.

Salic, A., & Mitchison, T. J. (2008). A chemical method for fast and sensitive detection of DNA synthesis in vivo. *Proceedings of the National Academy of Sciences of the United States of America*, *105*, 2415–2420.

Sirbu, B. M., Couch, F. B., & Cortez, D. (2012). Monitoring the spatiotemporal dynamics of proteins at replication forks and in assembled chromatin using isolation of proteins on nascent DNA. *Nature Protocols*, *7*, 594–605.

Sirbu, B. M., Couch, F. B., Feigerle, J. T., Bhaskara, S., Hiebert, S. W., & Cortez, D. (2011). Analysis of protein dynamics at active, stalled, and collapsed replication forks. *Genes & Development*, *25*, 1320–1327.

Sirbu, B. M., McDonald, W. H., Dungrawala, H., Badu-Nkansah, A., Kavanaugh, G. M., Chen, Y., et al. (2013). Identification of proteins at active, stalled, and collapsed replication forks using isolation of proteins on nascent DNA (iPOND) coupled with mass spectrometry. *The Journal of Biological Chemistry*, *288*, 31458–31467.

CHAPTER THREE

DNA Fiber Analysis: Mind the Gap!

Annabel Quinet, Denisse Carvajal-Maldonado, Delphine Lemacon, Alessandro Vindigni[1]
Saint Louis University School of Medicine, St. Louis, MO, United States
[1]Corresponding author: e-mail address: avindign@slu.edu

Contents

Abstract

Understanding the mechanisms of replication stress response following genotoxic stress induction is rapidly emerging as a central theme in cell survival and human disease. The DNA fiber assay is one of the most powerful tools to study alterations in replication fork dynamics genome-wide at single-molecule resolution. This approach relies on the ability of many organisms to incorporate thymidine analogs into replicating DNA and is widely used to study how genotoxic agents perturb DNA replication. Here, we review different approaches available to prepare DNA fibers and discuss important limitations of each approach. We also review how DNA fiber analysis can be used to shed light upon several replication parameters including fork progression, restart, termination, and new origin firing. Next, we discuss a modified DNA fiber protocol to monitor the presence of single-stranded DNA (ssDNA) gaps on ongoing replication forks. ssDNA gaps are very common intermediates of several replication stress response mechanisms,

Methods in Enzymology, Volume 591
ISSN 0076-6879
http://dx.doi.org/10.1016/bs.mie.2017.03.019

but they cannot be detected by standard DNA fiber approaches due to the resolution limits of this technique. We discuss a novel strategy that relies on the use of an ssDNA-specific endonuclease to nick the ssDNA gaps and generate shorter DNA fibers that can be used as readout for the presence of ssDNA gaps. Finally, we describe a follow-up DNA fiber approach that can be used to study how ssDNA gaps are repaired postreplicatively.

1. INTRODUCTION

Accurate DNA replication is essential to protect genome integrity and for the high-fidelity transmission of genomic information to daughter cells. Our genome is under constant attack by agents that arise from either normal metabolism or exposure to natural or artificial products in the environment. Cells have evolved several molecular pathways aimed at preserving the stability of perturbed replication forks and promoting their accurate restart in response to these endogenous and exogenous challenges (Berti & Vindigni, 2016; Zeman & Cimprich, 2014). These replication stress response pathways have become a central theme in genome stability, cell survival, and human disease (Berti & Vindigni, 2016; Menck & Munford, 2014; Roos, Thomas, & Kaina, 2015). For example, patients with genetic mutations in different replication stress response mechanisms are more cancer prone, highlighting the intimate relationship between replication stress and cancer (Gaillard, Garcia-Muse, & Aguilera, 2015; Menck & Munford, 2014). At the same time, highly proliferating cancer cells rely on replication stress response mechanisms to survive, making these ideal targets for cancer treatment (Gaillard et al., 2015; Roos et al., 2015).

DNA fiber analysis is a central technique to gain mechanistic insight into how genotoxic agents perturb replication fork dynamics genome-wide at single-molecule resolution (Bensimon et al., 1994; Jackson & Pombo, 1998; Merrick, Jackson, & Diffley, 2004; Michalet et al., 1997; Parra & Windle, 1993; Técher et al., 2013). In this assay, progressing replication forks are sequentially labeled with two consecutive thymidine analogs—e.g., 5-iodo-2′-deoxyuridine (IdU) and 5-chloro-2′-deoxyuridine (CldU) (Merrick et al., 2004). Immunostaining of each of these analogs allows for microscopic visualization of both tracts and to study how replication is perturbed genome wide by cumulative analysis of individual DNA molecules (Técher et al., 2013). This simple labeling scheme allows monitoring several key replication parameters including changes in replication fork progression, fork symmetry, and origin firing, as well as nucleolytic degradation of nascent DNA following replication stress induction (Vengrova &

Dalgaard, 2009). In this review, we describe different approaches that can be used to monitor these replication parameters by DNA fiber analysis and discuss the advantages and disadvantages of each approach.

One important limitation of the DNA fiber technique is that stretched DNA fibers have a stretching range from 2 to 3 kilobases (kb) per micrometer (μm), thereby limiting the resolution to few kb of DNA (Bensimon et al., 1994; Jackson & Pombo, 1998; Michalet et al., 1997). In addition, the DNA fiber technology does not allow distinction of the two newly replicated strands because they are both equally labeled with the thymidine analogs and they are "collapsed" in a single fiber upon DNA stretching. Because of these limitations, short single-stranded DNA (ssDNA) gaps present either on the leading or lagging strand cannot be detected by this approach (Bertolin, Mansilla, & Gottifredi, 2015). Small ssDNA regions are frequent intermediates of different replication stress response mechanisms. For example, they can be produced by the limited nuclease-dependent resection of stalled replication intermediates, necessary for fork restart (Hashimoto, Chaudhuri, Lopes, & Costanzo, 2010; Thangavel et al., 2015) or because of a faulty gap-filling mechanism, as discussed later (Edmunds, Simpson, & Sale, 2008; Elvers, Johansson, Groth, Erixon, & Helleday, 2011; Jansen, Tsaalbi-Shtylik, Hendriks, Verspuy, et al., 2009; Quinet et al., 2014). These ssDNA regions can be as small as 300–400 nucleotides (Lopes, Foiani, & Sogo, 2006) and can normally be detected only using higher resolution techniques such as electron microscopy (Vindigni & Lopes, 2016). Here, we also describe a modified DNA fiber protocol that employs an ssDNA-specific endonuclease to overcome the low-resolution limits of the DNA fiber approach and allow detection of small ssDNA gaps (Quinet et al., 2016).

2. PREPARATION OF DNA FIBERS

As mentioned earlier, the DNA fiber analysis exploits the ability of many organisms to incorporate halogenated pyrimidine nucleoside analogs into replicating DNA and provides a powerful tool to monitor genome-wide replication perturbations at single-molecule resolution. Further insights can be obtained by knocking down selected factors suspected to play a role during replication, repair, or postreplication repair (PRR) and by assessing how their loss affects the replication dynamics. Later, we illustrate three main techniques commonly used to extract, digest, and immobilize prelabeled DNA fibers prior immunodetection (Fig. 1). We also briefly discuss the procedures for image acquisition and data analysis.

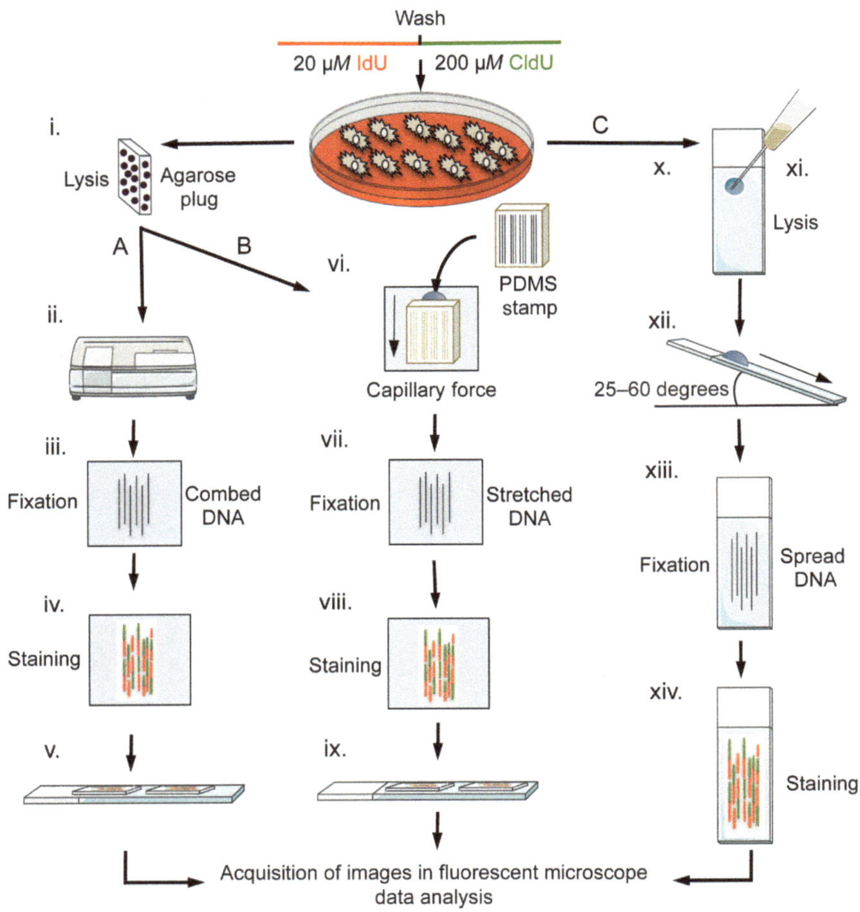

Fig. 1 Scheme of the three main approaches to obtain DNA fibers. In DNA combing (A) and stretching by *maRTA* (B), cells are resuspended and lysed in an agarose plug (i). (A) DNA is combed with a combing machine (ii) onto a silanized coverslip, fixed (iii), and immunostained (iv). Coverslips are mounted into a slide for microscopic visualization (v). (B) DNA is stretched using a PDMS patch containing small capillaries (vi). A drop of isolated DNA is added using a cut tip at one of the capillary ends of the PDMS patch, which stretches the DNA onto the silanized coverslip (vi) using capillary force. Afterward, the DNA is fixed (vii) and immunostained (viii). Samples are mounted onto a microscope slide (ix). (C) For DNA spreading, a drop of prelabeled cells is transferred to a positively coated microscope slide (x) and lysed (xi). Slide is then tilted at a 25–60 degrees angle to allow DNA spreading down the slide (xii). DNA is then fixed (xiii) and immunostained (xiv) DNA fibers are visualized through a fluorescent microscope and data analysis can be performed with ImageJ.

2.1 Pulse-Labeling Replication Forks In Vivo

Replication forks must be pulse labeled with two consecutive thymidine analogs—i.e., IdU followed by CldU (or vice versa) (Técher et al., 2013). For the purpose of this review we will designate IdU as the first label (red) and CldU as the second label (green). Typically, the first analog is added to the media to a final concentration of 10–25 μM. After careful wash, the second analog is supplemented to the media to a final concentration of 100–250 μM. *It is important to note that the concentration of the second analog (CldU) is 10-fold higher than the concentration of the first analog (IdU). This is to ensure that the second analog is able to displace any remaining IdU that was not removed during the washing steps.* The procedure of adding and washing the analogs from the cell culture must be performed relatively quickly to ensure accuracy of the incorporation timing as detailed later.

After labeling, the DNA molecules can be "combed," "stretched," or "spread" on a positively coated glass surface. The combing approach uses a controlled force to elongate and separate DNA fibers in solution on a positively coated glass surface. The force is tightly controlled using a specialized DNA combing machine that regulates the speed by which the DNA molecules are "combed" on the glass surface (Fig. 1A; Bensimon et al., 1994; Michalet et al., 1997). In the second approach, the DNA molecules are "stretched" using a microfluidic-assisted capillary force, also named *maRTA* (for microfluidic-assisted replication tract analysis) (Fig. 1B; Sidorova, Li, Schwartz, Folch, & Monnat, 2009). Alternatively, DNA molecules can be stretched using gravity to separate and "spread" the DNA on a positively charged microscope slide (Fig. 1C; Parra & Windle, 1993). Examples of images obtained with the three techniques are shown in Fig. 2. These three techniques are briefly described in the following paragraphs.

2.2 Preparation of DNA Fibers Using DNA Combing and maRTA

DNA combing and *maRTA* use the same DNA preparation protocol (Fig. 1i). Briefly, cells are resuspended in 1% low-melting agarose and transferred into a plug mold. Cells are then lysed using a lysis buffer (LB) containing detergent and proteinase K in order to remove proteins and cell debris and trap the isolated DNA in the agarose mesh of the plug. After 2 days of lysis, DNA plugs are washed, melted, and then digested by β-agarase to release the pure DNA in solution. After this procedure, the DNA is ready to be "combed" or "stretched" using either the DNA combing or *maRTA* approach as detailed later. Prior to "combing" or "stretching," glass

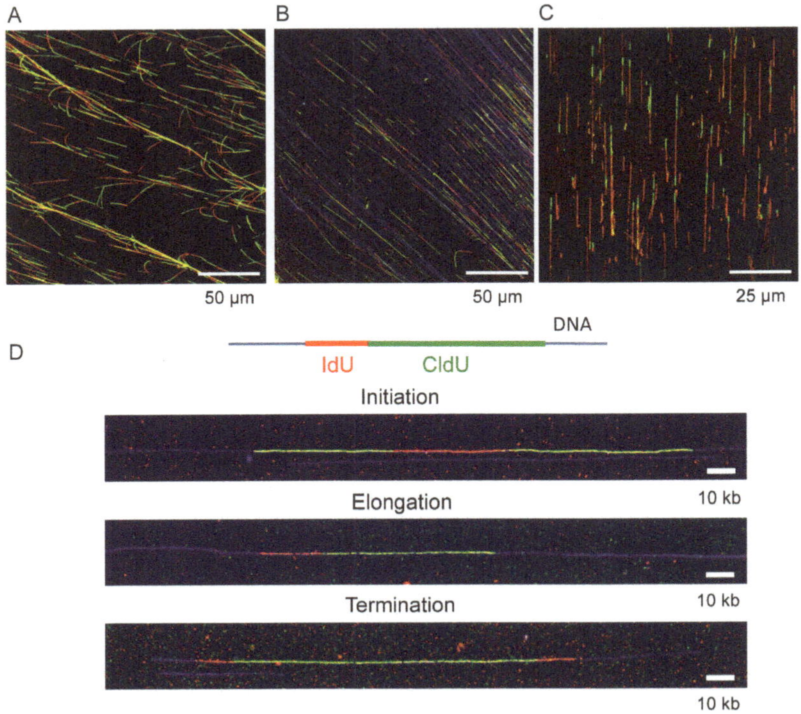

Fig. 2 Representative images of DNA fibers. DNA fibers obtained using the spreading technique without (A) or with (B) counterstaining of ssDNA. (C) Stretched DNA fibers obtained using *maRTA*. (D) Combed DNA fibers with DNA counterstaining. *Images of combed DNA fibers were kindly provided by Dr. Philippe Pasero and published in Sokol, A. M., Cruet-Hennequart, S., Pasero, P., & Carty, M. P. (2013). DNA polymerase η modulates replication fork progression and DNA damage responses in platinum-treated human cells.* Scientific Reports, 3, *3277. http://doi.org/10.1038/srep03277.*

coverslips must be cleaned and positively coated. For more details on the DNA preparation protocol, see Michalet et al. (1997), Sidorova et al. (2009), and Técher et al. (2013).

2.2.1 Preparation of Fibers Using DNA Combing Machine

In the DNA combing technique, also referred as "dynamic molecular combing," precoated coverslips are submerged into a solution containing the DNA and are then pulled out with a mechanical device, known as the "DNA combing machine" (Fig. 1A; Bensimon et al., 1994; Bianco et al., 2012; Gallo, Wang, Yip, & Brown, 2016; Michalet et al., 1997; Pasero, Bensimon, & Schwob, 2002). All combing machines comb the DNA onto coverslips at a very slow and constant speed. The receding air–water interface elongates and aligns the DNA in separate tracts. DNA attaches to the positively coated surface coating the entire coverslip.

The main advantage of the DNA combing technique is that it allows accurate determination of replication fork velocities because the force and speed of the DNA deposition is tightly controlled by the machine arm. The length of the DNA fibers can be accurately measured, because all DNA fibers are elongated by a constant factor of 2 kb/μm (Conti et al., 2007; Michalet et al., 1997). In addition, this technique is well suited to measure the interorigin distance, which is the distance between two adjacent two-color-labeled replication forks, and study new origin firing because the DNA combing technique aligns all molecules along a single axis, preventing crossing of DNA fibers (also see Section 3.2 for more details). One limitation of this technique is that it is time consuming and it allows preparation of only few samples at a time. Moreover, some DNA breakage might occur, in particular at fragile sites, thereby limiting the size of observable tracts to a few 100 kb in length (Bianco et al., 2012; Michalet et al., 1997). The force applied to comb the DNA (around 160 pN) is 10 times less than that required to break the covalent bonds within the molecule (Bensimon, Simon, Croquette, & Bensimon, 1995), therefore breakage likely occurs before combing of the DNA. Most breakages occur during the melting of the agarose plugs or while handling of the isolated DNA when pouring it into the combing reservoir. A modified protocol addresses this limitation by melting the plugs in the reservoir in β-agarase buffer containing 100 m*M* NaCl pH 6.0 and then adding the β-agarase to digest the remaining agarose (Kaykov, Taillefumier, Bensimon, & Nurse, 2016). This modified protocol reduces the mechanical shear that normally occurs while handling of the isolated DNA. In addition, the Na^+ ions in the buffer (100 m*M* NaCl) coat the backbone of the DNA and prevent any further breakage that may be caused by the force applied by the water–air interface. Using this modified protocol, the authors were able to obtain DNA fibers of up to 10 megabases in length, vastly expanding the amount information that can be obtained from each individual fiber (Kaykov et al., 2016).

2.2.2 Preparation of Fibers Using Capillary Force (or maRTA)

The *maRTA* approach uses capillary force to elongate and separate the DNA fibers on the positively coated coverslips, as described (Sidorova et al., 2009; Fig. 1B). Briefly, a PDMS (PolyDiMethylSiloxane) patch is prepared using a SU-8 master mold, in which a series of 100 capillaries are precarved using a laser. The capillaries have the width of a double-stranded DNA molecule.

This ensures that only one DNA molecule can fit in each capillary and greatly reduces the risk of crossing between DNA molecules. The PDMS patch is placed onto the silanized coverslip, and a 1-μL drop of DNA isolated from approximately 2000 cells is placed at one of the capillary ends. The DNA solution is slowly and evenly sucked into the patch by capillary force. After allowing the DNA to dry for 10 min, the PDMS patch is carefully lifted up and discarded. The DNA is then stretched on the surface of the slide. For a detailed protocol, refer to Sidorova et al. (2009).

Using this technique, the fibers are evenly stretched and separated on the glass surface, similarly to the DNA combing technique. For this reason, *maRTA* can also be used to study the same replication parameters monitored by DNA combing. However, an important limitation of this technique is that subtle differences during the drying of the PDMS patch can alter the stretching efficiency thereby preventing accurate measurements of replication fork velocity or changes in interorigin distances. Moreover, this technique is time consuming and requires considerable efforts to prepare the coverslips and the PDMS patches.

2.3 Preparation of DNA Fibers Using DNA Spreading

DNA spreading, first described by Parra and Windle (1993), uses gravity to separate and "spread" the DNA on a positively charged microscope slide (Fig. 1C). Briefly, after DNA labeling with the two analogs, cells are collected and resuspended in phosphate-buffered saline (PBS) at a concentration of 1000–1500 cells/μL. A 2-μL drop containing cells is transferred to the positively coated slide and lysed. The slides are carefully tilted at 25–60 degrees to allow a stream of DNA to travel slowly down the slide (Fig. 1C). For a more detailed protocol, refer to Section 4.1.

This technique allows a faster preparation of the DNA fibers compared to the two previous techniques, because cell lysis and spreading can be done immediately after labeling the cells with the thymidine analogs, drastically reducing the sample preparation time. This technique also requires the least amount of materials and preparation. As such, it allows for a higher throughput of samples and it is ideal for a quick survey of replication perturbations under different conditions. For these reasons, the DNA fiber by spreading approach is widely used to test how different genotoxic agents or specific genetic backgrounds perturb DNA replication. However, the spreading by gravity can lead to a nonuniform spreading of fibers that result in frequent crossings (Fig. 2). To circumvent this problem, it is crucial to select areas where the fibers do not significantly overlap during the image acquisition step. This limitation also complicates the analysis of more complex parameters, such as interorigin distances, or the frequency of termination and new origin firing events. In addition, in contrast to the two previous techniques, the DNA is not isolated and therefore cannot be stored.

2.4 Data Acquisition and Analysis

Images are acquired with a fluorescence microscope capable of detecting emissions from at least two different fluorophores, typically red and green (also see Section 4.1). Images are normally taken using 63 × or 100 × objective for optimal fiber resolution.

For data analysis, the length of each labeled DNA tract (IdU and CldU) should be measured using specialized software, such as ImageJ (Schneider, Rasband, & Eliceiri, 2012). A minimum of 150–200 individual tracts must be measured for a reliable estimation of the different replication parameters and to account for normal differences between progressing forks (Técher et al., 2013). Moreover, it is important to take pictures of fibers across the whole slide to rigorously measure replication fork dynamics, in particular when using the spreading technique to prepare the DNA fibers. The length of IdU and CldU tracts should be recorded in μm for each replication event. The measurements obtained for each tract can then be converted to kb. For DNA fibers obtained by molecular combing the conversion factor is 2 kb/μm (Michalet et al., 1997), whereas for DNA fibers obtained by the spreading a commonly used conversion factor is 2.59 kb/μm (Daigaku, Davies, & Ulrich, 2010; Jackson & Pombo, 1998). Graphical representations with column bars corresponding to averages are not typically used because they do not adequately address the heterogeneity of the fibers. Data are frequently plotted as scatter dot plots, or in box and whiskers plots. Statistical significance between two groups is normally assessed using the Mann–Whitney *U* test (unpaired and nonparametric).

3. BIOLOGICAL QUESTIONS AND LABELING SCHEMES

The DNA fiber assay allows monitoring of several replication parameters (Fig. 3) and how these are affected by the loss of a particular protein or upon treatment with replication stress-inducing agents, as detailed in the sections later.

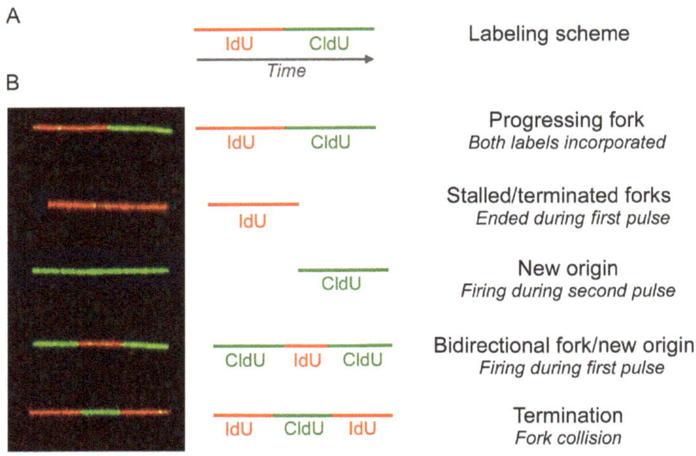

Fig. 3 Schematic of different replication events that can be studied DNA fiber analysis. (A) Labeling scheme: IdU is incorporated as the first analog, followed by CldU, incorporated as the second analog. (B) Replication parameters observed by DNA fiber assay and their interpretations.

3.1 Replication Fork Velocity and Symmetry

The DNA combing technique is ideal to measure fork velocity, because the machine arm tightly controls the speed and force of the combing and all DNA fibers are equally elongated (Michalet et al., 1997). The simplest labeling scheme to measure fork velocity is the consecutive labeling with two thymidine analogs (Fig. 3A). After measuring the length of both labeled tracts, the values are converted from μm into kb (see earlier), and fork speed is calculated by dividing the length of the tract by the labeling time (kb/min). Fork velocity can vary from 1 to 3 kb/min depending on the particular organism and cell type (Bianco et al., 2012; Conti et al., 2007; Jackson & Pombo, 1998; Técher et al., 2013).

Fork symmetry is the parameter that monitors the synchronized progression of two sister forks emanating from a given origin. Fork asymmetry refers to the scenario where the tracts of one of the two sister fork are longer than the other (Fig. 4A). Fork asymmetry can be measured from the ratio between the length of the two sister forks and used as a parameter to estimate the frequency of fork stalling/collapse (Conti et al., 2007; Tuduri et al., 2009). DNA combing is the technique of choice to study fork asymmetry because it aligns all DNA molecules on the same axis. For these experiments it is also crucial to counterstain the entire DNA filaments to confirm that two replication events belong to the same molecule and that DNA fibers are not

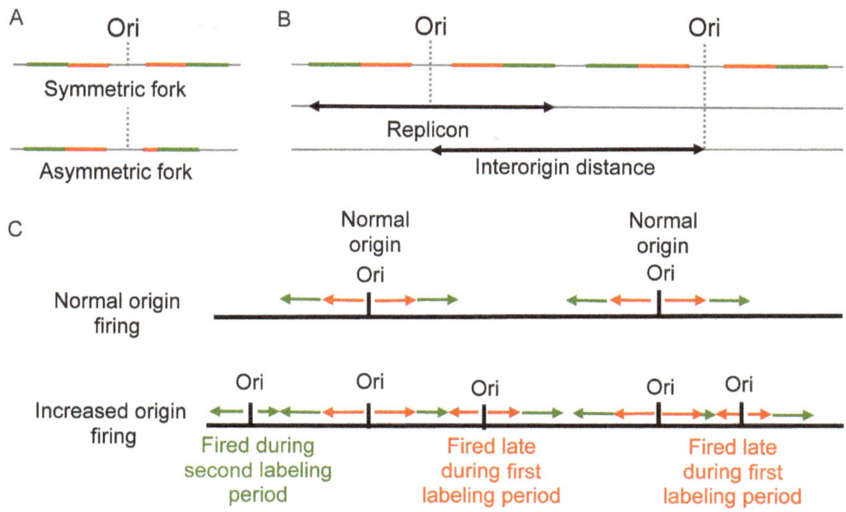

Fig. 4 Schematic representation of fork asymmetry, new origin firing, and interorigin distances. (A) Schematic representation of symmetric vs asymmetric forks. (B) Schematic representation of the labeled tracts of replication forks that emanate from fired origins. The origin is in the middle of two forks traveling in opposite directions. The replicon is the length end to end of two adjacent forks. The interorigin distance is the length between two adjacent origins. (C) Schematic representation of increased origin firing. In normal cells with canonical origin firing most of the tracts are *red–green*. Increased origin firing leads to licensing of new origins at different time points during the labeling which results in an increase in *green* only and *green–red–green* tracts as well as more fork collisions.

broken. Total ssDNA can be immunostained with specific primary antibodies commercially available (see Fig. 2B and D for representative images of DNA counterstaining and Section 4.1 for more details). Fork symmetry can also be estimated from DNA spreading experiments by measuring the length of the two green tracts in bidirectional forks that have a contiguous green–red–green signal (see Fig. 3B). Asymmetric forks will result in the green tract being shorter on one end of the bidirectional fork. However, bidirectional forks with a contiguous green–red–green signal are a relatively rare event, thereby impairing a proper quantification of fork asymmetry.

3.2 New Origin Firing and Interorigin Distance

DNA fiber analysis has largely contributed to the advancements in our understanding of new origin firing and replication clusters. In particular, DNA combing and *maRTA* allow monitoring of interorigin distances because they align DNA molecules along the same axis (Bensimon et al.,

1994; Sidorova et al., 2009). To assess the length of replicons and interorigin distances, cells are labeled sequentially with IdU followed by CldU (Fig. 3A). DNA fibers are prepared following the combing/stretching procedure described in Section 2.2 and immunostained alongside with DNA counterstaining. The length from end to end of two adjacent progressing forks traveling in the opposite direction (Fig. 4B) represents the length of the replicon, while the distance between the consecutive tracts that incorporated both labels corresponds to interorigin distance (Fig. 4B; Técher et al., 2013; Tuduri, Tourrière, & Pasero, 2010).

DNA fiber labeling can also be used to determine the frequency of origin firing events and whether origin firing is altered after treatment with genotoxic agents. An increase in the frequency of origin firing events leads to an increase in the number of green only tracts (origins that fired during the second labeling period), as well as an increase in the number of green–red–green tracts (origins that fired during the first labeling period) (Fig. 3B). In principle, all three stretching techniques can be used to measure the percentage of new origin firing, even though crossing fibers usually observed using the DNA spreading approach hamper this kind of analysis. By DNA combing and *maRTA*, an increase in new origin firing can be evaluated not only by measuring the percentage of new origin firing but also by measuring interorigin distances, expected to be shorter if more origins are firing (Fig. 4C; Técher et al., 2013; Tuduri et al., 2010).

3.3 Replication Fork Restart and Progression Upon Treatment With Genotoxic Agents

Before discussing how to study fork restart by DNA fiber, it is important to make a distinction between genotoxic agents that globally perturb all replication forks and genotoxic agents that might differentially perturb replication forks by inducing DNA lesions. For example, all replication forks are likely to be quickly and equally perturbed in the presence of hydroxyurea (HU), which transiently inhibits DNA synthesis by causing an imbalance in the deoxyribonucleotide pool (Poli et al., 2012), or the replicative polymerase inhibitor aphidicolin. Herein, we will call these agents "global replication-stalling drugs." On the other hand, DNA-damaging agents such as UV light, the inter- and intrastrand cross-linking agents cisplatin and mitomycin C (MMC), or the alkylating agent methylmethane sulfonate (MMS) cause DNA lesions randomly throughout the genome and forks in close proximity to the lesion might be differentially perturbed compared to replication forks that are distant.

DNA-damaging agents must be used at relatively high doses so that the density of DNA lesions is high enough to guarantee that a significant fraction of pulse-labeled forks is encountering the lesion. This is particularly important in the case of short treatments with drugs that have to be metabolized within the cell before damaging the DNA. For example, DNA fiber assays with cisplatin are normally performed using high concentrations of the drug (from 300 μ*M* to 2.5 m*M*) for 20–25 min, concomitantly with the second thymidine analog (Henry-Mowatt et al., 2003; Phillips & Sale, 2010). Other studies use a different scheme, where cisplatin is added alongside with the first tract for 30–45 min followed by pulse labeling with the second analog for additional 20–45 min. In this case, the cisplatin doses are slightly lower (25–40 μ*M*) (Povlsen et al., 2012; Schaaf et al., 2016). Alternatively, lower concentrations of DNA-damaging agents can be used by extending the timing of treatment. For example, Sokol and colleagues used a much lower dose of cisplatin (1.7 μ*M*) by doing a pretreatment for 24 h prior to thymidine analog incorporation (Sokol, Cruet-Hennequart, Pasero, & Carty, 2013). In conclusion, when using DNA-damaging drugs, two parameters have to be taken into account: drug concentration and timing of treatment. The choice of either depends on the biological question to be asked as well as on the particular cell type used for the experiments, as discussed in the section later.

3.3.1 Global Replication-Stalling Drugs

When using global replication-stalling drugs such as HU or aphidicolin, fork restart can be evaluated by treating the cells with the first analog, followed by addition of the drug for a specific amount of time and pulse labeling with the second analog after the removal of the drug (see labeling scheme of Fig. 5A). The second pulse should be relatively short (typically 15–20 min for human cells) to increase the chance of detecting any fork restart defect. The longer the second pulse, the more time it gives the forks to restart, thereby obscuring a possible defect in restart. Restarting forks are represented by replication tracts that have incorporated both analogs (IdU–CldU, CldU–IdU–CldU, and IdU–CldU–IdU), whereas stalled forks correspond to IdU only tracts (Fig. 5A). A defect or delay in fork restart will lead to a decrease in the percentage of restarting forks and a concomitant increase in the percentage of stalled forks compared to untreated cells.

A slightly modified version of the labeling scheme described earlier can be used to test whether a specific genotoxic agent impairs the ability of the fork to progress after restart. In this modified version, a longer second pulse

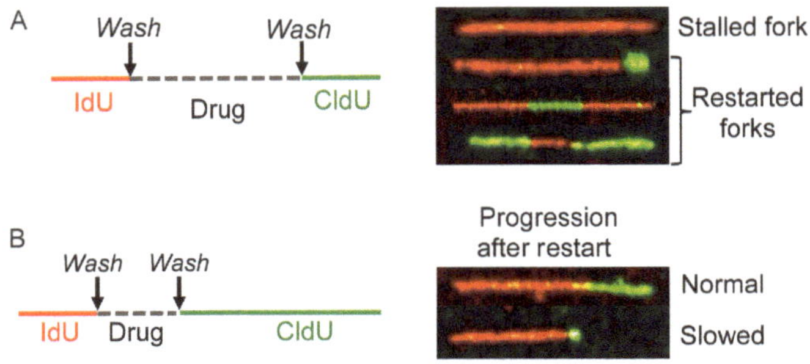

Fig. 5 Labeling scheme to evaluate replication fork restart and progression after treatment with a global replication-stalling agent. (A) Cells are first labeled with IdU analog (*red*) and then treated with a global replication-stalling agent followed by labeling with the CldU analog (*green*). (B) Fork progression after restart is evaluated with a longer pulse with the second analog CldU.

with CldU (40–120min, Fig. 5B) can be used to examine how forks progress following treatment with HU or aphidicolin (Ray Chaudhuri et al., 2016). By increasing the timing of CldU incorporation, forks have more time to restart and fork progression after restart can be more accurately evaluated by measuring CldU tracts or CldU/IdU ratio (Jones, Kotsantis, Stewart, Groth, & Petermann, 2014; Ray Chaudhuri et al., 2016).

3.3.2 DNA-Damaging Agents

In the case of DNA-damaging agents, studying fork restart is less straightforward because forks continue replicating until they encounter the lesion (Fig. 6). Consequently, it is difficult to distinguish between defects in fork restart or fork progression. Nevertheless, labeling schemes similar to the one described earlier (IdU-treatment-CldU) were successfully used to assess fork restart upon exposure to different DNA-damaging agents (Berti et al., 2013; Mourón et al., 2013; Thangavel et al., 2015). Fork progression upon DNA damage can be monitored using the same labeling scheme described for global replication-stalling drugs (timing of the second tract incorporation is typically from 20 to 120min) (Elvers et al., 2011; Jansen et al., 2014; Vallerga, Mansilla, Federico, Bertolin, & Gottifredi, 2015; Fig. 6A and B). Another strategy to assess fork progression is pretreating cells with DNA-damaging agent for 1–24h and then pulse-label with both analogs (Sokol et al., 2013). In this case, the drug concentration can be lower and defects in fork progression can be monitored by comparing the length of both tracts in the presence and absence of the drug (see Fig. 6C).

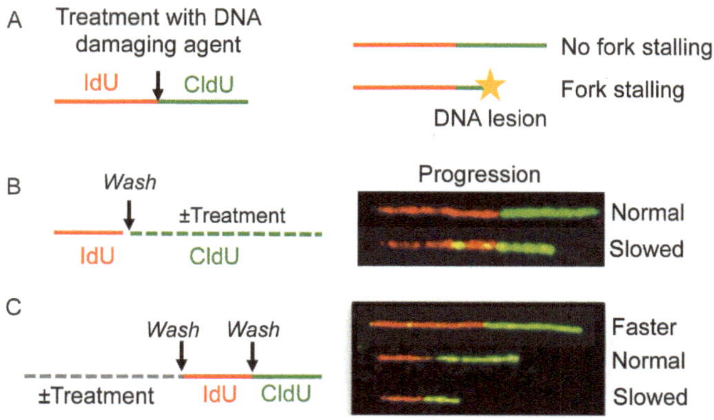

Fig. 6 Replication fork stalling upon treatment with a DNA-damaging agent. When cells are treated with a DNA-damaging agent prior to (A) or concomitant with (B) CldU incorporation, progressing forks that collide with the lesion (represented by the *yellow star*) might stall, generating a shorter *green* track compared to nonstalled forks. (C) Fork progression can also be monitored by treating cells prior to incorporation of both analogs.

3.4 Nucleolytic Degradation of Nascent DNA (or Resection)

Extensive nuclease-dependent degradation of nascent DNA has been detected by DNA fiber analysis in the absence of key *Fanconi anemia*/homologous recombination factors such as the *Br*east *ca*ncer susceptibility proteins 1 and 2 (Brca1 and Brca2) (Ray Chaudhuri et al., 2016; Schlacher et al., 2011). In order to detect extensive nucleolytic degradation of nascent DNA, the crucial aspect in the labeling scheme is the incorporation of one or both analogs prior drug treatment (Fig. 7). Shortening of the tract incorporated prior drug treatment relative to untreated cells reflects nuclease-dependent degradation of the stalled forks upon replication stress induction. This conclusion can be validated by inhibiting or depleting the nucleases involved in this process and confirming that their absence rescues the shortening phenotype (Ray Chaudhuri et al., 2016; Schlacher et al., 2011). Two labeling schemes are commonly used to study nucleolytic degradation. The first starts with the incorporation of the two analogs followed by drug treatment (Fig. 7A). If there is resection, the second tract (CldU) is expected to be shorter compared to IdU tracts. In the second, cells are pulse labeled with the first thymidine analog IdU (red label), followed by treatment with a selected replication inhibitor or DNA-damaging agent, and sequential or concomitant labeling with the second thymidine analog, CldU (green label) (Fig. 7B and C). In this case, the IdU length must be measured. An important caution to use in this analysis is that shortening of the first tract

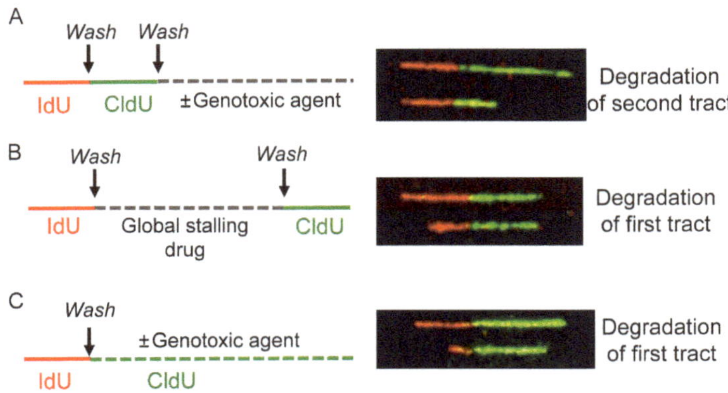

Fig. 7 Labeling schemes that evaluate nucleolytic degradation of newly synthesized DNA after treatment with a genotoxic agent. (A) Cells are treated with both analogs and then exposed to a genotoxic agent. In case of DNA degradation, the second tracts CldU (*green*) will be shorter. (B) Cells are treated with a global replication-stalling drug in between the two labeling schemes or (C) cells are treated with the first analog, and then the genotoxic agent (global stalling or DNA-damaging drug) is added prior to or concomitant with the second analog for an extended period of time. In these scenarios, DNA resection leads to shorter first tracts (IdU, in *red*).

should be measured only on forks characterized by contiguous IdU–CldU signals (and not on forks that have only the IdU label) to ensure that the shortening phenotype is indeed due to nucleolytic resection of stalled replication forks that can resume DNA synthesis and not to premature termination events (which would carry only the first label).

4. APPLICATION TO STUDY POSTREPLICATION REPAIR/ GAP-FILLING MECHANISM

Replication forks have several mechanisms to "skip" DNA lesions and continue replication, while the lesion is tolerated behind the fork, uncoupled from the replication machinery (Lehmann & Fuchs, 2006). This "skipping" mechanism leads to the formation of a ssDNA gap opposite to the damage on the daughter strand, also known as "daughter-strand gap" (Daigaku et al., 2010; Lopes et al., 2006; Meneghini, 1976; Rupp & Howard-flanders, 1968). The mechanism that leads to ssDNA gap formation on the leading strand is called de novo DNA priming (or repriming) (García-Gómez et al., 2013; Heller & Marians, 2006; Mourón et al., 2013). These gaps are then filled (or repaired) postreplicatively by gap-filling or PRR mechanisms (Daigaku et al., 2010; Diamant et al., 2012; Edmunds et al., 2008; Izhar, Ziv, Cohen, Geacintov, & Livneh, 2013; Karras & Jentsch, 2010; Quinet et al., 2016). These mechanisms cannot be studied using the conventional DNA fiber approach because the ssDNA gaps have a size ranging from 300 to 400 nucleotides, while the DNA fiber technique has a resolution of a few kb, as already discussed. In this section, we discuss modified versions of the DNA fiber approach that allow studying two aspects of the PRR pathway: (1) the presence of ssDNA gaps on ongoing forks at early time points and (2) the repair of these ssDNA gaps (or gap filling) at later time points.

4.1 Detection of ssDNA Gaps on Ongoing Forks

To directly evaluate the presence of ssDNA gaps on ongoing forks, we discuss a modified DNA fiber protocol based on the use of a ssDNA-specific enzyme, the S1 nuclease from *Aspergillus oryza* (Quinet et al., 2016). S1 nuclease is used to nick the ssDNA opposite to the gap, thus converting the ssDNA gap into a double-strand break (Schumacher, Menck, & Meneghini, 1983). A protocol describing the use of S1 nuclease in *maRTA* was also recently described (Welcsh, Kehrli, Lazarchuk, Ladiges, & Sidorova, 2016). In the presence of ssDNA gaps, treatment of exposed nuclei with S1 nuclease before DNA spreading leads to shorter CldU tracts (Fig. 8A and B).

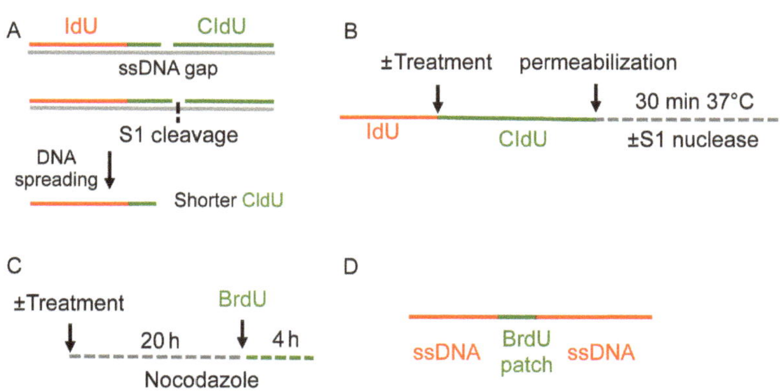

Fig. 8 Detection of ssDNA gaps on ongoing forks and postreplicative gap filling. (A) ssDNA gaps are not detected by DNA fiber analysis. Treatment with the ssDNA-specific S1 nuclease leads to cleavage at the gap and results in a shorter second label (*green*) in tracts. (B) Scheme of the protocol to detect ssDNA gaps using S1 nuclease. (C) Scheme of the procedure to monitor postreplicative gap filling. (D) Scheme of the expected tract in case of postreplicative gap filling.

It is important to note that ssDNA gaps form only after forks have "skipped" the lesion and therefore might not be detectable immediately after replication stress induction. For example, ssDNA gaps are detected only 60 min after UV exposure in repair-deficient cells (Quinet et al., 2016, 2014). Therefore, in order to detect ssDNA gaps, the labeling scheme must include a relative long pulse after or concomitant to treatment with a genotoxic agent.

4.1.1 Material, Solutions, and Reagents

- Microscope slides: positive charged (Denville Ultraclear Cat # M1021, for example)
- Slides jar
- Thymidine analogs: IdU (Sigma-Aldrich Cat # I7125) and CldU (Sigma-Aldrich Cat # C6891)
- PBS 1 ×
- LB: 200 m*M* Tris–HCl, pH 7.5, 50 m*M* EDTA, 0.5% SDS
- PBS-T: PBS 1 ×, 0.05% Tween-20
- PBS-T-BSA: PBS-T, 1% BSA (bovine serum albumin)
- PBS + 0.1% BSA
- CSK100 buffer: 100 m*M* NaCl, 10 m*M* MOPS, pH 7, 3 m*M* $MgCl_2$, 300 m*M* sucrose, 0.5% Triton X-100
- S1 nuclease (Invitrogen Cat # 18001016) *check exact concentration on the enzyme tube; prepare aliquots of 10 μL of prediluted S1 nuclease at 1/100 or 1/200 (in dilution buffer, provided with the enzyme) and store them at −20°C (use it once and discard)*
- S1 nuclease buffer: 30 m*M* sodium acetate, 10 m*M* zinc acetate, 5% glycerol, 50 m*M* NaCl, pH 4.6
- Methanol 3:1 acetic acid
- HCL 2.5 *M*
- Antibodies: rat anti-BrdU (Abcam Cat # Ab6326), mouse anti-BrdU (BD Biosciences Cat # 347580), antimouse IgG1 Alexa Fluor 547 (Invitrogen Cat # A21123), antirat Alexa Fluor 488 (Invitrogen Cat # A21470), mouse anti-ssDNA (Millipore Cat # MAB3034), antimouse IgG2 Alexa Fluor 647 (Invitrogen Cat # A21241)
- Prolong antifade (Invitrogen Cat # P36930)
- Fluorescent microscope with 63 × or 100 × objective with filters or lasers for the following wavelengths: 488 and 547 nm, as well as 647 nm (far-red) in case of ssDNA staining

4.1.2 Procedure

Plating: for each cell line, at least four conditions: ±treatment ± S1 nuclease.

Analogs incorporation and treatment with S1 nuclease:

- Incubate cells with IdU at 20 μ*M* for 20 min (precise). *This timing can be changed.*
- Wash twice with PBS.
- Treat cells.
- Incubate cells with CldU at 200 μ*M* for 60 min (precise). *To detect ssDNA gaps, the second analog must be incorporated long enough upon treatment (see earlier).*
- Wash with PBS.
- Permeabilize cells with CSK100 (enough volume to cover the well/plate): 10 min R.T. *At this time, only the nuclei are observable at the microscope; after this step, wash carefully to avoid detaching the nuclei.*
- Wash with PBS, carefully.
- Wash once with S1 nuclease buffer.
- Add S1 buffer with 20 U/mL S1: 30 min at 37°C. *Add only S1 buffer without the S1 nuclease as a control.*
- Remove the S1 buffer and add PBS + 0.1% BSA. *The low percentage BSA helps precipitating the nuclei.*
- Scrape the cells and put them in an appropriate annotated tube on ice. *Trypsin cannot be used here as it degrades the exposed nuclei.*
- Centrifuge at 7000 rpm for 5 min at 4°C.
- Remove supernatant by leaving the volume necessary to have a concentration of 1.0–2.0 × 10^3 cells/μL. *Nuclei cannot be counted, so this refers to the amount of cells prior to cell permeabilization.*
- Resuspend the pellet well. *The pellet here is harder to resuspend than usual.*
- Put on ice and start quickly the protocol.

Spreading DNA fibers:

- Mix samples well by pipetting up and down.
- Pipette 2 μL of each sample and place the drop near the top of the slide. To increase the amount of DNA that can be analyzed on a single slide, two drops may be added per slide. *Ensuring that the drops do not merge during any of the following steps is critical if using two drops.*
- Add 8 μL of LB and gently pipette up and down about five times to lyse the cells (avoid creating any bubbles that may shear the DNA); then, with the pipette expand the drop a few millimeters to the bottom to determine the path of the drop.

- Wait for 8 min to lyse the nuclei. *For standard spreading, cells are lysed with 6 μL of LB for 5 min. The volume of LB and the time of lysis can be adapted according to the cell type and the conditions of the room. In this protocol the nuclei are exposed and are prone to aggregate, therefore LB can vary from 6 to 8 μL and timing from 6 to 10 min.*
- Carefully tilt the slide at around 25–60 degrees angle and allow the drop to slowly travel down the slide (see Fig. 1C).
- Wait for the slide to dry (10–15 min).
- Fix DNA by placing the slides in the jar with freshly prepared methanol 3:1 acetic acid for 5 min.
- Let the slides dry.
- Store at 4°C or proceed with the staining.

Staining

- Wash the slides 2 × 5 min in PBS.
- Denature DNA with 1 h HCl 2.5 *M* at R.T.
- Wash slides 3 × 5 min in PBS.
- Block for 30–45 min at 37°C with BSA 5% previously warmed at 37°C.
- Remove the excess of liquid by gently tapping.
- Add 30 μL of primary antibodies: mouse anti-BrdU 1/20 (for IdU) and rat anti-BrdU (for CldU) 1/100 in PBS-T-BSA and add a coverslip on top without making too many bubbles.
- Incubate for 1 h 30 min at R.T. in a humid chamber.
- Place the slides in PBS, wait 1–2 min, and remove the coverslips.
- Wash 3 × 5 min with PBS-T, then place the slides in PBS.
- Remove the excess of liquid by gently tapping.
- Add 30 μL of secondary antibodies: antimouse IgG1 Alexa Fluor 546 and antirat Alexa Fluor 488 at 1/100 in PBS-T-BSA and add a coverslip on top.
- Incubate for 1 h at R.T. in a humid chamber in the dark.
- Place the slides in PBS, wait 1–2 min, and remove the coverslips.
- Wash 3 × 5 min with PBS-T, then place the slides in PBS.
  - o Optional: to ensure the DNA fibers are not broken during the preparation of the fibers or by an inappropriate action of the S1 nuclease, immunostain ssDNA as described in the following steps.
  - o Add 30 μL of mouse IgG2 anti-ssDNA at 1/200, place a coverslip on top, and incubate for 45 min at R.T. in a humid chamber in the dark.
  - o Place the slides in PBS, wait 1–2 min, and remove the coverslips; wash 3 × 5 min with PBS-T, then place in PBS.

- o Add 30 μL of antimouse IgG2 Alexa Fluor 647 at 1/200, place a coverslip on top, and incubate 45 min at R.T. in a humid chamber in the dark.
- o Place in PBS, wait 1–2 min, and remove the coverslips; wash 3 × 5 min with PBS-T, then place in PBS.

- Dry the slides.
- Add 20 μL of prolong gold antifade and add a coverslip on top without making bubbles.
- Let the slides dry at R.T. in the dark.
- Store at 4°C (or −20°C for longer conservation).

4.1.3 Data Analysis

Data analysis is performed as described in Section 2.4. Results can be reported in length of CldU tracts (second tracts). In the presence of ssDNA gaps, treatment with S1 nuclease will generate shorter CldU tracts. Alternatively, data can be reported as IdU/CldU tract ratios, if IdU tracts are not affected by treatment with the genotoxic agent. In the presence of S1 nuclease, IdU/CldU tract ratios are expected to be higher compared to controls (Fig. 8A).

4.2 Detection of Postreplication Repair Tracts (or Gap Filling)

A modified DNA fiber approach can also be used to detect filled gaps/PRR tracts in yeast and human cells (Daigaku et al., 2010; Quinet et al., 2016). In this case, the experiments must be performed on a later time point following treatment to allow gap-filling mechanisms to take place. Nocodazole is added after or concomitant to treatment to accumulate cells in the G2/M phase and avoid the following replication cycle. For the last 4 h of treatment, BrdU is added to cell culture media so that it can be incorporated during gap filling (Fig. 8C and D). The rationale of this strategy is based on the finding that gap filling occurs separately from replication in the late S and G2/M phases of the cell cycle, as discussed earlier.

ssDNA gaps activate the checkpoint in the G2/M phase (Callegari, Clark, Pneuman, & Kelly, 2010; Jansen, Tsaalbi-Shtylik, Hendriks, Gali, et al., 2009; Temviriyanukul et al., 2012; Wigan et al., 2012). Accumulation of cells in G2/M upon treatment with a selected genotoxic agent is a good indication of ssDNA gaps formation. Therefore, preliminary experiments can be performed to determine the time point in which the selected genotoxic agent induces a G2/M cell cycle arrest. This time point represents the best timing for the gap-filling assay. As mentioned previously, DNA fibers should be treated with relative high doses of DNA-damaging agents to ensure a significant density of DNA lesions. However, cell cycle analysis following treatment with high doses of the agent could mask an eventual accumulation of cells in G2/M due to S-phase arrest. Therefore, cell cycle analysis must be performed with relative low doses to determine the time point of G2-phase arrest upon treatment with the selected DNA-damaging agent. Next, gap filling can be monitored at this timing by DNA fiber using a two to five times higher dose of the same agent (Daigaku et al., 2010; Quinet et al., 2016).

4.2.1 Reagents

- Nocodazole (Sigma-Aldrich Cat # M1404)
- BrdU (5-bromo-2′-deoxyuridine, Sigma-Aldrich Cat # B5002) or CldU (Sigma-Aldrich Cat # C6891)
- Antibodies: mouse anti-ssDNA antibody (Millipore Cat # MAB3299), rat anti-BrdU (Abcam Cat # Ab6326), antimouse IgG1 Alexa Fluor 547 (Invitrogen Cat # A21123) or antimouse Alexa Fluor 594, antirat Alexa Fluor 488 (Invitrogen Cat # A21470)

4.2.2 Protocol

- Treat cells with a genotoxic agent.
- Add nocodazole at a final concentration of 100 ng/mL for 24 h. *This timing can be changed.*
- For the last 4 h, add BrdU to the medium at a final concentration of 10 μ*M*.
- Proceed with the preparation of DNA fibers by spreading (see protocol earlier for standard spreading).
- Proceed with immunostaining as described earlier with the following modifications:
  - o Primary antibodies: anti-ssDNA (1/40) and anti-BrdU (1/40) for 3–4 h at R.T. in a humid chamber.
  - o Secondary antibodies: antimouse and antirat at 1/100 for 1 h at R.T. in a humid chamber in the dark.

4.2.3 Data Analysis

Results are expressed as density of PRR tracts. For this purpose, ssDNA fibers only (with no continuous BrdU staining) should be evaluated in kb. As mentioned in Section 2.4, 1 μm of DNA fiber prepared by spreading corresponds to approximately 2.59 kb (Jackson & Pombo, 1998). The amount of clear BrdU patches (PRR tracts) overlapping the DNA should be quantified. Density of PRR tracts (kb^{-1}) = number of PRR tracts/length of ssDNA in kb (Daigaku et al., 2010; Quinet et al., 2016).

5. CONCLUDING REMARKS AND FUTURE DIRECTIONS

DNA fiber analysis is a powerful tool to monitor how several key replication parameters are affected by treatment with genotoxic agents or by the changes of the genetic background. When using genotoxic agents, it is crucial to distinguish agents that globally perturb replication forks from DNA-damaging agents that might differentially affect forks that are in close proximity to the lesions compared to replication forks that are distant. The type of agent used to challenge replication affects the choice of labeling scheme and conditions of treatment for DNA fiber analysis. In addition, the recent development of novel technologies to specifically detect DNA lesions opens new avenues to improve even further the potential applications of DNA fiber analysis in the replication field. For example, a recently developed quantum dot technology allows visualization of the cross-linked sites on genomic DNA (Huang et al., 2013). Using this technique, the authors elegantly demonstrated that replication forks that meet the cross-linked site are able to traverse the damaged site and continue replicating, although with some delay (Huang et al., 2013). Whether forks that are in close vicinity to DNA lesions are differentially perturbed compared to distant forks remains an open question in the field. To properly address this question, future studies should extend the quantum dot technology to other drugs or replication inhibitors by using custom-made antibodies or novel strategies to specifically detect DNA lesions on genomic DNA. Another current limitation of the DNA fiber technique is its relatively low resolution and its inability to detect short ssDNA discontinuities of the leading or lagging strand. Here, we described a novel approach that partially overcomes this limitation by using an ssDNA-specific endonuclease that cleaves ssDNA gaps leading to shorter thymidine-labeled tracts that can be detected by DNA fiber analysis. This protocol paves the way to study how and when ssDNA gaps are generated on newly replicated DNA. We envision that this protocol will be increasingly used in combination with other techniques—e.g., electron microscopy (Berti et al., 2013; Neelsen, Chaudhuri, Follonier, Herrador, & Lopes, 2014; Neelsen & Lopes, 2015; Thangavel et al., 2015; Zellweger et al., 2015), isolation of proteins on nascent DNA (iPOND)

(Sirbu, Couch, & Cortez, 2012; Sirbu et al., 2011), and chromatin immunoprecipitation (ChIP) sequencing (Lubelsky, MacAlpine, & MacAlpine, 2012; Ostrow, Viggiani, Aparicio, & Aparicio, 2015)—to study the biological function of replication intermediates containing ssDNA discontinuities and shed light onto the constantly growing number of DNA replication stress response mechanisms.

ACKNOWLEDGMENTS

We thank Joel Eissenberg and Carlos F.M. Menck for their careful reading of the manuscript and insightful comments. We also thank Carlos F.M. Menck for contributing to the development of the method to detect ssDNA gaps on ongóing forks and Philippe Pasero for supplying the DNA fibers images obtained with the combing machine. Work in the A.V. laboratory is supported by NIH Grant R01GM108648 and by DOD BRCP Breakthrough Award BC151728.

REFERENCES

Bensimon, A., Simon, A., Chiffaudel, A., Croquette, V., Heslot, F., & Bensimon, D. (1994). Alignment and sensitive detection of DNA by a moving interface. *Science*, *265*(5181), 2096–2098. http://doi.org/10.1126/science.7522347.

Bensimon, D., Simon, A. J., Croquette, V., & Bensimon, A. (1995). Stretching DNA with a receding meniscus: Experiments and models. *Physical Review Letters*, *74*(23), 4754–4757. http://doi.org/10.1103/PhysRevLett.74.4754.

Berti, M., Ray Chaudhuri, A., Thangavel, S., Gomathinayagam, S., Kenig, S., Vujanovic, M., et al. (2013). Human RECQ1 promotes restart of replication forks reversed by DNA topoisomerase I inhibition. *Nature Structural & Molecular Biology*, *20*(3), 347–354. http://doi.org/10.1038/nsmb.2501.

Berti, M., & Vindigni, A. (2016). Replication stress: Getting back on track. *Nature Structural & Molecular Biology*, *23*(2), 103–109. http://doi.org/10.1038/nsmb.3163.

Bertolin, A. P., Mansilla, S. F., & Gottifredi, V. (2015). The identification of translesion DNA synthesis regulators: Inhibitors in the spotlight. *DNA Repair*, *32*, 158–164. http://doi.org/10.1016/j.dnarep.2015.04.027.

Bianco, J. N., Poli, J., Saksouk, J., Bacal, J., Silva, M. J., Yoshida, K., et al. (2012). Analysis of DNA replication profiles in budding yeast and mammalian cells using DNA combing. *Methods*, *57*(2), 149–157. http://doi.org/10.1016/j.ymeth.2012.04.007.

Callegari, A. J., Clark, E., Pneuman, A., & Kelly, T. J. (2010). Postreplication gaps at UV lesions are signals for checkpoint activation. *Proceedings of the National Academy of Sciences of the United States of America*, *107*(18), 8219–8224. http://doi.org/10.1073/pnas.1003449107.

Conti, C., Sacca, B., Herrick, J., Lalou, C., Pommier, Y., & Bensimon, A. (2007). Replication fork velocities at adjacent replication origins are coordinately modified during DNA replication in human cells. *Molecular Biology of the Cell*, *18*(8), 3059–3067. http://doi.org/10.1091/mbc.E06-08-0689.

Daigaku, Y., Davies, A. A., & Ulrich, H. D. (2010). Ubiquitin-dependent DNA damage bypass is separable from genome replication. *Nature*, *465*(7300), 951–955. http://doi.org/10.1038/nature09097.

Diamant, N., Hendel, A., Vered, I., Carell, T., Reißner, T., De Wind, N., et al. (2012). DNA damage bypass operates in the S and G2 phases of the cell cycle and exhibits differential mutagenicity. *Nucleic Acids Research*, *40*(1), 170–180. http://doi.org/10.1093/nar/gkr596.

Edmunds, C. E., Simpson, L. J., & Sale, J. E. (2008). PCNA ubiquitination and REV1 define temporally distinct mechanisms for controlling translesion synthesis in the avian cell line DT40. *Molecular Cell, 30*(4), 519–529. http://doi.org/10.1016/j.molcel.2008.03.024.

Elvers, I., Johansson, F., Groth, P., Erixon, K., & Helleday, T. (2011). UV stalled replication forks restart by re-priming in human fibroblasts. *Nucleic Acids Research, 39*(16), 7049–7057. http://doi.org/10.1093/nar/gkr420.

Gaillard, H., Garcia-Muse, T., & Aguilera, A. (2015). Replication stress and cancer. *Nature Reviews. Cancer, 15*(5), 276–289. http://doi.org/10.1038/nrc3916.

Gallo, D., Wang, G., Yip, C. M., & Brown, G. W. (2016). Analysis of replicating yeast chromosomes by DNA combing. *Cold Spring Harbor Protocols, 2016*(2), 90–101. pdb.prot085118. http://doi.org/10.1101/pdb.prot085118.

García-Gómez, S., Reyes, A., Martínez-Jiménez, M. I., Chocrón, E. S., Mourón, S., Terrados, G., et al. (2013). PrimPol, an archaic primase/polymerase operating in human cells. *Molecular Cell, 52*, 541–553. http://doi.org/10.1016/j.molcel.2013.09.025.

Hashimoto, Y., Chaudhuri, A. R., Lopes, M., & Costanzo, V. (2010). Rad51 protects nascent DNA from Mre11-dependent degradation and promotes continuous DNA synthesis. *Nature Structural & Molecular Biology, 17*(11), 1305–1311. http://doi.org/10.1038/nsmb.1927.

Heller, R. C., & Marians, K. J. (2006). Replication fork reactivation downstream of a blocked nascent leading strand. *Nature, 439*(7076), 557–562. http://doi.org/10.1038/nature04329.

Henry-Mowatt, J., Jackson, D., Masson, J. Y., Johnson, P. A., Clements, P. M., Benson, F. E., et al. (2003). XRCC3 and Rad51 modulate replication fork progression on damaged vertebrate chromosomes. *Molecular Cell, 11*(4), 1109–1117. http://doi.org/10.1016/S1097-2765(03)00132-1.

Huang, J., Liu, S., Bellani, M. A., Thazhathveetil, A., Ling, C., deWinter, J. P., et al. (2013). The DNA translocase FANCM/MHF promotes replication traverse of DNA interstrand crosslinks. *Molecular Cell, 52*(3), 434–446. http://doi.org/10.1016/j.molcel.2013.09.021.

Izhar, L., Ziv, O., Cohen, I. S., Geacintov, N. E., & Livneh, Z. (2013). Genomic assay reveals tolerance of DNA damage by both translesion DNA synthesis and homology-dependent repair in mammalian cells. *Proceedings of the National Academy of Sciences of the United States of America, 110*(16), E1462–E1469. http://doi.org/10.1073/pnas.1216894110.

Jackson, D. A., & Pombo, A. (1998). Replicon clusters are stable units of chromosome structure: Evidence that nuclear organization contributes to the efficient activation and propagation of S phase in human cells. *Journal of Cell Biology, 140*(6), 1285–1295. http://doi.org/10.1083/jcb.140.6.1285.

Jansen, J. G., Temviriyanukul, P., Wit, N., Delbos, F., Reynaud, C. A., Jacobs, H., et al. (2014). Redundancy of mammalian Y family DNA polymerases in cellular responses to genomic DNA lesions induced by ultraviolet light. *Nucleic Acids Research, 42*(17), 11071–11082. http://doi.org/10.1093/nar/gku779.

Jansen, J. G., Tsaalbi-Shtylik, A., Hendriks, G., Gali, H., Hendel, A., Johansson, F., et al. (2009a). Separate domains of Rev1 mediate two modes of DNA damage bypass in mammalian cells. *Molecular and Cellular Biology, 29*(11), 3113–3123. http://doi.org/10.1128/MCB.00071-09.

Jansen, J. G., Tsaalbi-Shtylik, A., Hendriks, G., Verspuy, J., Gali, H., Haracska, L., et al. (2009b). Mammalian polymerase zeta is essential for post-replication repair of UV-induced DNA lesions. *DNA Repair, 8*(12), 1444–1451. http://doi.org/10.1016/j.dnarep.2009.09.006.

Jones, R. M., Kotsantis, P., Stewart, G. S., Groth, P., & Petermann, E. (2014). BRCA2 and RAD51 promote double-strand break formation and cell death in response to gemcitabine. *Molecular Cancer Therapeutics, 13*(10), 2412–2421. http://doi.org/10.1158/1535-7163.MCT-13-0862.

Karras, G. I., & Jentsch, S. (2010). The RAD6 DNA damage tolerance pathway operates uncoupled from the replication fork and is functional beyond S phase. *Cell*, *141*(2), 255–267. http://doi.org/10.1016/j.cell.2010.02.028.

Kaykov, A., Taillefumier, T., Bensimon, A., & Nurse, P. (2016). Molecular combing of single DNA molecules on the 10 megabase scale. *Scientific Reports*, *6*, 19636. Nature Publishing Group. http://doi.org/10.1038/srep19636.

Lehmann, A. R., & Fuchs, R. P. (2006). Gaps and forks in DNA replication: Rediscovering old models. *DNA Repair*, *5*(12), 1495–1498. http://doi.org/10.1016/j.dnarep.2006.07.002.

Lopes, M., Foiani, M., & Sogo, J. M. (2006). Multiple mechanisms control chromosome integrity after replication fork uncoupling and restart at irreparable UV lesions. *Molecular Cell*, *21*(1), 15–27. http://doi.org/10.1016/j.molcel.2005.11.015.

Lubelsky, Y., MacAlpine, H. K., & MacAlpine, D. M. (2012). Genome-wide localization of replication factors. *Methods*, *57*(2), 187–195. http://doi.org/10.1016/j.ymeth.2012.03.022.

Menck, C. F. M., & Munford, V. (2014). DNA repair diseases: What do they tell us about cancer and aging? *Genetics and Molecular Biology*, *37*(1 Suppl), 220–233. http://doi.org/10.1590/S1415-47572014000200008.

Meneghini, R. (1976). Gaps in DNA synthesized by ultraviolet light-irradiated WI38 human cells. *Biochimica et Biophysica Acta*, *425*(4), 419–427. http://doi.org/10.1016/0005-2787(76)90006-X.

Merrick, C. J., Jackson, D., & Diffley, J. F. X. (2004). Visualization of altered replication dynamics after DNA damage in human cells. *Journal of Biological Chemistry*, *279*(19), 20067–20075. http://doi.org/10.1074/jbc.M400022200.

Michalet, X., Rosemary, E., Fougerousse, F., Rousseaux, S., Schurra, C., Hornigold, N., et al. (1997). Dynamic molecular combing: Stretching the whole human genome for high-resolution studies. *Science*, *277*(5331), 1518–1523. http://doi.org/10.1126/science.277.5331.1518.

Mourón, S., Rodriguez-Acebes, S., Martínez-Jiménez, M. I., García-Gómez, S., Chocrón, S., Blanco, L., et al. (2013). Repriming of DNA synthesis at stalled replication forks by human PrimPol. *Nature Structural & Molecular Biology*, *20*(12), 1383–1389. http://doi.org/10.1038/nsmb.2719.

Neelsen, K. J., Chaudhuri, A. R., Follonier, C., Herrador, R., & Lopes, M. (2014). Visualization and interpretation of eukaryotic DNA replication intermediates in vivo by electron microscopy. *Methods in Molecular Biology*, *1094*, 177–208. http://doi.org/10.1007/978-1-62703-706-8_15.

Neelsen, K. J., & Lopes, M. (2015). Replication fork reversal in eukaryotes: From dead end to dynamic response. *Nature Reviews. Molecular Cell Biology*, *16*(4), 207–220. http://doi.org/10.1038/nrm3935.

Ostrow, A. Z., Viggiani, C. J., Aparicio, J. G., & Aparicio, O. M. (2015). ChIP-Seq to analyze the binding of replication proteins to chromatin. *Methods in Molecular Biology*, *1300*, 155–168. http://doi.org/10.1007/978-1-4939-2596-4_11.

Parra, I., & Windle, B. (1993). High resolution visual mapping of stretched DNA by fluorescent hybridization. *Nature Genetics*, *5*(1), 17–21. http://doi.org/10.1038/ng0993-17.

Pasero, P., Bensimon, A., & Schwob, E. (2002). Single-molecule analysis reveals clustering and epigenetic regulation of replication origins at the yeast rDNA locus. *Genes & Development*, *16*(19), 2479–2484. http://doi.org/10.1101/gad.232902.

Phillips, L. G., & Sale, J. E. (2010). The Werner's syndrome protein collaborates with REV1 to promote replication fork progression on damaged DNA. *DNA Repair*, *9*(10), 1064–1072. http://doi.org/10.1016/j.dnarep.2010.07.006.

Poli, J., Tsaponina, O., Crabbé, L., Keszthelyi, A., Pantesco, V., Chabes, A., et al. (2012). dNTP pools determine fork progression and origin usage under replication stress. *The EMBO Journal*, *31*(4), 883–894. http://doi.org/10.1038/emboj.2011.470.

Povlsen, L. K., Beli, P., Wagner, S. A., Poulsen, S. L., Sylvestersen, K. B., Poulsen, J. W., et al. (2012). Systems-wide analysis of ubiquitylation dynamics reveals a key role for PAF15 ubiquitylation in DNA-damage bypass. *Nature Cell Biology*, *14*(10), 1089–1098. http://doi.org/10.1038/ncb2579.

Quinet, A., Martins, D. J., Vessoni, A. T., Biard, D., Sarasin, A., Stary, A., et al. (2016). Translesion synthesis mechanisms depend on the nature of DNA damage in UV-irradiated human cells. *Nucleic Acids Research*, *44*(12), 5717–5731. http://doi.org/10.1093/nar/gkw280.

Quinet, A., Vessoni, A. T., Rocha, C. R. R., Gottifredi, V., Biard, D., Sarasin, A., et al. (2014). Gap-filling and bypass at the replication fork are both active mechanisms for tolerance of low-dose ultraviolet-induced DNA damage in the human genome. *DNA Repair*, *14*(1), 27–38. http://doi.org/10.1016/j.dnarep.2013.12.005.

Ray Chaudhuri, A., Callen, E., Ding, X., Gogola, E., Duarte, A. A., Lee, J.-E., et al. (2016). Replication fork stability confers chemoresistance in BRCA-deficient cells. *Nature*, *535*(7612), 382–387. http://doi.org/10.1038/nature18325.

Roos, W. P., Thomas, A. D., & Kaina, B. (2015). DNA damage and the balance between survival and death in cancer biology. *Nature Reviews Cancer*, *16*(1), 20–33. http://doi.org/10.1038/nrc.2015.2.

Rupp, W. D., & Howard-flanders, P. (1968). Discontinuities in the DNA synthesized in an excision-defective strain of Escherichia coli following ultraviolet irradiation. *Journal of Molecular Biology*, *31*(2), 291–304. http://doi.org/10.1016/0022-2836(68)90445-2.

Schaaf, L., Schwab, M., Ulmer, C., Heine, S., Murdter, T. E., Schmid, J. O., et al. (2016). Hyperthermia synergizes with chemotherapy by inhibiting PARP1-dependent DNA replication arrest. *Cancer Research*, *76*(10), 2868–2875. http://doi.org/10.1158/0008-5472.CAN-15-2908.

Schlacher, K., Christ, N., Siaud, N., Egashira, A., Wu, H., & Jasin, M. (2011). Double-strand break repair-independent role for BRCA2 in blocking stalled replication fork degradation by MRE11. *Cell*, *145*(4), 529–542. http://doi.org/10.1016/j.cell.2011.03.041.

Schneider, C. A., Rasband, W. S., & Eliceiri, K. W. (2012). NIH Image to ImageJ: 25 Years of image analysis. *Nature Methods*, *9*(7), 671–675.

Schumacher, R. I., Menck, C. F., & Meneghini, R. (1983). Sites sensitive to S1 nuclease and discontinuities in DNA nascent strands of ultraviolet irradiated mouse cells. *Photochemistry and Photobiology*, *37*(6), 605–610. http://doi.org/10.1111/j.1751-1097.1983.tb04528.x.

Sidorova, J. M., Li, N., Schwartz, D. C., Folch, A., & Monnat, R. J., Jr. (2009). Microfluidic-assisted analysis of replicating DNA molecules. *Nature Protocols*, *4*(6), 849–861. http://doi.org/10.1038/nprot.2009.54.

Sirbu, B. M., Couch, F. B., & Cortez, D. (2012). Monitoring the spatiotemporal dynamics of proteins at replication forks and in assembled chromatin using isolation of proteins on nascent DNA. *Nature Protocols*, 7(3), 594–605. http://doi.org/10.1038/nprot.2012.010.

Sirbu, B. M., Couch, F. B., Feigerle, J. T., Bhaskara, S., Hiebert, S. W., & Cortez, D. (2011). Analysis of protein dynamics at active, stalled, and collapsed replication forks. *Genes & Development*, *25*(12), 1320–1327. http://doi.org/10.1101/gad.2053211.

Sokol, A. M., Cruet-Hennequart, S., Pasero, P., & Carty, M. P. (2013). DNA polymerase η modulates replication fork progression and DNA damage responses in platinum-treated human cells. *Scientific Reports*, *3*, 3277. http://doi.org/10.1038/srep03277.

Técher, H., Koundrioukoff, S., Azar, D., Wilhelm, T., Carignon, S., Brison, O., et al. (2013). Replication dynamics: Biases and robustness of DNA fiber analysis. *Journal of Molecular Biology*, *425*(23), 4845–4855. http://doi.org/10.1016/j.jmb.2013.03.040.

Temviriyanukul, P., van Hees-Stuivenberg, S., Delbos, F., Jacobs, H., de Wind, N., & Jansen, J. G. (2012). Temporally distinct translesion synthesis pathways for ultraviolet light-induced photoproducts in the mammalian genome. *DNA Repair, 11*(6), 550–558. http://doi.org/10.1016/j.dnarep.2012.03.007.

Thangavel, S., Berti, M., Levikova, M., Pinto, C., Gomathinayagam, S., Vujanovic, M., et al. (2015). DNA2 drives processing and restart of reversed replication forks in human cells. *Journal of Cell Biology, 208*(5), 545–562. http://doi.org/10.1083/jcb.201406100.

Tuduri, S., Crabbé, L., Conti, C., Tourrière, H., Holtgreve-Grez, H., Jauch, A., et al. (2009). Topoisomerase I suppresses genomic instability by preventing interference between replication and transcription. *Nature Cell Biology, 11*(11), 1315–1324. http://doi.org/10.1038/ncb1984.

Tuduri, S., Tourrière, H., & Pasero, P. (2010). Defining replication origin efficiency using DNA fiber assays. *Chromosome Research, 18*(1), 91–102. http://doi.org/10.1007/s10577-009-9098-y.

Vallerga, M. B., Mansilla, S. F., Federico, M. B., Bertolin, A. P., & Gottifredi, V. (2015). Rad51 recombinase prevents Mre11 nuclease-dependent degradation and excessive PrimPol-mediated elongation of nascent DNA after UV irradiation. *Proceedings of the National Academy of Sciences of the United States of America, 112*(48), E6624–E6633. http://doi.org/10.1073/pnas.1508543112.

Vengrova, S., & Dalgaard, J. Z. (Eds.), (2009). In *DNA replication* (vol 521). Totowa, NJ: Humana Press. http://doi.org/10.1007/978-1-60327-815-7.

Vindigni, A., & Lopes, M. (2016). Combining electron microscopy with single molecule DNA fiber approaches to study DNA replication dynamics. *Biophysical Chemistry*. http://doi.org/10.1016/j.bpc.2016.11.014 (Epub ahead of print).

Welcsh, P., Kehrli, K., Lazarchuk, P., Ladiges, W., & Sidorova, J. (2016). Application of the microfluidic-assisted replication track analysis to measure DNA repair in human and mouse cells. *Methods, 108*, 99–110. http://doi.org/10.1016/j.ymeth.2016.04.029.

Wigan, M., Pinder, A., Giles, N., Pavey, S., Burgess, A., Wong, S., et al. (2012). A UVR-induced G2-phase checkpoint response to ssDNA gaps produced by replication fork bypass of unrepaired lesions is defective in melanoma. *The Journal of Investigative Dermatology, 132*, 1681–1688 (1523–1747 (Electronic)). http://doi.org/10.1038/jid.2012.41.

Zellweger, R., Dalcher, D., Mutreja, K., Berti, M., Schmid, J. A., Herrador, R., et al. (2015). Rad51-mediated replication fork reversal is a global response to genotoxic treatments in human cells. *Journal of Cell Biology, 208*(5), 563–579. http://doi.org/10.1083/jcb.201406099.

Zeman, M. K., & Cimprich, K. A. (2014). Causes and consequences of replication stress. *Nature Cell Biology, 16*(1), 2–9. http://doi.org/10.1038/ncb2897.

CHAPTER FOUR

Comet-FISH for Ultrasensitive Strand-Specific Detection of DNA Damage in Single Cells

Manas Mondal, Jia Guo[1]
Biodesign Institute & School of Molecular Sciences, Arizona State University, Tempe, AZ, United States
[1]Corresponding author: e-mail address: jiaguo@asu.edu

Contents

Abstract

The genome integrity of living organisms is constantly threatened by endogenous cellular metabolic processes and environmental agents. To quantify these low, physiologically relevant levels of DNA damage, a single-cell gel electrophoresis (comet) combined with strand-specific fluorescence in situ hybridization (FISH)-based approach has been developed. This approach enables the quantification of low levels of specific DNA lesions in each strand of the selected sequence at the single-molecule sensitivity, as well as in the genome overall in single cells. In this method, the percentage of DNA in the comet tail is used to quantify lesions in the genome overall. Lesions in the respective strands of the designated sequence are analyzed using strand-specific FISH probes. These probes targeting the 3′ and 5′ termini of the selected sequence are conjugated with two distinct fluorophores. Following the comet-FISH assay, the two termini of the designated sequence are visualized as two spots with different colors, under a fluorescence microscope. Separated spots indicate a damage strand, while adjacent or

Methods in Enzymology, Volume 591
ISSN 0076-6879
http://dx.doi.org/10.1016/bs.mie.2017.03.023

colocalized spots imply an intact strand. Any DNA lesions or DNA modifications, which can be converted into strand breaks enzymatically or chemically, can be quantified by this method. The comet-FISH approach described here can be applied to the study of the molecular mechanisms of various repair pathways, as well as in drug screening to develop inhibitors for specific repair pathways.

1. INTRODUCTION

To maintain genome integrity, living organisms must remove lesions in their DNA induced by environmental agents and endogenous cellular metabolic processes (Ciccia & Elledge, 2010). Repair of transcriptionally active sequences could be more urgent than that in silent domains of the genome, as accurate transcription is critical for cell function and survival. Transcription-coupled repair (TCR), a subpathway of nucleotide excision repair (NER), is dedicated to the removal of DNA damage from the transcribed strands of active genes (Bohr, Smith, Okumoto, & Hanawalt, 1985; Hanawalt & Spivak, 2008; Mellon, Spivak, & Hanawalt, 1987; Spivak, 2016). The southern blot (Bohr et al., 1985; Mellon et al., 1987) and ligation-mediated polymerase chain reaction (LM-PCR) (Tornaletti & Pfeifer, 1994)-based approaches have been applied to study TCR of pyrimidine dimers. Although these techniques are powerful tools to study strand-specific DNA repair, they are limited by the requirements of high doses of DNA-damaging agents to induce detectable number of lesions, which the cells or the organisms under study may not tolerate.

To study strand-specific repair of low, physiologically relevant levels of DNA damage, the single-cell gel electrophoresis (comet) combined with strand-specific fluorescence in situ hybridization (FISH)-based method has been developed (Guo, Hanawalt, & Spivak, 2013; Spivak, 2015a). This approach enables the quantification of low levels of specific DNA lesions in each strand of the selected sequence, as well as in the genome overall. In the comet assay (Fig. 1A), cells are embedded in low-melting agarose gel on a microscopic slide and are mildly lysed to break down membranes and remove histones and other soluble proteins. Subsequently, cells are incubated with glycosylases or endonucleases, which convert the lesions into strand breaks specifically at the lesion sites. Upon electrophoresis under alkaline conditions, intact supercoiled DNA attached to nuclear matrix form the "head" of the comet, while DNA-containing breaks become unwound and migrate toward the positive electrode, appearing as the comet "tail."

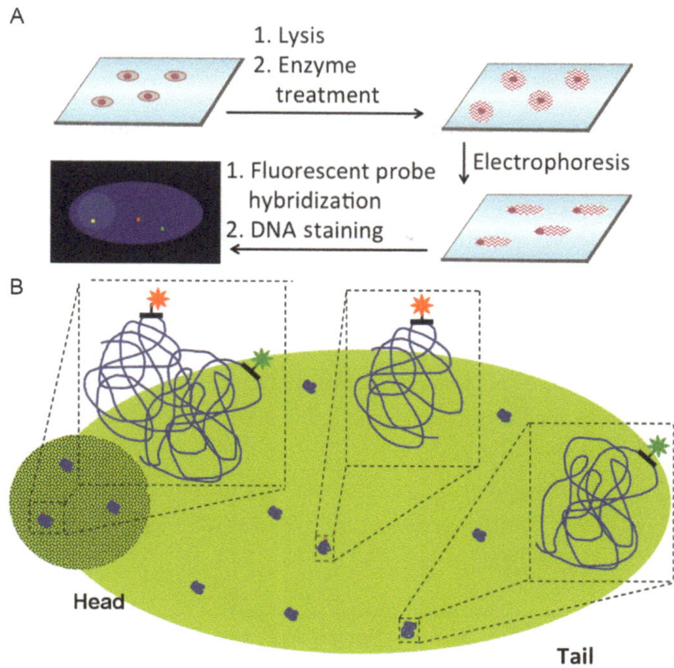

Fig. 1 Comet-FISH with strand-specific probes. (A) Following DNA-damaging treatment, cells are lysed, incubated with glycosylases or endonucleases, and subjected to electrophoresis. Staining the bulk of the DNA permits the analysis of GGR; hybridization of strand-specific probes to the termini of the selected DNA segments allows the quantification of TCR. (B) Schematic representation of comet-FISH. Separated *green* and *red* FISH signals indicate a damaged DNA strand; adjacent *green* and *red* FISH signals suggest an intact DNA strand. *Adapted from Guo, J., Hanawalt, P. C., & Spivak, G. (2013). Comet-FISH with strand-specific probes reveals transcription-coupled repair of 8-oxoGuanine in human cells.* Nucleic Acids Research, 41, *7700–7712.*

The percentage of DNA in the comet tail reflects the total number of strand breaks in individual cells (Spivak, 2010) and is thus used to quantify repair in the genome overall, or global genomic repair (GGR).

To quantify repair in the respective strands of the selected sequence, the 3′ and 5′ termini of the strand are hybridized with strand-specific FISH probes carrying two different fluorophores, such as Alexa 488 and Alexa 594 (Fig. 1B). Assuming that single-stranded DNA in the gel has flexible 3D structures, the 3′ and 5′ termini of an intact strand should be within a short distance. For example, the theoretical end-to-end distance of the single-stranded 146 kb ataxia telangiectasia mutated (ATM) gene is less than 3 μm (Rechendorff, Witz, Adamcik, & Dietler, 2009). Thus, under a fluorescence microscope, the 3′ and 5′ termini of the ATM strand should be visualized as two spots adjacent to each other or colocalized. For the strand containing a break, upon electrophoresis the 3′ and 5′ termini of the strand will migrate to different locations. Consequently, when imaged under a fluorescence microscope, two well-separated spots will be observed. In this way, the comet-FISH approach described here enables the quantification of strand-specific DNA repair in designated sequences with single-molecule sensitivity.

This comet-FISH technique has been validated by documenting TCR of low levels of cyclobutane pyrimidine dimers in the ATM gene in human skin fibroblasts. Applying this approach, it has also been demonstrated that 8-oxoGuanine is preferentially repaired in the transcribed strand of the ATM gene, and the hOGG1, CSB, XPA, and UVSSA proteins, as well as elongating RNA polymerase II, are required for this process (Guo et al., 2013). The detailed protocol of the comet-FISH approach used in these studies is described later.

2. DESIGN AND SYNTHESIS OF STRAND-SPECIFIC FISH PROBES

2.1 Gene or Sequence Selection

First, the gene of interest should be expressed throughout the cell cycles, so that the template strand is actively transcribed in every cell in the sample. The nontranscribed strand of the selected gene should not serve as a template for transcription of other genes. Additionally, the selected gene should not have any matrix attachment sites, as these sites may interfere with the migration of the attached DNA fragments during gel electrophoresis. Finally, to detect low levels of DNA damage, such as 1–10 lesions per 10^6 nucleotides, in a significant percentage of DNA strands, and also to minimize the instances of more than 1 lesion per strand, the size of the selected gene should be around 100–300 kb.

2.2 Synthesis of Probes by PCR

To achieve optimal gel penetration and ideal hybridization specificity and efficiency, the FISH probes are preferred to be ~250 bases in length (Rapp, Hausmann, & Greulich, 2005). On each probe, the fluorophores should be at least 10 bases apart to avoid self-quenching (Levsky, Shenoy, Pezo, & Singer, 2002). For these reasons, to obtain detectable FISH signals, approximately 40–50 different probes targeting one terminus of the selected gene should be synthesized. The generated probes are purified and verified by agarose gel electrophoresis before strand separation and fluorophore coupling. PCR primers can be designed using Gene Fisher, http://bibiserv.techfak.uni-bielefeld.de/genefisher2/. Biotinylated forward primers and natural reverse primers are utilized during PCR to allow separation of the probes for transcribed strands and nontranscribed strands. To generate amino groups on PCR products for subsequent fluorophore conjugation, amino-modified dUTP is combined with the four natural nucleotides for PCR (Fig. 2).

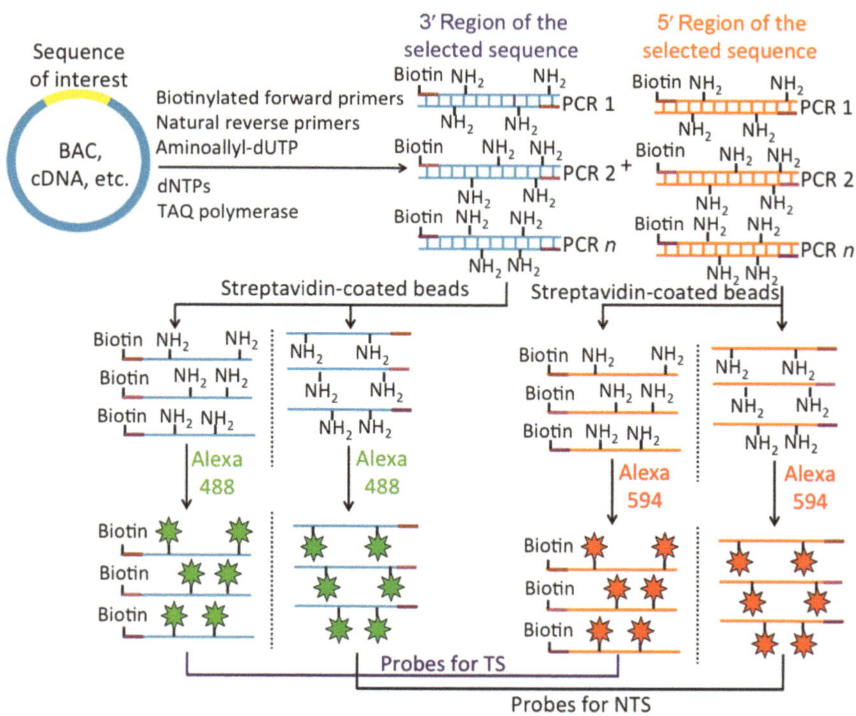

Fig. 2 Design and synthesis of strand-specific FISH probes. The 3′ and 5′ regions of the selected sequence are amplified by PCR using biotinylated forward primers, natural reverse primers, and aminoallyl-dUTP. Then, the biotinylated strands and the nonbiotinylated strands of the generated PCR products are separated using streptavidin-coated beads. The FISH probes targeting the 3′ and 5′ regions of the selected gene are conjugated with two differenct fluorophores, such as Alexa 488 and Alexa 594. Finally, all the biotinylated probes are mixed as probes for the transcribed strand (TS) of the designated sequence, while all the nonbiotinylated probes are combined as probes for the nontranscribed strand (NTS) of the designated sequence. *Adapted from Guo, J., Hanawalt, P. C., & Spivak, G. (2013). Comet-FISH with strand-specific probes reveals transcription-coupled repair of 8-oxoGuanine in human cells.* Nucleic Acids Research, 41, *7700–7712.*

1. Prepare 90 μL reagent mixture for PCR; several PCR reactions can be run simultaneously with GeneAmp PCR system 2400 (Perkin Elmer)
   400 pmol of each of the primers
   20 nmol of each dATP, dCTP, and dGTP
   6.7 nmol of dTTP
   13.3 nmol of aminoallyl-dUTP
   2.5 U of Taq polymerase
   10 μL of 10 × CoralLoad PCR buffer (Qiagen)
   Adjust total volume to 90 μL using DNAse- and RNAse-free water.
2. Add 90 μL of reagent mix to the PCR reaction tube containing 10 μL of template DNA (BAC clone, genomic library, etc.). The optimal concentration of the template DNA should be determined experimentally.
3. Perform PCR at 94°C for 1 min, 55°C for 1 min, 72°C for 1 min 30 cycles, followed by the last extension at 72°C for 10 min.
4. Purify PCR products using PCR purification kit (Qiagen).

2.3 Purification of Single-Stranded FISH Probes

1. Incubate 8 μg of combined PCR products with 40 μg of streptavidin-coated beads at room temperature for 30 min.
2. Use a magnet to capture beads pellet and discard the supernatant.
3. Incubate beads with 20 μL of 0.1 *M* NaOH at room temperature for 10 min. Save the supernatant containing nonbiotinylated DNA.
4. Denature streptavidin with a solution containing 10 m*M* ethylenediaminetetraacetic acid (EDTA) and 95% formamide at 90°C for 3 min. Save the supernatant containing the biotinylated DNA.
5. Purify both biotinylated and nonbiotinylated probes using nucleotide removal kit (Qiagen). Elute the probes with pure water.
6. Characterize the separated single-stranded probes by 1.5% agarose gel electrophoresis. Biotinylated probes should migrate slightly slower than nonbiotinylated probes. Double-stranded PCR products should have higher mobility compared to single-stranded probes (Fig. 3A).
7. Calculate DNA concentration by absorbance at 260 nm. Single-stranded probes can be stored at −20°C for 12 months.

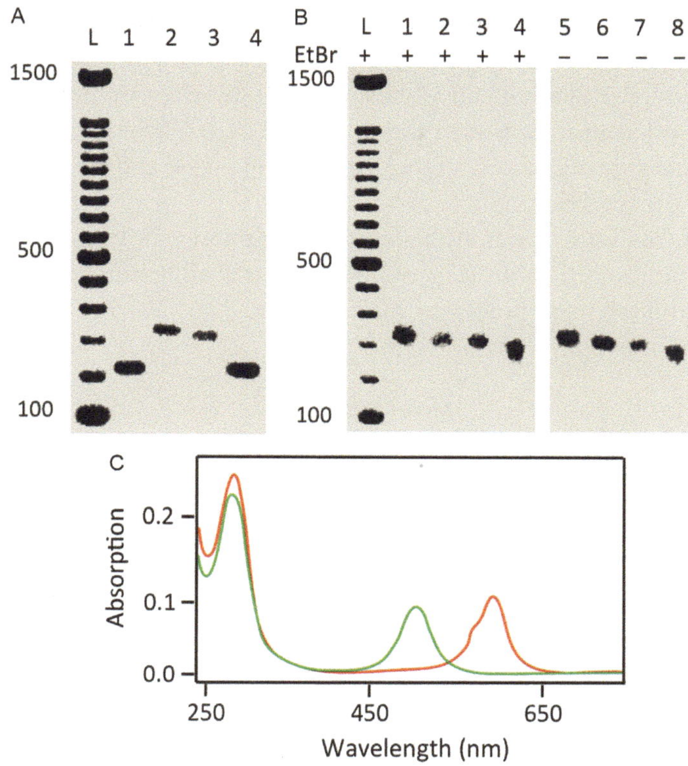

Fig. 3 (A) Agarose gel containing the double-stranded PCR product (lane 1), biotinylated single-stranded DNA segments (lane 2), nonbiotinyaled single-stranded DNA segments (lane 3), and annealed double-stranded DNA segments (lane 4). (B) Agarose gel with strand-specific FISH probes stained with ethidium bromide (*left*) or unstained (*right*). (C) Absorption spectra of strand-specific FISH probes labeled with Alexa 488 (*green*) and Alexa 594 (*red*). *Adapted from Guo, J., Hanawalt, P. C., & Spivak, G. (2013). Comet-FISH with strand-specific probes reveals transcription-coupled repair of 8-oxoGuanine in human cells.* Nucleic Acids Research, 41, *7700–7712.*

2.4 Labeling of Single-Stranded Probes With Fluorophores

To distinguish the 3′ and 5′ termini of the selected gene under a fluorescence microscope, probes targeting the two termini should be coupled to fluorophores with distinct colors, such as Alexa 488 or Alexa 594 (Fig. 2).

1. Incubate combined 1 μg of amino-modified single-stranded probes with 3 μmol of $NaHCO_3$, ~20 μg of *N*-hydroxysuccinimide ester-modified fluorophores, and 2 μL of dimethyl sulfoxide (DMSO) in a total volume of 10 μL at room temperature for 2 h.

2. Purify fluorescently labeled probes with a nucleotide removal kit. Elute the probes with pure water.
3. Characterize fluorescently labeled probes by electrophoresis in 1.5% standard agarose gels; stain the gel with 3 μg/mL ethidium bromide to compare the migration of DNA (Fig. 3B lanes 1–4) and the fluorophores (Fig. 3B lanes 5–8).
4. Measure absorbance at the maximum absorption wavelength of the fluorophore and the DNA (Fig. 3C). Calculate labeling number of the generated fluorescent probes by the equation:

$$N_L = \frac{A_{dye} \times \varepsilon_{base} \times 100}{\left(A_{base} - \alpha \times A_{dye}\right) \times \varepsilon_{dye}}$$

N_L is the number of fluorophore moieties per 100 bases. A_{dye} is the absorption of fluorescently labeled probes at the maximum absorption wavelength of the fluorophore. A_{base} is the absorption of fluorescently labeled probes at 260 nm. ε_{dye} and ε_{base} are the molar extinction coefficients of the fluorophore and base, respectively. α is the ratio of absorbance of the fluorophore at 260 nm to that at the maximum absorbance wavelength. The optimal labeling number should be around five for maximum fluorophore densities and minimum self-quenching. The fluorescent probes can be stored at −20°C for 12 months.

3. COMET ASSAY

1. Coat microscope slides with 1% agarose and store at room temperature for at least 2 weeks.
2. Harvest cells for the single-cell electrophoresis (comet) assay, wash with PBS, and suspend in PBS at 2×10^5 cells/mL.
3. Add 85 μL of the mixture containing a 1:1 ratio of the cell solution and 1.2% low-melting point agarose onto each agarose-coated microscopic slides (VWR). Prepare enough slides for hybridization with probes for each of the complementary strands of the sequence of interest.
4. Cover slides with coverslips and place at 4°C for 30 min.
5. Remove coverslips and incubate the cells in lysis solution [2.5 *M* NaCl, 100 m*M* EDTA, 10 m*M* Tris–HCl, pH 10, with DMSO and 1% Triton X-100] at 4°C overnight.
6. To convert DNA lesions into strand breaks, incubate the agarose-embedded cells with appropriate glycosylases or endonucleases.

7. Place slides in a gel electrophoresis tank and incubate with cold electrophoresis buffer (300 m*M* NaOH and 1 m*M* Na_2EDTA, pH > 13) for 40 min. Conduct electrophoresis at 24 V and 300 mA for 30 min.
8. Remove slides from the electrophoresis apparatus.
9. Wash three times with neutralizing buffer (0.4 *M* Tris, pH 7.5) for 5 min each at 4°C.
10. Incubate slides with 100% ethanol at 4°C for 30 min.
11. Place slides in 0.5 *M* NaOH at room temperature for 25 min.
12. Dehydrate slides with 70%, 85%, and 95% ethanol in water for 5 min each at room temperature.
13. Leave slides in the dark overnight to air-dry. Slides can be stored for several days.

4. HYBRIDIZATION

1. Ethanol precipitate human Cot-1 DNA, evaporate, dissolve in water with the final concentration of 1 μg/μL, aliquot, and store at −20°C. Human Cot-1 DNA hybridizes to repetitive DNA sequences, thus reducing the nonspecific binding of the FISH probes.
2. Mix 10 μg of human Cot-1 DNA with 20 ng of pooled single-stranded fluorescent probes for each slides. Vacuum-dry the mixture completely.
3. Dissolve the probe mixture in 10 μL hybridization buffer [50% formamide, 10% dextran sulfate, and 2× saline sodium citrate buffer (300 m*M* NaCl and 30 m*M* sodium citrate, pH 7.0)].
4. Keep hybridization mixture in ice. Set up a water bath or heat block at 73°C.
5. Heat the probe mixture at 73°C for 5 min and then incubate the mixture at 37°C for 20 min.
6. Prewarm dried comet slides in a humidified chamber to 37°C for 5 min.
7. Add 10 μL of hybridization mixture at the center of prepared comet slides and cover with 24 × 60 mm coverslip.
8. Incubate the slide at 37°C overnight in a humidified chamber.
9. Wash slides twice with 50% formamide, 2× saline sodium phosphate EDTA (SSPE) buffer (300 m*M* NaCl, 20 m*M* NaH_2PO_4, and 2 m*M* EDTA at pH 7.4) at 37°C for 15 min.
10. Wash once with 2× SSPE at 37°C for 15 min.
11. Wash once with 1× SSPE at room temperature for 10 min.

5. COUNTERSTAINING AND IMAGING

1. Prepare a staining solution with 1 μg/mL 4′,6-diamino-2-phenylindole (DAPI) in 1 × PBS. This solution can be stored in the dark and used repeatedly for more than 1 year.
2. Incubate slides with the staining solution at room temperature in the dark for 15 min.
3. Dehydrate slides by dipping quickly in 100% ethanol and let air-dry the slides in the dark.
4. Subsequently add 20 μL of Prolong Gold antifade reagent and cover with 24 × 60 mm coverslip; cure for 1 h in the dark.
5. Store at 4°C in the dark until ready to image.

6. ANALYSIS OF STRAND-SPECIFIC REPAIR

1. Locate comets using the DAPI filter and the 20 × or 40 × magnification lens. Switch to the 100 × lens with immersion oil and the appropriate filters to image the strand-specific FISH probes. Each terminus should be visualized as a single spot under the fluorescence microscope. Minimize photobleaching of fluorophores by limiting the exposure time.
2. Superimpose the images of FISH probes targeting the 3′ and 5′ termini of the selected gene. Well-separated spots corresponding to the 3′ and 5′ termini indicate damaged DNA strands. Adjacent or colocalized spots corresponding to the 3′ and 5′ termini suggest intact DNA strands (Fig. 4).
3. Add the total number of breaks in each comet and calculate the percentage of the damaged strands among all the analyzed strands. For example, if there are 9 breaks in 30 cells with 2 alleles in each cell, 15% of the strands are damaged.

7. ANALYSIS OF GGR

1. Use the 40 × magnification lens and the DAPI filter to capture images of whole comets.
2. Import comet images into a software program for analysis, such as public software including ImageJ (http://rsbweb.nih.gov/ij/), NIH Image

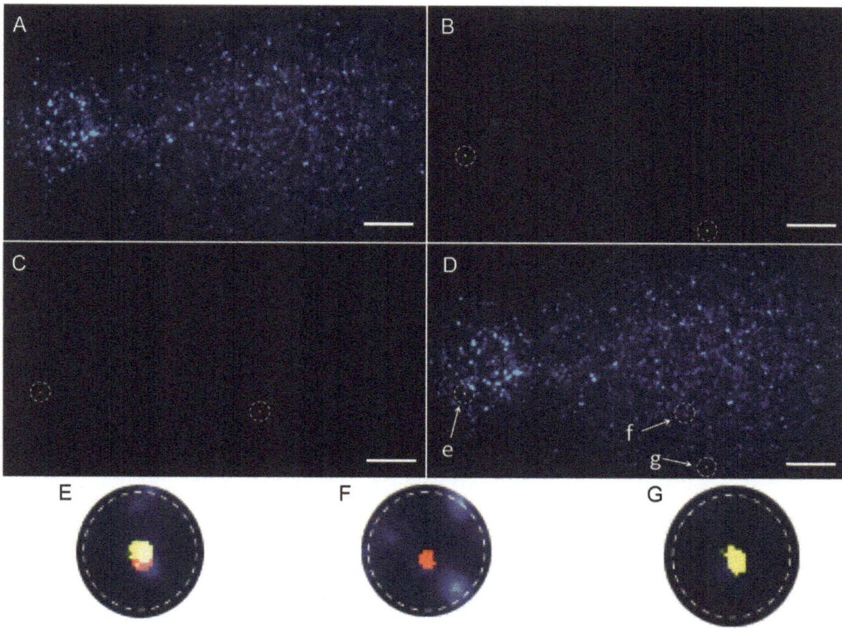

Fig. 4 Representative comet-FISH images showing (A) the bulk DNA stained with DAPI, (B) Alexa 488-labeled probes hybridized to the 3′ termini of the selected stand, (C) Alexa 594-labeled probes hybridized to the 5′ termini of the selected stand, and (D) an overlay of (A–C) (scale bars, 5 μm). (E–G) Enlargements of the FISH signals shown in (D). *Adapted from Guo, J., Hanawalt, P. C., & Spivak, G. (2013). Comet-FISH with strand-specific probes reveals transcription-coupled repair of 8-oxoGuanine in human cells.* Nucleic Acids Research, 41, *7700–7712.*

(http://rsb.info.nih.gov/nih-image/), and CASP (http://www.casplab.com), or commercial packages including Comet Assay IV (Perceptive Instruments) and Komet (Andor).

3. Quantify the fluorescence intensities in the heads and tails of comets, and also in a dark area to measure fluorescence background. Some programs discussed earlier automatically identify the comet heads and tails, subtract backgrounds, and analyze signals. We use ImageJ to perform these tasks manually. The percentage of DNA in comet tails is calculated for 30–100 comets on each slide.

8. CONCLUSIONS

The comet-FISH method described here has the following advantages: (i) this approach allows the quantification of global repair and strand-specific repair of lesions at low, physiologically relevant levels, which are approximately 100 times lower than those in other typical studies (Spivak, Cox, & Hanawalt, 2009). (ii) With the FISH probes of ~250 bases in length, the ideal gel penetration, hybridization efficiency, and specificity are achieved. (iii) Applying multiple FISH probes to hybridize to the termini of the selected sequence, this method eliminates the time-consuming signal amplification step used in previous comet-FISH approaches (Horváthová, Dusinská, Shaposhnikov, & Collins, 2004). (iv) By detecting damaged strands at the single-molecule sensitivity in individual cells, this approach requires fewer cells than other methods to study TCR.

This comet-FISH approach allows the study of the induction and removal of any DNA lesions or DNA modifications, which can be converted into strand breaks enzymatically or chemically. These lesions or modifications include single-strand breaks (Caldecott, 2008), double-strand breaks (O'Driscoll & Jeggo, 2006), lesions removed by NER (Spivak, 2015b) or base excision repair pathways (Kim & Wilson, 2012), ribonucleotides incorporated into DNA (Reijns et al., 2012), and 5-formylcytosine and 5-carboxylcytosine involved in DNA demethylation (Nabel & Kohli, 2011). By labeling the FISH probes with multispectral fluorophores (Dai, Guo, Teo, & Kool, 2011; Guo, Wang, Dai, Teo, & Kool, 2011; Wang, Guo, Ono, & Kool, 2012) or cleavable fluorophores (Guo, 2016; Mondal, Liao, Xiao, Eno, & Guo, 2017), or through reiterative hybridization (Xiao & Guo, 2015), lesions in each strand of multiple designated genomic regions can be quantified simultaneously in single cells. This comet-FISH approach will have wide applications in the study of the molecular mechanisms of various repair pathways, as well as in drug screening to develop inhibitors for specific repair pathways.

ACKNOWLEDGMENTS

This work is supported by grants from Arizona State University and Arizona State University/Mayo Clinic seed grant (ARI-219693).

REFERENCES

Bohr, V. A., Smith, C. A., Okumoto, D. S., & Hanawalt, P. C. (1985). DNA repair in an active gene: Removal of pyrimidine dimers from the DHFR gene of CHO cells is much more efficient than in the genome overall. *Cell*, *40*, 359–369.

Caldecott, K. W. (2008). Single-strand break repair and genetic disease. *Nature Reviews. Genetics*, *9*, 619–631.

Ciccia, A., & Elledge, S. J. (2010). The DNA damage response: Making it safe to play with knives. *Molecular Cell*, *40*, 179–204.

Dai, N., Guo, J., Teo, Y. N., & Kool, E. T. (2011). Protease probes built from DNA: Multispectral fluorescent DNA-peptide conjugates as caspase chemosensors. *Angewandte Chemie (International Edition in English)*, *50*, 5105–5109.

Guo, J. (2016). System and method for iterative detection of biological molecules. US patent application 20160054308A1.
Guo, J., Hanawalt, P. C., & Spivak, G. (2013). Comet-FISH with strand-specific probes reveals transcription-coupled repair of 8-oxoGuanine in human cells. *Nucleic Acids Research*, *41*, 7700–7712.
Guo, J., Wang, S., Dai, N., Teo, Y. N., & Kool, E. T. (2011). Multispectral labeling of antibodies with polyfluorophores on a DNA backbone and application in cellular imaging. *Proceedings of the National Academy of Sciences of the United States of America*, *108*, 3493–3498.
Hanawalt, P. C., & Spivak, G. (2008). Transcription-coupled DNA repair: Two decades of progress and surprises. *Nature Reviews. Molecular Cell Biology*, *9*, 958–970.
Horváthová, E., Dusinská, M., Shaposhnikov, S., & Collins, A. R. (2004). DNA damage and repair measured in different genomic regions using the comet assay with fluorescent in situ hybridization. *Mutagenesis*, *19*, 269–276.
Kim, Y.-J., & Wilson, D. M. (2012). Overview of base excision repair biochemistry. *Current Molecular Pharmacology*, *5*, 3–13.
Levsky, J. M., Shenoy, S. M., Pezo, R. C., & Singer, R. H. (2002). Single-cell gene expression profiling. *Science*, *297*, 836–840.
Mellon, I., Spivak, G., & Hanawalt, P. C. (1987). Selective removal of transcription-blocking DNA damage from the transcribed strand of the mammalian DHFR gene. *Cell*, *51*, 241–249.
Mondal, M., Liao, R., Xiao, L., Eno, T., & Guo, J. (2017). Highly multiplexed single-cell in situ protein analysis with cleavable fluorescent antibodies. *Angewandte Chemie (International Edition in English)*, *56*, 2636–2639.
Nabel, C. S., & Kohli, R. M. (2011). Demystifying DNA demethylation. *Science*, *333*, 1229–1230.
O'Driscoll, M., & Jeggo, P. A. (2006). The role of double-strand break repair—Insights from human genetics. *Nature Reviews. Genetics*, *7*, 45–54.
Rapp, A., Hausmann, M., & Greulich, K. O. (2005). The comet-FISH technique: A tool for detection of specific DNA damage and repair. *Methods in Molecular Biology*, *291*, 107–119.
Rechendorff, K., Witz, G., Adamcik, J., & Dietler, G. (2009). Persistence length and scaling properties of single-stranded DNA adsorbed on modified graphite. *The Journal of Chemical Physics*, *131*, 95103.
Reijns, M. A. M., Rabe, B., Rigby, R. E., Mill, P., Astell, K. R., Lettice, L. A., et al. (2012). Enzymatic removal of ribonucleotides from DNA is essential for mammalian genome integrity and development. *Cell*, *149*, 1008–1022.
Spivak, G. (2010). The comet-FISH assay for the analysis of DNA damage and repair. *Methods in Molecular Biology*, *659*, 129–145.
Spivak, G. (2015a). New developments in comet-FISH. *Mutagenesis*, *30*, 5–9.
Spivak, G. (2015b). Nucleotide excision repair in humans. *DNA Repair*, *36*, 13–18.
Spivak, G. (2016). Transcription-coupled repair: An update. *Archives of Toxicology*, *90*, 2583–2594.
Spivak, G., Cox, R. A., & Hanawalt, P. C. (2009). New applications of the Comet assay: Comet-FISH and transcription-coupled DNA repair. *Mutation Research, Reviews in Mutation Research*, *681*, 44–50.
Tornaletti, S., & Pfeifer, G. P. (1994). Slow repair of pyrimidine dimers at p53 mutation hotspots in skin cancer. *Science*, *263*, 1436–1438.
Wang, S., Guo, J., Ono, T., & Kool, E. T. (2012). DNA polyfluorophores for real-time multicolor tracking of dynamic biological systems. *Angewandte Chemie (International Edition in English)*, *51*, 7176–7180.
Xiao, L., & Guo, J. (2015). Multiplexed single-cell in situ RNA analysis by reiterative hybridization. *Analytical Methods*, *7*, 7290–7295.

CHAPTER FIVE

Examining DNA Double-Strand Break Repair in a Cell Cycle-Dependent Manner

Janapriya Saha, Shih-Ya Wang, Anthony J. Davis[1]
University of Texas Southwestern Medical Center, Dallas, TX, United States
[1]Corresponding author: e-mail address: anthony.davis@utsouthwestern.edu

Contents

Abstract

DNA double-strand breaks (DSBs) are deleterious DNA lesions that must be properly repaired to maintain genome stability. Agents, generated both exogenously (environmental radiation, dental X-rays, etc.) and endogenously (reactive oxygen species, DNA replication, V(D)J recombination, etc.), induce numerous DSBs every day. To counter these DSBs, there are two major repair pathways in mammalian cells, nonhomologous

Methods in Enzymology, Volume 591
ISSN 0076-6879
http://dx.doi.org/10.1016/bs.mie.2017.03.012

end joining (NHEJ) and homologous recombination (HR). NHEJ directly mediates the religation of the broken DNA molecule and is active in all phases of the cell cycle. HR directs repair via the use of a homologous DNA sequence as a template and is primarily active in only S/G2 phases owing to the availability of a DNA template via a sister chromatid. As NHEJ and HR are active in multiple cell cycle phases, there is significant interest in how a cell chooses between the two DSB repair pathways. Therefore, it is essential to utilize assays to study DSB repair that can distinguish between the two DSB repair pathways and the different phases of the cell cycle. In this chapter, we describe methods to measure the contribution of DNA repair pathways in different phases of the cell cycle. These methods are simple, can be applied to most mammalian cell lines, and can be used as a broad utility to monitor cell cycle-dependent DSB repair.

1. INTRODUCTION

The human genome is constantly under attack from a variety of agents that generate tens of thousands of DNA lesions per day. The most deleterious of these lesions is the DNA double-strand break (DSB). Two major pathways direct repair of DSBs in mammalian cells, homologous recombination (HR) and nonhomologous end joining (NHEJ) (Goodarzi & Jeggo, 2013; Hoeijmakers, 2001; Jackson & Bartek, 2009; Schipler & Iliakis, 2013). HR drives DSB repair by using a homologous DNA sequence as a template to guide error-free restoration of the DNA molecule. Since an accessible homologous template is found on a sister chromatid, error-free HR is believed to be primarily active in mid-S phase to early G2 phase of the cell cycle. NHEJ functions by directly religating the two broken DNA strands. As NHEJ does not require a homologous template, it is not restricted to a particular cell cycle phase. It should be noted that there is also an alternative end-joining (Alt-EJ) pathway, which is believed to primarily be a backup pathway for both HR and NHEJ. Alt-EJ typically utilizes microhomologies distant from the DSB site to drive repair (Schipler & Iliakis, 2013).

Since there are multiple DSB repair processes, a cell must properly choose the specific pathway to repair a broken DNA molecule. The cell cycle phase likely plays a role in this process as HR is primarily active in mid-S to early G2 phase of the cell cycle. However, NHEJ is also active in these cell cycle phases and thus there must be a process that assists the cell in choosing the appropriate DSB repair pathway. In particular, due to the high replication activity and the formation of single-ended replication fork-associated breaks in S phase and the critical G2 phase preceding the subsequent division in M phase, error-free repair of DSBs in S/G2 is paramount. Importantly, it has been shown that the majority of breaks are still repaired by NHEJ in early S phase with activities transitioning to the HR pathway from mid-S phase (Karanam, Kafri, Loewer, & Lahav, 2012). Thus, it is also important to distinguish and demarcate different subphases within the S phase to decipher DNA repair activity and pathway contributions accurately.

In this chapter, we will describe protocols that can be used to examine DSB repair processes in a cell cycle-specific manner. These methods were originally developed by other groups and later modified by us and utilized in various publications (Davis et al., 2015; Davis, So, & Chen, 2010; Lee et al., 2016; Shao et al., 2012). The protocols include: examining real-time dynamics of repair proteins localizing and dissociating from DSBs (Jackson & Bartek, 2009); immunofluorescence-based methods to monitor NHEJ, DNA end resection, and ongoing HR (Schipler & Iliakis, 2013); and determining overall repair capacity (Goodarzi & Jeggo, 2013).

2. DYNAMICS OF REPAIR PROTEINS TO LASER-GENERATED DSBS

The cellular response to DSBs initiates with the recognition of the ends of the broken DNA molecule. This DSB recognition results in the recruitment of a significant number of factors to the DSB site and the surrounding area. In this section, we will describe a technique that utilizes a microlaser system to generate DSBs coupled with live-cell microscopy to examine the recruitment and dynamics of a yellow fluorescent protein (YFP)-tagged protein to DSBs. To allow differentiation of cells in S phase and non-S phase, DsRed-tagged PCNA is monitored, as PCNA shows a faint and even distribution in non-S phase cells and forms a distinct punctate patterning in S phase (Fig. 1) (Shao et al., 2012). Here, we will outline the recruitment and kinetics of the NHEJ factor Ku80 to laser-generated DSBs, but this technique has also been successfully used to examine the localization and dynamics of other repair proteins to laser-generated DSBs, including DNA-PKcs, ATM, MDC1, and MRE11 (Davis & Chen, 2010; Kim et al., 2005; So, Davis, & Chen, 2009). It should be noted that the method outlined here is specific for our laser/microscope setup and appropriate adjustments will have to be made for other setups.

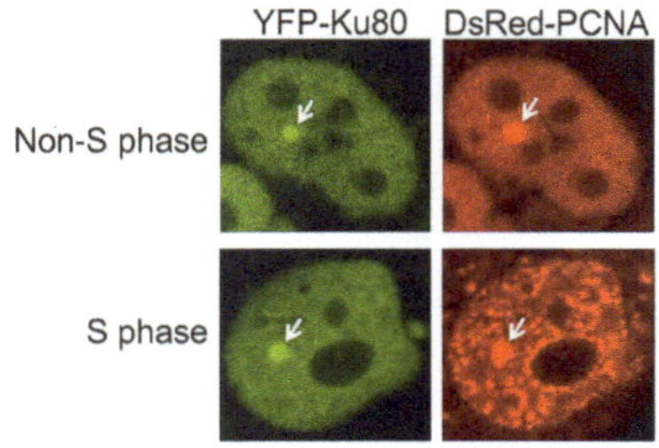

Fig. 1 Differentiating S and non-S phase cells. YFP-tagged Ku80 and DsRed-tagged PCNA were transiently expressed in the Ku80-deficient CHO cell line Xrs5. S phase and non-S phase cells are differentiated by the localization pattern of DsRed-PCNA. Localization of Ku80 and PCNA to laser-generated DSBs is marked by a *white arrow*.

2.1 Transient Expression of YFP-Tagged Ku80 and DsRed-Tagged PCNA

2.1.1 Equipment

- Amaxa Biosystems Nucleofector II (Lonza)

2.1.2 Buffers and Reagents

- Full-length wild-type Ku80 cDNA was subcloned into a modified pcDNA3 vector that carries an YFP tag upstream of the multiple cloning site to generate YFP-tagged Ku80 (YFP-Ku80).
- Full-length wild-type PCNA cDNA was subcloned into pDsRed-Monomer-C1 vector that carries DsRed tag upstream of the multiple cloning site to generate DsRed-tagged PCNA (DsRed-PCNA).
- U2OS or HT1080 cells. It should be noted that these experiments are also performed in rodent cell lines deficient for the YFP-tagged protein of interest. For example, when testing YFP-Ku80, experiments can be performed in the K80-deficient Chinese Hamster Ovary (CHO) cell line Xrs5 or Xrs6 or Ku80-deficient mouse embryonic fibroblasts (MEFs).
- Alpha-minimum Eagle's medium (Fisher, #SH3026502) supplemented with 10% fetal bovine serum and fetal calf serum (1:1 ratio).
- CO_2-independent medium (Invitrogen, #19045088).
- Transfection reagent such as Lipofectamine 2000 reagent (Invitrogen) or electroporation with Nucleofector (Lonza).
- 35-mm Glass-bottom culture dishes (MatTek, P35G-0-14-C).

2.1.3 Procedure

1. Cotransfect the YFP-Ku80 and DsRed-PCNA constructs via Nucleofector or Lipofectamine 2000 following the manufacturer's

protocol into U2OS or HT1080 cells. If using Nucleofector, use protocol X-001 with solution V to electroporate YFP-Ku80 and DsRed-PCNA into U2OS and HT1080 cells. For rodent cells, use protocol U-27 or U-23 with solution T.

2. After transfection, split the cells and plate 1×10^5 cells onto a 35-mm glass-bottom culture dish, allow to attach, and grow for 48h. Check expression via microscopy.

2.1.4 Notes

1. To amplify the Ku80 constructs, transform 50–100ng of plasmid into DH5α competent cells, and then plate on LB^{+AMP} plate for overnight incubation at 37°C. Pick a single colony (choose small, not large colonies) and grow in 200mL LB^{+AMP} in a shaker incubator at 37°C for 24h. Next day, purify the plasmid DNA by the Qiagen midi-prep kit according to manufacturer's instructions. A typical yield for the Ku80 cDNA-containing vectors is approximately 300μg. It should be noted that expression constructs large in size, such as those that contain full-length cDNAs of DNA-PKcs, ATM, 53BP1, and BRCA1, tend to recombine during DNA amplification; therefore, *Escherichia coli* strains deficient in endonucleases and recombination (e.g., XL10) are recommended to improve the quality of the DNA preparation.
2. We typically use cells that are stably expressing the YFP-tagged protein of interest in laser microirradiation assays as this allows for more consistent experimental results. To make stable cell lines, linearize the vector with a restriction enzyme and verify the complete digestion via agarose gel electrophoresis. Purify the linearized vector by phenol extraction (twice) followed by ethanol precipitation and wash with 70% ethanol and allow the pellet to air dry. Resuspend the pellet in TE or water and measure the DNA concentration. Perform transfection as described. Following transfection, plate 1×10^4 cells on 100mm (4 total dishes) with normal growth medium. 24h posttransfection, add the selection drug G418 (500 μg/mL) and incubate the cells for 10 days. Pick single colonies and check expression of YFP-Ku80 via microscopy (YFP signal) and Western blot analysis (antibodies that recognize the YFP tag and/or Ku80).

2.2 Microscope and Laser-Irradiation Setup

2.2.1 Equipment

- An inverted Axiovert 200M microscope equipped with a Plan-Apochromat 63×/NA 1.40 oil immersion objective and an AxioCam

HRm digital camera (Carl Zeiss MicroImaging, Inc.) for time-lapse imaging. Computer with Carl Zeiss AxioVision 4.8.2 software.

- Laser microirradiation unit. Spectra-physics nitrogen laser (Andor Technology, VSL-337ND-S): The Spectra-Physics VSL-337ND-S nitrogen laser emits 4-ns pulses in the UV at 337 nm. The pulse repetition rate may be varied from 1 to 60 Hz with a pulse energy of up to 300 μJ. The nitrogen laser is directly coupled to the epifluorescence path of the microscope through a Micropoint Laser Illumination and Ablation System (Photonic Instruments, Inc.) and focused through a Plan-Apochromat 63 × oil immersion objective. The nitrogen wavelength can be changed from 337 to 365 nm by BPBD 365 dye solution. Power output is controlled by the Micropoint System ranged from 1% to 89%.
- Digital temperature control system with a heating insert P (PeCon, #0426.100) and a Tempcontrol 37-2 control unit (PeCon, #0503.000).

2.2.2 Procedure

1. Switch on temperature control at least 15 min prior to use for live-cell imaging (37°C).
2. Turn on Stage Controller/MCU28.
3. Turn on laser via the key switch. Set repetition rate to 12 or 4 o'clock position depending on the experiment.
4. Switch on Micropoint and set number to 75 for DSB induction.
5. Turn on microscope and the EXFO: UV lamp.
6. Select Objective on 63 × (oil immersion).
7. Use 70/30 beam splitter for laser damage along with GFP filter on the UV adaptor.
8. Filter set selection: select #5 filter for laser damage (FITC_TexasRed).
9. Replace normal growth medium with heated CO_2-independent medium for microirradiation experiment.
10. Apply a drop of oil on the objective and lower objective with "focus down" bottom before placing the sample on it.
11. Attach the 35-mm dish onto the heated dish holder to stabilize the dish and then place this onto the heating insert.
12. Focus a cell and put "Cross mark" (only on the right side-eye) in the center of the nucleus.
13. Set up imaging parameter utilizing AxioVision software. The parameters utilized for multidimensional acquisition are:
    a. "Dye"—YFP
    b. "Exposure"—"fixed"—"400 ms"

c. "Hardware setting"
   i. During acquisition—"workgroup: FITC_TexasRed"
   ii. After acquisition—"workgroup: shutter closed"
d. For initial accumulation (up to 10 min) utilize "Time lapse setting"
   i. Interval—20 s
   ii. Cycles—32
e. For 2 h kinetics, utilize a hand-held timer and take images at specific time points. Typically, images are acquired before irradiation and then 3, 10, 20, 30, 60, 90, and 120 min postirradiation

2.3 DSB Repair Kinetics With Laser Microirradiation

2.3.1 Procedure

1. DSBs are induced in the nucleus of cultured cells by microirradiation with a pulsed nitrogen laser (Spectra-Physics; 365 nm, 10 Hz pulse). Set the output of the laser power at 75% of the maximum.
2. Change the complete alpha-MEM media with CO_2-independent medium heated to 37°C.
3. Identify multiple cells expressing YFP-Ku80 and DsRed-PCNA. DsRed-PCNA is used to differentiate S and non-S phase cells as PCNA forms a distinct punctate pattern in S phase, whereas it shows faint and even expression in non-S phase cells (Fig. 1). Select at least 10 cells in S phase and non-S phase.
4. Initial accumulation (10 min) of YFP-Ku80 to laser-generated DSBs.
   a. Start image acquisition before laser microirradiation to obtain an image of the unirradiated cell.
   b. After first image acquisition, induce DSBs in cell nuclei by microirradiation with pulsed nitrogen laser allowing 8 laser pulses to hit a defined region of a single nucleus.
   c. As described in Section 2.2.2, Note 13, the defined parameters in AxioVision software will acquire images before and after irradiation via a 400-ms exposure time.
   d. Capture images every 20 s for the duration of the 10-min time course.
   e. Convert signal intensities of accumulated YFP at the microirradiated site into a numerical value by the use of the Carl Zeiss AxioVision software (see below).
5. Kinetics of YFP-Ku80 at laser-induced DSBs (2 h).
   a. Apply the same procedure as described above for 2-h microirradiation and time-lapse imaging except capture images at defined time points (3, 10, 20, 30, 60, 90, and 120 min) post microirradiation.

b. Set a fixed exposure time to capture all images before and after irradiation.

c. Convert signal intensities of accumulated YFP at the microirradiated site into a numerical value by the use of the Carl Zeiss AxioVision software (see below).

2.4 Calculation of Relative Fluorescent Intensity for Protein Recruitment Kinetics

2.4.1 Procedure

1. Open AxioVision software and the appropriate image set for each individual cell.
2. Circle damage site (D_t, white circles in Fig. 2) and a background area (B_t, yellow circles in Fig. 2) at each time point. The circle of the DNA

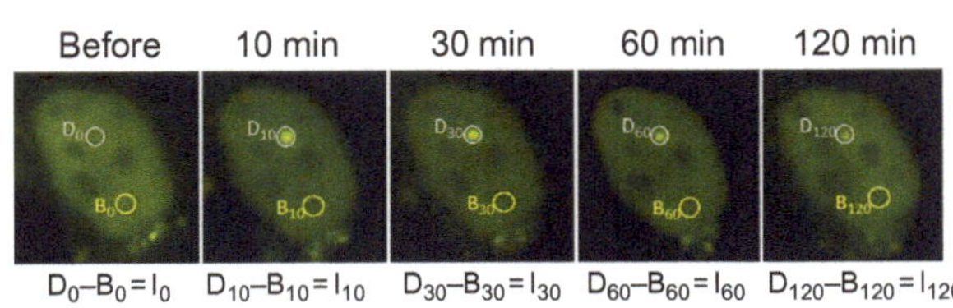

Normalization:

$NI_0 = (I_0/B_0)^*D_0$

$NI_{10} = (I_{10}/B_{10})^*D_0$

$NI_{30} = (I_{30}/B_{30})^*D_0$

$NI_{60} = (I_{60}/B_{60})^*D_0$

$NI_{120} = (I_{120}/B_{120})^*D_0$

Relative intensity:

$RFI_0 = (NI_0 - NI_0)/(NI_{10} - NI_0) = 0$

$RFI_{10} = (NI_{10} - NI_0)/(NI_{10} - NI_0) = 1$

$RFI_{30} = (NI_{30} - NI_0)/(NI_{10} - NI_0) < 1$

$RFI_{60} = (NI_{60} - NI_0)/(NI_{10} - NI_0) < 1$

$RFI_{120} = (NI_{120} - NI_0)/(NI_{10} - NI_0) < 1$

Fig. 2 Calculation of relative fluorescence intensity for recruitment kinetics. A panel of images showing the localization of YFP-tagged Ku80 to laser-generated DSBs in a 2-h time course. Relative intensities of YFP-tagged Ku80 at sites of laser damage at each time point is calculated by using the following equations. D_t, fluorescence intensity accumulated at the damaged spot at a given time; B_t, background fluorescence intensity at an undamaged spot a given time; I_t, fluorescence intensity of the damaged spot at a given time; NI_t, normalized fluorescence intensity of the damaged spot at a given time; RFI_t, relative fluorescence intensity of the damaged spot at a given time compared to that of maximum normalized fluorescence intensity.

damage site (accumulation site of YFP-Ku80) is typically set at 1 μm, but may need to be adjusted based on the size of the spot. The background site is typically 2 μm.

3. Obtain densitometric mean (fluorescence of the spot) via the AxioVision software, which is based on the fluorescence intensity and area size. Perform all calculations by utilizing Microsoft Excel. The densitometric means are placed in Excel for calculations.
4. To eliminate the background fluorescence, laser-induced fluorescence intensity accumulated at the damaged spot D_t is subtracted with the background fluorescence intensity B_t at an undamaged spot in the same nuclei at each time point (Fig. 2). The fluorescence intensity $I_t = D_t - B_t$.
5. Compensate nonspecific photobleaching and UV lamp output fluctuation by normalizing the absolute fluorescence intensity $I_{(t)}$ accumulation at the damaged spot of each time point based on background intensity prior to laser damage using the formula: normalized fluorescence intensity $NI_t = (I_t/B_t)*D_0$, where B_t represents the undamaged site background intensity of each time point, and D_0 represents the intensity of the damaged spot prior to irradiation.
6. Calculate relative fluorescence intensity (RFI) by using the formula: $RFI_t = (NI_t - NI_0)/(NI_{max} - NI_0)$, where NI_0 means normalized fluorescence intensity of the damaged spot prior to laser irradiation and NI_{max} is the maximum normalized fluorescence intensity of the damaged spot.
7. Each data point is the average of 10 independent measurements.
8. Average the relative intensity to make a graph (Fig. 3).

2.5 Enhancement of Laser Microirradiation With DNA Photosensitizer

1. DNA photosensitizers, such as Hoechst 33258 and bromodeoxyuridine (BrdU), can also be used to assist in increasing the fluorescent signal due to increased DNA damage.
2. For sensitization with Hoechst 33258, change to CO_2-independent medium, then add 1 μL of 10 mg/mL Hoechst 33258 in 2.5 mL medium and incubate the cells for 10 min before initiating the experiment.
3. For BrdU sensitization, treat cells in regular medium with 10 μ*M* BrdU for 16–24 h, then change to CO_2-independent medium, followed by laser microirradiation.

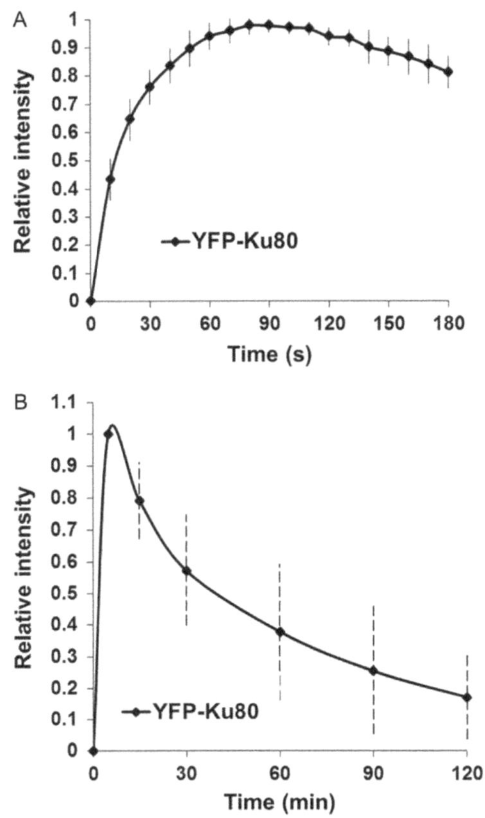

Fig. 3 Localization and kinetics of YFP-tagged Ku80 to laser-generated DSBs. Initial localization (A) and 2 h kinetics (B) of relative fluorescence intensity of YFP-tagged Ku80 to laser-induced DSB in Xrs5 cells.

3. CELL CYCLE-SPECIFIC IMMUNOFLUORESCENCE ASSAYS TO EXAMINE NHEJ, DNA END RESECTION, AND ONGOING HR

A powerful tool utilized in the DSB repair field is an immunofluorescence-based technique to monitor DSB-induced foci formation and resolution of repair proteins. This technique uses immunodetection of specific proteins in response to DNA damage and can be coupled with prelabeling cells with EdU (5-ethynyl-2′-deoxyuridine) to allow differentiation of cell cycle stages. Here, we will describe methods to examine NHEJ, DNA end resection, and ongoing HR in a cell cycle-specific manner using immunofluorescence coupled with microscopy.

3.1 Pulse-Labeling Cells With EdU to Allow Differentiation of Cell Cycle Stages

3.1.1 Buffers and Reagents

- Falcon™ Culture Slides (4 chambers) (Fisher Scientific, Catalog no: 08-774-209).
- Fisherfinest™ Premium Cover Glasses (50 × 20 mm) (Fisher Scientific, Catalog no: 12-548-5E).
- Click-iT® EdU Alexa Fluor® 555 Imaging Kit (ThermoFisher Scientific, Catalog no: C10338).
- U2OS, HT1080, or rodent cell lines.

3.1.2 Procedure

1. Seed an appropriate number of cells on the Falcon culture slides (typically 10–50 × 10^3 cells) and allow to grow for 24–48 h. Make sure the cells are healthy and well dispersed.
2. Prepare a 2× working solution of EdU from the Click-iT kit (Component A) in complete medium.
3. Prewarm the 2× EdU solution, then add an equal volume of the 2× EdU solution to the volume of the media currently on the cells. Replacing all of the media is not recommended as this could affect the rate of cell proliferation. Protect from light.
4. The final concentration of EdU typically used is 50 μ*M* (35 μ*M* for rodent cell lines).
5. Incubate for 30 min to allow the incorporation of EdU into replicating DNA.
6. Following the incubation with EdU, wash the cells 2× with complete medium and then replace with fresh medium. The cells can now be exposed to a DNA damaging agent and the experiment performed. EdU detection will be performed in the middle of the procedures outlined in Sections 3.2 and 3.3. Fig. 4 shows a representation of EdU labeling.

3.2 Monitoring NHEJ in G1 Phase of the Cell Cycle

3.2.1 Equipment

- Gamma irradiator (Cs^{137})

3.2.2 Buffers and Reagents

- Phosphate-buffered saline (PBS), pH 7.4.
- 4% Paraformaldehyde solution (PFA): Dissolve 4 g of paraformaldehyde in 50 mL of water and 1 mL of 1 *M* NaOH (heat at 65°C in a water bath

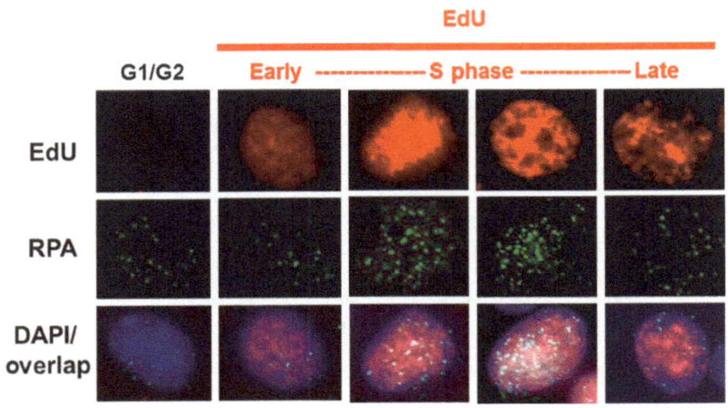

Fig. 4 EdU labeling allows identification of cells in S phase. Panel of images depicting cells in different cell cycle stages, in particular, cells in early-, mid-, and late-S phase via differences in nuclear EdU staining. The images also include an overlay of nuclear EdU staining with RPA foci and DAPI stain.

until powder is completely dissolved). Cool to room temperature. Add 10 mL of 10 × PBS. Adjust the pH to 7.4 using HCl. Make up the volume to 100 mL with water. Filter through 0.2 μm filter. Store at −20°C in aliquots.

- Triton X-100 (0.5%) in 1 × PBS, store at 4°C and use within 2–3 weeks.
- 5% Normal goat serum in 1 × PBS, store at 4°C and use within 2–3 weeks.
- Wash buffer (1% BSA in 1 × PBS), store at 4°C and use within 2–3 weeks.
- Click-iT® EdU Alexa Fluor® 555 Imaging Kit (ThermoFisher Scientific, Catalog no: C10337).
- Vectashield with DAPI (Vector Labs, Catalog no: H-1200).
- Anti-53BP1 rabbit polyclonal antibody (Santa Cruz, Catalog no: sc-22760).
- Anti-DNA-PKcs (phospho S2056) rabbit polyclonal antibody (Abcam, Catalog no: ab18192).
- FITC-488-antirabbit secondary antibody.

3.2.3 Procedure

1. Follow the pulse labeling of cells with EdU protocol from Section 3.1.
2. Expose cells to 1 Gy of γ-rays and place in incubator.
3. At different time points after irradiation (30, 60, 120, 240, and 480 min post-IR), wash cells twice with ice-cold 1 × PBS. (Remove the buffer with pipette to avoid cell loss.)

4. Add 4% paraformaldehyde (in 1 × PBS) to fix cells and allow to incubate for 20 min at room temperature.
5. Wash cells five times with ice-cold 1 × PBS (final two washes are 5-min washes with rocking). See Section 3.2.4, Note 1.
6. Incubate the cells in ice-cold 0.5% Triton X-100 (in PBS) on ice for 10 min.
7. Wash the cells five times with ice-cold 1 × PBS (final two washes are 5-min washes with rocking).
8. Incubate the cells in blocking solution (5% goat serum in 1 × PBS) for 2 h or overnight. See Section 3.2.4, Note 1.
9. Wash the cells once with ice-cold 1 × PBS.
10. Add the DNA-PKcs phospho-2056 (1:200 dilution) or 53BP1 (1:500 dilution) rabbit polyclonal antibody diluted in 5% normal goat serum in 1 × PBS to the cells. See Section 3.2.4, Note 2.
11. Incubate the cells with the antibody at room temperature for 2–4 h.
12. Wash the cells three times with ice-cold wash buffer (each wash for 5 min with rocking).
13. Perform the Click-iT reaction for EdU detection following the manufacturer's established protocol. In total, the Click-iT reaction typically takes 60–90 min.
14. Once the Click-iT reaction is completed, wash the cells five times with ice-cold wash buffer.
15. Incubate the cells with FITC-488-antirabbit secondary antibody (diluted 1:1000) in 1% BSA and 2.5% normal goat serum in 1 × PBS for 1 h at room temperature (in the dark).
16. Wash the cells five times with ice-cold wash buffer (each wash for 5 min with rocking).
17. After the last wash, remove the entire wash buffer completely and allow the cells to air dry.
18. Remove the chamber partition and mount the cells in VectaShield mounting medium containing DAPI.
19. Acquire images and count foci (see Section 3.4).

3.2.4 Notes

1. At this step, the cells can be stored at 4°C for a few days.
2. G1 phase nuclei can be distinguished from S and G2 phase nuclei by their lack of EdU staining, low intensity of DAPI staining, and smaller size of the nuclei.
3. Both DNA-PKcs phospho-2056 and 53BP1 focus formation can be used to monitor NHEJ and either can be used in an individual experiment.

3.3 Monitoring DNA End Resection or Ongoing HR in Mid-S Phase of the Cell Cycle

3.3.1 Equipment

- Gamma irradiator (Cs^{137})

3.3.2 Reagents and Buffers

- Phosphate-buffered saline (PBS), pH 7.4.
- 4% Paraformaldehyde (PFA).
- Triton X-100 (0.5%) in PBS.
- 5% Normal goat serum in PBS.
- Wash buffer (1% BSA in 1 × PBS).
- Click-iT® EdU Alexa Fluor® 555 Imaging Kit (ThermoFisher Scientific, Catalog no: C10338).
- Vectashield with DAPI (Vector Labs, Catalog no: H-1200).
- Anti-RPA2 mouse monoclonal antibody (Millipore, Catalog no: NA19L).
- Anti-Rad51 rabbit polyclonal antibody (Santa Cruz, Catalog no: sc-8349).
- Extraction Buffer (CSK Buffer): 10 m*M* HEPES, pH 7.4, 300 m*M* sucrose, 100 m*M* NaCl, 3 m*M* $MgCl_2$, 0.1% Triton X-100. Prepare fresh.

3.3.3 Procedure

1. Follow the pulse labeling of cells with EdU protocol from Section 3.1.
2. Expose cells to 8 Gy of γ-rays and place in incubator.
3. At different time points (Mock, 2, 4, 8, and 12 h) post-IR, wash the cells twice with ice-cold 1 × PBS.
4. Add ice-cold Extraction Buffer (CSK Buffer) to the cells and incubate for 7–8 min on ice.
5. Wash the cells five times with ice-cold 1 × PBS (remove the buffer with pipette to avoid cell loss).
6. Fix the cells with 4% paraformaldehyde (in 1 × PBS) for 20 min at RT.
7. Wash the cells five times with ice-cold 1 × PBS (final two washes are 5-min washes with rocking).
8. Incubate the cells in ice-cold 0.5% Triton X-100 (in PBS) on ice for 10 min.
9. Wash the cells five times with ice-cold 1 × PBS (final two washes are 5-min washes with rocking).

10. Incubate the cells in blocking solution (5% goat serum in 1 × PBS) for 2 h or overnight. See Section 3.2.3, Note 1.
11. Wash the cells once with ice-cold 1 × PBS.
12. Add the RPA2 (1:500 dilution) mouse monoclonal antibody or Rad51 (1:800 dilution) rabbit polyclonal antibody diluted in 5% normal goat serum in 1 × PBS to the cells. Incubate for 2–3 h.
13. Incubate the cells with the antibody at room temperature for 2–4 h.
14. Wash the cells three times with ice-cold wash buffer (each wash for 5 min with rocking).
15. Perform the Click-iT reaction for EdU detection following the manufacturer's established protocol. In total, the Click-iT reaction typically takes 60–90 min.
16. Once the Click-iT reaction is completed, wash the cells five times with ice-cold wash buffer.
17. Incubate the cells with FITC-488-antimouse secondary antibody for RPA (diluted 1:1000) or FITC-488-antirabbit secondary antibody for Rad51 (diluted 1:1000) in 1% BSA and 2.5% normal goat serum in 1 × PBS for 1 h at room temperature (in the dark).
18. Wash the cells five times with ice-cold wash buffer (each wash for 5 min with rocking).
19. After the last wash, remove the entire wash buffer completely and allow the cells to air dry.
20. Remove the chamber partition and mount the cells in VectaShield mounting medium containing DAPI.
21. Acquire images and count foci (see Section 3.4). Fig. 4 shows a representation of EdU labeling and RPA foci staining.

3.4 Quantification of Foci

3.4.1 Equipment

- Fluorescence or confocal microscope
- Imaris or ImageJ image analysis software

3.4.2 Procedure

- Acquire images using fluorescence or confocal microscope with a magnification of 60 × or higher.
- For mid-S phase cells, select cells with even pan-nuclear EdU stained nuclei. See Fig. 4.

- Preferentially, acquire Z stacks and count foci on maximum intensity projection (all Z sections merged) images so that all foci are visualized at the same plane.
- Foci numbers per nuclei can be enumerated using Imaris image analysis software or ImageJ software.

4. DETERMINATION OF DNA REPAIR CAPACITY IN DIFFERENT PHASES OF THE CELL CYCLE

Colony formation or cell survival assays test cell proliferation after being challenged with a DNA damaging agent. It is a well-established and accurate method to determine the sensitivity of cells to DNA damaging agents like ionizing radiation and chemotherapeutic agents. Typically, cell survival assays are performed with asynchronous cell populations (Munshi, Hobbs, & Meyn, 2005), which may skew the results as cells have differential radiosensitivities throughout the cell cycle (more radioresistant in S phase and more radiosensitivity in mitosis), and it does not allow for the ability to compare radioresponses in different phases of the cell cycle. Thus, in order to accurately determine radiosensitivity of all subpopulations within a specific cell type, survival assays with synchronized population of cells are imperative. Here, we describe a method to amalgamate a simple synchronization method utilizing double-thymidine block (Bootsma, Budke, & Vos, 1964), where cell cycle progression is hindered by excess thymidine leading to feedback inhibition of nucleotide synthesis, with a cell survival assay to examine radiosensitivity of cells in G1 and S phase of the cell cycle (Lee et al., 2016).

4.1 Cell Synchronization Utilizing Double-Thymidine Block

4.1.1 Buffers and Reagents

- Thymidine (T9250, Sigma Aldrich).
- Complete DMEM (D6429, Sigma Aldrich) (with 10% fetal bovine serum plus penicillin–streptomycin).
- 1 × PBS, pH 7.4.
- Cells of interest. This protocol has been optimized specifically for mouse ear fibroblasts and MEFs and should be modified for other cell lines.

4.1.2 Procedure

1. Plate cells in DMEM at 30% confluency 1 day prior to thymidine treatment.

2. First thymidine block. Add thymidine to the cells to a final concentration of 2 m*M* and incubate for 8 h.
3. Wash the cells three times with 1 × PBS.
4. Add complete DMEM for 4 h to release the cells from the thymidine block.
5. Second thymidine block. Add thymidine to the cells to a final concentration of 2 m*M* and incubate for 12 h. This should result in the cells to be fully synchronized at the G1/S phase border. For G1 phase survival assay, cells can be used directly at this stage (see Section 4.1.3, Note 2). Take a small aliquot of cells for cell cycle analysis (see Section 4.2). For collecting S phase cells for survival assay, proceed to step 6.
6. Wash the cells 3 × with 1 × PBS.
7. Add complete DMEM to release the cells from the thymidine block and incubate at 37°C for 2.5 h to achieve a maximum population of cells in the mid- to late-S phase of the cell cycle. See Section 4.1.3, Note 3.
8. Place the cells on ice to stop cell cycle progression.
9. Take a small aliquot of cells for cell cycle analysis (see Section 4.2).

4.1.3 Notes

1. The synchronization protocol described here is specific for rodent fibroblasts with a doubling time of 12–14 h.
2. Before attempting synchronization, it is essential to know the doubling time of the experimental cell line. Most common cell lines have established protocols for synchronization that can be readily used. If not available, BrdU incorporation assay should be performed to estimate the duration of different phases of the cell line and then synchronization attempted.
3. For G1 phase synchronization, approximately 75%–80% cells are in G1/S border using thymidine double block method.
4. For S phase synchronization, typically, approximately 73%–76% of the cells will be in S phase, 14%–16% in G1 phase, and 10%–11% in G2 phase of the cell cycle.

4.2 Cell Cycle Analysis by Propidium Iodide Staining Followed by Flow Cytometry

4.2.1 Equipment

- Flow cytometer (AMNIS FlowSight® Imaging Flowcytometer)
- Gamma irradiator (Cs^{137})
- Biohazard hood

4.2.2 Buffers and Reagents

- 70% Ethanol (in DI water)
- 1 × PBS, pH 7.4
- Propidium Iodide (PI) (50 μg/mL) in 1 × PBS
- Ribonuclease A (RNase A) (100 μg/mL) in 1 × PBS

4.2.3 Procedure

1. Harvest cells by trypsinization or scraping and wash with 1 × PBS.
2. Fix cells by adding ice-cold 70% ethanol (approx. 1 mL 70% ethanol to 1 million cells). Add dropwise to the cell pellet while vortexing to ensure fixation of all cells and minimize clumping.
3. Incubate for at least 30 min at 4°C to ensure full fixation. See Section 4.2.4, Note 1.
4. Wash the fixed cells twice in 1 × PBS.
5. Spin at 2000 rpm and discard the supernatant. See Section 4.2.4, Note 2.
6. Add 50 μL of 100 μg/mL RNase to the cells. This will ensure that only the DNA will be stained by the PI.
7. Add 400 μL of PI per million cells and mix well.
8. Incubate the cells for 5–10 min in a 37°C water bath or 30 min at room temperature.
9. Analyze samples by flow cytometry and measure at least 10,000 single cells. See Section 4.2.4, Note 3.

4.2.4 Notes

1. Once fixed, the samples can be stored for several weeks at 4°C or −20°C.
2. Care should be taken to avoid cell loss when discarding the supernatant, especially the 70% ethanol fixation step.
3. The cells can be directly analyzed in the PI/RNaseA solution and thus there is no need to wash the cells.

4.3 Survival Assay With Various Fractions of Synchronous Cells Obtained From Double-Thymidine Block Method

4.3.1 Buffers and Reagents

- Synchronized cells (see Section 4.1).
- Complete DMEM (with 10% fetal bovine serum plus penicillin–streptomycin).
- Trypsin–EDTA, to make single-cell suspensions from monolayer cultures. Store at 4°C.
- 60-mm Tissue culture dishes.

- 15-mL Conical tubes.
- 1 × PBS, pH 7.4.
- 0.5% Crystal Violet (made in 100% methanol). Store at room temperature in a dark plastic bottle.

4.3.2 Procedure

1. Label the 60-mm dishes and 15-mL conical tubes for each sample. One will need 60-mm dishes in triplicate and two cell numbers will be plated for each dose of radiation, i.e., 6 total dishes per dose (see Section 4.3.3, Note 1). Label the bottom of each dish and not on the lid as the lids will be discarded during staining.
2. Add 5 mL of growth medium to the flasks and keep them aside in a hood.
3. Make sure that the cells are in single-cell suspension and obtain an accurate cell count using an automated cell counter or a hemocytometer.
4. Count the cells and divide them into 5 dishes corresponding to 0, 1, 2, 4, 6 Gy of IR, while still maintaining on ice.
5. Irradiate the dishes marked 1–6 Gy with the corresponding doses of IR.
6. After the irradiation treatment, make serial dilutions to obtain the number of cells to be plated for each radiation dose. Plate the cells in triplicate and place in the incubator for 10–12 days to allow colony formation. For example, plate 50 and 100 cells in triplicate for 0 Gy control.
7. Remove the medium from the plates and add 0.5% Crystal Violet stain to the dishes. Allow to sit for 3–4 min.
8. Remove the Crystal Violet stain and rinse the stained plates upside down in a pan with lukewarm water. Rinsing the plates upside down in the pan prevents the colonies from loosening and washing off.
9. Let the plates air dry overnight and then count the colonies using a dissecting microscope under a magnified field. A cluster of blue-staining cells is considered a colony if it comprises at least 50 cells. Note the numbers for both the A and B dilutions and make a chart.
10. Use the nonirradiated control cells to obtain a plating efficiency. Average the three colony counts for each dilution A and B and divide the mean by the number of cells plated. This will give the PE:

$$\mathrm{PE} = \frac{\text{number of colonies counted}}{\text{number of cells plated}} \times 100$$

11. Following determination of PE, calculate the fraction of cells surviving a given treatment. First, normalize all the plating efficiencies of the treated samples to that of the control unirradiated plates, considering that to be 100%. The surviving fraction (SF) is determined by dividing the PE of the treated cells by the PE of the controls, and then multiplying by 100:

$$\mathrm{SF} = \frac{\text{PE of treated sample}}{\text{PE of control}} \times 100$$

12. Plot the data on an Excel spreadsheet with the dose of radiation on the x-axis and survival on the y-axis.
13. A representative survival curve is shown in Fig. 5.

4.3.3 Notes

1. To determine the number of cells to be plated for each radiation dose, a preliminary experiment must be performed where increasing number of

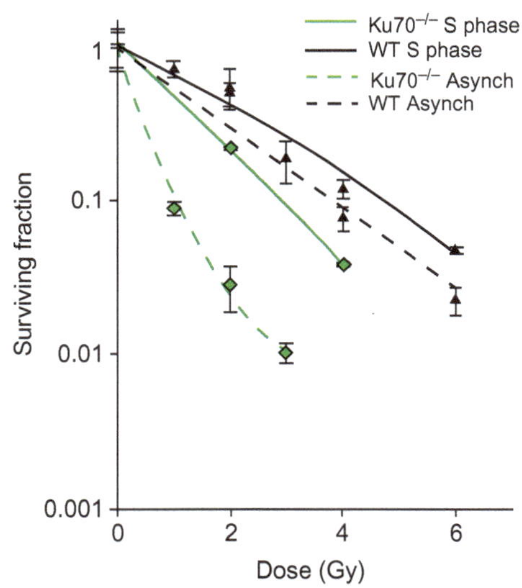

Fig. 5 Cell cycle-specific IR survival assay. Colony formation assays were performed to compare radiation sensitivity of Ku70$^{-/-}$ mouse fibroblasts or Ku70$^{-/-}$ fibroblasts complemented with Ku70 wild type in synchronized S phase or as an asynchronous cell population. The cell lines were left cycling or synchronized by double-thymidine block method and then released. Subsequently, the cells were irradiated at the indicated doses and plated for analysis of survival and colony-forming ability. An increase in radioresistance was observed in S phase of Ku70$^{-/-}$ cells as compared to its asynchronous counterpart. Error bars denote S.D. Note that the Ku70$^{-/-}$ mouse fibroblasts have higher rates of homologous recombination due to the absence of classical NHEJ machinery.

cells are plated for each dose of radiation, which yields distinct cell colonies for counting, preferably less than 100 colonies per dish. Typically, we plate X and 2X with the cell number being previously determined by colony-forming assays.

2. Staining plates with Crystal Violet stain is easy, but care must be taken not to get it on one's clothes, because it is difficult to remove. It is suggested that a laboratory coat and double gloves be used for the staining procedure.
3. If possible, it is a good idea to dedicate an incubator to clonogenic cell survival. This avoids unnecessary bumping of colonies by other users when they open and close the door of the incubator. As the colonies grow, bumping the incubator or shelf can cause the cells to detach and settle as new colonies, thereby leading to an increase in the colony count and erroneous results. This is especially true for CHO cells.

ACKNOWLEDGMENTS

The authors are supported by the National Institutes of Health grants (CA162804 and CA092584).

REFERENCES

Bootsma, D., Budke, L., & Vos, O. (1964). Studies on synchronous division of tissue culture cells initiated by excess thymidine. *Experimental Cell Research*, *33*, 301–309.

Davis, A. J., & Chen, D. J. (2010). A role for ATM kinase activity and Mre11 in microhomology-mediated end-joining. *Cell Cycle*, *9*, 3147–3148.

Davis, A. J., Chi, L., So, S., Lee, K. J., Mori, E., Fattah, K., et al. (2015). BRCA1 modulates the autophosphorylation status of DNA-PKcs in S phase of the cell cycle. *Nucleic Acids Research*, *42*, 11487–11501.

Davis, A. J., So, S., & Chen, D. J. (2010). Dynamics of the PI3K-like protein kinase members ATM and DNA-PKcs at DNA double strand breaks. *Cell Cycle*, *9*, 2529–2536.

Goodarzi, A. A., & Jeggo, P. A. (2013). The repair and signaling responses to DNA double-strand breaks. *Advances in Genetics*, *82*, 1–45.

Hoeijmakers, J. H. (2001). Genome maintenance mechanisms for preventing cancer. *Nature*, *411*, 366–374.

Jackson, S. P., & Bartek, J. (2009). The DNA-damage response in human biology and disease. *Nature*, *461*, 1071–1078.

Karanam, K., Kafri, R., Loewer, A., & Lahav, G. (2012). Quantitative live cell imaging reveals a gradual shift between DNA repair mechanisms and a maximal use of HR in mid S phase. *Molecular Cell*, *47*, 320–329.

Kim, J. S., Krasieva, T. B., Kurumizaka, H., Chen, D. J., Taylor, A. M., & Yokomori, K. (2005). Independent and sequential recruitment of NHEJ and HR factors to DNA damage sites in mammalian cells. *The Journal of Cell Biology*, *170*, 341–347.

Lee, K. J., Saha, J., Sun, J., Fattah, K. R., Wang, S. C., Jakob, B., et al. (2016). Phosphorylation of Ku dictates DNA double-strand break (DSB) repair pathway choice in S phase. *Nucleic Acids Research*, *44*, 1732–1745.

Munshi, A., Hobbs, M., & Meyn, R. E. (2005). Clonogenic cell survival assay. *Methods in Molecular Medicine*, *110*, 21–28.

Schipler, A., & Iliakis, G. (2013). DNA double-strand-break complexity levels and their possible contributions to the probability for error-prone processing and repair pathway choice. *Nucleic Acids Research*, *41*, 7589–7605.

Shao, Z., Davis, A. J., Fattah, K. R., So, S., Sun, J., Lee, K. J., et al. (2012). Persistently bound Ku at DNA ends attenuates DNA end resection and homologous recombination. *DNA Repair (Amst)*, *11*, 310–316.

So, S., Davis, A. J., & Chen, D. J. (2009). Autophosphorylation at serine 1981 stabilizes ATM at DNA damage sites. *The Journal of Cell Biology*, *187*, 977–990.

CHAPTER SIX

Base Excision Repair Variants in Cancer

Carolyn G. Marsden*, Julie A. Dragon*, Susan S. Wallace*, Joann B. Sweasy*,†,1

*The Markey Center for Molecular Genetics, University of Vermont, Burlington, VT, United States

†Yale University School of Medicine, New Haven, CT, United States

[1]Corresponding author: e-mail address: joann.sweasy@yale.edu

Contents

Abstract

Base excision repair (BER) is a key genome maintenance pathway that removes endogenously damaged DNA bases that arise in cells at very high levels on a daily basis. Failure to remove these damaged DNA bases leads to increased levels of mutagenesis and chromosomal instability, which have the potential to drive carcinogenesis. Next-generation sequencing of the germline and tumor genomes of thousands of individuals has uncovered many rare mutations in BER genes. Given that BER is critical for genome maintenance, it is important to determine whether BER genomic variants have functional phenotypes. In this chapter, we present our in silico methods for the identification and prioritization of BER variants for further study. We also provide detailed instructions and commentary on the initial cellular assays we employ to dissect potentially important phenotypes of human BER variants and highlight the strengths and weaknesses of

Methods in Enzymology, Volume 591
ISSN 0076-6879
http://dx.doi.org/10.1016/bs.mie.2017.03.003

our approaches. BER variants possessing interesting functional phenotypes can then be studied in more detail to provide important mechanistic insights regarding the role of aberrant BER in carcinogenesis.

1. INTRODUCTION

DNA damage occurs as a result of normal metabolism resulting in oxidized, deaminated, and alkylated bases at a rate of at least 30,000 lesions per cell per day (Freidberg, Wood, Walker, & Siede, 2006). Base excision repair (BER) removes the majority of these lesions. The simplest and most common form of BER is short-patch BER, which is initiated by one of several different DNA glycosylases, each having preferences for specific types of lesions (for a comprehensive review, see Wallace, Murphy, & Sweasy, 2012). Monofunctional DNA glycosylases recognize DNA lesions and catalyze the hydrolysis of the *N*-glycosyl bond that releases the damaged base and generates an abasic site. The abasic site is nicked at its 5′side by the apurinic/apyrimidinic endonuclease 1 (APE1), leaving a 3′OH and a 5′deoxyribose phosphate (dRP). DNA polymerase β (Pol β) fills in the single-nucleotide gap and catalyzes removal of the dRP group with its associated lyase activity. Bifunctional DNA glycosylases, which usually recognize and remove oxidative lesions, have an associated lyase activity that cleaves the DNA backbone, leaving either a phosphate group or an α,β-unsaturated aldehyde attached to the 3′end of the DNA. Phosphate groups at the 3′end of the DNA are removed by polynucleotide kinase (PNKP). The α,β-unsaturated aldehydes are processed by APE1 endonuclease, which creates a 3′OH that is recognized by Pol β, which fills in the single-nucleotide gap. The X-ray cross-complementing 1 (XRCC1)/Ligase IIIα complex catalyzes ligation of the DNA ends. Long-patch BER is thought to be a minor DNA repair pathway in human cells and takes place if the 5′sugar is modified or if the lyase activity of Pol β is defective (for reviews, see Balakrishnan & Bambara, 2013; Wallace et al., 2012). In this case, Pol β performs strand displacement synthesis to initiate long-patch BER, and 2–12 nucleotides are added by DNA polymerases δ and ε, resulting in the generation of a 5′DNA flap. The flap is removed by Flap endonuclease 1 (FEN1), and DNA ligase 1 or XRCC1/Lig3α seal the nick.

BER is critical for maintaining genomic stability. The inability to remove endogenous DNA damage can lead to the accumulation of point mutations that likely arise as a result of error prone lesion bypass by replicative DNA polymerases, DNA polymerase λ or β, or translesion polymerases (for an excellent review see Bacolla, Cooper, & Vasquez, 2014). Aberrant processing of base damage by mutant enzymes in the BER pathway may also lead to the accumulation of BER intermediates, including single- and double-strand breaks, which can result in chromosomal instability (for example, see Yamtich, Nemec, Keh, & Sweasy, 2012).

Little is known about the relationship between single-nucleotide polymorphisms (SNPs) and/or somatic variants in genes encoding for BER proteins and the etiology of human cancer. Recently, a linkage to colorectal and perhaps multiple cancers was identified for a truncation within the *NTHL1* gene (Rivera, Castellsague, Bah, van Kempen, & Foulkes, 2015; Weren et al., 2015). This DNA glycosylase recognizes and removes oxidized pyrimidines (Asagoshi et al., 2000; Aspinwall et al., 1997; Eide et al., 2001; Ikeda et al., 1998). Previous to this, it was shown that SNPs within the *MUTYH* gene predispose individuals to MUTYH-associated polyposis and perhaps other types of cancer (for detailed reviews, see David, O'Shea, & Kundu, 2007; Wallace et al., 2012). MUTYH recognizes and removes adenine that has been inserted opposite 8-oxoguanine or FapyG (Au, Cabrera, Miller, & Modrich, 1988; Lu, Tsai-Wu, & Cillo, 1995; McGoldrick, Yeh, Solomon, Essigmann, & Lu, 1995; Pope & David, 2005; Slupska et al., 1996; Slupska, Luther, Chiang, Yang, & Miller, 1999). Mutations in *MUTYH* result in an inability of the protein to efficiently remove adenine opposite 8-oxoguanine, resulting in the accumulation of GC to TA transversions within the adenomatous polyposis coli (*APC*) gene, leading to its inactivation and also within the *KRAS* gene, leading to its activation (Al-Tassan et al., 2002; Boparai et al., 2008).

There is general agreement among cancer geneticists that there is missing hereditability underlying the disease of cancer (Schork, Murray, Frazer, & Topol, 2009). This missing hereditability could result from multiple variants each having small effects that are additive or rare variants that have large phenotypic effects, referred to as "the rare variant hypothesis." Because current methods of statistical analysis of the association of rare variants to the etiology of human cancer are underpowered, other methods must be considered. Our approach has been to characterize the function of coding SNPs and tumor-associated variants in specific human BER genes using a combined biological and biochemical approach. Using this approach, we have shown that expression of rare human germline or somatic BER genetic variants in human cells results in genomic instability and cellular transformation. These include human genetic variants in the *NTHL1* (Galick et al., 2013) and thymine DNA glycosylase (*TDG*) genes (Sjolund et al., 2014), as well as in the *POLB* (Donigan, Hile, Eckert, & Sweasy, 2012; Donigan, Sun, et al., 2012; Lang, Dalal, Chikova, DiMaio, & Sweasy, 2007; Lang, Maitra, Starcevic, Li, & Sweasy, 2004; Murphy, Donigan, Jaeger, & Sweasy, 2012; Nemec, Donigan, Murphy, Jaeger, & Sweasy, 2012; Nemec, Murphy, Donigan, & Sweasy, 2014; Sweasy et al., 2005; Yamtich et al., 2012) and *XRCC1* (Sizova, Keh, Taylor, & Sweasy, 2015) genes. Our work suggests that inherited or somatically acquired genetic alteration in the BER pathway has potential to drive carcinogenesis.

In this chapter, we describe our approach and methods for the characterization of human genetic BER variants. We initiate our studies with identification and prioritization of variants for further study. The variants are then expressed in human cells and assessed for their abilities to induce genomic instability and cellular transformation. Variants that induce these phenotypes undergo thorough characterization to uncover specific mechanisms underlying genomic instability and cellular transformation, the details of which can be found in our previously published work.

2. IDENTIFICATION AND PRIORITIZATION OF BASE EXCISION REPAIR VARIANTS USING IN SILICO METHODS

We identified variants in silico based on public and private sequence data and linked them to features that reflect their significance for human health, for example, predicted impact on the protein function and frequency in the general population, in order to prioritize variants for experimental studies (Fig. 1).

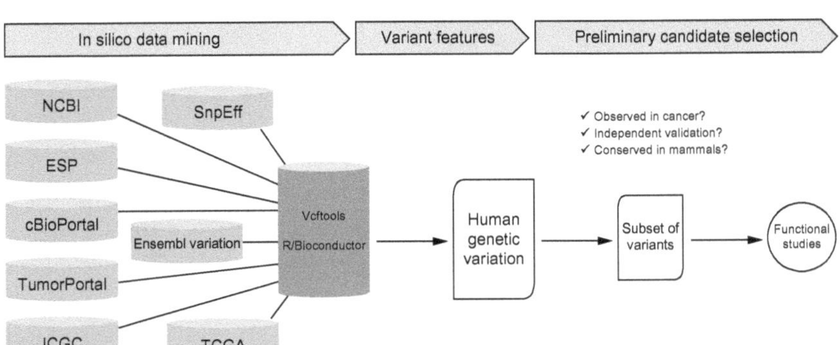

Fig. 1 Structure and operation of the base excision repair variant identification pipeline. See text for details.

2.1 In Silico Data Analysis

For the in silico analysis, all known variants in BER genes are assembled from multiple databases. We use publicly available open source bioinformatics tools [SnpEff (Cingolani et al., 2012), Ensembl Variation (Chen et al., 2010), vcftools (Danecek et al., 2011), and R/Bioconductor packages (Gentleman et al., 2004)] to interface with a variety of databases including Ensembl (Chen et al., 2010; Yates et al., 2016) and cBioPortal (Gao et al., 2013) using the R/Bioconductor cgdsr package, over 200 restricted access TCGA germline whole exome sequence samples from tumor–normal pairs from patients with invasive breast cancer (dbGaP accession phs000178.v5.p5), the International Cancer Genome Consortium (ICGC), National Center for Biotechnology Information (NCBI), the NHLBI Exome Sequencing Project (ESP), and TumorPortal (Lawrence et al., 2014). We combine the information from across these resources, and this comprehensive body of evidence about the variants is merged into a single matrix file that can be easily sorted and filtered with Microsoft Excel.

2.2 Variant Features

The Microsoft Excel worksheet generated from in silico analysis includes information on both germline and somatic variants. The germline variants have unique identifiers from NCBI's dbSNP database and the somatic variants have unique identifiers from the Catalog of Somatic Mutations in Cancer (COSMIC). Amino acid alterations are annotated as is each type of mutation, namely, frameshift, stop-gained, missense, or nonsense mutations. These variant BER proteins are then assessed by PolyPhenII (Adzhubei et al., 2010) to predict whether they are damaging, meaning that they have the potential to alter protein function. Ancestry corrected allele frequencies are also annotated in the Excel spreadsheet, as well as alternative transcript identifiers and associated PubMed identifiers for supporting literature. Variants are then filtered by a combination of features. We are generally interested in the BER variants that are predicted to be damaging to the protein product and have a minor allele frequency $<5\%$.

2.3 Preliminary Evaluation of Variants

A variant selected for experimental studies must be supported by evidence that it (1) reflects a correct variant call, (2) is infrequent in the general population, (3) is predicted to have a protein functional consequence, and (4) is either a germline or a somatic mutation identified in a human tumor, or both. Evidence bearing on the confidence of the variant call is evaluated, including independent validation, sequencing quality metrics such as the quality score for the nucleotide call and of the mapping of sequence reads to the reference genome, as well as the depth of sequencing coverage of each variant. Interpretation of protein functional consequences is based on the collection of information and inspection by crystallographers and biochemists. Variants chosen for further study include those that are within important and conserved functional domains of proteins with amino acid alterations that would be predicted to affect protein function. For example, we identified the *NTHL1* D239Y germline variant as altering a potentially important amino acid residue involved in the catalytic function of this DNA glycosylase. Subsequent studies demonstrated that the NTHL1 D239Y protein was not only catalytically inactive and but that it induced genomic instability and cellular transformation upon heterozygous expression in human epithelial cells (see below for additional details) (Galick et al., 2013).

3. CELLULAR CHARACTERIZATION OF BER VARIANTS

To accurately assess the potential for a BER variant to drive carcinogenesis, the cellular characterization of BER variants should be performed in a physiologically relevant system. The system described here enables the expression of BER variants of interest in human cells at levels roughly equivalent to those of the endogenous protein (as shown in Fig. 2A) to generate a biologically relevant system for the study of BER variants. Using this method, the mechanisms by which BER variants induce cellular transformation and genomic instability are informative, directly relating to the etiology and progression of human disease. For cellular characterization of BER variants, the cDNA of the variant is stably expressed in immortal but nontransformed human cells. We have experience using the MCF10A or HME-1 mammary epithelial cells (for example, see Galick et al., 2013; Nemec et al., 2016). Here we will describe our work with MCF10A, which is a nontransformed, immortalized mammary epithelial cell line that is near diploid with a stable karyotype. Although the cell line does contain some karyotype abnormalities (including an extra chromosome), MCF10A cells are not able to form tumors in immunocompromised mice (Yoon et al., 2002) and cannot form colonies in soft agar (Soule et al., 1990). Therefore, the MCF10A cell line is extremely useful to study the effects of BER variants on premalignant progression.

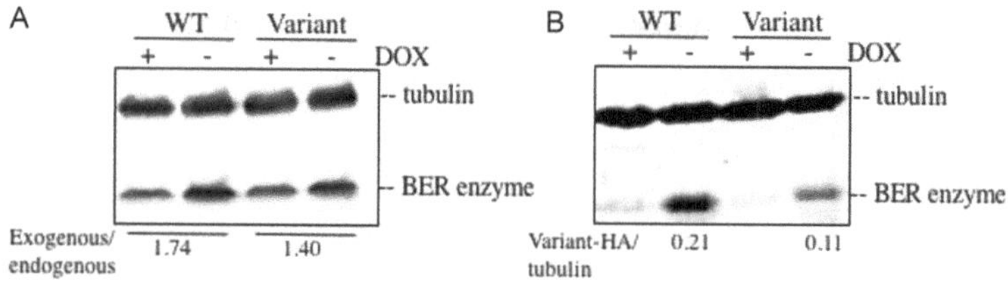

Fig. 2 Western blot to quantify expression of exogenous WT and variant protein expression. (A) The level of exogenous protein expression above endogenous protein expression is determined using a primary antibody against the BER enzyme of interest. The signal produced in lysates from cells not cultured with doxycycline (exogenous + endogenous protein) is divided by the signal produced in lysates from cells cultured with doxycycline (endogenous protein only) to calculate the fold increase of protein expression as a result of exogenous protein expression. To properly characterize the cellular effects of a human somatic or germline BER variant, it is important to maintain expression levels of the exogenous protein that are close to the levels of the endogenous protein. (B) Expression of WT and variant is detected using a primary antibody against the HA epitope tag. Expression is normalized by dividing the HA signal by the tubulin signal.

3.1 Subcloning in the pRVY-tet Vector

Cellular expression of BER variants of interest is achieved using the pRVY-tet vector. The pRVY-tet vector is a retroviral vector encoding ampicillin resistance for selective bacterial expression and hygromycin resistance for selective expression in mammalian cells. The pRVY-tet vector conveniently provides tetracycline-controlled gene expression of the BER variants of interest. In the TET-off (or tTA-dependent) system, a tetracycline transactivator fusion protein (tTA), composed of a tetracycline repressor (TetR) and the C-terminal domain of VP_{16} (virion protein 16), is unable to bind to a tetracycline response element (TRE) located upstream of the promoter when tetracycline (or doxycycline) is added to the cell culture media, thereby turning off gene expression (Gossen & Bujard, 1992). A major benefit of the TET-off system is the reversible and rapid induction of gene expression upon the removal of tetracycline (or doxycycline) from the medium. However, "leakiness" can occur, which results in detectable levels of gene expression even in the presence of tetracycline or doxycycline. Furthermore, tetracycline can cause toxicity in mammalian cells at higher concentrations, so titration of tetracycline is necessary to achieve optimal inhibition of gene expression while minimizing toxicity. We generally use doxycycline (an analog of tetracycline) to regulate gene expression because it has lower toxicity in mammalian cells, a known half-life of 24h, and because it binds to tTA with high affinity, therefore reducing the likelihood of unintended gene expression (or "leakiness"). The TET-off feature of the pRVY-tet vector is particularly useful for expression of BER variants that may reduce cellular viability and/or cellular proliferation making study difficult with continuous expression.

The easiest and most time-effective method for generating pRVY-tet constructs expressing BER variants of interest is to first subclone the WT sequence of the BER enzyme into the pRVY-tet vector and then perform site-directed mutagenesis to generate the desired mutation. Since the pRVY-tet vector does not contain a multiple cloning site, the restriction enzymes available for subcloning are limited to the *Not*I and *Bam*HI sites, thereby requiring a multistep approach to subcloning into the pRVY-tet vector. In contrast, site-directed mutagenesis is a relatively simple method involving PCR amplification with substitution of the desired nucleotide/s followed by digestion of the parental vector. In addition, if an epitope tag needs to be added to the BER enzyme of interest in order to distinguish the endogenous and exogenously expressed proteins, then the sequence encoding the tag to be used is included in the primer design during subcloning. It is important to consider the structural and catalytic constraints of an epitope tag on the BER enzymes of interest. Furthermore, it is imperative to perform functional assays using the tagged BER enzyme to ensure that the epitope tag does not interfere with its catalytic functions. BER enzymes studied in our laboratory have been successfully tagged with either the FLAG or human influenza hemagglutinin (HA) epitope tags without affecting catalytic activity (Galick et al., 2013).

3.1.1 Equipment

- PCR thermal cycler
- 37°C water bath
- PCR cleanup kit
- Standard electrophoresis equipment
- Gel extraction kit
- Imaging equipment
- Spectrophotometer

3.1.2 Buffers and Reagents

- High-fidelity DNA polymerase
- PCR buffer
- dNTP stock solution
- Primers

- *Not*I
- *Bam*HI
- Restriction enzyme buffer
- DNA ligase
- DNA ligase buffer
- LB broth and LB agar
- Ampicillin (100 μg/mL final)

3.1.3 Procedure

1. Amplify sequence of BER enzyme of interest by PCR, using primers designed to add *Not*I and *Bam*HI restriction sites to 5′- and 3′ends, respectively.
2. Isolate the amplified sequence using PCR cleanup kit.
3. Digest the PCR product from step 1 and the pRVY vector with *Not*I and *Bam*HI for 1 h at 37°C.
4. Run the digested vector on a 1% agarose gel. Cut out the appropriate band.
5. Gel purify the digested vector backbone.
6. Use a PCR cleanup kit to isolate the digested PCR product.
7. Determine the DNA concentration of the digested and purified pRVY vector and PCR product.
8. Perform a ligation reaction using 1:1, 1:2, and 1:4 digested pRVY vector (25 ng): digested PCR product, DNA ligase, and DNA ligase buffer at 16°C overnight or at room temperature for 30 min^{-1} h.
9. Transform ligation reaction into competent *Escherichia coli* (such as DH5α).
10. Plate onto prewarmed LB/ampicillin plates.
11. Isolate plasmid DNA from the colonies and sequence the insert to confirm its presence.
12. Upon successful subcloning of WT sequence of BER enzyme of interest into the pRVY-tet vector, perform site-directed mutagenesis to generate the BER variant following the manufacturer's instructions (see Note 4).

3.1.4 Notes

1. A two-step PCR reaction may be needed to successfully amplify the sequence of interest, particularly if your primers are long. If this is the case, use an appropriate annealing temperature (T_A) for the sequence in the primers that overlaps with the sequence to be amplified (this

would be the 8–12 base pairs that anneal with the target sequence only, not the T_A for the entire primer sequence) for 5 cycles, followed by 30 cycles using a T_A appropriate for the entire length of the primers.
2. Check for amplification by agarose gel electrophoresis before continuing.
3. We prefer to the New England Biolabs Monarch PCR purification/gel purification kit. We have found that the amount of DNA recovered is higher than other commercially available kits.
4. There are two commercially available site-directed mutagenesis kits that have been successful in our laboratory: the Q5® site-directed mutagenesis kit from New England Biolabs and the QuikChange II XL site-directed mutagenesis kit from Agilent. Primer design is quite different between the two kits, and therefore, it is very important to design the primers using the appropriate primer design software for each kit. We most commonly use the Q5® site-directed mutagenesis kit from New England Biolabs.

3.2 Generation of Stable MCF10A Pools and Clones Expressing the BER Variant of Interest

A major benefit of the pRVY expression system is the moderate level of expression of the exogenous protein of interest in the cells. As shown in Fig. 2B, we measure the levels of exogenous protein expression by Western blotting using a primary antibody against the epitope tag, which in this case is the HA tag. However, in order to compare the level of exogenous protein expression in relation to endogenous protein expression levels, we also blot with an antibody against the BER enzyme of interest as shown in Fig. 2A. It is also important to use pools and/or clones that exhibit similar expression levels of the exogenous WT enzyme and BER variant of interest so that any phenotypic changes detected cannot be attributed to an artifact as a result of higher expression levels of the variant enzyme than the WT enzyme. Some BER variants may not express well in cells, as seen in Fig. 2B where expression of the variant is about half of WT expression levels. Lower expression of a variant compared to the WT enzyme may mask a possible cellular phenotype, which must be kept in mind during subsequent analyses. However, detection of a phenotype can be considered a positive result in these circumstances.

When performing downstream assays, there is a choice between using pools of cells or clones. Retroviral transduction results in random integration of the introduced sequence into the genome. Each cell will have a different number of integrations of the construct as well as different integration locations within the genome. Therefore, when selecting a pool of cells with stable integration, the resulting population will be heterogeneous and expression levels will reflect the heterogeneity of integration. In contrast, a clone will contain a population of cells that arose from a single cell and therefore all cells in that population will be homogeneous for the number and location of integrations. Although cellular transformation due to an integration event is not an issue when working with pools of cells, the heterogeneity of expression in pools may mask a cellular phenotype. In contrast, the homogeneous expression in clones makes it easier for cellular phenotypes to emerge; however the possibility of transformation due to an integration event needs to be addressed by screening at least 8–10 different clones with stable expression of the WT or variant enzyme. We typically select both pools and clones but perform our initial characterization assays using pools of cells. If the results using pools are inconsistent or hard to interpret, we will then employ clones for our assays.

3.2.1 Equipment

- Biosafety cabinet
- CO_2 incubator (5% CO_2/37°C)

3.2.2 Buffers and Reagents

- 100 mm tissue culture plates
- T25 and T75 tissue culture flasks
- 0.25 M calcium chloride (3.68 g into 100 mL ddH_2O), sterile filtered
- 2× HEBS (12 m*M* Dextrose, 50 m*M* HEPES, 10 m*M* KCl, 280 m*M* NaCl, 1.5 m*M* $Na_2HPO_4 \cdot 2H_2O$), pH adjusted to 7.05 with 10 N NaOH, sterile filtered
- 10 μg of pRVY plasmid expressing WT and BER variant of interest
- 10 μg of pVSV-G plasmid
- 0.45 μm pore filter
- 10–30 mL syringe (without a needle)
- Dulbecco's Modified Eagle's Medium, high glucose
- Fetal bovine serum (FBS)
- Penicillin/streptomycin (P/S) (100× stock)
- Polybrene (4 mg/mL stock)
- Hygromycin B (50 mg/mL stock)
- Dulbecco's Modified Eagle Medium: Nutrient Mixture F-12 (DMEM/F-12)

- Horse serum
- Insulin, recombinant, 4 mg/mL stock (Life Technologies)
- Epidermal growth factor (EGF), 100 µg/mL stock (Peprotech)
- Hydrocortisone, 1 mg/mL stock (Sigma)
- Cholera Toxin from *Vibrio cholera*, 1 mg/mL stock (Sigma)
- MCF10A complete growth media (DMEM/F-12, 5% horse serum, 20 ng/mL EGF, 0.5 µg/mL hydrocortisone, 100 ng/mL cholera toxin, 10 µg/mL recombinant insulin, 1% P/S)

3.2.3 Procedure

1. Plate GP2-293 cells at 2.5×10^6 cells per 100 mm plate in DMEM 10% FBS/1% P/S for each pRVY vector and one control plate (no pRVY vector included in transfection) 18–24 h prior to transfection.
2. Replace media with fresh DMEM 10% FBS/1% P/S about 1 h prior to transfection.
3. Combine 500 µL of 2 × HEBS, 10 µg of pRVY vector, and 10 µg of pVSV-G vector in a sterile prelabeled tube (one tube per pRVY expression vector).
4. Add dropwise 500 µL of 0.25 *M* $CaCl_2$ with aeration to the 2 × HEBS/DNA solution.
5. Add dropwise the $CaCl_2$/2 × HEBS/DNA solution to the GP2-293 cells.
6. Incubate at 37°C/5% CO_2 for 6–8 h.
7. Replace the media with fresh DMEM 10% FBS/1% P/S and incubate overnight at 37°C/5% CO_2.
8. The next day, change the media again. See Note 4 below.
9. 48 h posttransfection begin selection with 200 µg/mL Hygromycin B. If the cells are >80–85% confluent, split at a ratio of 1:4 and begin selection 24 h later.
10. Select for stable integration for 5–8 days until all cells on the control plate are dead and visible colonies have formed on pRVY-transfected plates.
11. *Generation of high titer virus*: Plate the selected GP2-293 cells at 2.5×10^6 cells per 100 mm plate 18–24 h prior to transfection.
12. Replace the media with fresh DMEM 10% FBS/1% P/S about 1 h prior to transfection.
13. Perform transfection as described in steps 3–8 using 10 µg of pVSV-G vector only for each pRVY-expressing GP2-293 cell line.

14. Plate 5×10^5 MCF10A cells in T25 flasks in MCF10A complete growth media 24 h prior to infection. Plate two to three flasks for each stable GP2-293 cell line generated, as well as one flask as a control for selection (no virus-containing media + polybrene).
15. Change media on MCF10A cells about 1 h prior to infection.
16. For the infection: collect media 72 h after transfection (contains high titer virus) and filter through a 0.45-μ*M* pore filter.
17. Add polybrene to the filtered virus-containing media at a final concentration of 8 μg/mL.
18. Replace media on plated MCF10A cells in T25 flasks with 3 mL of virus-containing media or nonvirus-containing media + polybrene (as a control for selection).
19. Incubate MCF10A cells with virus-containing media at 37°C/5% CO_2, rocking the flasks every 30 min.
20. Add 2 mL MCF10A complete media and incubate overnight at 37°C/5% CO_2.
21. The next day, replace the media with MCF10A complete growth media.
22. For the generation of pools with stable expression, begin selection 48 h after infection with 200 μg/mL hygromycin B. Add 2 μg/mL of doxycycline to prevent expression of BER enzymes during selection.
23. For the generation of clones, serially dilute the infected cells and plate several dilutions in 100 mm plates. Begin selection with 200 μg/mL hygromycin B the next day.
24. Once selection is complete, change media to MCF10A complete media containing 15 μg/mL hygromycin B to maintain selective pressure and 2 μg/mL of doxycycline to prevent expression of the BER variant of interest until prepared to plate for downstream assays.
25. Expression of BER enzyme/variant of interest is then measured by Western blot. Plate each pool or clone with and without doxycycline. Collect cell lysates for Western blot.

3.2.4 Notes

1. Test for precipitate for every new stock of $CaCl_2$ and 2 × HEBS before transfection.
2. The pVSV-G construct expresses a viral envelope protein that is necessary for virus production and viral infection.
3. Add $CaCl_2$ dropwise using a pipetman while adding aeration using a Pasteur pipet and pipet aid.

4. Check for precipitate 20 min after adding $CaCl_2$/2× HEBS/DNA solution to the GP2-293 cells.
5. At the point prior to selection of transfected GP2-293 cells, the cells are producing low titer virus. Low titer virus can be used to infect MCF10A cells with reasonable infectivity. However, we typically select GP2-293 cells with stable integration of pRVY construct and perform a second transfection with the pVSV-G vector to generate high titer virus.
6. A viral titer assay can be performed to calculate the colony-forming units (CFU)/mL using 3T3 mouse fibroblasts if optimization is needed.
7. Do not attempt selection with the cells at >80–85% confluence. Split the cells 1:4 and begin selection 24 h later.
8. The virus-containing media can be diluted into MCF10A complete media if the CFU was determined to be high or if lower integration events are desired. In this case, 1:2 and 1:4 dilutions of MCF10A complete media: virus-containing media is a good place to start.
9. Infection of MCF10A cells can be performed at 32°C/5% CO_2 to increase virus viability.
10. If the infected MCF10A cells are confluent 48 h after infection, split each T25 flask into a T75 flask and then begin selection the next day.
11. We have performed a dose titration with hygromycin B for MCF10A cells and found 200 μg/mL hygromycin B to be the optimal concentration for selection. If selection is not efficient or the entire cell population dies, then determine the optimal selection concentration for your own purposes.
12. Doxycycline should be added every 48 h to maintain inhibition of gene expression.

3.3 Cellular Transformation in Human Cells

The MCF10A cell line is a nontransformed human epithelial cell line that is used to measure cellular transformation upon expression of a BER enzyme of interest (the Brugge laboratory website is an excellent source for general MCF10A protocols: http://brugge.med.harvard.edu/protocols). The potential of a BER variant of interest to transform human cells is initially assessed by measuring changes in proliferation and anchorage-independent growth. Other characteristic cellular properties not described here but that emerge as a result of transformation including cellular migration and invasive potential (for example, see Galick et al., 2013) can also be assessed following these initial studies. MCF10A cells cannot grow in soft agar and therefore serve as an excellent cell line to assess the ability of a BER variant to induce anchorage-independent growth. We have adapted more high-throughput "kit-based" assays to test both proliferation and anchorage-independent growth for the purpose of efficient characterization of several BER variants at the same time. However, if the use of a kit is not preferred, then refer to the notes section where alternative methods are described briefly. The general methodology is similar between the two approaches with the differences mainly including plate size, cell number, and end-point measurement (as is the case for the proliferation assay). The CYQUANT® NF cell proliferation assay measures DNA content to quantify proliferation; therefore the end-point measurement is relative fluorescence units as opposed to cell number which is the end-point measurement if you choose to count cells manually. Anchorage-independent growth is measured by counting the number of colonies formed whether you use the Cytoselect™ 96-well cell transformation assay or the alternative soft agar assay. The main difference between the two soft agar assays described is the composition of agar used. Proliferation and anchorage-independent growth are measured starting at passage 2 and then measured every other passage (i.e., passage 4, passage 6, passage 8, etc.). If using pools, you can begin with one WT pool and one variant pool. If you are using clones, you should begin with at three to five WT clones and three to five variant clones. We have detected differences in anchorage-independent growth between WT- and BER variant-expressing cells as early as passage 6 and as late as passage 14. Each time you plate cells for proliferation and anchorage-independent growth you should also plate cells for a Western blot to monitor expression of the exogenous WT and variant proteins in the cells over passages. For BER variants, we have found that some of the hallmarks of cellular transformation, including increased cellular proliferation and anchorage independent growth, occur as a function of the numbers of passages of the cells and have suggested that BER is a tumor-suppressor mechanism (Sweasy, Lang, & DiMaio, 2006). During the growth and passaging of the cells, aberrant BER will occur if the variant under study is deficient in its ability to remove and process DNA damage. Aberrant BER can occur as a result of deficient removal of damaged bases by DNA glycosylases, inefficient end remodeling, or backbone incision by enzymes including PNKP and APE1, slow, defective, or error-prone gap filling by Pol β, defective scaffolding of the BER enzyme complex by XRCC1, or inefficient ligation. Each of these events has the potential to result in point mutations in growth control genes (for an example, see

Donigan, Hile, et al., 2012) and/or genomic instability (for an example, see Nemec et al., 2016), but the cells must replicate and divide in order for the phenotype to manifest itself.

3.3.1 Equipment

- Plate reader (480/520 nm)
- Light microscope
- Water bath (40°C)
- Microwave
- 0.22-μm pore filter, bottle top

3.3.2 Buffers and Reagents

- MCF10A complete media
- CYQUANT® NF cell proliferation assay (Invitrogen)
- 2 × DMEM/F12 (Sigma)
- 2 × MCF10A complete media (10% horse serum, 40 ng/mL EGF, 1.0 μg/mL hydrocortisone, 200 ng/mL cholera toxin, 20 μg/mL recombinant insulin, 2% P/S)
- Cytoselect™ 96-well cell transformation assay (Cell Biolabs)

3.3.3 Procedure

1. *Proliferation assay*: Plate 10^3 cells expressing either the WT or the variant enzyme per well in triplicate in a 96-well plate in MCF10A complete media without hygromycin B (see Fig. 3A for example of plate setup).
2. Add media to the wells to adjust the total volume per well to 200 μL.
3. Begin measuring proliferation 48 h after plating by following the manufacturers' instructions.
4. Measure proliferation twice a day while cells are in log growth (we typically measure twice a day beginning on day 4 until day 6.5).
5. *Anchorage-independent growth*: Turn on the water bath and set to 40°C.
6. Prepare 2 × DMEM/F12 as per the manufacturers' instructions. Sterile filter using a bottle top 0.22-μm pore filter and store at 4°C.
7. Prepare 2 × MCF10A complete media (for recipe see buffers and reagents) using the prepared 2 × DMEM/F12 prepared in the previous step.
8. Prepare the 1.2% agar solution as described by the manufacturer.
9. Boil the 1.2% agar solution in the microwave until completely dissolved.
10. Place in water bath until cools slightly.
11. Warm 2 × MCF10A complete media in same water bath.

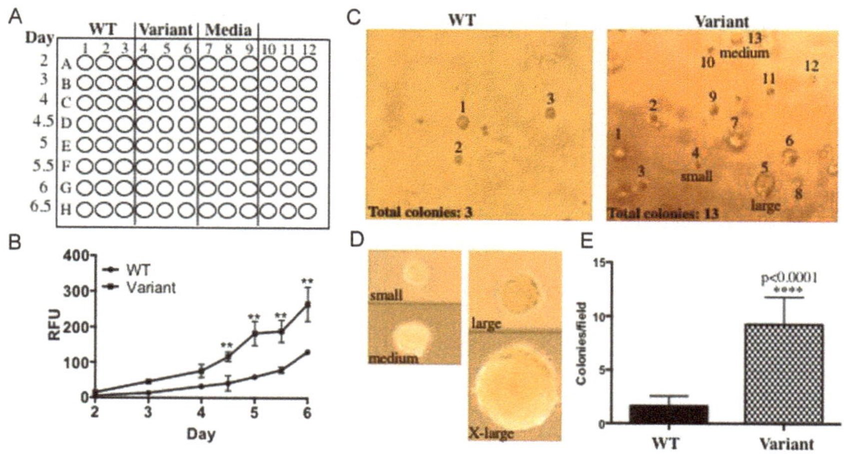

Fig. 3 (A) An example a 96-well plate setup for the proliferation assay. Plate WT-expressing cells in the first three columns, variant-expressing cells in the next three columns, and media only in the next *three columns* in *rows A–H*. Cells will be counted in triplicate (*each column*) over time (*each row*). The media only wells serve as a control for background. (B) As seen in the graph, the CYQUANT® NF cell proliferation assay in a 96-well format measures proliferation during early log phase growth. Significant differences have been observed during early log phase growth in cells expressing a BER variant as shown in the graph. To generate a complete growth curve, the 6-well plate format with manual counting would be a better option. (C–E) Representative fields of view of a soft agar assay using the Cytoselect™ 96-well cell transformation assay for a WT enzyme and a BER variant. In each field of view, the number of colonies is provided and the colonies that were included in the count have been labeled. Colony size can vary; however, colonies ranging from small to extra (X)-large are counted in the graph shown (E). Varying colony sizes are labeled in the variant field of view as well as in the panel below (D). If larger colonies are consistently formed by a BER variant, colonies of different sizes can be scored separately. To score the soft agar assay in the 96-well format, five fields of view are counted for each of four wells containing cells expressing either the WT enzyme or the BER variant. As shown in (E), colonies per field of view are graphed on the *Y*-axis.

12. Calculate the volume of media: agar solution required for the bottom agar, considering that 50 μL per well is needed and that each cell line will be plated in quadruplicate, including at least two wells without cells to serve as a negative control.
13. Combine 1.2% agar solution with 2× MCF10A complete media at a 1:1 ratio in a 15-mL conical tube.
14. Plate 50 μL of the bottom agar solution per well. Avoid creating bubbles in the agar when plating.
15. Tap the side of the plate gently to evenly distribute the agar: media solution.

16. Place at 4°C for 30 min.
17. Remove the plate from 4°C and place in the incubator for 15 min.
18. Melt the 1.2% agar in the microwave again and put in the water bath to cool. Place the 2 × MCF10A complete media in the water bath as well.
19. Prepare your cells: dilute 2 × 10^5 cells in 1 mL of 2 × MCF10A complete media.
20. Combine 125 μL of 2 × 10^5 cells, 125 μL 2 × MCF10A complete media, and 125 μL of 1.2% agar. Add 75 μL per well of the (cell: media:agar solution) to the wells with the solidified bottom agar.
21. Place the plate at 4°C for 15 min to allow the agar to solidify.
22. Add 100 μL of MCF10A complete media (without hygromycin) to each well and place in incubator. (Be sure you add 1 × growth media, not the 2 × media you have been using throughout this protocol).
23. Colonies should start forming in about 3–4 days. Count colonies about 8–10 days after plating.
24. Count 4–5 randomly chosen fields per well using a light microscope. See Fig. 3C and D for representative fields and colonies.

3.3.4 Notes

1. Do not pipet less than 1.5 μL of cells when plating or pipet error will be high and results will be inconsistent.
2. As an alternative method to using the CYQUANT® NF cell proliferation assay, plate 10^4 cells expressing WT and variant enzyme in triplicate in 6-well plates, then count the cells 48 h after plating for a total of 8 days using a cell counter or a hemocytometer.
3. The Cytoselect™ 96-well cell transformation assay is sold as a kit containing multiple components; however, we only use the Cytoselect™ Agar powder for this assay. We prepare our own 2 × media because MCF10A cells grow in DMEM/F12 not DMEM. However, if you are using cells that grow in DMEM, then prepare the 2 × DMEM provided in the kit. The other reagents provided are for agar solubilization, cell lysis, and cell quantification. We manually count the number of colonies that form in the agar.
4. You will need to move fast when plating the agar/media solutions so that the agar does not solidify before you finish plating.
5. Be sure that the agar has cooled to about 40–45°C before combining with cells. The cells will lyse when combined with the agar/media solution if it is too hot.

6. Similar to the proliferation assay, anchorage-independent growth can be measured without using the Cytoselect™ 96-well cell transformation assay. The alternative assay utilizes 1.0% Noble agar (Affymetrix) for the bottom agar and 0.7% Noble agar for the top agar, plated in a 6-well plate in triplicate for each cell line. For this type of assay, plate 10^4 cells per well using the same method described earlier, and simply adjust the volumes. The major difference is that colonies will be counted 30 days after plating instead of 8–10 days.
7. We recommend that cells from each passage be preserved in liquid nitrogen in order to perform the assays described later and also in case of contamination.

3.4 Genomic Instability and Mutagenesis

MCF10A cells expressing a BER variant that exhibit increased cellular proliferation and anchorage-independent growth, indicating cellular transformation, are then further interrogated for genomic instability and mutagenesis. While a few BER variants have exhibited increased proliferation and increased anchorage-independent growth, many BER variants exhibit increased anchorage-independent growth without increased proliferation. Increased proliferation may be counteracted by increased apoptosis or decreased cell viability and therefore is a more complex and less direct assessment of cellular transformation. Changes in anchorage-independent growth are a more direct reflection of cellular transformation and therefore are a key phenotype for continuing with characterization of a BER variant. To characterize the effects of a BER variant on genomic stability, we measure the amount and types of chromosomal aberrations using metaphase spread preparations and we quantify the presence of micronucleus formation to assess chromosome damage. Increased chromosomal aberrations are an indication of structural changes in chromosomes and signify genomic instability. Micronuclei arise from lagging chromosomes (either whole chromosomes or fragments) during mitosis. Micronuclei are a reflection of chromosome damage, such as chromosome breakage or whole chromosome loss, and therefore increased levels of micronuclei are another biomarker for genomic instability. Scoring chromosomal aberrations and micronuclei can often be subjective based on the quality of the images or cellular staining and the investigator that is scoring. It is highly recommended that the slides be coded prior to imaging and/or scoring to avoid any bias. It is also imperative that chromosomal aberrations and micronuclei are scored in cells at passages prior to those in which significant levels of anchorage-independent growth are observed, as described earlier. The purpose of these assays is to understand the mechanisms driving transformation. Genomic instability is an inherent feature of cells that are already transformed; therefore, no conclusions can be drawn as to whether or not the BER variant induced genomic instability or if increased genomic instability contributed to cellular transformation if transformed cells are utilized in these assays. In addition to DNA damage (i.e., breaks) and structural chromosomal aberrations, a BER variant may induce cellular transformation by increasing mutagenesis mainly in the form of point mutations. The ouabain assay is utilized to measure mutagenesis. Ouabain binds to and inhibits the Na^+/K^+-ATPase sodium potassium ion pump, leading to accumulation of intracellular sodium and eventual cell death. Mutations in the Na^+/K^+-ATPase sodium potassium ion pump can result in ouabain resistance; therefore, increased ouabain resistance serves as a marker for increased mutagenesis. The purpose of the assays outlined in this section is to determine whether the expression of a BER variant induces genomic instability or increased mutagenesis, providing key information that will guide preliminary mechanistic studies to determine the molecular mechanisms contributing to the cellular phenotypes observed.

3.4.1 Equipment

- Centrifuge
- 37°C water bath
- Clamp for pipetman
- Glass slides and coverslips
- Coplin staining jar
- Microscope (500/526 filter for DNA visualization)
- Two Hand Tally Counters
- Counter pen (Thermo Scientific)

3.4.2 Buffers and Reagents

- MCF10A complete media
- KaryoMax® Colcemid™ (10 μg/mL stock) (ThermoFisher)
- Carnoy's fixative (3:1 methanol:acetic acid), prechilled to −20°C
- 1 × PBS
- 0.05% trypsin
- Potassium chloride (Gibco, 0.075 *M*), prewarmed to 37°C
- Mounting media with DAPI (we use Sigma Fluoroshield™ with DAPI)
- Cytochalasin B (Cyt-B) in ddH_2O

- Methanol
- Acridine orange in PBS
- Ouabain octahydrate (Sigma)
- 0.5% Crystal Violet (500 mg diluted in 100 mL 80% methanol)

3.4.3 Procedure

1. *Metaphase spreads*: Plate 5 × 10^5 cells of each MCF10A pool (or 3–5 clones of each line) in four 100 mm plates.
2. When cells reach 50–65% confluency, add KaryoMax® Colcemid™ to each plate at a final concentration of 0.1 μg/mL.
3. Incubate cells in KaryoMax® Colcemid™ for 3–4 h.
4. Label 50 mL tubes for each MCF10A pool
5. After the KaryoMax® Colcemid™ incubation, collect the media from the plates into the 50 mL tubes.
6. Wash plates with 5 mL of 1 × PBS and collect into the same 50 mL tubes containing the collected media.
7. Add 4 mL of 0.05% trypsin to each plate and incubate at 37°C/5% CO_2 for 15–20 min, until the rest of the cells detach from the plate.
8. Collect the cells into the same 50 mL tube.
9. Centrifuge the cells at 1000 rpm for 5 min.
10. Aspirate supernatant leaving about 500 μL left. Resuspend cells by flicking the tube. Avoid pipetting the cells from this point on.
11. Wash cells in 10 mL of 1 × PBS.
12. Centrifuge at 1000 rpm for 5 min.
13. Aspirate supernatant leaving about 500 μL. Resuspend cells by flicking the tube.
14. Add 10 mL of prewarmed KCl (75 m*M*) to the cells. Add the first 3 mL dropwise slowly while gently mixing, and add the last 7 mL a little faster while still mixing.
15. Incubate the cells in KCl for 30 min in a 37°C water bath.
16. Add 10–15 drops of prechilled (−20°C) Carnoy's fixative while gently mixing.
17. Centrifuge the cells at 1000 rpm for 5 min.
18. Remove the supernatant by pipetting, taking care not to disturb the cell pellet, leaving about 500 μL of liquid. Resuspend the cells by flicking the tube gently.
19. Slowly add 10 mL of prechilled Carnoy's fixative dropwise while gently mixing. It should take about 2 min to add the first 2 mL of Carnoy's fixative.

20. Incubate at room temperature for 10 min.
21. Centrifuge the cells at 1000 rpm for 5 min.
22. Repeat fixation (steps 18–21) 3–4 times.
23. During the fixation steps, prepare the glass slides for later by rinsing with Carnoy's fixative and allow to dry. Also, place a container (such as a tip box cover or a plastic microcentrifuge box top) with wet Kimwipes at 37°C to prepare a pseudo-humidified chamber.
24. After centrifugation in the final fixation step, remove the supernatant as before leaving an appropriate amount of fixative based on the approximate number of cells presumed in the cell pellet. This volume is typically between 0.5 and 2 mL of fixative. If the cells are overly dense or overly sparse after dropping the first slide, the volume can be adjusted.
25. Place your P20 pipetman into a clamp setup and place on a shelf above your bench such that the distance between the pipetman and the bench surface is about 3–4 ft. See Fig. 4A for a picture of our setup.
26. Place the prepared glass slide into a humidified chamber, with the frosted side end of the slide on the edge of the chamber and the other end of the slide inside the chamber such that the slide angled.
27. Line up the slide under your pipetman, drop 10–15 μL of cells onto the slide. Wait about a minute to allow the cells to spread out on the slide.
28. Check the spread under a light microscope (20 × objective). In Fig. 4B and C, cells in the metaphase stage with well spread chromosomes are annotated by (*). The spread in Fig. 4B is a little sparse; however, cells are fairly well dispersed. The spread in Fig. 4C has several areas of clustered cells (indicated by the arrows), emphasizing the importance of step 16 in the procedure. If you do not add several drops of fixative prior to centrifugation after the KCl incubation, the cells can stick together. If the cells are sparse, add 5–6 mL of Carnoy's fixative, centrifuge, and resuspend the cells in a smaller volume. If the cells are closely packed, then increase the volume, drop another 10–15 μL of cells, and check again under the microscope.
29. Allow slides to dry for 24–48 h.
30. Add one to two drops of mounting media containing DAPI and place coverslip on slide.
31. Image metaphase spreads 24 h after mounting.
32. Score metaphase spreads for aberrations. We typically score fragments, fusions, and breaks. Fig. 4D shows "normal" chromosomes in a metaphase spread from a pool of MCF10A cells expressing a WT BER enzyme. In contrast, the metaphase spread from cells expressing a

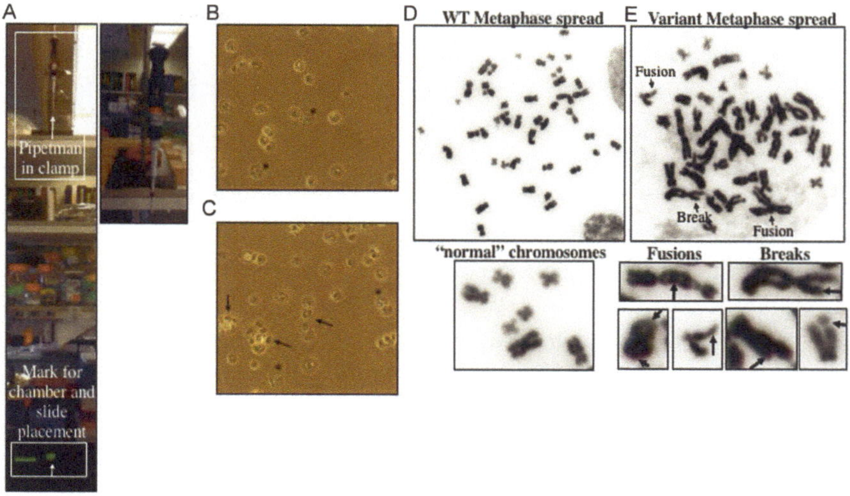

Fig. 4 (A) Picture of the pipetman clamp setup for dropping cells for metaphase preparation and micronucleus analysis. The pipetman is held in a clamp that is placed on a shelf about 3 ft. above the bench surface. A mark is placed on the bench top for appropriate placement of the slides. (B) Bright-field images using a 20 × objective of metaphase spreads prior to mounting. Cells in metaphase are indicated by (*). The *arrows* point to clumps of cells stuck together. Checking the slides after dropping cells serves both to adjust the volume if the cells are too sparse or confluent and to get an idea of how many cells are in metaphase prior to mounting. (D)–(E) Representative image of metaphase spreads from a cell expressing WT or BER variant enzyme taken using a 100 × objective on an Olympus BX50 research microscope with a Qlmaging Retiga 2000R digital camera. The WT metaphase spread (D) contains "normal" chromosomes without aberrations. However, the BER variant metaphase spread (E) contains multiple chromosomes with aberrations, including breaks and fusions, as indicated in the *image*. Additional representative chromosomes with fusions or breaks are included below (E) to serve as a guide for scoring aberrations.

BER variant in Fig. 4E contains chromosomes with aberrations including fusions and breaks as annotated. Representative chromosomes with fusions or breaks are included later the metaphase spread to serve as a reference. Fragments, which are not represented in the images, are acentric pieces of DNA.

33. *Micronucleus assay*: Plate 5 × 10^5 cells of each MCF10A pool (or three to five clones of each line) in three 100 mm plates.
34. When cells reach 50–65% confluency, add 6 μg/mL of Cyt-B per plate for 24 h at 37°C/5% CO_2.
35. After 24 h, trypsinize cells and collect into a 50-mL conical tube.
36. Centrifuge at 1000 rpm for 5 min.

37. Resuspend cells in 7 mL of 0.075 *M* KCl.
38. Incubate in KCl in a 37°C water bath for 10 min.
39. After 10 min, add 3 mL of methanol to each tube and incubate for 1 h at room temperature.
40. Centrifuge the cells at 1000 rpm for 5 min.
41. Fix cells twice in Carnoy's fixative as described earlier.
42. Drop cells as described earlier for metaphase preparation.
43. Stain slides in 10 μg/mL acridine orange for 10–20 min in a coplin jar wrapped in tin foil to protect from the light.
44. Briefly rinse slides with water. Keep the slides in a coplin jar in water until ready to score.
45. Slides should be scored immediately after staining using a microscope with a 500/520 excitation/emission filter. Take a slide out of the coplin jar in the dark and cover with a coverslip using the residual water as an aqueous buffer between the coverslip and the slide.
46. With hand tally counters in each hand score binucleated cells with one counter and binucleated cells with micronuclei with the other counter. Fig. 5 shows a representative image of a binucleated cell with a distinct single micronucleus. For more detailed information on scoring micronuclei, refer to *the in vitro micronucleus technique* (Fenech, 2000). Score 1000 binucleated cells per slide.
47. *Ouabain resistance assay*: You will need to plate the following for each MCF10A pool in 100 mm plates:

Untreated cells

- 10^3 (cells)
- 10^2

Ionizing radiation (IR) only

- 8 GY: 10^4, 10^3, 10^2

Ouabain only

- 50 n*M* ouabain: 10^4, 5×10^4
- 100 n*M* ouabain: 5×10^4, 10^5

Ouabain + IR

- 50 n*M* ouabain + 8 GY IR: 10^6, 7.5×10^5, 5×10^5
- 100 n*M* ouabain + 8 GY IR: 10^6, 7.5×10^5, 5×10^5

48. The following day, add 50 or 100 n*M* ouabain to the ouabain only plates. Also, irradiate the IR only and IR + ouabain plates. IR serves as a positive control for this assay.
49. 72 h after IR exposure, begin treatment with ouabain for IR + ouabain plates.

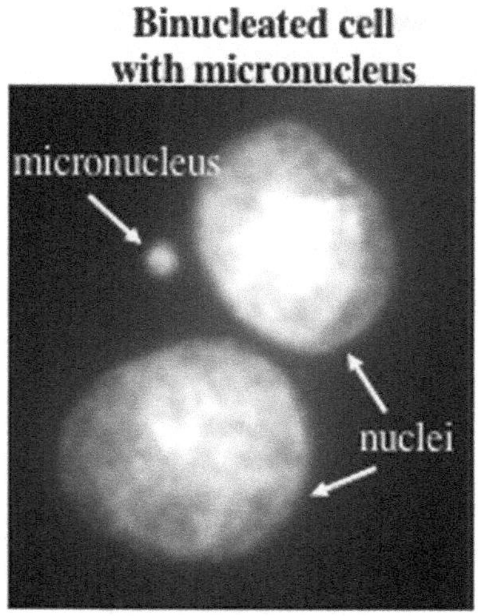

Fig. 5 Representative image of a binucleated cell >with a micronucleus.

50. Change the media on the ouabain and IR+ ouabain plates every 3–4 days.
51. Colonies that can be scored will form within about 8–10 days on the untreated plates and IR only plates. Colonies that can be scored will form within about 2–4 weeks on the ouabain and IR+ ouabain plates.
52. To accurately count colonies, use a counting pen. The plates will be marked with black marker after counting colonies; therefore take images of the plates prior to scoring.
53. To calculate the mutation frequency, the total number of colonies growing on ouabain is divided by the total number of colonies grown in the absence of ouabain.

3.4.4 Notes

1. If cells are more than 75% confluent you should replate and start again. KaryoMax® Colcemid™, also known as demecolcine, inactivates spindle fiber formation, thus arresting cells in metaphase. If the cells are not in log phase growth, then incubation with KaryoMax® Colcemid™ will not have any effect on the number of cells arrested in metaphase.
2. Shorter incubations in KaryoMax® Colcemid™ will reduce the number of cells in metaphase at the time of collection; however, longer incubations will condense the chromosomes, which makes scoring aberrations more difficult. For MCF10A cells, we have found

incubation with KaryoMax® Colcemid™ for 3–4 h is optimal, but this is likely cell type dependent.

3. Cells in mitosis become more rounded and therefore readily detach from the plate without trypsin. Therefore, it is important to collect all washes to ensure optimal recovery of cells in metaphase.
4. Incubation in potassium chloride will swell the cells so that when dropped onto glass slides, the chromosomes will spread nicely. The length of incubation is highly cell type dependent. We have optimized 30 min in KCl for MCF10A cells; however, if working with another cell line, you will need to optimize the length of incubation.
5. It is important to add 10–15 drops of Carnoy's fixative prior to centrifugation to keep the cells from sticking together.
6. Incubation at 4°C for 30 min is also acceptable.
7. If you cannot setup a pipetman in a clamp, you can also hold the pipetman while dropping your cells onto the slides. However, it is more difficult to consistently get the drops onto the slide.
8. Different protocols recommend varying distances between the pipetman and the slide. For MCF10A cells, we found that dropping the cells from heights of 3–4 ft. produces excellent metaphase spreads that are easily scored for aberrations.
9. If exposing the cells to damaging agents, then add the Cyt-B 24 h after IR or H_2O_2 exposure or treat cells for 24 h with the DNA damaging agent (i.e., cisplatin), then add Cyt-B upon removal of the agent.
10. Incubation in KCl is shorter for the micronucleus assay, as compared to preparation of metaphase spreads, and while it is desirable for the cells to swell, it is important that the nuclei and, if present, micronuclei remain in close proximity for scoring purposes. The optimal time in our hands is 10 min in KCl, but this is a step that may need to be optimized for each cell line and condition.
11. We have found that acridine orange produces high background if mounted in mounting media (such as Aquapolymount or antifade mounting media).
12. If the slide starts to dry out while imaging, place back in water and reposition the coverslip.
13. The dilutions listed are a starting point based on our results. The number of cells plated may need to be adjusted, particularly in the ouabain + IR (or any damaging agent) conditions.
14. We suggest trying 50 and 100 n*M* when beginning the assay. Once an optimal concentration of ouabain is determined, then perform future

assays using only that optimal concentration. The optimal concentration of ouabain will depend on the mutation frequency of the cells, the cytotoxicity of ouabain, and the combined cytotoxicity of ouabain and the chosen damaging agent.

15. The dose of irradiation may need to be adjusted. The dose recommended (8 GY) generates mutations but is also cytotoxic. In combination with the cytotoxicity of ouabain, it may decrease the number of scorable colonies. The dose can be decreased to 4 GY, which will be less cytotoxic but will also generate fewer mutations and therefore may decrease the numbers of cells that acquire ouabain resistance (also leading to fewer colonies).
16. IR is just a suggested damaging agent to generate mutations. Other damaging agent can be used in place of IR, such as hydrogen peroxide or menadione.
17. The counting pen marker dries out quickly. If the black marker on the counter pen dries up and stops working, count colonies with the counter pen in one hand and a marker in the other in order to mark the colonies you have already counted.

3.5 Preliminary Mechanistic Analysis

The information gathered about a BER variant from the studies performed thus far should be used as a guide for the design of additional studies to determine the mechanisms contributing to the cellular outcomes. For instance, if expression of a BER variant increases the levels of chromosomal aberrations and micronuclei, as determined by the studies in the previous section, then one possibility is that the BER variant does not remove polymerase-blocking lesions from DNA, thereby disrupting replication. Disruption of replication can be tested indirectly by measuring DNA double-strand breaks (DSBs) that arise during S-phase as a result of collapsed replication forks and directly by measuring the initiation and progression of sites of replication using the DNA fiber assay (for examples, see chapter "DNA Fiber Analysis: Mind the Gap!" by Quinet et al. and Nemec et al., 2016). A phosphorylated form of the histone variant H2A.X (γH2A.X) is used as a marker to measure DSBs and propidium iodide (PI) is used to measure DNA content to delineate G1, S, and G2/M phases of the cell cycle by flow cytometry. Using this method, the percentage of cells positive for γH2A.X in S-phase can be compared between WT-expressing and BER variant-expressing cells. The basic principles of flow cytometry and flow cytometric data analysis will not be described here, but more information can be obtained from Adan, Alizada, Kiraz, Baran, and Nalbant (2017). The DNA fiber assay utilizes the incorporation of halogenated thymidine analogs into the DNA during replication in the cell allowing visualization of sites of DNA replication. The DNA fiber assay provides information about the efficiency of replication by permitting the measurement of mean replication tract length or replication fork speed as well as the measurement of replication structures including elongating replication forks and stalled replication forks. Described here are only two assays that can be used to tease out the mechanisms driving transformation. Depending on the specific function of your BER variant of interest and the results of the assays already performed, another avenue of investigation may be more appropriate.

3.5.1 Equipment

- Microcentrifuge
- Microcentrifuge tubes
- Flow cytometer
- Rocker
- Software for flow cytometric analysis (we use FlowJo software)
- Angled platform
- Staining jar
- Glass slides

3.5.2 Buffers and Reagents

- PBS
- MCF10A complete media
- DMEM/F12 (no serum) supplemented with 1% P/S
- 0.05% trypsin
- Halt protease/phosphatase inhibitor tablets (Roche)
- EDTA (100 m*M* stock)
- 1% paraformaldehyde with 5 m*M* EDTA diluted in PBS
- 70% ethanol (prechilled −20°C)
- Flow buffer: 1% bovine serum albumin (BSA) (w/v), 0.1% Triton X-100, 5 m*M* EDTA diluted in PBS
- Mouse monoclonal anti-phospho-Histone H2A.X (Ser139) (Millipore, clone JBW301), 1/300 dilution in flow buffer
- AlexaFlour® 488 goat anti-mouse IgG (H + L) antibody (Invitrogen), 1/500 dilution in flow buffer
- BD PI/RNase Staining Buffer supplemented with 5 m*M* EDTA

- 44 μm pore filter mesh
- 12 × 75mm round-bottom polystyrene tubes, nonsterile without caps
- Iododeoxyuridine (IdU), 5 m*M* dissolved in DMSO (Sigma)
- Chlorodeoxyuridine (CIdU), 2.5 m*M* dissolved in ddH_2O (Sigma)
- Methanol:acetic acid fixative (3:1)
- 2.5 *M* HCl
- Fiber lysis solution: 50 m*M* EDTA, 0.5% SDS, 200 m*M* Tris–HCl pH 7.5 diluted in ddH_2O.
- 5% BSA (w/v) in PBS
- Rat monoclonal anti-BrdU primary antibody (Abcam, ab6326)
- Mouse monoclonal anti-BrdU primary antibody (BD biosciences, 347580)
- AlexaFlour® 647 goat anti-rat IgG (H + L) antibody (Invitrogen)
- Mounting media (No DAPI)

3.5.3 Procedure

1. Plate 1×10^5 cells per well in triplicate for each cell line in 6-well plates in MCF10A complete growth media. You will need three wells for each line for 0, 2, 4, and 8 h time-points as well as one well for each cell line for an unstained control, and PI only control.
2. The next day, remove the media and wash the cells 2 × with PBS to remove any residual serum in the wells.
3. Add serum-free DMEM/F12 media (SFM) containing 1% pen-strep to each well. Incubate the cells in SFM for at least 24 h to halt the cells in G_0/G_1.
4. Replace SFM with MCF10A complete growth media 16–18 h prior to exposing cells to a damaging agent (i.e., IR).
5. Prior to exposing cells to damaging agent, prepare 0.05% trypsin with 5 m*M* EDTA and one protease/phosphatase inhibitor tablet per 10 mL of trypsin. Warm in a 37°C water bath.
6. Expose cells to DNA-damaging agent 16 h postrelease from serum deprivation.
7. Place 0 h plates on ice until collection. Put the rest of the plates back into the incubator.
8. *For each time-point collection*, wash the cells with PBS (unstained and PI only plates can be collected at any time point).
9. Trypsinize cells with 0.05% trypsin, 5 m*M* EDTA, protease/ phosphatase inhibitors prepared earlier until the cells are just

detaching from the plate. Be careful to minimize the time the cells are in trypsin.

10. While the cells are trypsinizing, label microcentrifuge tubes for each sample containing 500 μL of PBS with 5 m*M* EDTA and place on ice.
11. Once the cells have detached, immediately pipet the cells into the labeled microcentrifuge tubes on ice.
12. Centrifuge cells in a microcentrifuge that is in a cold room or refrigerated (4°C) at 190 × *g* for 5 min.
13. Using a vacuum aspirator, remove the supernatant carefully without disturbing the cell pellet, leaving about 50 μL of supernatant to avoid losing cells.
14. Add 500 μL of 1% paraformaldehyde with 5 m*M* EDTA diluted in PBS and vortex gently to resuspend to cells.
15. Incubate cells in fixative for 20 min on ice.
16. Centrifuge samples at 4°C at 5200 × *g* for 5 min. Aspirate supernatant leaving about 50 μL of supernatant to avoid losing cells.
17. Add 1 mL of cold (−20°C) 70% EtOH to tubes and vortex gently to resuspend cells. Incubate for at least 30 min at −20°C (see Note 5 below).
18. Centrifuge samples at 5200 × *g* for 5 min at 4°C. Aspirate supernatant leaving about 50 μL of supernatant.
19. Add 1 mL of flow buffer and vortex gently to resuspend cells. Incubate on ice for 15 min to rehydrate the cells.
20. Centrifuge samples at 5200 × *g* for 5 min. Aspirate the supernatant leaving about 50 μL of supernatant.
21. Add 100 μL of diluted (1/300) mouse monoclonal anti-phospho-Histone H2A.X primary antibody to each sample. Vortex gently to resuspend.
22. Incubate overnight on a slant on a rocker in a cold room (4°C).
23. The next day, add 1 mL flow buffer to samples and vortex gently. Incubate on ice for 10 min.
24. Centrifuge samples at 5200 × *g* for 5 min. Aspirate supernatant leaving about 50 μL.
25. *All steps starting from this point must be performed in the dark*. Add 100 μL of diluted (1/500) AlexaFlour® 488 goat antimouse secondary antibody to samples and vortex gently.
26. Incubate at room temperature for 1 h with gentle shaking.
27. Add 1 mL flow buffer and vortex gently. Incubate at room temperature for 10 min.

28. Centrifuge samples at 5200 × *g* for 5 min. Aspirate the supernatant leaving about 50 μL.
29. Add 450 μL of PI/RNase buffer with 5 m*M* EDTA to samples. Vortex gently to mix.
30. Incubate at room temperature for 15 min.
31. Pipet through 44 μm pore filter mesh into 12 × 75 mm round-bottom polystyrene tubes. Keep samples in the dark and at 4°C until analysis.
32. Fig. 6 provides a brief overview of cell cycle/γH2A.X gating analysis. However, for a comprehensive, detailed protocol describing gating and analysis of γH2A.X labeling refer to Kataoka, Bindokas, Duggan, Murley, and Grdina (2006) and Tanaka et al. (2007). If you have a flow cytometry facility at your institution, that would be the best resource for learning about flow cytometry.
33. *DNA fiber assay*: Plate 1 × 10^5 cells for each cell line in 35 mm plates in MCF10A complete growth media.
34. Culture the cells until about 40–50% confluent.
35. Add IdU to the medium at a final concentration of 25 μ*M*. Place plates back into the incubator for 20 min.
36. During the IdU incubation, prepare 250 μ*M* (final concentration) of CldU in MCF10A complete growth media for the second pulse.
37. Aspirate the media and wash the cells three times with PBS.
38. Add MCF10A complete media containing 250 μ*M* CldU. Place back into the incubator for 20 min.
39. After 20 min, trypsinize the cells and perform a cell count.
40. Resuspend the cells in PBS at 1 × 10^6–2.5 × 10^6 cells/mL (which will equal 2000–5000 cells in 2 μL in the next step). Place cells on ice.
41. Prepare the slides and the setup to spread the DNA fibers. Label glass slides for each sample in pencil.
42. Place the labeled glass slides that will be spotted first into a container that will eventually be angled (see Fig. 7A).
43. Setup a surface that will allow the container with the glass slides to be tilted to a 15-degree angle to allow the fibers to spread along the slide. See Fig. 7A–E for an example setup. Once everything is set up, you are ready to spot your cells onto the slides.
44. Spot 2 μL of cell suspension onto labeled glass slides.
45. Air dry the cell suspension for 5 min or until the volume is greatly reduced but not dry.
46. Pipet 7 μL of fiber lysis solution into the cell suspension and gently mix with the pipet tip.

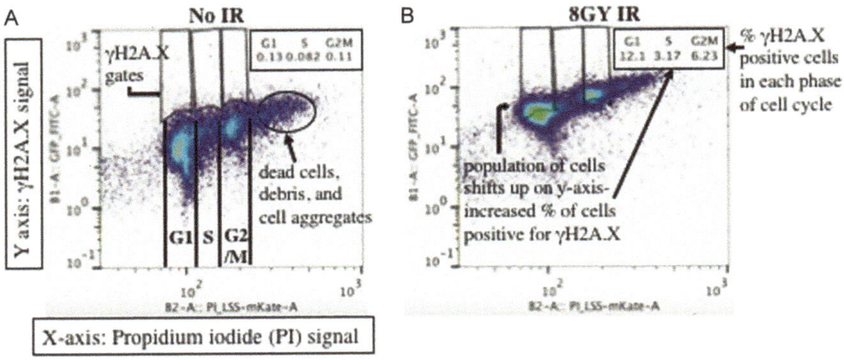

Fig. 6 (A)–(B) Sample 2D dot plots of MCF10A cells synchronized using serum deprivation and then either untreated (No IR) or exposed to 8 GY of IR. In these dot plots, the PI signal is plotted on the *X*-axis. PI binds to double-stranded DNA by intercalating between base pairs and therefore provides a measurement of DNA content in the cells. DNA content is then used to identify cell populations in G1, S, and G2/M phases of the cell cycle. The cell populations being analyzed in these dot plots were not previously filtered for live single-cell populations using forward and side scatter. By gating for the live, single-cell population using forward and side scatter, you will mostly eliminate the population clustered to the right of cells in G2/M (A), which includes cell doublets and aggregates, dead cells and debris. The γH2A.X signal is plotted on the *Y*-axis. To determine the percentage of γH2A.X-positive cells in your samples, draw gates that delineate G1, S, and G2/M phases of the cell cycle. Use your untreated sample plots (A) to determine placement of the gates, serving as a baseline for γH2A.X expression. Copy and paste those gates from the untreated sample plots into the treated sample plots. The software will measure the percentage of cells within the entire cell population for each phase of the cell cycle (i.e., the number cells, referred to as events, that were measured during data collection on the flow cytometer) that is positive for γH2A.X expression based on the gates provided.

47. Incubate for 2 min to allow for the cells to lyse.

48. After 2 min of lysis, tilt the slides to a 15-degree angle. Watch carefully as the solution spreads down the slide. If the solution is traveling rapidly down the slide, decrease the angle to 10 degrees; however, if the solution is not moving fast enough, the angle can be increased to 25–30 degrees. It should take about 30 s^{-1} min for the solution to reach the bottom of the slide. The speed at which the solution spreads down the slide will significantly impact the dispersal of the DNA fibers along the slide.

49. Place the slides horizontally to allow to dry. Once dry a thin opaque line will be visible along the slide. Mark the beginning of the stretched fibers with a pencil to help with locating the fibers under the microscope. The line will not be visible after staining.

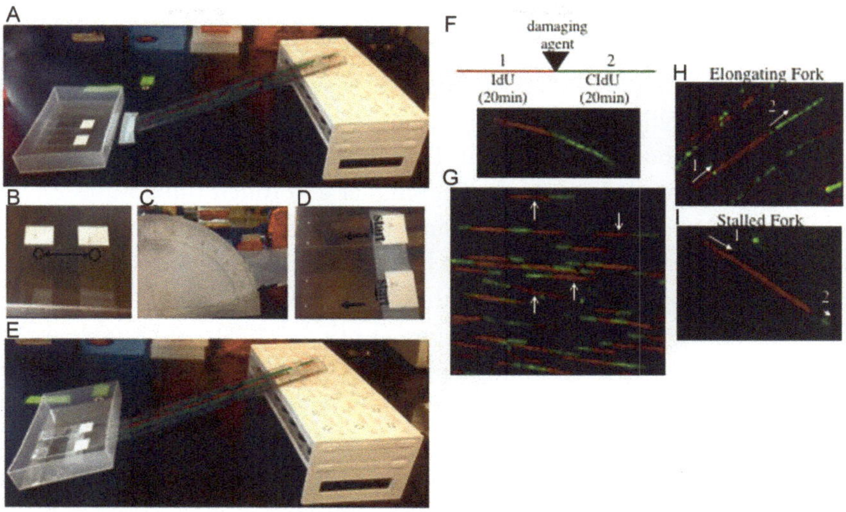

Fig. 7 (A) An image of our setup for the lysis step in the DNA fiber assay through to spreading the DNA fibers onto the slides. The slides are in a container that will sit flat and even on the bench. The *ruler* and *rack* (turned upside down) serves as a platform for tilting the slides in the container to a 15-degree angle after lysis. (B) An image of slides with 2 μL of cells spotted at the top of the slide. (C, E) After lysis, the container is moved up the ruler until the bottom edge of the container (corresponding to the bottom of the slides) reaches a 15-degree angle. (D) The *lines* at the top of the slide mark the start of the DNA fibers and the *arrows* indicate the movement of the DNA fibers as they move down the slide. This is a critical step in the DNA fiber assay. How the fibers move down the slide will significantly impact the quality of images for analysis. Slide the container holding the lysed cells slowly up the ladder (or whatever you use to make an angled surface) watching the movement of the fiber solution carefully. The fibers should spread slowly down the slide. If the fiber solution is moving down the slide rapidly, lower the angle to a 10-degree angle or until the movement is slowed. (F) A schematic of the experiment and an image of a representative DNA fiber with incorporation of IdU in the first pulse (which is *red* in the image) and then incorporation of CldU in the second pulse (which is *green* in the image). To measure the effect of a damaging agent on fork progression or fork collapse, the cells are treated with a damaging agent after the first pulse with IdU. (G) A representative image with well-separated, elongating fibers. The *arrows* indicate a few of the elongating fibers in the image that can be used to measure mean tract length. (H) A representative image of an elongating fiber indicating the progression of replication: (1) labels the first pulse with IdU and (2) labels the second pulse with CldU. Therefore, replication was moving from *left* to *right* in this fiber. (I) A representative image of a stalled replication fork. IdU is incorporated during the first pulse; however, CldU is not incorporated or there is initial incorporation without continuation (as seen in this *image*). Elongating replication forks can be used to measure mean tract length. Replication structures, including elongating and stalled forks, can also be scored.

50. Immerse the slides in methanol:acetic acid (3:1) in a glass staining jar for 10 min.
51. Wash the slides with distilled H_2O and then immerse the slides in 2.5 *M* HCl for 2.5 h.
52. Wash the slides three times in PBS for 5 min each.
53. Remove the slides from the staining jar and touch the bottom of the slide to a kimwipe to remove an excess PBS.
54. Place the slides horizontally in a humidified chamber and add 100 µL of 5% BSA to each slide. Cover the slides carefully with a coverslip and incubate at room temperature for 20 min.
55. Dilute the primary antibodies in 5% BSA at the following concentrations: mouse anti-BrDU 1:25; rat anti-BrDU 1:200.
56. Gently slide the coverslip to the bottom of the slide to remove. If the coverslip sticks to the slide rehydrate the slide in PBS until the coverslip becomes loose and can be removed with ease.
57. Add 50 µL of the diluted primary antibody solution to each slide. Cover with a coverslip carefully to avoid creating any bubbles or air pockets between the coverslip and slide.
58. Incubate at room temperature for 1.5 h.
59. During this incubation, dilute the secondary antibodies at 1/1000 in 5% BSA and keep in the dark at 4°C until ready to use.
60. Remove the coverslip as before. Wash the slides three times in PBS for 5 min each.
61. *All steps from this point forward should be performed in the dark*: Add 50 µL of the diluted secondary antibodies and carefully cover with a coverslip as before.
62. Incubate in the dark for 2.5 h.
63. Remove the coverslips as before and wash the slides three times in PBS for 5 min each.
64. Add a drop of mounting media (such as Prolong Gold or Vectashield without DAPI) to the slides and apply a coverslip. Gently press the coverslip, pushing out any bubbles between the coverslip and the slide. Remove any excess mounting media along the sides of the slide with a kimwipe.
65. Image the DNA fibers using a 60 × objective. Use the pencil mark made in step 49 as a guide to find the DNA fibers on the slide. Typically, there will be a main fiber bundle, but the fibers in the main bundle are too entangled and therefore cannot be analyzed. Move away from the main fiber bundle down the slide to find areas with nicely separated

fibers. Fig. 7G is a great example of a field of view to image for analysis. The fibers are well separated and easily visible. Also, most of the fibers in Fig. 7G have a single stretch of IdU incorporation and a single stretch of CldU incorporation, which is important for analysis. If you can find fields of view similar to Fig. 7G, then 10 pictures of each slide will suffice. If you find all your slides are extremely crowded with DNA fibers and you cannot identify images that can be analyzed, in the next assay you will want to dilute your cells labeled with IdU and CldU with unlabeled cells at a ratio that is appropriate to get images for analysis (for instance 200 μL labeled cells:800 μL unlabeled cells or 500 μL labeled cells:500 μL unlabeled cells). For more information on the DNA fiber assay, see Chapter X by Vindigni et al.. A step-by-step video of the procedure can be found at the following URL: http://www.jove.com/video/3255/visualization-dna-replication-vertebrate-model-system-dt40-using-dna.

3.5.4 Notes

1. The easiest way to set up the experiment for collection is to plate three wells of WT-expressing cells and three wells of BER variant-expressing cells in one 6-well plate for each time point.
2. Residual serum in the wells will result poor synchronization.
3. MCF10A cells enter early S-phase about 16 h after release from serum deprivation. Therefore, you want to time release from serum deprivation with treatment with a damaging agent 16–18 h later. Cells will be collected at 0, 2, 4, and 8 h posttreatment; therefore, we release the cells from serum deprivation in the late afternoon and treat with a damaging agent the next morning.
4. The addition of EDTA to reagents in this protocol is meant to prevent cells from aggregating or sticking together.
5. The antibody binds to a H2A.X that has been phosphorylated at serine 139 forming what is known as gamma H2A.X (γH2A.X); therefore, it is imperative that the protease/phosphatase inhibitor tablet you choose to use contains phosphatase inhibitors, not just protease inhibitors.
6. Once the cells are in 70% EtOH, you can store your samples at −20°C for up to 1 month.
7. To filter the cells, you can also use the Falcon® 5 mL Round-Bottom Polystyrene Test Tube with a Cell Strainer Snap Cap. The cell strainer is a 35-μm mesh filter that can be detached from the round-bottom tube once cells are filtered through by centrifugation. It is a more

convenient method however also more expensive and sometimes allows cell aggregates through probably because of the force applied during centrifugation.

8. It is important to filter the cells prior to analyzing on the flow cytometer to remove large cell aggregates. If you do not perform this step, there is a good chance you will clog the sample injection port and the machine will need to be repaired prior to running any more samples.
9. DNA fiber assay: Similar to metaphase preparation, the cells should be in log phase growth to maximize the number of cells in S-phase when pulsing the cells with the thymidine analogs.
10. Washing the plates with PBS is a critical step to ensure complete removal of IdU from the cells. Incomplete washing of the cells will result in overlapping signals in the DNA fiber due to incorporation of both IdU and CIdU in the second pulse. The fiber will appear yellow instead of green and therefore cannot be scored.
11. Do not prepare more than five slides at a time. While adding lysis solution to the cells, you will not be able to move fast enough to avoid other slides drying out in the meantime. Once the cell solution dries out, lysis will be incomplete and the DNA will not spread down the slide as desired.
12. Once the slides have been fixed in methanol:acetic acid (3:1), they can be stored at 4°C until the next day.
13. We make a humidified chamber using a slide box and place wet paper towels in the center below the slides; however, any container that allows the slides to lay flat horizontally with a source of humidity will work.
14. We have found that a 100× objective restricts too much light and makes the fibers difficult to image.
15. Microfluidics and other types of instruments can be purchased to spread DNA fibers on the slides.

REFERENCES

Adan, A., Alizada, G., Kiraz, Y., Baran, Y., & Nalbant, A. (2017). Flow cytometry: Basic principles and applications. *Critical Reviews in Biotechnology*, *37*, 163–176.

Adzhubei, I. A., Schmidt, S., Peshkin, L., Ramensky, V. E., Gerasimova, A., Bork, P., et al. (2010). A method and server for predicting damaging missense mutations. *Nature Methods*, 7, 248–249.

Al-Tassan, N., Chmiel, N. H., Maynard, J., Fleming, N., Livingston, A. L., Williams, G. T., et al. (2002). Inherited variants of MYH associated with somatic G:C→T:A mutations in colorectal tumors. *Nature Genetics*, *30*, 227–232.

Asagoshi, K., Yamada, T., Okada, Y., Terato, H., Ohyama, Y., Seki, S., et al. (2000). Recognition of formamidopyrimidine by Escherichia coli and mammalian thymine glycol glycosylases. Distinctive paired base effects and biological and mechanistic implications. *275*, 24781–24786.

Aspinwall, R., Rothwell, D. G., Roldan-Arjona, T., Anselmino, C., Ward, C. J., Cheadle, J. P., et al. (1997). Cloning and characterization of a functional human homolog of Escherichia coli endonuclease III. *Proceedings of the National Academy of Sciences of the United States of America, 94*, 109–114.

Au, K. G., Cabrera, M., Miller, J. H., & Modrich, P. (1988). Escherichia coli mutY gene product is required for specific A-G----C.G mismatch correction. *Proceedings of the National Academy of Sciences of the United States of America, 85*, 9163–9166.

Bacolla, A., Cooper, D. N., & Vasquez, K. M. (2014). Mechanisms of base substitution mutagenesis in cancer genomes. *Genes (Basel), 5*, 108–146.

Balakrishnan, L., & Bambara, R. A. (2013). Flap endonuclease 1. *Annual Review of Biochemistry, 82*, 119–138.

Boparai, K. S., Dekker, E., Van Eeden, S., Polak, M. M., Bartelsman, J. F., Mathus-Vliegen, E. M., et al. (2008). Hyperplastic polyps and sessile serrated adenomas as a phenotypic expression of MYH-associated polyposis. *Gastroenterology, 135*, 2014–2018.

Chen, Y., Cunningham, F., Rios, D., McLaren, W. M., Smith, J., Pritchard, B., et al. (2010). Ensembl variation resources. *BMC Genomics, 11*, 293.

Cingolani, P., Platts, A., Wang, L. L., Coon, M., Nguyen, T., Wang, L., et al. (2012). A program for annotating and predicting the effects of single nucleotide polymorphisms, SnpEff. *Fly, 6*, 80–92.

Danecek, P., Auton, A., Abecasis, G., Albers, C. A., Banks, E., DePristo, M. A., et al. (2011). The variant call format and VCFtools. *Bioinformatics, 27*, 2156–2158.

David, S. S., O'Shea, V. L., & Kundu, S. (2007). Base-excision repair of oxidative DNA damage. *Nature, 447*, 941–950.

Donigan, K. A., Hile, S. E., Eckert, K. A., & Sweasy, J. B. (2012b). The human gastric cancer-associated DNA polymerase beta variant D160N is a mutator that induces cellular transformation. *DNA Repair (Amst), 11*, 381–390.

Donigan, K. A., Sun, K. W., Nemec, A. A., Murphy, D. L., Cong, X., Northrup, V., et al. (2012a). Human POLB gene is mutated in high percentage of colorectal tumors. *The Journal of Biological Chemistry, 287*, 23830–23839.

Eide, L., Luna, L., Gustad, E. C., Henderson, P. T., Essigmann, J. M., Demple, B., et al. (2001). Human endonuclease III acts preferentially on DNA damage opposite guanine residues in DNA. *Biochemistry, 40*, 6653–6659.

Fenech, M. (2000). The in vitro micronucleus technique. *Mutation Research, 455*, 81–95.

Freidberg, E. C., Wood, R. D., Walker, G. C., & Siede, W. (2006). *DNA repair and mutagenesis* (2nd ed.). Washington, DC: ASM Press.

Galick, H. A., Kathe, S., Liu, M., Robey-Bond, S., Kidane, D., Wallace, S. S., et al. (2013). Germ-line variant of human NTH1 DNA glycosylase induces genomic instability and cellular transformation. *Proceedings of the National Academy of Sciences of the United States of America, 110*, 14314–14319.

Gao, J., Aksoy, B. A., Dogrusoz, U., Dresdner, G., Gross, B., Sumer, S. O., et al. (2013). Integrative analysis of complex cancer genomics and clinical profiles using the cBioPortal. *Science Signaling, 6*, pl1.

Gentleman, R. C., Carey, V., Bates, D. M., Bolstad, B., Dettling, M., Dudoit, S., et al. (2004). Bioconductor: Open software development for computational biology and bioinformatics. *Genome Biology, 5*, R80.

Gossen, M., & Bujard, H. (1992). Tight control of gene expression in mammalian cells by tetracycline-responsive promoters. *Proceedings of the National Academy of Sciences of the United States of America, 89*, 5547–5551.

Ikeda, S., Biswas, T., Roy, R., Izumi, T., Boldogh, I., Kurosky, A., et al. (1998). Purification and characterization of human NTH1, a homolog of Escherichia coli endonuclease III. Direct identification of Lys-212 as the active nucleophilic residue. *The Journal of Biological Chemistry*, *273*, 21585–21593.

Kataoka, Y., Bindokas, V. P., Duggan, R. C., Murley, J. S., & Grdina, D. J. (2006). Flow cytometric analysis of phosphorylated histone H2AX following exposure to ionizing radiation in human microvascular endothelial cells. *Journal of Radiation Research*, *47*, 245–257.

Lang, T., Dalal, S., Chikova, A., DiMaio, D., & Sweasy, J. B. (2007). The E295K DNA polymerase beta gastric cancer-associated variant interferes with base excision repair and induces cellular transformation. *Molecular and Cellular Biology*, *27*, 5587–5596.

Lang, T., Maitra, M., Starcevic, D., Li, S. X., & Sweasy, J. B. (2004). A DNA polymerase beta mutant from colon cancer cells induces mutations. *Proceedings of the National Academy of Sciences of the United States of America*, *101*, 6074–6079.

Lawrence, M. S., Stojanov, P., Mermel, C. H., Robinson, J. T., Garraway, L. A., Golub, T. R., et al. (2014). Discovery and saturation analysis of cancer genes across 21 tumour types. *Nature*, *505*, 495–501.

Lu, A. L., Tsai-Wu, J. J., & Cillo, J. (1995). DNA determinants and substrate specificities of Escherichia coli MutY. *The Journal of Biological Chemistry*, *270*, 23582–23588.

McGoldrick, J. P., Yeh, Y. C., Solomon, M., Essigmann, J. M., & Lu, A. L. (1995). Characterization of a mammalian homolog of the Escherichia coli MutY mismatch repair protein. *Molecular and Cellular Biology*, *15*, 989–996.

Murphy, D. L., Donigan, K. A., Jaeger, J., & Sweasy, J. B. (2012). The E288K colon tumor variant of DNA polymerase beta is a sequence specific mutator. *Biochemistry*, *51*, 5269–5275.

Nemec, A. A., Bush, K. B., Towle-Weicksel, J. B., Taylor, B. F., Schulz, V., Weidhaas, J. B., et al. (2016). Estrogen drives cellular transformation and mutagenesis in cells expressing the breast cancer-associated R438W DNA polymerase lambda protein. *Molecular Cancer Research*, *14*, 1068–1077.

Nemec, A. A., Donigan, K. A., Murphy, D. L., Jaeger, J., & Sweasy, J. B. (2012). Colon cancer-associated DNA polymerase beta variant induces genomic instability and cellular transformation. *The Journal of Biological Chemistry*, *287*, 23840–23849.

Nemec, A. A., Murphy, D. L., Donigan, K. A., & Sweasy, J. B. (2014). The S229L colon tumor-associated variant of DNA polymerase beta induces cellular transformation as a result of decreased polymerization efficiency. *The Journal of Biological Chemistry*, *289*, 13708–13716.

Pope, M. A., & David, S. S. (2005). DNA damage recognition and repair by the murine MutY homologue. *DNA Repair (Amst)*, *4*, 91–102.

Rivera, B., Castellsague, E., Bah, I., van Kempen, L. C., & Foulkes, W. D. (2015). Biallelic NTHL1 mutations in a woman with multiple primary tumors. *The New England Journal of Medicine*, *373*, 1985–1986.

Schork, N. J., Murray, S. S., Frazer, K. A., & Topol, E. J. (2009). Common vs. rare allele hypotheses for complex diseases. *Current Opinion in Genetics & Development*, *19*, 212–219.

Sizova, D. V., Keh, A., Taylor, B. F., & Sweasy, J. B. (2015). The R280H X-ray cross-complementing 1 germline variant induces genomic instability and cellular transformation. *DNA Repair (Amst)*, *31*, 73–79.

Sjolund, A., Nemec, A. A., Paquet, N., Prakash, A., Sung, P., Doublie, S., et al. (2014). A germline polymorphism of thymine DNA glycosylase induces genomic instability and cellular transformation. *PLoS Genetics*, *10*, e1004753.

Slupska, M. M., Baikalov, C., Luther, W. M., Chiang, J. H., Wei, Y. F., & Miller, J. H. (1996). Cloning and sequencing a human homolog (hMYH) of the Escherichia coli

mutY gene whose function is required for the repair of oxidative DNA damage. *Journal of Bacteriology*, *178*, 3885–3892.
Slupska, M. M., Luther, W. M., Chiang, J. H., Yang, H., & Miller, J. H. (1999). Functional expression of hMYH, a human homolog of the Escherichia coli MutY protein. *Journal of Bacteriology*, *181*, 6210–6213.
Soule, H. D., Maloney, T. M., Wolman, S. R., Peterson, W. D., Jr., Brenz, R., McGrath, C. M., et al. (1990). Isolation and characterization of a spontaneously immortalized human breast epithelial cell line, MCF-10. *Cancer Research*, *50*, 6075–6086.
Sweasy, J. B., Lang, T., & DiMaio, D. (2006). Is base excision repair a tumor suppressor mechanism? *Cell Cycle*, *5*, 250–259.
Sweasy, J. B., Lang, T., Starcevic, D., Sun, K. W., Lai, C. C., Dimaio, D., et al. (2005). Expression of DNA polymerase {beta} cancer-associated variants in mouse cells results in cellular transformation. *Proceedings of the National Academy of Sciences of the United States of America*, *102*, 14350–14355.
Tanaka, T., Huang, X., Halicka, H. D., Zhao, H., Traganos, F., Albino, A. P., et al. (2007). Cytometry of ATM activation and histone H2AX phosphorylation to estimate extent of DNA damage induced by exogenous agents. *Cytometry Part A*, *71*, 648–661.
Wallace, S. S., Murphy, D. L., & Sweasy, J. B. (2012). Base excision repair and cancer. *Cancer Letters*, *327*, 73–89.
Weren, R. D., Ligtenberg, M. J., Kets, C. M., de Voer, R. M., Verwiel, E. T., Spruijt, L., et al. (2015). A germline homozygous mutation in the base-excision repair gene NTHL1 causes adenomatous polyposis and colorectal cancer. *Nature Genetics*, *47*, 668–671.
Yamtich, J., Nemec, A. A., Keh, A., & Sweasy, J. B. (2012). A germline polymorphism of DNA polymerase beta induces genomic instability and cellular transformation. *PLoS Genetics*, *8*, e1003052.
Yates, A., Akanni, W., Amode, M. R., Barrell, D., Billis, K., Carvalho-Silva, D., et al. (2016). Ensembl 2016. *Nucleic Acids Research*, *44*, D710–D716.
Yoon, D. S., Wersto, R. P., Zhou, W., Chrest, F. J., Garrett, E. S., Kwon, T. K., et al. (2002). Variable levels of chromosomal instability and mitotic spindle checkpoint defects in breast cancer. *The American Journal of Pathology*, *161*, 391–397.

CHAPTER SEVEN

Fluorescence-Based Reporters for Detection of Mutagenesis in *E. coli*

Melissa Standley, Jennifer Allen, Layla Cervantes, Joshua Lilly, Manel Camps[1]
University of California—Santa Cruz, Santa Cruz, CA, United States
[1]Corresponding author: e-mail address: mcamps@ucsc.edu

Contents

Methods in Enzymology, Volume 591
ISSN 0076-6879
http://dx.doi.org/10.1016/bs.mie.2017.03.013

Abstract

Mutagenesis in model organisms following exposure to chemicals is used as an indicator of genotoxicity. Mutagenesis assays are also used to study mechanisms of DNA homeostasis. This chapter focuses on detection of mutagenesis in prokaryotes, which boils down to two approaches: reporter inactivation (forward mutation assay) and reversion of an inactivating mutation (reversion mutation assay). Both methods are labor intensive, involving visual screening, quantification of colonies on solid media, or determining a Poisson distribution in liquid culture. Here, we present two reversion reporters for in vivo mutagenesis that produce a quantitative output, and thus have the potential to greatly reduce the amount of test chemical and labor involved in these assays. This output is obtained by coupling a TEM β lactamase-based reversion assay with GFP fluorescence, either by placing the two genes on the same plasmid or by fusing them translationally and interrupting the N-terminus of the chimeric ORF with a stop codon. We also describe a reporter aimed at facilitating the monitoring of continuous mutagenesis in mutator strains. This reporter couples two reversion markers, allowing the temporal separation of mutation events in time, thus providing information about the dynamics of mutagenesis in mutator strains. Here, we describe these reporter systems, provide protocols for use, and demonstrate their key functional features using error-prone Pol I mutagenesis as a source of mutations.

1. INTRODUCTION

Mutagenesis following exposure to chemicals is used to detect genotoxicity, which is an indicator of potential to cause cancer and birth defects (Biran et al., 2010; Krewski et al., 2010). Mutagenesis assays are also used as a readout to study processes of DNA replication, DNA repair, and DNA damage tolerization (Ahluwalia & Schaaper, 2013; Curti, McDonald, Mead, & Woodgate, 2009; Oller, Fijalkowska, Dunn, & Schaaper, 1992; Paul, Million-Weaver, Chattopadhyay, Sokurenko, & Merrikh, 2013).

This chapter focuses on direct detection of mutagenesis in prokaryotes. Genotoxicity can also be monitored in other ways. It can be detected indirectly, through transcriptional fusion of a reporter gene to a promoter that is induced following DNA damage such as *alkA* or *nrdA*, and genes belonging to the SOS response (*umuDC*, *sulA*, *recN*, *recA*) (reviewed in Biran et al., 2010). Genotoxicity can also be detected physically through visualization of DNA damage (breaks or rearrangements) (Solanky & Haydel, 2012). In addition to prokaryotic systems, a variety of eukaryotic organisms, notably yeast, *Drosophila*, and mouse, have also been used as reporters for genotoxicity. These eukaryotic systems have been reviewed extensively elsewhere (Gaivao & Sierra, 2014; Lynch et al., 2011; Nohmi, Suzuki, & Masumura, 2000; Vogel & Nivard, 2003).

Relative to indirect methods of mutagenesis detection, mutagenesis assays have the advantage of being more specific because they detect changes in DNA sequence rather than DNA damage-induced alterations in gene expression. Relative to eukaryotic model systems, bacterial assays are fast and cheap, although they cannot report on nonconserved targets (cytoskeleton, nucleotide excision repair to name two), nor account for bioactivation as accurately. Still, direct mutagenesis assays in bacteria constitute one of three assays mandated for demonstration of safety for compounds in the pipeline for clinical development (the other two being a eukaryotic cell culture one and an animal test) and, because of their relative low cost, are typically the first ones used to explore the safety of a compound (Krewski et al., 2010; Lynch et al., 2011).

Mutagenesis detection in prokaryotes boils down to two approaches: reporter inactivation (forward mutation assay) and reversion of an inactivating mutation (reversion mutation assay). Both are labor intensive, involving visual screening, quantification of colonies on solid media, or obtaining a Poisson distribution for growth in a large number of parallel cultures.

Forward mutation assays are based on the inactivation of a reporter. Reporters can produce colorimetric, luminescent, fluorescent, or electrochemical signals (reviewed in Biran et al., 2010). Inactivation can be the result of a variety of mutations. Thus, compared to reversion assays, forward mutation assays detect events that are more frequent, which allows screening. Forward mutation assays also provide a more accurate representation of the range of genetic changes induced by the relevant mutagen because they are not dependent on specifically predetermined mutations. In some cases, the readout for these assays is a selection, greatly increasing sensitivity. RpoB (RNA polymerase) is an example, as mutations in a variety of loci produce resistance to the antibiotic rifampin (Curti et al., 2009; Severinov, Soushko, Goldfarb, & Nikiforov, 1994). AraD is another example. The cells used in this assay have a mutation in the *araD* gene, which leads to accumulation of a toxic intermediate when arabinose is present. Mutations upstream of *araD* that inactivate the operon prevent the metabolism of arabinose, making cells resistant to arabinose (Whong, Stewart, & Ong, 1981).

Reversion assays detect the return to wild type of an inactivating mutation in a predetermined site, typically through a selection (auxotrophy, antibiotic resistance, FACS sorting, etc.). Availability of selection for reversion assays increases their sensitivity, but their dependence on specific mutations at predetermined sites makes them susceptible to sequence context effects and limits the range of genetic changes that can be detected.

The *Ames Test* was one of the first of these assays to be described and still by far the most widely used prokaryotic testing method, in part because it is mandatory for regulatory compliance. This assay is based on the reversion of a mutation preventing the biosynthesis of histidine, producing colonies on solid agar in the presence of trace amounts of histidine (Mortelmans & Zeiger, 2000). A set of six strains have been developed to detect a broad range of point mutations and frameshifts. Two variations of the *Ames Test* facilitate high-throughput formatting and reduce the amount of sample needed: *Mini-Ames* (which follows the standard Ames Test protocol, except at 1/5 the size) (Flamand, Meunier, Meunier, & Agapakis-Causse, 2001) and the *Ames Fluctuation Test* (which is performed in liquid culture, with growth detected through a chromophore) (Fluckiger-Isler et al., 2004).

Reversion assays based on LacZ (Cupples & Miller, 1989) and TEM β-lactamase have also been described (Schmid, Arndt, & Reifferscheid, 2003; Suzuki, Suzuki, Tashiro, Saito, & Umeno, 2007), and one of the latter includes a set of six point mutations reporting on each type of point mutation that is possible in double-stranded DNA (Suzuki et al., 2007). Reversion, however, produces a binary output, i.e., growth vs no growth. This means that the generation of a single data point requires fine-tuning of the dose and of the dilution to obtain countable colonies (on solid plates), or a number of positive wells that follows a Poisson distribution (in liquid).

A special type of reversion assay is the papillation assay, which is used to detect alterations in mutagenesis rates in vivo (Oller, Fijalkowska, & Schaaper, 1993). This assay is based on a mutation in the gal2K gene, which makes cells unable to ferment galactose. Cells are grown on MacConkey-galactose plates, producing white colonies. Spotting the surface of these colonies, each colored papilla (sectors) represents a microcolony derived from a single Gal+ mutant capable of galactose fermentation. The output is semi-quantitative, though, as it depends on mutation events occurring early enough to allow for visual detection (Schaaper, 1996).

Here, we present two sets of reporters for in vivo mutagenesis that produce a quantitative output, and thus have the potential to greatly reduce the amount of test chemical and labor involved in these assays. This output is obtained by coupling a TEM β-lactamase-based reversion assay with GFP fluorescence, either by placing the two genes on the same plasmid or by fusing them translationally.

As mentioned earlier, mutator strains, i.e., strains consistently exhibiting an elevated mutation frequency, can be identified by their ability to produce sectored colonies (for reviews on mutator strains, see Marinus, 2010; Miller & Michaels, 1996). There are some indications that mutation rates in these strains are not constant, as there is a counterselection against high mutation rates due to the deleterious effects of mutations and possibly because of additional physiological adaptations to the stress caused by accumulation of deleterious mutations. In addition, studying the dynamics of mutagenesis in mutator strains using reporters is difficult because mutations can inactivate the reporter regardless of its forward or reversion status with a probability that grows exponentially with the number of mutations present. Here, we describe a third reporter system aimed at facilitating monitoring mutagenesis in mutator strains. This reporter couples two reversion assays: TEM β-lactamase reversion and GFP reversion. This double set of markers allows the detection of sequential hits, separating mutation events in time and thus facilitating the detection of changes in mutation rates over time.

Here, we describe these reporter systems, provide protocols for use, and demonstrate their key functional features using error-prone Pol I mutagenesis as a source of mutations.

2. DESCRIPTION OF THE REPORTER SYSTEMS

Our plasmid-based reporter systems are diagrammed in Fig. 1. Figure legends describe the annotation for each of these reporters in detail. The sequences in MacVector and FASTA formats are provided as supplemental materials.

We selected TEM-1 β-lactamase, a gene that confers resistance to carbenicillin, as one of the reversion markers. TEM-1 was inactivated through mutations in the S70 position, the serine residue that polarizes the carbonyl group of the β-lactam amide bond in the β-lactam ring of β-lactamase antibiotics (Matagne, Lamotte-Brasseur, & Frere, 1998) and is completely intolerant to amino acid changes (Firnberg, Labonte, Gray, & Ostermeier, 2014). We engineered point mutations at this position to be one nucleotide away from a serine-coding codon so that we can detect each of the six pairs of nucleotide substitutions that are possible in duplex DNA. Table 1 shows each of the six reporter codons, the single nucleotide changes that produce a serine codon and what that "nearest" codon is. The primers used to introduce these point mutations at the S70 position are

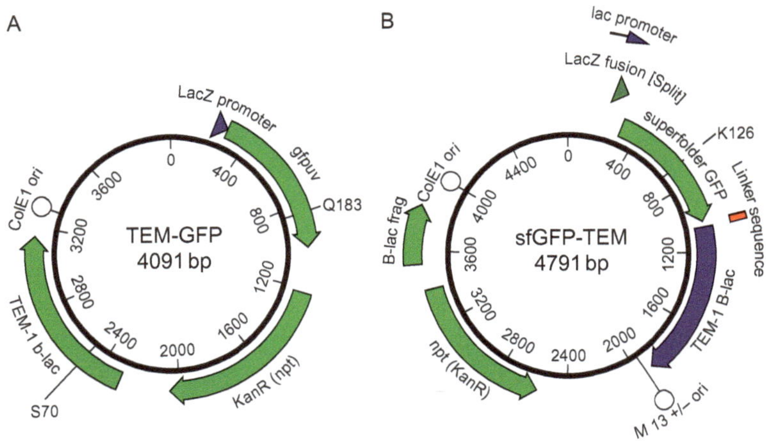

Fig. 1 Reporter constructs. (A) TEMrev-GFP reporter. Main features: Lac promoter: 143–172; LacZ fusion: 217–288; Cycle 3 GFP: 289–1005; Kan. Res.: 1219–2004; β-lactamase: 2291–3151 (2492–2494 S70X reversion reporter); ColE1-like (pMB1) plasmid origin of replication 3299–4091. In the TEMrev-GFPrev variant, the Q183R mutant codon is at positions 835–837. The mutant codon is CGA (R), which requires a G → A transition to revert to CCA (Q). (B) sfGFP-TEM reporter. pMB1 ori 3999–479; sfGFP lac promoter 143–172; lacZ fusion 223–259, sfGDP: 201–1014; 12 aa serine/glycine-rich linker 1015–1050; lactamase 1051–1911; the K126stop codon is at positions 579–81; M13 ori: 1953–2462; Kan resistance (opposite orientation) 3400–2575; lactamase fragment 3553–3851.

Table 1 S70 Reporter Strategy

Original Codon	Original AA	Reporter Codon	Reporter AA	Mutation Needed for Reversion	Reversion Codon
AGC	S	CCA	P	C:G → T:A	TCA
AGC	S	ACA	T	A:T → T:A	TCA
AGC	S	AGA	R	A:T → C/T:G/A	AGC or AGT
AGC	S	TGA	*	G:C → C:G	TCA
AGC	S	AAC	N	A:T → G:C	AGC
AGC	S	CGC	R	C:G → A:T	AGC

*Stop codon.

described in Table S1 in the online version at http://dx.doi.org/10.1016/bs.mie.2017.03.013.

The six reporters are: S70P, which detects C:G → T:A mutations; S70T, which reports A:T → T:A mutations; S70R1, which reports A:T → C:G and A:T → T:A mutations; S70stop, which detects G:C → C:G mutations; S70N, which detects A:T → G:C mutations; and S70R2, which detects C: G → A:T mutations. Note that this approach was previously described by Suzuki et al. (2007), except that two of the six S70 mutations used were different: S70A (GCT) for G:C → T:A mutations and S70stop2 (TAA) for A:T → C:G.

2.1 TEMrev-GFP

We coupled a GFP fluorescent marker to the S70X TEM β-lactamase reporter set described above. Cycle 3 GFP, a variant of GFP optimized for fluorescence in *Escherichia coli* (Crameri, Whitehorn, Tate, & Stemmer, 1996), is placed on the same plasmid downstream of TEM so that it is cotranscribed with it (TEMrev-GFP series, Fig. 1A). This way, growth in the presence of carbenicillin can be detected through fluorescence, a signal with much higher sensitivity and wider dynamic range than turbidity. GFP does not require cell lysis, thus facilitating monitoring growth over time.

2.2 sfGFP-TEM

Another way in which we coupled reversion of TEM to GFP was through a translational fusion, placing GFP in the same open reading frame as β-lactamase, with a 12 serine and glycine-rich linker in between (sfGFP-TEM, Fig. 1B).

To obtain a functional translational fusion, we had to use superfolder GFP (sfGFP), an evolved form of cycle 3 GFP with two additional mutations selected for robustness to translational fusions (Pedelacq, Cabantous, Tran, Terwilliger, & Waldo, 2006). In addition, we found that sfGFP is only functional if located at the N-terminus, not at the C-terminus. Even this fusion is not stable in all strains: while it emits a high level of fluorescence in Top10 cells, we found that it produced inconsistent fluorescence JS200 cells and no fluorescence in AB1157 and its GW7101 (*ada alkB::CAT*) derivative.

The introduction of a stop codon truncating GFP expression inactivates both GPF and β-lactamase, resulting in a nonfluorescent, carbenicillin-sensitive phenotype. As a proof of principle, we introduced a TGA stop codon at three different positions (Q69, K113, and K126). As expected, the reversion of the stop codon at these positions resulted in the acquisition of both fluorescence and carbenicillin resistance (stop codon reversions for sfGFP-TEM K126stop shown in Fig. 2).

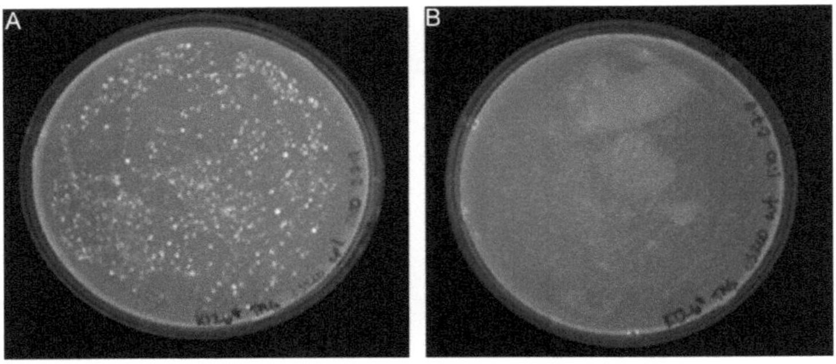

Fig. 2 sfGFP-TEM K126stop reporter on solid plates. Following Pol I mutagenesis, JS200 cells were plated on LB carbenicillin, and exposed to UV light in order to reveal GFP fluorescence. (A) Cells expressing LF-Pol I. (B) Cells expressing WT-Pol I (control).

2.3 TEMrev-GFPrev

The goal of this reporter is to facilitate monitoring mutagenesis in mutator strains in a more quantitative manner than papillation assays. TEMrev-GFPrev is identical to the TEMrev-GFP reporter described earlier, but it carries a mutant GFP inactivated through a point mutation replacing Q183 with an R (CAA to CGA), which reverts with C:G → T:A mutations (Fig. 1A). The method is diagrammed in Fig. 3. Colonies are first grown on carbenicillin to identify S70 reversion events. A few nonfluorescent carbenicillin-resistant colonies are picked, and grown in liquid culture. The plasmid DNA from these cultures is recovered and retransformed into a readout strain (Top10 or DH5α) to identify R183Q (fluorescent) revertants (Fig. 3). Under these conditions, fluorescent colonies have to be the result of a mutation event that occurred after carbenicillin reversion, unless the reversion was already present in one of the copies of the plasmid pool when the first mutation occurred. This alternate explanation can be ruled out if the frequency of reversion is lower than one divided by the copy number of the reporter plasmid.

3. EXPERIMENTAL VALIDATION

To validate our reporter system, we used error-prone Pol I plasmid replication, as previously described in detail by our group (Alexander et al., 2014; Troll, Alexander, Allen, Marquette, & Camps, 2011). This system is based on the expression of an error-prone variant of DNA polymerase I (low-fidelity Pol I or LF-Pol I) in JS200, a *polA12* (temperature-sensitive) strain of *E. coli* (Camps & Loeb, 2005). LF-Pol I bears three mutations synergistically decreasing its replication fidelity: I709N in motif A (broadening

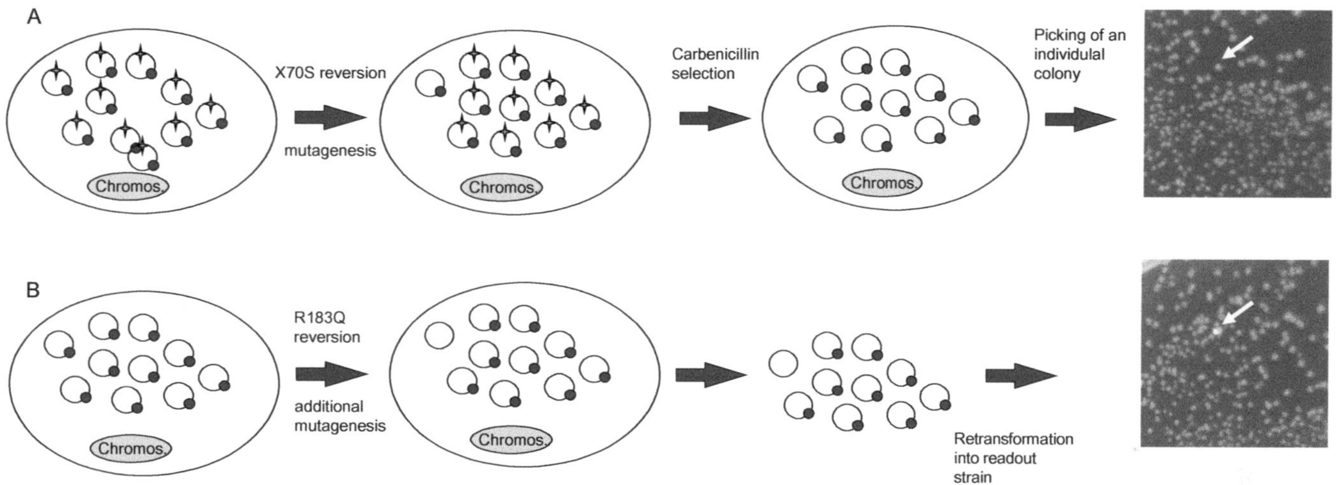

Fig. 3 Method for detection of two mutations separated in time. Colonies representing reversions in TEM1 β-lactamase are expanded under restrictive (mutagenic) conditions, their plasmid pools recovered through miniprep and retransformed. The two reversion reporters are shown as *star* (S70X) and *circle* (Q183R). (A) Detection of first mutation: once reversion at the S70X site occurs, under selective pressure, the reversion events get amplified, representing a majority of the plasmid population, and leading to carbenicillin resistance. (B) Detection of second mutation: single, carbenicillin-resistant colonies (*white arrow*) are grown. Plasmids from these cultures are recovered. Reversions at the Q183 site of the GFP reporter are detected by retransformation of recovered plasmids into a readout strain, producing fluorescent colonies on a background on nonfluorescent ones (*white arrow*). Reversion cannot have already been present in one of the copies of the plasmid pool when the first mutation occurred if the frequency of reversion is lower than one divided by the copy number of the reporter plasmid (in this example, one in 10).

its active site), A759R in motif B (favoring its closed conformation), and D424A (inactivating its proofreading domain) (Camps, Naukkarinen, Johnson, & Loeb, 2003). Shift of this strain to 37°C makes JS200 cells dependent on the activity of LF-Pol I for survival. Pol I performs ColE1 plasmid replication (Troll et al., 2014) and processes of Okazaki fragments during lagging-strand replication in both plasmid and chromosomal DNA (Camps & Loeb, 2005; Troll et al., 2014).

Overnight culture under restrictive conditions (37°C) leads to an increased mutation frequency by over three orders of magnitude in ColE1 plasmids, about 1 nucleotide substitution per 1.5 kb. This is true for most of the plasmid sequence, where Pol I appears to be competing with Pol III (Troll et al., 2014). These loads are higher in areas replicated exclusively by Pol I: the 150 nucleotides immediately downstream of the RNA/DNA switch (leading-strand synthesis by Pol I), ~500 nucleotides upstream of the RNA/DNA switch (gap-filling of lagging-strand synthesis by Pol I), and ~20 nt patches corresponding to areas of Okazaki fragment processing by Pol I (Allen et al., 2011; Troll et al., 2011). It is worth noting that LF-Pol I is partially dominant in vivo, as expression of this polymerase still produces Col E1 plasmid mutagenesis at permissive temperature in *polA12* cells or at 37°C in *polA* WT strains, albeit with a three- to fivefold lower frequency relative to *polA12* cells at restrictive temperature (Alexander et al., 2014).

In terms of mutation spectrum, we were able to estimate the mutation frequency of LF-Pol I on a single strand in vivo (Troll et al., 2014). The vast majority of mutations (>95%) are point mutations and can be grouped in four groups: most frequent: C → T transitions (60%); frequent: A → G and A → T (20% and 10% of the total, respectively); rare: G → T, G → A, and G → T; and extremely rare: T → C, T → A, A → C, and C → G. Note that, based on the very low frequency of T → C transitions observed in vivo relative to the frequency of these transitions reported for proofreading-deficient Pol I in vitro, mismatch repair appears to be intact in these cells (Troll et al., 2014). Given that our reporter detects mutations in double-stranded DNA, i.e., in pairs of complementary mutations, we expect the following ranking based on the frequency: C:G → T:A > A:T → G:C ~ A:T → T:A > G:C → T:A > T:A → G:C > G:C → C:G.

3.1 Mutagenesis

We transformed our six TEMrev-GFP reporters, a TEM-GFP positive control, and a negative control not bearing the TEM1 gene (supplemental file and Fig. S1 in the online version at http://dx.doi.org/10.1016/bs.mie.2017.03.013) into JS200 *E. coli* cells expressing LF-Pol I (Camps et al., 2003). As an additional control, we also transformed these reporter and control plasmids into a JS200 cells expressing WT-Pol I. After recovery at 30°C, cells were plated onto LB agar plates prewarmed to 37°C containing kanamycin, thus switching our transformants to restrictive conditions. Mutagenesis occurred during growth overnight at 37°C.

3.2 Readout on Solid Media

Transformants produced a high density of colonies (near-lawn). These colonies were harvested from the plate into ~1.5 mL of LB broth. Absorbance at 600 nm was determined to normalize the washes to $OD_{600} = 1$. These normalized stocks were used to plate kanamycin (at further dilution of $1{:}10^7$) and carbenicillin plates at different dilutions, depending on reversion frequencies (between neat and $1{:}10^3$ dilutions). This time, plates were incubated overnight at 30°C (i.e., under permissive conditions) to minimize additional mutagenesis. Following incubation, the number of colonies on each plate was counted, and this number was used to calculate the reversion rate for each reporter (Fig. 4). Interestingly, fluorescence was not uniform across all carbenicillin-resistant colonies, possibly due to the presence of additional mutations affecting GFP expression and/or function (not shown).

For the sfGFP-TEM K126stop reporter, plating a transformation of JS200 cells expressing LF-Pol I at the restrictive temperature produced a semilawn of carbenicillin-resistant, fluorescent colonies (Fig. 2A). Sequencing of 10 these colonies showed point mutations at the stop codon in all cases, producing an L (three times), W (three times), Q (twice), and Y (twice). In eight of these cases, the WT signal was still detectable, suggesting that the plasmid carrying the K126 point mutations had not replaced all the copies of the original K126 stop reporter. Cells expressing WT-Pol I had practically no colonies (Fig. 2B).

3.3 Readout in Liquid Media

Following mutagenesis, plate washes were normalized to $OD_{600} = 1$ and used to 96-well plates in a 1:20 dilution. The plates were deep-well round-bottom plates with glass beads (to facilitate oxygenation) and a final volume of 1 mL (see Section 6). The plates were then covered with AirPore breathable sheets, in order to protect against cross-contamination and evaporation effects, while still allowing for microbe growth under

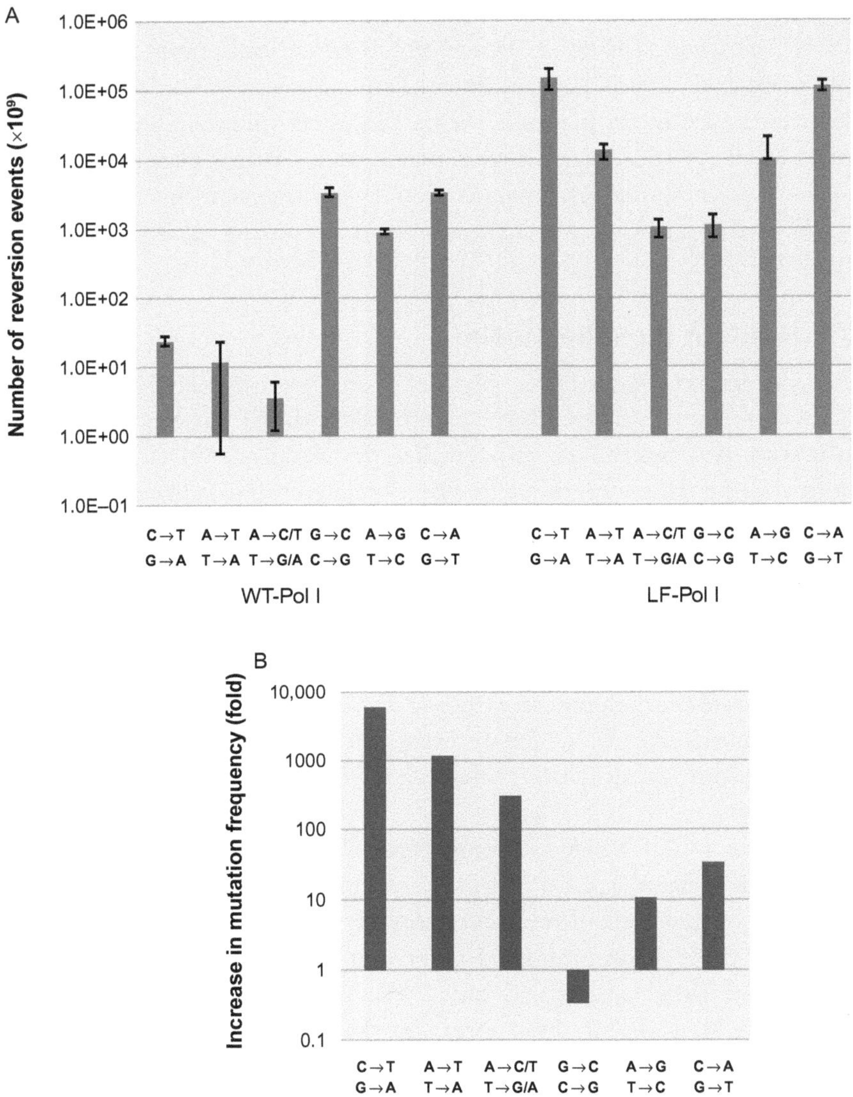

Fig. 4 LF-Pol I reversion profile. The set of six TEMrev-GFP reporters was transformed in JS200 cells expressing LF-Pol I, and error-prone reporter plasmid replication was performed as described in Section 6. As a control, the same reporters were transformed into JS200 cells expressing WT-Pol I. (A) Original reversion frequencies, in log scale. *Error bars* represent standard deviation of triplicates. Reversion frequencies for LF-Pol I samples were normalized for viability (these cells exhibited a 50% decrease in viability) and for plasmid copy number (at least 10-fold lower than WT) (Camps et al., 2003). (B) Reversion frequency relative to control cells expressing WT polymerase I (fold, in log scale).

aerobic conditions, and grown at 30°C shaking at 325 rpm. At different time-points 200 μL of each culture were transferred to a set of black-walled flat-bottomed 96-well microtiter plates and kept at 4°C. At the end of the experiment, we determined growth by reading absorbance at 600 nm. Fluorescence readings were also obtained on a fluorescence-enabled spectrophotometric plate reader, with excitement $\lambda = 395$ nm and emission $\lambda = 509$ nm. Results were then used to plot growth kinetics curves for each construct under carbenicillin selection. Fig. 5 shows the growth and fluorescence emission kinetics for two of our reporters, S70P, which detects C:G → T:A mutations (Fig. 5A and C), and S70R1, which detects A:T → C:G and A:T → T:A mutations (Fig. 5B and D).

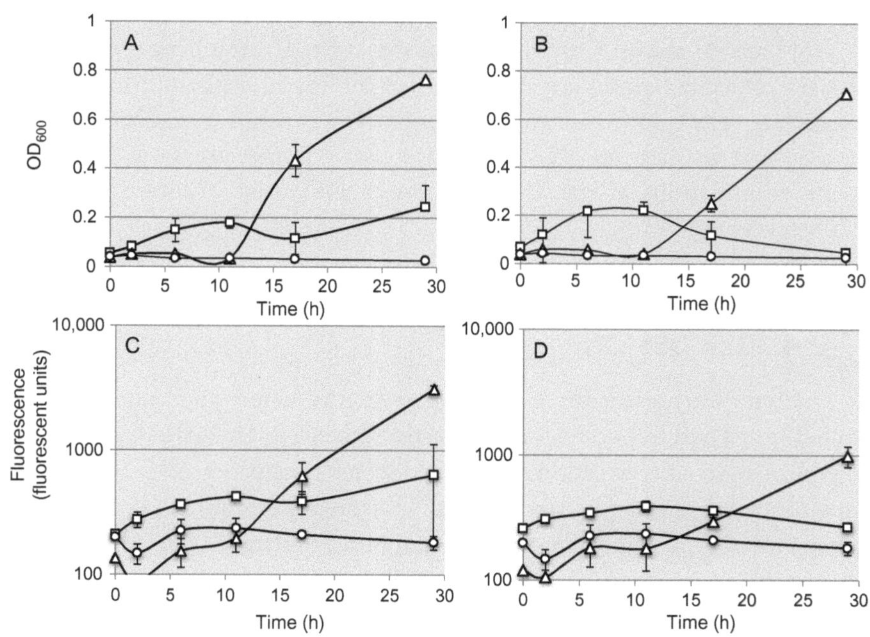

Fig. 5 Mutagenesis assay in 96-well format. Cells bearing two sample reporters, S70P (which detects C:G → T:A transitions) or S70R1 (which detects A:T → C:G and A:T → T:A mutations) underwent error-prone plasmid replication as described in Section 6, recovered by washing the plates, and inoculated into 96 deep-well plates to a final OD_{600} of 0.05. At different time-points (shown in the *x*-axis), samples were drawn and kept at 4°C. After completion of the time-course, fluorescence and optical density (OD_{600}) were measured. LF-Pol I mutagenesis, *triangles*; WT-Pol I control, *squares*; negative control with no β-lactamase gene, *circles*. (A) S70P reporter, OD_{600}. (B) S70R1 reporter, OD_{600}. (C) S70P reporter, fluorescence, in log scale. (D) S70R1 reporter, fluorescence, in log scale. *Error bars* represent standard deviation between duplicates.

3.4 Continuous Mutagenesis Detection

Colonies expressing LF-Pol I and bearing the TEMrev-GFPrev S70P reporter were plated under restrictive conditions as described in Section 3.2 but at a higher dilution in order to obtain individual carbenicillin-resistant colonies. Three nonfluorescent, carbenicillin-resistant colonies were picked, and grown in liquid culture under restrictive conditions. The DNA from these cultures was recovered and retransformed into DH5α cells to identify R183Q (fluorescent) revertants (Fig. 3B diagram). The results of these experiments are listed in Table 2. Using the GFP reporter to quantify "second hit" mutagenesis, we found a mutation frequency of 137 in 10^6 cells, in line with frequencies seen by carbenicillin reversion (Fig. 6). Three colonies were also grown under permissive conditions and retransformed into reporter cells. Here, we find a frequency that is 3.7-fold lower, about 37 fluorescent colonies per 10^6 transformants. Given that the average plasmid copy number for the reporter plasmid in LF-Pol-expressing cells is less than 10 plasmids per cell (Camps et al., 2003), these results confirm that the observed fluorescent colonies are most likely the result of mutations at the 183 position of GFP that occurred after the P70S reversion.

4. DISCUSSION

Here, we present three reporter systems to detect and quantify point mutations. These reporters are on a plasmid bearing a pMB1 (ColE1-like) plasmid origin of replication. This has several advantages over a chromosomal location: (1) a plasmid reporter increases the number of targets for mutagenesis by at least one order of magnitude, since ColE1 plasmids are multicopy plasmids (Cesareni, Helmer-Citterich, & Castagnoli, 1991; Million-Weaver & Camps, 2014); (2) the fact that plasmids are present in multiple copies also allows amplification of reporter signal through selection; and (3) a plasmid reporter facilitates exposure to mutagens ex vivo; in this scenario, transformation would be performed only to obtain a readout. It needs to be kept in mind, though, that most mutagens require biotransformation (Guengerich, 2000), restricting the applications of such an ex vivo assay.

The first reporter system, TEMrev-GFP, is based on the reversion of an inactivating mutation in TEM β-lactamase and cotranscription of GFP. A similar approach was previously reported by Schmid et al., although in

Table 2 GFP Reversion Frequencies

Pol I	Temperature (°C)	Total Colonies Screened ($\times 10^4$)	Total Number of Transformations	Average Transformation Efficiency	Number of Fluorescent Colonies	Frequency of GFP Reversion ($\times 10^6$)	S70P Reversion Frequency ($\times 10^6$)
WT	30	60.4	6	10.1	0	<1.7	N/A
WT	37	18.1	6	3.0	3	16.5	0.004
LF	30	13.4	8	1.6	5	37.2	N/A
LF	37	5.1	8	0.6	7	137.0	180.9

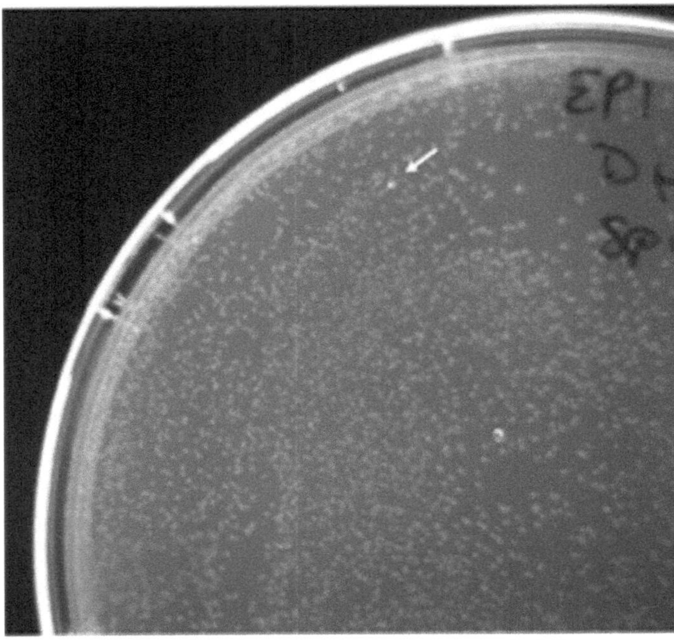

Fig. 6 Q183R reporter: R183Q reversion on a plate containing 9400 colonies is shown. Transformed plasmids were recovered following expansion in liquid culture of a non-fluorescent carbenicillin-resistant colony grown under restrictive (37°C) conditions (diagrammed in Fig. 3).

their case reversion of TEM β-lactamase was coupled to β-galactosidase (Schmid et al., 2003).

One limitation of reversion-based reporter systems is that they look at a limited number of sequence changes in a predetermined site. To be able to characterize the complete spectrum of point mutations, we generated a panel of mutations in the S70 of β-lactamase position reporting for all possible point mutations. A similar approach that has been previously described also for TEM β-lactamase (Suzuki et al., 2007) and for LacZ (Cupples & Miller, 1989). The spectrum results obtained through these sets of reversion reporters is still subject to sequence context and strand bias effects, though.

The point mutation spectrum profile of LF-Pol I that we describe here (Fig. 4) is consistent with the one previously reported by Suzuki et al. (2007). Compared with the profile that our group generated previously based on sequencing (Troll et al., 2014), only one pair exhibits a lower reversion frequency than expected: A:T → G:C. This could be the result of sequence context-dependent effects (Lee, Popodi, Tang, & Foster, 2012; Rogozin & Pavlov, 2003), which can be in part due to differential efficiency of mismatch repair (Ahluwalia & Schaaper, 2013). Our observation that S70R1, which detects A:T → C:G and A:T → T:A mutations, produces fewer reversions than S70T, which detects A:T → T:A alone directly confirms the impact of local sequence context on mutation rates. Overall, then, profiling mutation spectrum using our TEMrev-GFP mutation set gives a general idea of which type of point mutations are favored, particularly if there is a strong bias for a specific type, but the spectrum obtained is not as reliable as that obtained by sequencing a forward mutagenesis reporter or genomic sequence because it does not sample a variety of sequence contexts.

When the frequency of mutagenesis is expressed as the ratio of LF-Pol I vs WT-Pol I, one of the values is below 1 (G:C → C:G, Fig. 4B). This could be due to a conservative estimate of difference in plasmid copy number between WT-expressing and LF-Pol I-expressing cells (10-fold) or to a different mutagenic mechanism operating in WT-Pol I expressing cells, as suggested by the high mutation frequency seen in these cells, which is between 10- and 100-fold above that of other reporters. In any case, LF-Pol I appears to induce a very low frequency of G:C → C:G, consistent with our previous estimate of LF-Pol I mutation spectrum based on extensive sequencing (Troll et al., 2014).

It is also worth noting that two of the six reporters in the set produce between 10- and 100-fold higher background mutation frequencies in the control strain expressing WT-Pol I relative to the other three (Fig. 4A). The two reporters are S70stop and S70T, which report G:C → C:G and A:T → T:A, respectively. We ignore the reason for this increased background in mutation frequency for two transversions in our reporter, as C:G → T:A transitions predominate instead in the spontaneous mutation spectrum of *E. coli* (Lee et al., 2012) and Pol I produces predominantly transitions as well (Curti et al., 2009). The replication fidelity of Pol I can be modulated by pols II and IV, though, which could result in an increased number of transversions (Curti et al., 2009).

GFP facilitates quantification of growth in vivo, producing a much stronger signal than turbidity. Note that in Fig. 5, the results of fluorescence (panels C and D) are shown in a logarithmic scale and that the ratio of background to signal is at least two orders of magnitude higher. Further, for a given time-point, the result of GFP fluorescence is quantitative. Our reversion assays report that LF-Pol I produces more C:G → T:A mutations relative to A:T → C:G and A:T → T:A mutations. In Fig. 5, the fluorescent signal is much stronger at 30 h time-point for the C:G → T:A reporter, so a measurement at this time-point would have been proportional to mutation frequency determined by plating. Relative to other quantitative reporters, GFP has several advantages that make it ideal for continuous measurement in culture: (1) it is highly stable; (2) no addition of an external substrate is necessary; (3) no cell lysis is required; and (4) it is less susceptible to substrate interference. However, in genotoxicity sensors based on fusions with DNA-damage-inducible promoters, GFP has been found to have low sensitivity compared to enzyme-based reporters, even when GFP variants producing enhanced signal through mutagenesis were used (Hakkila, Maksimow, Karp, & Virta, 2002). Here, we rely on two levels of signal amplification: one is plasmid, which is present in multiple copies, increasing GFP expression; the other one is growth: TEM β-lactamase gives revertants a dramatic growth advantage, increasing the number of cells present in liquid culture at a given time in a mutator relative to a control (see Fig. 5A and B).

Our sfGFP-TEM K126stop reporter is based on generating a translational fusion between GFP and TEM β-lactamase and interrupting translation through a stop codon or a frameshift. Reversion of this codon results in both resistance to carbenicillin and fluorescence (Fig. 2A). A translational fusion represents a new approach for quantitative detection of reversions that should further increase the signal-to-noise ratio for GFP fluorescence, as cells without a reversion are not even fluorescent. Further, this cassette can be placed in any desired location and orientation. However, this fusion protein appears not to be very stable and only works in some strains of *E. coli*.

Finally, our TEMrev-GFPrev reporter represents an alternative to papillation assays for the characterization of mutator strains. The main advantage is that the output in this case is quantitative rather than semiquantitative, allowing head-to-head comparisons between different mutators and growth conditions. Different inactivating GFP mutations can be introduced, depending on the mutagenic profile of the mutator strain. The chromophore-containing cyclized hexapeptide (residues 64–69) is a good target for inactivating mutations, with narrow tolerance to alternative amino acids (Tse et al., 2016). Table S2 in the online version at http://dx.doi.org/10.1016/bs.mie.2017.03.013 shows a list of 12 mutations inactivating cycle 3 GFP fluorescence in *E. coli* outside the active site hexapeptide that can also be explored as possible reversion sites.

Position 183 is located within a 9-Å shell around the chromophore without directly contacting it. A C:G → T:A mutation, which is by far the predominant nucleotide substitution introduced by LF-Pol I, reverts 183R back to Q. We confirmed that reversion to Q is the predominant mutation in fluorescent colonies: nine out of nine fluorescent colonies sequenced produced this reversion. We do not know if other amino acid substitutions are allowed at this site, but C and A substitutions are not tolerated either because of a dramatic destabilizing effect on GFP (Jain & Ranganathan, 2004).

A decreased plasmid copy number for LF-Pol I cells compared to WT-expressing cells, previously reported in Camps et al. (2003) and factored in Fig. 4, is the likely cause for the decreased transformation efficiency of DNA recovered from these cells (Table 2).

The mutation frequency for "second hit" mutagenesis that we obtained using our GFP reversion reporter is slightly lower relative to that of the first round of mutagenesis (137, compared to 181 in 10^6 cells, Table 2) although given that two different reporters were used to measure "first hit" and "second hit" mutagenesis and that the readout for carbenicillin resistance involved two rounds of growth on solid plate (at 37°C and then at 30°C), whereas GFP reversion only involved one round, it is hard to draw any conclusions regarding the frequency of mutagenesis of the second hit vs the first hit other that the two results are consistent between each other. We also found a significant amount of revertants in our WT-expressing control grown at 37°C (16 in 10^6 cells) but not at 30°C (Table 2) and do not have an explanation at the moment for this observation.

For DNA from colonies grown under permissive conditions, we find a frequency that is 3.7-fold lower relative to DNA from colonies grown under restrictive conditions, about 37 fluorescent colonies per 10^6 transformants. This result aligns well with a previous report of partial dominance for LF-Pol I (Alexander et al., 2014). Given that the average plasmid copy number for the reporter plasmid in LF-Pol-expressing cells is less than 10 plasmids per cell (Camps et al., 2003), the very low rate of reversion that we observed confirms that the fluorescent colonies that we see are most likely the result of mutations at the 183 position of GFP that occurred after the P70S reversion.

This means that LF-Pol I-expressing cells continue to generate mutations after one passage in culture. Our TEMrev-GFPrev reporter system should be of use to fine-tune LF-Pol I-expressing cells and other existing mutator strains such as XL-1 red (Muteeb & Sen, 2010), the MP6 mutagenesis system (Badran & Liu, 2015), or strains with altered dNTP pools (Tse et al., 2016) to identify conditions supporting constant mutation rates over time.

5. MATERIALS

5.1 Transformation and Mutagenesis

1. Competent cells
   a. Top10
   b. JS200-pHSG_WT
   c. JS200-pHSG_LF
2. ColE1 vectors
   a. TEMrev-GFP
   b. TEMrev-GFPrev
   c. sfGFP-TEM K126stop
   d. pGFPuv_KanR and TEM-GFP as positive and negative controls
3. 500 mL centrifuge bottles
4. Eppendorf centrifuge 5810 R (Eppendorf)
5. 50 mL conical tubes (Fisher Scientific, Cat.# 1443222)
6. 15 mL culture tubes (E&K Scientific, Cat.# EK-62262)
7. LB broth (Fisher Scientific, Cat.# BP1426-2)
8. LB agar (Fisher Scientific, Cat.# BP1425-2)
9. 100 mm × 15 mm disposable Petri dishes (Fisher Scientific, Cat.# FB0875713)
10. Kanamycin solution, 30 mg/mL, store at −20°C
11. Kanamycin (30 μg/mL) LB agar and broth
12. 1.5 mL microfuge tubes (E&K Scientific, Cat.# 280150)
13. TropiCooler, Model 260014 (Boekel Scientific)
14. MaxQ 4000 shaker/incubator (Barnstead International)
15. Water-jacketed incubator (Forma Scientific)

5.2 Washing Plates

1. Kanamycin (30 μg/mL) LB broth
2. Plate spinner
3. Plate spreader
4. Ethanol (200 proof)
5. Bunsen burner
6. Spectrophotometer cuvettes (Fisher Scientific, Cat.# 14955127)
7. 1.5 mL microfuge tubes (E&K Scientific, Cat.# 280150)
8. BioMate 3 Spectrophotometer (Thermo Scientific)

5.3 Readout (Plates)

1. Kanamycin (30 μg/mL) LB agar and broth
2. Carbenicillin (100 μg/mL) LB agar and broth
3. 1.5 mL microfuge tubes (E&K Scientific, Cat.# 280150)
4. Plate spreader
5. Plate spinner
6. Ethanol (200 proof)
7. Bunsen burner
8. Water-jacketed incubator (Forma Scientific)
9. UV light

5.4 Readout (Liquid Culture Assay)

1. Kanamycin (30 μg/mL) LB broth
2. Carbenicillin (100 μg/mL) LB broth
3. 3 mm diameter glass beads (Sigma-Aldrich, Cat.# Z265926)
4. AirPore tape sheets (Qiagen, Cat.# 2017-10-RP)
5. 96-well round-bottomed deep-well plates (Fisher Scientific, Cat.# 10011-944)
6. 96-well flat-bottomed black-walled plates (Fisher Scientific, Cat.# 82050-744)
7. Microtiter plate lids (Fisher Scientific, Cat.# 82050-829)
8. MaxQ 4000 shaker/incubator (Barnstead International)
9. SpectraMax M2e Fluorometric and Spectrophotometric plate reader. Dual monochromators, absorbance 200–1000 nm and excitation 250–850 nm (Molecular Devices)

5.5 Plasmid Recovery

1. 15 mL culture tubes (E&K Scientific, Cat.# EK-62262)
2. 1.5 mL microfuge tubes (E&K Scientific, Cat.# 280150)
3. NucleoSpin® Plasmid (NoLid) kit (Macherey-Nagel, Cat.# 740499.250)

5.6 Sequencing Plasmids of Interest

1. NanoDrop ND-1000 Spectrophotometer for DNA quantification (Thermo Scientific)
2. 0.6 mL microfuge tubes (E&K Scientific, Cat.# 280060-S)
3. MacVector version 12.7.5 for sequence analysis (MacVector Inc.)

6. METHODS

6.1 Transformation and Mutagenesis

6.1.1 *Preparation of Competent Cells for Chemical Transformation*

1. Prepare a 5-mL overnight culture in a 15-mL culture tube in LB media for the strain of interest. If necessary, include selective antibiotic in the media for the desired strain.
2. Expand this culture into a sterile 1-L Erlenmeyer flask containing 500 mL of LB media with selective antibiotic.
3. Incubate this flask at 30°C or 37°C (depending on the cell line; incubate JS200 cell lines at 30°C, all others at 37°C) with shaking (225 rpm) until exponential phase is reached ($OD_{600} = 0.4$–0.6).
4. Chill the flask containing the cells on ice for 20 min.[a]
5. Transfer the liquid cultures to 500 mL plastic centrifuge bottles and centrifuge at 4000 rpm for 20 min at 4°C.
6. Pour off supernatant, and resuspend the cell pellet in 50 mL of chilled calcium chloride solution (100 m*M* $CaCl_2$, 10 m*M* HEPES, 15% Glycerol, pH 7).
7. Transfer the resuspended cells into a 50-mL conical tube.
8. Centrifuge the cells at 4000 rpm for 20 min at 4°C.
9. Pour off supernatant, and resuspend the cell pellet in 50 mL of chilled calcium chloride solution, and centrifuge the cells at 4000 rpm for 20 min at 4°C.
10. Repeat step 9.
11. After third wash with calcium chloride solution, pour off supernatant, and resuspend the cells in 5 mL of chilled calcium chloride solution.
12. Keep cells on wet ice and use immediately, or aliquot into 1.5 mL microfuge tubes and place on dry ice for storage at −80°C.

6.1.2 *Chemical Transformation*

1. Pipette 40 μL of chemically competent cells into 1.5 mL microfuge tube per transformation.
2. Pipette 100 ng of plasmid DNA into the tube containing the competent cells and mix well by pipetting up and down.
3. Incubate on ice for 30 min.

[a] From this point forward, cells must be kept chilled at all times for best results.

4. Heat-shock the cells at 42°C for 90 s on TropiCooler block.
5. Place cells back on ice for 5 min.
6. Add 1 mL of LB broth to the microfuge tube containing the cells and DNA.
7. Allow the cells to recover for 30 min–1 h at 30°C or 37°C[b] with shaking (225 rpm).
8. Plate transformed cells by spreading 100 μL with sterile plate spreader onto prewarmed LB agar plates containing 30 μg/mL kanamycin.
9. Allow the cells to grow overnight at either 30°C or 37°C.

6.1.3 Mutagenesis

1. For ColE1-like ori plasmids which have been transformed into JS200-LF-Pol I expressing strains, incubate at 37°C to induce mutagenesis.[c]
2. Use the same plasmids transformed into JS200-pHSG_WT as control.

6.2 Washing Plates

1. Observe plate for "near-lawn," high density of colonies.
2. Place plate on a plate spinner.
3. Add 1 mL LB broth directly to the plate surface containing bacterial growth.
4. Use sterile plate spreader to collect colonies into LB broth.
5. Tilt plate slightly of collect broth containing harvested colonies into one area, and transfer as much as possible into 1.5 mL microfuge tube.
6. Repeat steps 3–5. Collect second wash into the same 1.5 mL microfuge tube.[d]
7. Dilute plate washes 1:20 directly in spectrophotometer cuvettes (950 μL media +50 μL plate wash), and mix by pipetting.
8. Measure OD_{600} of diluted plate wash using the BioMate 3 spectrophotometer, and multiply the measurement by 20 to obtain the actual OD_{600} of the undiluted plate wash.
9. Normalize all plate washes to $OD_{600} = 1$ prior to readout experiments.

---

[b] JS200 transformants need to be grown at 30°C if permissive conditions are desired; for mutagenesis or in the case of nontemperature-sensitive strains, cells need to be grown at 37°C.
[c] Prewarming plates to 37°C prior to plating cells is essential for efficient mutagenesis.
[d] Typical recovery per 2 mL of LB is ~1.5 mL of plate wash.

6.3 Readout (Plates)

1. Prewarm LB agar plates containing 30 μg/mL kanamycin and LB agar plates containing 100 μg/mL carbenicillin in incubator. For JS200 transformants, 30°C (permissive temperature) should be used.
2. Plate 100 μL of plate washes (all washes plated to both LB agar plates containing 30 μg/mL kanamycin and LB agar plates containing 100 μg/mL carbenicillin) at appropriate dilutions to yield countable colonies.[e]
3. Incubate overnight. For JS200 transformants, 30°C (permissive temperature) should be used.
4. Determine the number of colonies on each plate.
5. Use counts to determine CFU/mL of OD normalized cultures on each type of selective media.
6. Determine fraction of TEM β-lactamase S70 revertants in 10^9 cells by the formula:

$$\%\ \text{Reversion} = (\text{CFU/mL(Carb)}/\text{CFU/mL(Kan)}) * 10^9$$

6.4 Readout (Liquid Culture Assay)

1. Aseptically place one sterile 3 mm diameter glass bead into each well of a 96-well deep-well round-bottomed plate using sterilized forceps.[f]
2. Transfer 950 μL of LB broth containing 30 μg/mL kanamycin to one well and 950 μL of LB broth containing 100 μg/mL carbenicillin to another well for each construct to be tested at each time-point.[g]
3. Inoculate wells with 50 μL of 1:10 diluted plate washes (final inoculation $OD_{600} = 0.05$), leaving one row of noninoculated media as negative control.[h]
4. Cover with AirPore tape sheet.

[e] Experimenter must estimate dilution needed to achieve a total number of colonies on plate between 50 and 500. This may require some trial and error. Our results indicate that a dilution factor of 10^7 is effective for all constructs and controls on kanamycin plates and positive controls (WT β-lactamase) on carbenicillin plates and no dilution for negative controls on carbenicillin plates. For reporter constructs on carbenicillin plates, dilutions may vary depending on the expected reversion frequency, but generally range between a 1:10 dilution and a dilution factor of 10^3. Be sure to take note of the dilution factor used for each construct plated, as this will be used to calculate CFU/mL (see Section 6.3, step 6).

[f] Sterilize forceps by dipping in ethanol and holding over flame until red hot.

[g] It is recommended to have triplicates at each time point.

[h] Inoculate different time points and different constructs and controls into separate wells. Inoculate each culture into both kanamycin wells and carbenicillin wells. Leave at least three wells on each plate uninoculated, to be used as blanks during spectrophotometry/fluorimetry.

5. Remove 200 μL of T0 time-point to 96-well black-walled clear-bottomed flat-bottomed plates, cover with sterile plate lid, and place at 4°C.
6. Cover deep-welled plate with sterile plate lid and place in incubator at 30°C or 37°C[b] with shaking (325 rpm).[i]
7. At appropriate time-points, remove 200 μL of culture from preassigned well to 96-well black-walled plates and return culture to shaking incubator.[j]
8. Between time-points, the black-walled plates should be stored at 4°C, and the deep-welled plates should be incubated at the appropriate temperature with shaking.
9. At the last time-point, remove culture and uninoculated blank wells to black-walled plates.
10. Read OD_{600} and fluorescence (ex. 395 nm, em. 509 nm) from black-walled plates on SpectraMax M2e Fluorometric and Spectrophotometric plate reader.
11. Plot OD_{600} vs time and fluorescence vs time for constructs under both kanamycin selection and carbenicillin selection to estimate relative rates of TEM β-lactamase S70 reversion.

6.5 Plasmid Recovery

1. Pick reversion colonies from LB agar plates containing 100 μg/mL carbenicillin generated in Section 6.3 and inoculate into 3 mL of LB broth containing 100 μg/mL carbenicillin.
2. Grow cultures overnight under restrictive conditions (37°C) with shaking (225 rpm).
3. Harvest cells by centrifugation at 11,000 × *g* for 1 min, pour off supernatant, and isolate plasmid DNA (miniprep) based on manufacturer's instructions.

6.6 Sequencing Plasmids of Interest

1. Quantify plasmid DNA yield and purity using NanoDrop Spectrophotometer
    a. Open NanoDrop software (ND-1000, version 3.8.1) and select nucleic acid quantification.

[i] For cell lines growing at 30 °C, culture growth will be slower. Plan time points accordingly.

[j] Take 200 μL from fresh wells at each time point. Do not resample wells that have already been sampled at previous time points because the stirring of the culture that occurs when a sample is taken alters growth on a microtiter plate. Sample well by stabbing the micropipette tip through the AirPore sheet. Use caution to avoid disturbing/cross-contaminating wells containing later time points or blanks.

b. Place 2 μL of purified water on cleaned pedestal and lower arm to initialize spectrophotometer.
c. Place 2 μL of elution buffer on cleaned pedestal and lower arm to blank spectrophotometer.
d. Place 2 μL of sample to be quantified on cleaned pedestal and lower arm to measure the absorption spectrum between 220 and 350 nm.

2. Transfer 0.5–1 μg of plasmid DNA to 0.6 μL microfuge tube with appropriate label.
3. Transfer 10 μL of 5 μ*M* sequencing primer to 0.6 μL microfuge tube with appropriate label.
4. Send plasmid DNA and sequencing primer to sequencing facility.
5. Assemble and analyze sequences using the program MacVector.

Supplementary data to this article can be found online at http://dx.doi.org/10.1016/bs.mie.2017.03.013.

ACKNOWLEDGMENTS

This work was supported by R01ES019625-01A1 to M.C., and by funds from UCSC's Office of Research for support of intellectual property.

REFERENCES

Ahluwalia, D., & Schaaper, R. M. (2013). Hypermutability and error catastrophe due to defects in ribonucleotide reductase. *Proceedings of the National Academy of Sciences of the United States of America*, *110*, 18596–18601.

Alexander, D. L., Lilly, J., Hernandez, J., Romsdahl, J., Troll, C. J., & Camps, M. (2014). Random mutagenesis by error-prone pol plasmid replication in Escherichia coli. *Methods in Molecular Biology*, *1179*, 31–44.

Allen, J. M., Simcha, D. M., Ericson, N. G., Alexander, D. L., Marquette, J. T., Van Biber, B. P., et al. (2011). Roles of DNA polymerase I in leading and lagging-strand replication defined by a high-resolution mutation footprint of ColE1 plasmid replication. *Nucleic Acids Research*, *39*, 7020–7033.

Badran, A. H., & Liu, D. R. (2015). Development of potent in vivo mutagenesis plasmids with broad mutational spectra. *Nature Communications*, *6*, 8425.

Biran, A., Yagur-Kroll, S., Pedahzur, R., Buchinger, S., Reifferscheid, G., Ben-Yoav, H., et al. (2010). Bacterial genotoxicity bioreporters. *Microbial Biotechnology*, *3*, 412–427.

Camps, M., & Loeb, L. A. (2005). Critical role of R-loops in processing replication blocks. *Frontiers in Bioscience*, *10*, 689–698.

Camps, M., Naukkarinen, J., Johnson, B. P., & Loeb, L. A. (2003). Targeted gene evolution in Escherichia coli using a highly error-prone DNA polymerase I. *Proceedings of the National Academy of Sciences of the United States of America*, *100*, 9727–9732.

Cesareni, G., Helmer-Citterich, M., & Castagnoli, L. (1991). Control of ColE1 plasmid replication by antisense RNA. *Trends in Genetics*, *7*, 230–235.

Crameri, A., Whitehorn, E. A., Tate, E., & Stemmer, W. P. (1996). Improved green fluorescent protein by molecular evolution using DNA shuffling. *Nature Biotechnology*, *14*, 315–319.

Cupples, C. G., & Miller, J. H. (1989). A set of lacZ mutations in Escherichia coli that allow rapid detection of each of the six base substitutions. *Proceedings of the National Academy of Sciences of the United States of America, 86*, 5345–5349.

Curti, E., McDonald, J. P., Mead, S., & Woodgate, R. (2009). DNA polymerase switching: Effects on spontaneous mutagenesis in Escherichia coli. *Molecular Microbiology, 71*, 315–331.

Firnberg, E., Labonte, J. W., Gray, J. J., & Ostermeier, M. (2014). A comprehensive, high-resolution map of a gene's fitness landscape. *Molecular Biology and Evolution, 31*, 1581–1592.

Flamand, N., Meunier, J., Meunier, P., & Agapakis-Causse, C. (2001). Mini mutagenicity test: A miniaturized version of the Ames test used in a prescreening assay for point mutagenesis assessment. *Toxicology In Vitro, 15*, 105–114.

Fluckiger-Isler, S., Baumeister, M., Braun, K., Gervais, V., Hasler-Nguyen, N., Reimann, R., et al. (2004). Assessment of the performance of the Ames II assay: A collaborative study with 19 coded compounds. *Mutation Research, 558*, 181–197.

Gaivao, I., & Sierra, L. M. (2014). Drosophila comet assay: Insights, uses, and future perspectives. *Frontiers in Genetics, 5*, 304.

Guengerich, F. P. (2000). Metabolism of chemical carcinogens. *Carcinogenesis, 21*, 345–351.

Hakkila, K., Maksimow, M., Karp, M., & Virta, M. (2002). Reporter genes lucFF, luxCDABE, gfp, and dsred have different characteristics in whole-cell bacterial sensors. *Analytical Biochemistry, 301*, 235–242.

Jain, R. K., & Ranganathan, R. (2004). Local complexity of amino acid interactions in a protein core. *Proceedings of the National Academy of Sciences of the United States of America, 101*, 111–116.

Krewski, D., Acosta, D., Jr., Andersen, M., Anderson, H., Bailar, J. C., 3rd, Boekelheide, K., et al. (2010). Toxicity testing in the 21st century: A vision and a strategy. *Journal of Toxicology and Environmental Health. Part B, Critical Reviews, 13*, 51–138.

Lee, H., Popodi, E., Tang, H., & Foster, P. L. (2012). Rate and molecular spectrum of spontaneous mutations in the bacterium Escherichia coli as determined by whole-genome sequencing. *Proceedings of the National Academy of Sciences of the United States of America, 109*, E2774–E2783.

Lynch, A. M., Sasaki, J. C., Elespuru, R., Jacobson-Kram, D., Thybaud, V., De Boeck, M., et al. (2011). New and emerging technologies for genetic toxicity testing. *Environmental and Molecular Mutagenesis, 52*, 205–223.

Marinus, M. G. (2010). DNA methylation and mutator genes in Escherichia coli K-12. *Mutation Research, 705*, 71–76.

Matagne, A., Lamotte-Brasseur, J., & Frere, J. M. (1998). Catalytic properties of class A beta-lactamases: Efficiency and diversity. *The Biochemical Journal, 330*(Pt. 2), 581–598.

Miller, J. H., & Michaels, M. (1996). Finding new mutator strains of Escherichia coli—A review. *Gene, 179*, 129–132.

Million-Weaver, S., & Camps, M. (2014). Mechanisms of plasmid segregation: Have multicopy plasmids been overlooked? *Plasmid, 75*, 27–36.

Mortelmans, K., & Zeiger, E. (2000). The Ames Salmonella/microsome mutagenicity assay. *Mutation Research, 455*, 29–60.

Muteeb, G., & Sen, R. (2010). Random mutagenesis using a mutator strain. *Methods in Molecular Biology, 634*, 411–419.

Nohmi, T., Suzuki, T., & Masumura, K. (2000). Recent advances in the protocols of transgenic mouse mutation assays. *Mutation Research, 455*, 191–215.

Oller, A. R., Fijalkowska, I. J., Dunn, R. L., & Schaaper, R. M. (1992). Transcription-repair coupling determines the strandedness of ultraviolet mutagenesis in Escherichia coli. *Proceedings of the National Academy of Sciences of the United States of America, 89*, 11036–11040.

Oller, A. R., Fijalkowska, I. J., & Schaaper, R. M. (1993). The Escherichia coli galK2 papillation assay: Its specificity and application to seven newly isolated mutator strains. *Mutation Research*, *292*, 175–185.
Paul, S., Million-Weaver, S., Chattopadhyay, S., Sokurenko, E., & Merrikh, H. (2013). Accelerated gene evolution through replication-transcription conflicts. *Nature*, *495*, 512–515.
Pedelacq, J. D., Cabantous, S., Tran, T., Terwilliger, T. C., & Waldo, G. S. (2006). Engineering and characterization of a superfolder green fluorescent protein. *Nature Biotechnology*, *24*, 79–88.
Rogozin, I. B., & Pavlov, Y. I. (2003). Theoretical analysis of mutation hotspots and their DNA sequence context specificity. *Mutation Research*, *544*, 65–85.
Schaaper, R. M. (1996). Suppressors of Escherichia coli mutT: Antimutators for DNA replication errors. *Mutation Research*, *350*, 17–23.
Schmid, C., Arndt, C., & Reifferscheid, G. (2003). Mutagenicity test system based on a reporter gene assay for short-term detection of mutagens (MutaGen assay). *Mutation Research*, *535*, 55–72.
Severinov, K., Soushko, M., Goldfarb, A., & Nikiforov, V. (1994). RifR mutations in the beginning of the Escherichia coli rpoB gene. *Molecular & General Genetics*, *244*, 120–126.
Solanky, D., & Haydel, S. E. (2012). Adaptation of the neutral bacterial comet assay to assess antimicrobial-mediated DNA double-strand breaks in Escherichia coli. *Journal of Microbiological Methods*, *91*, 257–261.
Suzuki, T., Suzuki, K., Tashiro, Y., Saito, K., & Umeno, D. (2007). Probing the mutation spectrum in E. coli. *Nucleic Acids Symposium Series*, *53*, 289–290.
Troll, C., Alexander, D., Allen, J., Marquette, J., & Camps, M. (2011). Mutagenesis and functional selection protocols for directed evolution of proteins in *E. coli*. *Journal of Visualized Experiments*. https://www.jove.com/video/2505/mutagenesis-functional-selection-protocols-for-directed-evolution?sectionid=0.
Troll, C., Yoder, J., Alexander, D., Hernandez, J., Loh, Y., & Camps, M. (2014). The mutagenic footprint of low-fidelity Pol I ColE1 plasmid replication in E. coli reveals an extensive interplay between Pol I and Pol III. *Current Genetics*, *60*, 123–134.
Tse, L., Kang, T. M., Yuan, J., Mihora, D., Becket, E., Maslowska, K. H., et al. (2016). Extreme dNTP pool changes and hypermutability in dcd ndk strains. *Mutation Research*, *784-785*, 16–24.
Vogel, E. W., & Nivard, M. J. (2003). Model systems for studying germ cell mutagens: From flies to mammals. *Advances in Experimental Medicine and Biology*, *518*, 99–114.
Whong, W. Z., Stewart, J., & Ong, T. (1981). Use of the improved arabinose-resistant assay system of Salmonella typhimurium for mutagenesis testing. *Environmental Mutagenesis*, *3*, 95–99.

CHAPTER EIGHT

Sequencing DNA for the Oxidatively Modified Base 8-Oxo-7,8-Dihydroguanine

Aaron M. Fleming, Yun Ding, Cynthia J. Burrows[1]
University of Utah, Salt Lake City, UT, United States
[1]Corresponding author: e-mail address: burrows@chem.utah.edu

Contents

Methods in Enzymology, Volume 591
ISSN 0076-6879
http://dx.doi.org/10.1016/bs.mie.2017.03.004

Abstract

The DNA base guanine (G) can be oxidatively modified to 8-oxo-7,8-dihydroguanine (OG). Extraction of genomic DNA followed by nuclease digestion and mass spectrometry analysis has found OG is present at background levels of ~1 out of 10^6 Gs; however, this approach cannot determine the locations for the OGs in the genome. Thus, in this methods report, we outline three different methods (A, B, and C) for sequencing OG in DNA. Method A sequences OG by utilizing the base excision repair pathway to delete the OG nucleotide from the DNA that is then detected by Sanger sequencing as a deletion signature. Method B sequences OG by harnessing the base excision repair pathway to convert OG to an unnatural DNA base pair followed by Sanger sequencing to locate the unnatural base pair indicating where OG was located. Method C (i.e., OG-Seq) takes genomic DNA sheared to ~150 bps followed by selectively biotinylating the OG-containing fragments for affinity purification and enrichment of the OG-modified strands. The OG-modified fragments are sequenced on a next-generation sequencing platform to locate OG on the genomic scale with a resolution of ~150 bps. The methods outlined are then compared and contrasted allowing researchers to select the one that best suits their experimental goals.

1. INTRODUCTION

In DNA, the heterocyclic bases can be oxidatively modified by reactive oxygen species to yield many products (Cadet, Wagner, Shafirovich, & Geacintov, 2014; Delaney, Jarem, Volle, & Yennie, 2012; Fleming & Burrows, 2016). A prominent oxidative compound observed is 8-oxo-7,8-dihydroguanine (OG), a primary two-electron oxidation product of guanine (G). Cell-based experiments suggest that OG causes mutations suspected in disease initiation and progression (Al-Tassan et al., 2002); additionally, OG can modulate gene expression (Fleming, Ding, & Burrows, 2016; Pan et al., 2016; Pastukh et al., 2015). In vitro polymerase bypass and cell studies that knocked out proteins responsible for OG removal from the genome have identified that OG causes G → T transversion mutations due to OG Hoogsteen base pairing with adenine (Al-Tassan et al., 2002; Shibutani, Takeshita, & Grollman, 1991). Confirmation of the cellular mutation results was achieved by synthesizing OG site specifically into a plasmid and allowing replication of the plasmid to occur in prokaryotic cells followed by sequence analysis of the replicons (Wood, Esteve, Morningstar, Kuziemko, & Essigmann, 1992). This method verified that OG induces G → T transversion mutations. Recently, the modified base OG has been suspected in modulating gene expression on the basis of in vitro and in vivo studies (Fleming et al., 2016; Hailer-Morrison, Kotler, Martin, & Sugden, 2003; Moore, Toomire, & Strauss, 2013; Pan et al., 2016; Pastukh et al., 2015; Ramon et al., 1999; Tornaletti, Maeda, Kolodner, & Hanawalt, 2004). For instance, in vitro studies with OG located in transcription factor binding sequences identified a position-dependent impact on protein binding that could possibly alter gene expression (Hailer-Morrison et al., 2003; Moore et al., 2013; Ramon et al., 1999). When OG was synthesized in a plasmid in the template strand of the coding region for a reporter protein, the presence of OG negatively impacted protein expression in mammalian cells (Allgayer, Kitsera, Bartelt, Epe, & Khobta, 2016). In contrast, when OG was located in the *TNFα* or *VEGF* promoters in mammalian cells, gene activation was observed that was suspected to result from coordination of DNA repair with gene regulatory factors (Fleming et al., 2016; Pan et al., 2016; Pastukh et al., 2015). These studies indicate that OG has many roles in biological processes; therefore, to better understand these roles, sequencing DNA for OG is essential.

The presence of OG in the genome is traditionally determined by extraction of genomic DNA followed by nuclease digestion and mass spectrometry analysis which has found a background level of ~1 OG out of every 10^6 Gs (Gedik & Collins, 2005); however, this approach cannot determine the locations of the OGs in the genome. Thus, sequencing the genome for the thousands of OGs present will allow the identification of their locations to better understand the mutagenesis and gene regulatory effects of OG. Implementation of OG sequencing on the genomic scale requires a high-throughput method or sequencing for specific genomic loci. Additionally, synthetic incorporation of OG into plasmids requires OG sequencing to confirm the synthesis was successful; this can be achieved with targeted OG sequencing. Direct sequencing for OG is achievable by SMRT or nanopore sequencing methods on native DNA (Clark, Spittle, Turner, & Korlach, 2011; Schibel et al., 2010); however, a drawback to these methods is that the DNA cannot be PCR amplified and retain the OG location during this step. This methods paper outlines three approaches developed in our laboratory for sequencing OG in DNA (Ding, Fleming, & Burrows, 2017; Riedl, Ding, Fleming, & Burrows, 2015; Riedl, Fleming, & Burrows, 2015). The three methods described allow PCR amplification of the OG-containing DNA. In two of the three methods, the location of OG is marked prior to PCR amplification, allowing single-nucleotide resolution of the OG location (Methods A and B). In the third method, fragments of genomic DNA containing OG are purified away from the other fragments and then sequenced (Method C). The three methods are compared and contrasted to inform researchers for selecting the method that best suits the goals of their experiments. Other research groups have sequenced OG from genomic DNA via enrichment with an OG-binding antibody with a resolution ranging from $\sim 10^2$ to 10^6 bps (Ohno et al., 2006; Pastukh et al., 2015; Yoshihara, Jiang, Akatsuka, Suyama, & Toyokuni, 2014); the antibody-based methods will not be described in these protocols.

2. METHOD A: SEQUENCING OG BY CONVERSION TO A DELETION SIGNATURE

2.1 Theory

Sequencing OG by conversion to a deletion signature harnesses enzymes of the base excision repair pathway to delete OG from the DNA strand, leading to a deletion signature at the OG site by Sanger sequencing (Riedl, Fleming, et al., 2015). The DNA glycosylase formamidopyrimidine DNA glycosylase (Fpg) removes OG from duplex DNA by its *N*-glycosylase and AP-lyase activities (Fig. 1, Step I) (David & Williams, 1998). The action of Fpg yields a gapped site with 5′ and 3′ phosphates. Next, endonuclease IV (endo IV) catalyzes the removal of the 5′ phosphate to furnish a ligatable gap (Fig. 1, Step II) (David & Williams, 1998). The processed gapped site is a substrate for T4 DNA ligase that furnishes a ligated product strand one nucleotide shorter resulting from removal of the OG (Fig. 1, Step III). After exponential PCR amplification of the duplex (Fig. 1, Step IV), the amplicons are subjected to Sanger sequencing to produce a chromatogram that has two peaks out of register by one nucleotide starting at the OG site (Fig. 1, Step V) (Riedl, Fleming, et al., 2015). The two peaks are observed in the chromatogram because the OG-containing strand is one nucleotide shorter after processing than the complementary strand, and both are amplified during PCR prior to Sanger sequencing. This approach is capable of sequencing one OG or two OGs in the same strand separated by at least five nucleotides (Riedl, Fleming, et al., 2015), as this is the closest Fpg can recognize and operate efficiently on two substrates (Cunniffe, Lomax, & O'Neill, 2007). It is conceivable that more than two OGs could be sequenced via this method, although this concept has yet to be verified.

2.2 Equipment

- PCR thermal cycler
- Sanger sequencing facility

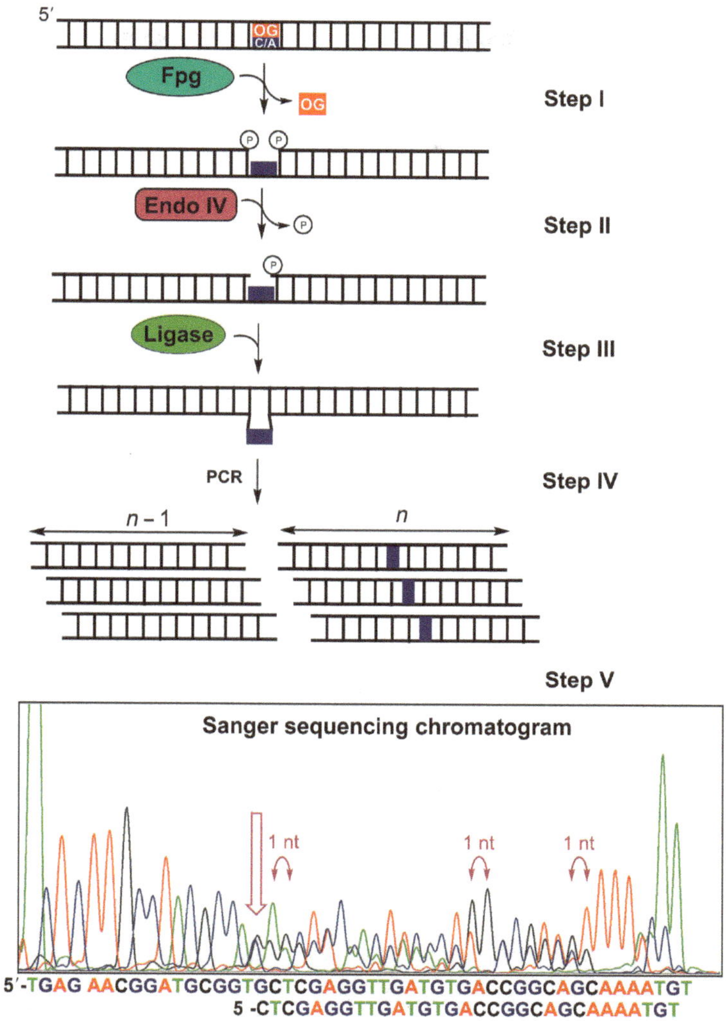

Fig. 1 Outline for sequencing OG via generation of a characteristic deletion signature in a Sanger sequencing chromatogram.

2.3 Materials

- Fpg (8000 units/mL; New England Biolabs)
- Endo IV (10,000 units/mL; New England Biolabs)
- T4 DNA ligase (400,000 units/mL; New England Biolabs)
- Reaction buffer (25 m*M* HEPES, pH 7.5; 10 m*M* $MgCl_2$; 5 m*M* KCl; 1 m*M* DTT; and 1 m*M* EDTA)
- ATP (2 m*M* stock in ddH_2O; New England Biolabs)

- Phusion® High-Fidelity DNA polymerase (2000 units/mL; New England Biolabs)
- dNTP solution mix (10 m*M* of each dNTP; New England Biolabs)
- PCR primers (forward and reverse in 8 μ*M* stock solutions)
- ddH_2O (autoclaved)
- DMSO
- Agarose
- TAE buffer (1 × = 40 m*M* Tris, 20 m*M* acetic acid, and 1 m*M* EDTA)
- Zymoclean™ Gel DNA Recovery kit
- Ethidium bromide (10 mg/mL)

2.4 Notes

- This procedure can be implemented on linear DNA duplexes or plasmid DNA. The method outlined in this report is for plasmid DNA.
- The example provided had OG synthetically incorporated in the pBR322 plasmid by a protocol outlined in the original publication of this method (Riedl, Fleming, et al., 2015).
- The PCR primers are designed to start polymerase extension at least 70 nucleotides away from the OG site in the plasmid. This distance is necessary for the Sanger sequencing chromatogram to provide reliable sequencing reads in the region of interest.

2.5 Procedure

(A) Suspend 5 ng of OG-containing plasmid DNA in 8.4 μL of reaction buffer in a 0.2-mL PCR tube.

(B) Thermally equilibrate the sample at 37°C in a PCR thermal cycler for 10 min prior to initiation of the reaction.

(C) After thermal equilibration at 37°C, add 5 units of Fpg (0.6 μL) and 10 units of endo IV (1.0 μL) to achieve a final volume of 10 μL. Allow the reaction to proceed for 30 min to generate a ligatable gap at the OG site (Fig. 1, Steps I and II).

(D) Following the 30-min reaction, change the thermal cycler temperature to 25°C and allow the sample to thermally equilibrate for 10 min at the new temperature.

(E) Once the sample is thermally equilibrated at 25°C, add ATP to a concentration of 83 μ*M* (0.5 μL) and 600 units of T4 DNA ligase (1.5 μL) to achieve a final volume of 12 μL. Next, allow the reaction to incubate for 2 h at 25°C (Fig. 1, Step III).

(F) After ligation of the gap, conduct exponential PCR to produce two product duplexes that differ in length by one nucleotide, in which the missing nucleotide in the shorter duplexes was the original site of OG (Fig. 1, Step IV).

(F.1) PCR conditions

- 2.0 μL processed plasmid reaction mixture
- 4.0 μL Phusion® GC buffer
- 0.4 μL dNTP solution mixture
- 0.6 μL DMSO
- 2.5 μL forward primer
- 2.5 μL reverse primer
- 1.0 μL Phusion® High-Fidelity DNA polymerase
- 7.0 μL ddH_2O
- Total volume = 20 μL

(F.2) PCR thermal cycler method

- Initiate the PCR with a single denaturing step for 2 min at 95°C
- Next, conduct 20 cycles of PCR with the following settings
  - Denaturation = 95°C for 45 s
  - Annealing = 55°C for 30 s
  - Extension = 72°C for 1.5 min
- After 20 cycles of PCR, conduct a final extension for 5 min at 72°C

(G) Once the PCR is complete, the product duplexes are separated from the plasmid template by agarose gel electrophoresis followed by removal of the PCR product strands from the gel with a Zymoclean™ Gel DNA Recovery kit.

(G.1) Agarose gel electrophoresis

- Prepare an agarose gel with ethidium bromide (0.5 μg/mL) following your laboratory's established protocol for the plasmid that is to be sequenced.
- After loading the samples into the agarose gel, conduct electrophoresis in 1 × TAE buffer using power setting appropriate for the plasmid analyzed.

(G.2) PCR product purification from the agarose gel

- Utilize a UV light box to locate and cut the PCR product band from the gel.
- Extract the product DNA using a Zymoclean™ Gel DNA Recovery kit following the manufacturer's protocol.

Additional information: Gel DNA extraction can be performed using other kits or established protocols in your laboratory.

(H) The purified PCR product is then submitted for standard Sanger sequencing. Because Sanger sequencing is not routinely conducted in most laboratories, refer to your sequencing facility for their instructions on how to prepare and submit the samples to be analyzed.

(I) After Sanger sequencing, the data are visually inspected to locate the OG site(s). In the Sanger sequencing chromatogram, the start of two peaks appearing out of register by one nucleotide reveals the location of the OG (Fig. 1, Step V). If more than one OG nucleotides are in the strand analyzed, refer to the original publication for analysis to identify the position of each OG (Riedl, Fleming, et al., 2015).

3. METHOD B: SEQUENCING OG BY CONVERSION TO AN UNNATURAL DNA BASE PAIR

3.1 Theory

This approach for sequencing OG harnesses the base excision repair pathway to convert a duplex with an OG-containing base pair into a PCR-amplifiable unnatural DNA base pair (e.g., dNaM:d5SICS; Fig. 2A) (Riedl, Ding, et al., 2015). Post PCR amplification, the amplicons are sequenced to reveal the location of the unnatural base pair, and hence OG. This method utilizes the *N*-glycosylase and AP-lyase activities of Fpg for removal of the OG nucleoside to yield a gapped site with phosphates on the 5′ and 3′ sides (Fig. 2B, Step I) (David & Williams, 1998). Removal of the 5′ phosphate is catalyzed by endo IV to generate a competent gap for introduction of an unnatural nucleotide (Fig. 2B, Step II). Next, the gapped site is filled with an unnatural nucleotide by Klenow fragment of DNA polymerase I deficient in exonuclease activity (Kf exo$^-$; Fig. 2B, Step III). The unnatural nucleotide introduced at the gap site includes dNaMTP, dMMO2TP, or d5SICSTP that was developed in the Romesberg laboratory (Fig. 2A) (Malyshev et al., 2012; Malyshev, Seo, Ordoukhanian, & Romesberg, 2009; Seo, Malyshev, Lavergne, Ordoukhanian, & Romesberg, 2011). Polymerase insertion of the unnatural nucleotide into the gap yields a nick on the 3′ side of the incorporated unnatural nucleotide base paired with a canonical nucleotide (i.e., A or C). The nick is then sealed with T4 DNA ligase to furnish an intact duplex (Fig. 2B, Step IV). Upon exponential PCR amplification of the duplex labeled with an unnatural

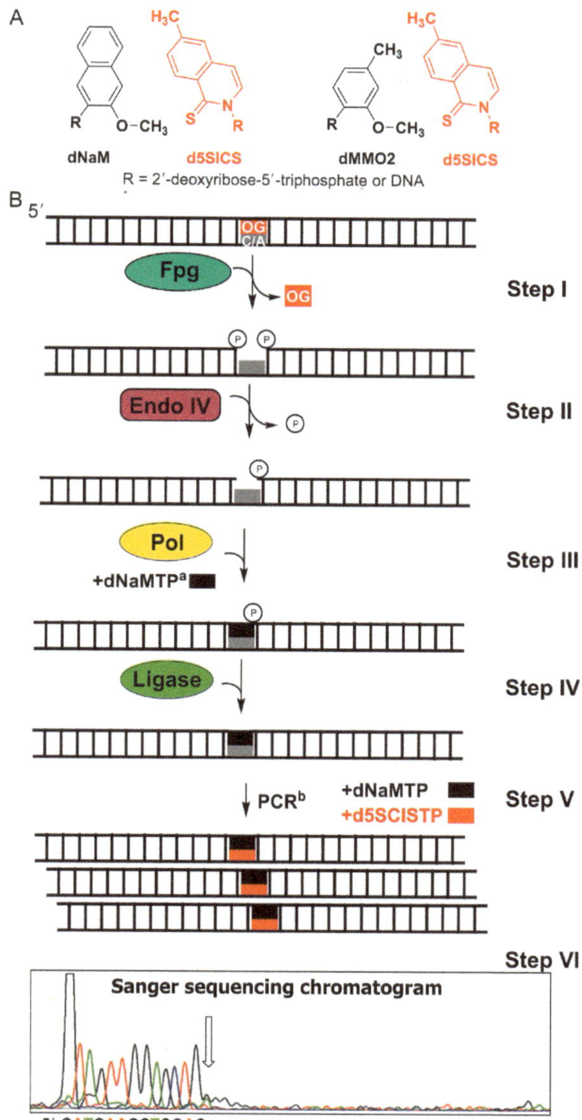

Fig. 2 Method outline to replace an OG-containing base pair with a dNaM:d5SICS or dMMO2:d5SICS unnatural DNA base pairs. (A) The dNaM:d5SICS and dMMO2:d5SICS structures. (B) Steps in the method to replace an OG-containing base pair with the dNaM:d5SICS unnatural DNA base pair. [a]For clarity only dNaMTP is shown in the figure; however, this step can also be achieved by allowing the polymerase to insert d5SICSTP. [b]For the sake of brevity, amplicons resulting from the unnatural nucleotide-containing strand are shown and sequenced. In practice, both strands from Step IV will amplify yielding a mixture of two duplexes, one with the unnatural base pair and the other without. A method to purify the unnatural base pair-containing duplex away from the other duplex is presented in a note below.

nucleotide in the presence of unnatural complementary nucleotides, new amplicons are generated bearing an unnatural base pair (Fig. 2A) at the site of OG (Fig. 2B, Step V). The unnatural DNA nucleotides developed by Romesberg and coworkers are PCR amplifiable with high efficiency and retention in the presence of natural dNTPs (Malyshev et al., 2009). Prior to sequencing, the unnatural base pair-containing duplex must be purified from the other duplex, which is outlined below. Sequencing for the dNaM:d5SICS or dMMO2:d5SICS base pair is traditionally achieved by Sanger sequencing to locate a strong stop in the sequencing chromatogram (Fig. 2B, Step VI). The strong stop is observed because the unnatural base pair that occurs where OG was originally positioned in the DNA cannot be sequenced past by Sanger sequencing that only looks for canonical nucleotides.

3.2 Equipment

- PCR thermal cycler
- Sanger sequencing facility

3.3 Materials

- Fpg (8000 units/mL; New England Biolabs)
- Endo IV (10,000 units/mL; New England Biolabs)
- T4 DNA ligase (400,000 units/mL; New England Biolabs)
- Reaction buffer (25 m*M* HEPES, pH 7.5; 10 m*M* $MgCl_2$; 5 m*M* KCl; and 1 m*M* EDTA)
- ATP (2 m*M* stock in ddH_2O; New England Biolabs)
- dNaMTP (500 μ*M* stock in ddH_2O)
- d5SICSTP (500 μ*M* stock in ddH_2O)
- $dMMO2^{SSBIO}TP$ (500 μ*M* stock in ddH_2O)
- One *Taq*® DNA polymerase (5000 units/mL; New England Biolabs)
- Klenow fragment deficient in exonuclease activity (5000 units/mL; New England Biolabs)
- dNTP solution mix (10 m*M* of each dNTP; New England Biolabs)
- PCR primers (forward and reverse in 8 μ*M* solutions)
- ddH_2O (autoclaved)
- DMSO
- DTT (300 m*M* stock in ddH_2O)
- TAE buffer (1 × = 40 m*M* Tris, 20 m*M* acetic acid, and 1 m*M* EDTA)
- Agarose

- Zymoclean™ Gel DNA Recovery kit
- Ethidium bromide (10 mg/mL)
- Streptavidin-coated magnetic beads

3.4 Notes

- This procedure can be implemented on linear DNA duplexes or plasmid DNA. The method outlined in this report is for plasmid DNA.
- The example provided had OG synthetically incorporated site specifically in the pBR322 plasmid by a protocol outlined in the original publication of this method (Riedl, Ding, et al., 2015).
- The PCR primers are designed to start polymerase extension at least 70 nucleotides away from the OG site in the plasmid. This distance is necessary for the Sanger sequencing chromatogram to provide reliable sequencing reads in the region of interest.
- The dNaM and d5SICS nucleosides can be purchased from Berry & Associates and need to be converted to the nucleotide triphosphates by literature protocols to utilize this OG-sequencing protocol (Malyshev et al., 2014, 2009).
- During PCR amplification, each member strand of the duplex will amplify (Fig. 2, Step IV), one has the unnatural DNA nucleotide and the other does not. For Sanger sequencing, the amplicon with the unnatural DNA base pair must be purified away from the amplicon with only canonical base pairs. This can be achieved by adding an unnatural nucleotide triphosphate functionalized with a biotin and a linker containing a disulfide during PCR amplification (e.g., dMMO2^{SSBIO}TP, Fig. 3A) (Seo et al., 2011). The functionalized unnatural nucleotide triphosphate allows affinity purification of the duplex with the unnatural base pair via streptavidin (STP)-coated magnetic beads, followed by release from the beads by reduction of the disulfide bond in the linker with DTT (Fig. 3B). At present, the dMMO2^{SSBIO}TP nucleotide required in this step must be synthesized by literature methods (Seo et al., 2011).

3.5 Procedure

(A) Suspend 50 ng of OG-containing plasmid in 8.4 μL of reaction buffer in a 0.2-mL PCR tube.

(B) Next, thermally equilibrate the PCR tube in a thermal cycler at 37°C for 10 min.

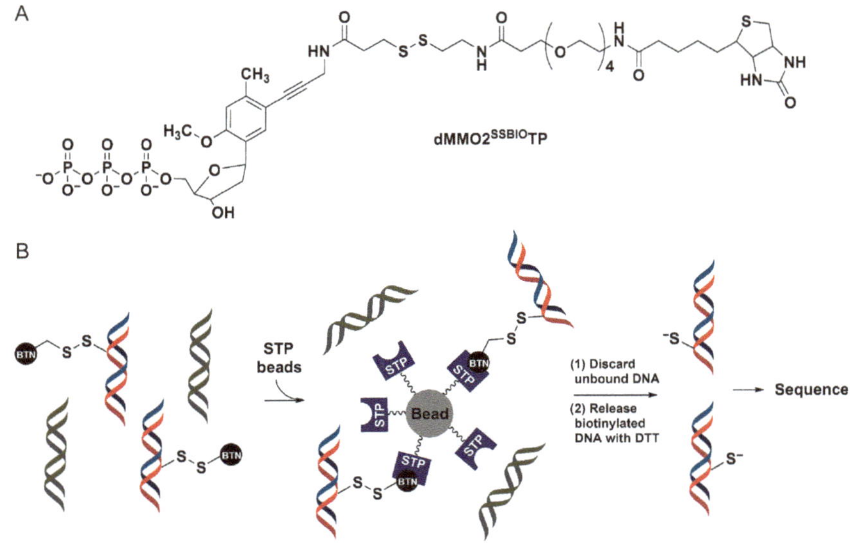

Fig. 3 Functionalized unnatural nucleotide utilized for selective purification of duplex DNA with an unnatural base pair at the site OG originally occupied. (A) Structure of dMMO2^{SSBIO}TP. (B) Scheme for affinity purification of duplexes with the dMMO2^{SSBIO} nucleotide. *BTN*, biotin; *STP*, streptavidin.

(C) After thermal equilibration, add 5 units of Fpg (0.6 μL) and 10 units of endo IV (1.0 μL) and allow the reaction to incubate for 30 min at 37°C. This enzyme combination will remove the OG nucleotide to furnish a gapped site with a 3′ OH on the 5′ side of the gap and a 5′ phosphate on the 3′ side of the gap (Fig. 2B, Steps I and II).

(D) After conversion of OG to a competent gapped site, add d5SICSTP to a concentration of 150 μ*M* (3 μL) and 7 units of Kf exo$^-$ (1.4 μL). Allow the reaction to progress at 37°C for 1 h (Fig. 2B, Step III). *Additional information:* This reaction can also be conducted with dNaMTP using identical reaction conditions as reported with the d5SICSTP.

(E) After polymerase insertion of the unnatural nucleotide into the gapped site, the reaction is quenched by heating at 90°C for 10 min in a preheated water bath.

(F) After quenching Kf exo$^-$, change the thermal cycler temperature to 25°C and allow the mixture to thermally equilibrate for 10 min.

(G) Once the sample is thermally equilibrated at 25°C, add ATP to achieve a concentration of 63 μ*M* (0.5 μL) and 600 units of T4 DNA ligase (1.5 μL) with a final volume of 16.5 μL. Next, allow the reaction to incubate for 16 h during which the nicked site with an unnatural nucleotide replacing OG is sealed (Fig. 2B, Step IV).

(H) After ligation, exponential PCR is conducted to yield two product duplexes, one of which has an unnatural DNA base pair and the other does not. To allow purification of the unnatural base pair-containing duplex, the PCR is conducted in the presence of d5SICSTP and $dMMO2^{SSBIO}TP$ to yield the $d5SICS{:}dMMO2^{SSBIO}$ base pair (Figs. 2A and 3).

(H.1) Conditions for the PCR amplification reaction.

- 2.0 μL processed plasmid reaction mixture
- 4.0 μL One*Taq*® reaction buffer
- 0.2 μL dNTP solution mixture
- 4.0 μL d5SICSTP
- 4.0 μL $dMMO2^{SSBIO}TP$
- 2.5 μL forward primer
- 2.5 μL reverse primer
- 0.5 μL One*Taq*® DNA polymerase
- 0.3 μL ddH_2O (total volume = 20 μL)

(H.2) Settings for the PCR thermal cycler.

- Conduct an initial denaturation step for 2 min at 95°C.
- Next, conduct 20 cycles of PCR using the following method.
  - Denaturation = 95°C for 45 s
  - Annealing = 55°C for 30 s
  - Extension = 72°C for 4 min
- After 20 cycles of PCR, conduct a final extension for 5 min at 72°C.

(I) The presence of $dMMO2^{SSBIO}TP$ allows affinity purification of the PCR-generated duplex with an unnatural base pair via STP-coated magnetic beads (Fig. 3A). Follow the manufacturer's protocol for proper use of the STP-coated magnetic beads during the purification step. After affinity purification of the duplexes is completed, release the DNA from the beads by reduction of the disulfide bond in the linker between the biotin and DNA with 30 m*M* DTT for 30 min at room temperature (Fig. 3B).

(J) After completion of the previous step, submit the sample for standard Sanger sequencing. Because Sanger sequencing is not routinely conducted in most laboratories, refer to your sequencing facility for their instructions on how they want the sample prepared.

(K) After Sanger sequencing, the data are visually inspected to reveal the location of the unnatural DNA base pair by observation of an abrupt stop in the sequencing chromatogram (Fig. 2, Step VI). The stop is observed because Sanger sequencing requires an additional PCR step conducted without the unnatural nucleotides. Consequently, sequencing OG by conversion to an unnatural base pair allows detection of one OG per strand by Sanger sequencing.

4. METHOD C: SEQUENCING OG BY OG-SEQ

4.1 Theory

The OG-Seq protocol was developed and implemented on mouse genomic DNA after enriching OG-containing fragments by OG-selective biotinylation (Fig. 4A) and affinity purification (Ding et al., 2017). The enriched fragments were then sequenced on a next-generation sequencing platform (NGS, Fig. 4B and C) (Ding et al., 2017). The fragment size will determine the resolution, which OG is sequenced in the genome; in our case, the fragments were ~150 bps. Selective biotinylation of OG is possible because the reduction potential of OG is sufficiently low such that only OG is oxidized with the mild one-electron oxidant K_2IrBr_6 (Ding et al., 2017). Upon two-electron oxidation of OG, a reactive electrophilic intermediate is formed that reacts with a nucleophilic primary amine analog of biotin ($BTN\text{-}NH_2$) to yield a stable covalent adduct (Fig. 4A) (Hosford, Muller, & Burrows, 2004; Xue & Greenberg, 2007). The duplex fragments containing the biotin-adducted OG are then enriched and purified by STP-coated magnetic beads. In this version of OG-Seq, the complementary strands to the biotinylated OG are released from the beads by NaOH denaturation and submitted for Illumina NGS. The NGS data are analyzed by aligning the reads to a reference genome to identify mapped reads that form regions of enrichment when compared to an input control experiment without OG enrichment (Fig. 4C). Further analysis of the sequencing reads can be tailored using any bioinformatic pipeline that is suitable for the analysis of these NGS data types.

4.2 Equipment

- DynaMag™-Spin magnet (ThermoFisher)
- S2 Focused-ultrasonicator (Covaris)
- Qubit Fluorometer (ThermoFisher)

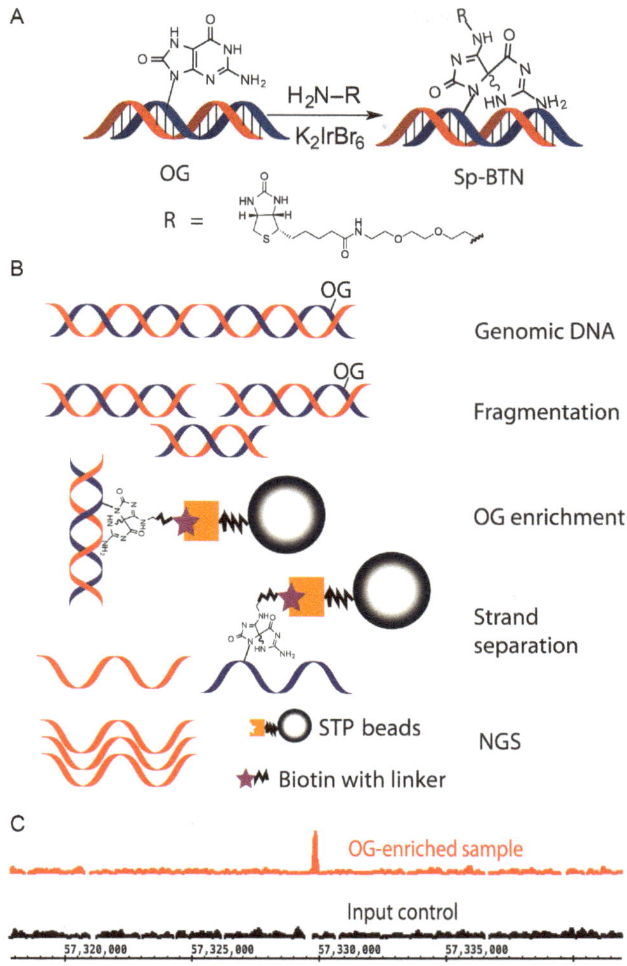

Fig. 4 Method overview for implementation of OG-Seq. (A) Mechanism for selectively labeling OG with biotin (BTN). (B) Sequence of steps to prepare ~150-mer DNA strands enriched in OG for submission to NGS. (C) Example data showing a region in the mouse genome where OG was found fivefold enriched.

- NanoDrop 1000 (ThermoFisher)
- Rotisserie mixer
- NGS facility

4.3 Materials

- Genomic DNA (~30 μg)
- ddH_2O (autoclaved)

- Desferal (100 m*M* solution in ddH_2O, Sigma Aldrich)
- Butylated hydroxytoluene (BHT, 100 m*M* solution in DMSO, Sigma Aldrich)
- Centricon-10 Concentrators (Sigma Aldrich)
- Micro Bio Spin P-6 gel columns (Bio-Rad)
- Qubit ssDNA assay kit (ThermoFisher)
- ssDNA/RNA clean & concentrator kit (Zymo Research)
- DNeasy blood and tissue kit (Qiagen)
- PCR clean-up kit (Mo Bio)
- Amine-PEG_2-Biotin ($BTN-NH_2$, 500 m*M* solution in ddH_2O, ThermoFisher)
- K_2IrBr_6 (100 m*M* solution in ddH_2O, Alfa Aesar)
- NaP_i buffer (100 m*M* solution at pH 8.0)
- Dynabeads MyOne Streptavidin C1 (Invitrogen)
- B&W buffer A (2 ×, 10 m*M* Tris, pH 7.5; 1 m*M* EDTA; 2 *M* NaCl; 0.1% Tween-20)
- B&W buffer B (2 ×, 10 m*M* Tris, pH 7.5; 1 m*M* EDTA; 2 *M* NaCl)
- Buffer 1 (150 m*M* NaCl; 0.01 *M* Tris, pH 7.0)
- TE buffer (10 m*M* Tris, pH 8.0; 1 m*M* EDTA)
- NaOH (150 m*M*)
- Tris buffer (10 m*M* solution at pH 8.0)
- Accel-NGS® 1S Plus DNA Library Kit (Swift Biosciences)

4.4 Notes

- OG-Seq was developed to sequence OG in genomic DNA. In the method outlined, the genomic DNA was derived from mouse embryonic fibroblast cells. The details of how these cells were grown can be found in the original publication (Ding et al., 2017); however, this method can be applied to any genomic DNA sample of interest, as long as you have enough DNA for implementation of this method (~30 μg for samples containing ~1 OG per 10^6 Gs).
- The OG-enriched strands obtained by the method below were sequenced on an Illumina Hi-Seq platform, although any NGS platform would be suitable for data collection.
- Bioinformatic analysis requires an input control that is comprised of genomic DNA that was sheared to the desired length and sequenced but not treated with the enrichment steps.

4.5 Procedure

(A) If mammalian cells are the source of OG for application of OG-Seq, obtain ~10^7 cells (~30 μg of genomic DNA) grown under conditions of interest to your research. Pellet the cells by centrifugation via standard laboratory methods. *Additional information*: If studies are conducted on other cells, harvest enough cells to obtain ~30 μg of genomic DNA if OG is present at ~1 OG per 10^6 Gs.

(B) Utilize a genomic DNA extraction kit, such as a DNeasy blood and tissue kit, to obtain pure genomic DNA following the manufacturer's protocol. Before commencing the extraction, add BHT and desferal to the solutions of the DNA extraction kit to a final concentration of 100 μ*M* for each compound. *Additional information*: The desferal binds iron ions and BHT is a reductant to prevent additional OG formation in the DNA during genomic DNA harvesting (Taghizadeh et al., 2008).

(C) Depending on the genomic DNA extraction kit used, the volume obtained will vary. Measure the genomic DNA concentration by either NanoDrop or Qubit™ analysis following the protocols for the instrument selected.

(D) The genomic DNA is then sheared by sonication to fragments of ~150 bps. In our studies, a Covaris S2 ultrasonicator was used that requires sample sizes of 5 μg per 130 μL of Tris buffer with 100 μ*M* of BHT and desferal present. Use the concentration obtained in the previous step to adjust the concentrations accordingly prior to sonication. Sonicate the samples with the settings set to the following values: duty cycle = 10%, intensity = 5, cycles/burst = 200, and time = 430 s. *Additional information*: If using a different ultrasonicator or shearing method, consult the manufacturer's protocol to obtain sheared DNA fragments of the desired length. In our published studies, ~150 bp lengths were the shortest that could be reliably obtained. If a different shearing method is selected, the resolution of the OG-Seq will be the average length of the sheared DNA.

(E) After sonication, exchange the residual Tris buffer from the previous steps with NaP_i buffer (100 m*M* pH 8.0) using a Micro Bio Spin P-6 gel column following the manufacturer's protocol. This step will also concentrate the sample to a volume of 20 μL. *Additional information*: It is important to exchange the Tris buffer with NaP_i buffer prior to the next step. During oxidation of OG an electrophile is formed that should react with the BTN-NH_2 nucleophile; however, Tris is a primary amine, and if it is present, it will compete with BTN-NH_2 for reacting with OG that would cause a loss in the enrichment of OG-containing fragments upon STP affinity purification.

(F) Take the concentrated sample and add 100 μL of 100 m*M* NaP_i (pH 8.0) buffer with BTN-NH_2 (5 μL or 20 m*M*) followed by heat the sample to 75°C for 10 min. After thermal equilibration, add K_2IrBr_6 (6.3 μL or 5 m*M*) and allow the react to progress for 1 h.

(G) After the biotinylation of OG, the excess BTN-NH_2 and K_2IrBr_6 are removed with a PCR clean-up kit following the manufacturer's protocol. Adjust the volume to 125 μL with Tris buffer.

(H) Use STP-coated magnetic beads (Dynabeads) to extract the DNA fragments biotinylated at OG. The steps for extraction are as follows.

(H.1) Take 100 μL of the Dynabeads for each sample to be analyzed and wash them with 1 mL of 1 × B&W buffer A four times. Between each wash step, use the DynaMag™-Spin magnet to retain the Dynabeads.

(H.2) Using the retained Dynabeads from the previous step, wash them with 1 mL of 1 × B&W buffer B using a DynaMag™-Spin magnet to retain the Dynabeads after the wash.

(H.3) Using the retained Dynabeads from the previous step, wash them with 1 mL of buffer 1 using a DynaMag™-Spin magnet to retain the Dynabeads after the wash.

(H.4) Resuspend the washed Dynabeads in 125 μL of B&W buffer B.

(H.5) Add the 125 μL of OG-biotinylated sample from step G to the washed beads from step H.4 and allow them to incubate overnight at 4°C on a rotisserie shaker.

(H.6) After the overnight incubation, the Dynabeads are washed twice with 1 mL of 1 × B&W buffer B using a DynaMag™-Spin magnet to retain the Dynabeads after the wash.

(H.7) The sample is then washed for a final time with 50 μL of TE buffer.

(H.8) To release the complementary strands to the bound OG-biotinylated strands, the beads are incubated in 150 m*M* NaOH at 20°C for 30 min.

(H.9) Separate the supernatant with the complementary strands from the beads.

(H.10) Use a single-strand DNA/RNA concentrator kit to elute the supernatant containing the complementary strands into 10 μL of ddH_2O following the manufacturer's protocol.

(H.11) Determine the sample concentration by Qubit analysis prior to submission of the sample to NGS to ensure that you have enough material for sequencing. The amount of DNA required is dependent on the NGS platform and you should consult your sequencing facility to determine the amounts required.

(I) Most laboratories do not possess an NGS instrument; therefore, refer to your NGS facility for further sample preparation prior to sequencing. If using an Illumina platform, the NGS library is constructed with an Accel-NGS® 1S Plus DNA Library Kit.

(J) The data are initially analyzed by aligning the reads to a reference genome using NovoAlign. The enriched peaks from the alignment are called from the mapped reads by MACS 2.0 using the input sample as a control (Zhang et al., 2008). *Additional information*: The bioinformatic analysis of the data can be further tailored using other pipelines that fit your experimental needs.

5. COMPARISON AND CONTRAST OF THE THREE OG-SEQUENCING METHODS

5.1 Sequencing Platforms

The three OG-sequencing methods outlined (A, B, and C) can be applied using different sequencing platforms. In Method A, OG is sequenced by deletion that is best observed in a Sanger sequencer because this platform will generate the two peaks out of register required for OG determining the location of OG (Fig. 1, Step V and Table 1). Utilization of Method A on an NGS platform would be problematic because deletion signatures are common sequencing errors that would require very high sequencing depth to reliably identify if they were real (Goodwin, McPherson, & McCombie, 2016). At present, Method B that sequences OG by conversion to an unnatural DNA base pair (e.g., d5SICS:dNaM; Fig. 2A) would be conducted on a Sanger sequencer (Fig. 2B, Step VI and Table 1); however, d5SCIS and dNaM were recently sequenced using the MspA protein nanopore (Craig et al., 2015). This opens the possibility of sequencing these unnatural nucleotides by Oxford Nanopore's MinION high-throughput

Table 1 Comparisons of Three OG-Sequencing Methods

	Method A: Sequencing OG by Formation of a Deletion Signature	**Method B: Sequencing OG by Conversion to an Unnatural Base Pair**	**Method C: Sequencing OG by OG-Seq**
Suitable sequencing platforms	Sanger	Sanger and NGS[a]	NGS
DNA contexts	Targeted sequencing of plasmids or genomic loci	Targeted sequencing of plasmids or genomic loci and the genomic scale[a]	Genomic scale
Resolution of sequenced OG	Single nucleotide	Single nucleotide	~150 bps
Ability to detect more than one OG per stand	Yes	No and yes[a]	No
Custom chemical synthesis required	No	Yes	No

[a]The ability to sequence OG by conversion to an unnatural DNA base pair on an NGS platform, the genomic scale, and more than one OG per strand analyzed is achievable as soon as the d5SICS, dNaM, and dMMO2 unnatural DNA bases can be sequenced on a commercial NGS platform. The concept of sequencing d5SICS and dNaM with the MspA protein nanopore has been demonstrated (Craig et al., 2015).

NGS platform that uses a protein nanopore (Jain et al., 2015); however, this has yet to be achieved. Lastly, sequencing OG by OG-Seq (i.e., Method C) utilizes affinity purification for enrichment of randomized OG-containing, genomic DNA fragments by selective biotinylation of the OG. This approach yields a large mixture of sequences to be analyzed that is best suited for high-throughput NGS platforms (Fig. 4B and Table 1).

5.2 DNA Contexts for Sequencing OG

The three sequencing methods outlined can be applied to different DNA sources. Method A requires PCR workup followed by Sanger sequencing, and therefore, this approach would best be applied to targeted sequences of established regions (Riedl, Fleming, et al., 2015). Method A would best confirm synthesis of plasmids with a site-specific OG or to interrogate single, specific loci in a genome (Table 1). On the same hand, Method B that finds

OG by conversion to an unnatural DNA base pair followed by PCR amplification (Riedl, Ding, et al., 2015), at present, would be applied to confirm synthesis of plasmids with a site-specific OG or for interrogation of single, specific loci in a genome. However, when sequencing d5SCIS or dNaM by commercial nanopore NGS technology is realized, OG-sequencing by this approach could be applied on the genomic scale (Table 1). Method C was developed and is applicable for sequencing OG on the genomic scale (Table 1) (Ding et al., 2017).

5.3 Resolution for Sequencing OG

Methods A and B achieve OG sequencing at the single-nucleotide resolution (Table 1) (Riedl, Ding, et al., 2015; Riedl, Fleming, et al., 2015). In contrast, Method C (i.e., OG-Seq) has a resolving power equal to the shearing size and the published version of this method had a resolution of ~150 bps (Table 1) (Ding et al., 2017). We anticipate new iterations of OG-Seq to emerge in the future with the power to sequence OG at singe-nucleotide resolution.

5.4 Ability to Sequence More Than One OG Per Strand

Method A detects OG by conversion to a deletion signature that is initiated with the DNA glycosylase Fpg and completed by the next steps of the base excision repair process; therefore, more than one OG can be detected per strand, as long as the OGs are spaced more than five nucleotides apart for Fpg to find and operate on both substrates (Table 1). On the other hand, Method B, at present, can only be used to detect a single OG per strand analyzed (Table 1); however, this limitation will be lifted once NGS sequencing for unnatural DNA nucleotides are achieve on a commercial instrument such as the Pacific Biosciences or Oxford Nanopore platforms. In the present version of Method C (i.e., OG-Seq) only one OG per 150-mer strand can be sequenced (Table 1).

5.5 Custom Chemical Synthesis Requirements

All materials needed to implement Method A or C can be obtained from commercial suppliers. Also, the sequencing facilities required to implement Method A or C for OG sequencing can be found at most research institutions (Table 1). In contrast, Method B requires custom synthesis of the unnatural dNTPs d5SICSTP, dNaMTP, and/or dMMO2^{SSBIO}TP, while the sequencing facility to conduct this method can be found at most research institutions (Table 1).

ACKNOWLEDGMENTS

Financial support for development of these methods was obtained from the National Institutes of Health (R01 GM093099), the National Cancer Institute (R01 CA090689), and a seed grant from the Nuclear Control Program at Huntsman Cancer Institute at the University of Utah (P30 CA042014). We thank the DNA/Peptide core facility at the University of Utah for synthesizing the oligomers and the Genomics core facility at the University of Utah for conducting the sequencing experiments; both are supported in part by a National Cancer Institute Cancer Center Support Grant (P30 CA042014). We are grateful to Dr. Jan Riedl for his hard work in developing Methods A and B.

REFERENCES

Allgayer, J., Kitsera, N., Bartelt, S., Epe, B., & Khobta, A. (2016). Widespread transcriptional gene inactivation initiated by a repair intermediate of 8-oxoguanine. *Nucleic Acids Research*, *44*, 7267–7280.

Al-Tassan, N., Chmiel, N. H., Maynard, J., Fleming, N., Livingston, A. L., Williams, G. T., et al. (2002). Inherited variants of MYH associated with somatic G:C–>T:A mutations in colorectal tumors. *Nature Genetics*, *30*, 227–232.

Cadet, J., Wagner, J. R., Shafirovich, V., & Geacintov, N. E. (2014). One-electron oxidation reactions of purine and pyrimidine bases in cellular DNA. *International Journal of Radiation Biology*, *90*, 423–432.

Clark, T. A., Spittle, K. E., Turner, S. W., & Korlach, J. (2011). Direct detection and sequencing of damaged DNA bases. *Genome Integrity*, *2*, 2–10.

Craig, J. M., Laszlo, A. H., Derrington, I. M., Ross, B. C., Brinkerhoff, H., Nova, I. C., et al. (2015). Direct detection of unnatural DNA nucleotides dNaM and d5SICS using the MspA nanopore. *PLoS One*, *10*, e0143253.

Cunniffe, S. M. T., Lomax, M. E., & O'Neill, P. (2007). An AP site can protect against the mutagenic potential of 8-oxoG when present within a tandem clustered site in *E. coli*. *DNA Repair*, *6*, 1839–1849.

David, S. S., & Williams, S. D. (1998). Chemistry of glycosylases and endonucleases involved in base-excision repair. *Chemical Reviews*, *98*, 1221–1262.

Delaney, S., Jarem, D. A., Volle, C. B., & Yennie, C. J. (2012). Chemical and biological consequences of oxidatively damaged guanine in DNA. *Free Radical Research*, *46*, 420–441.

Ding, Y., Fleming, A. M., & Burrows, C. J. (2017). Sequencing the mouse genome for the oxidatively modified base 8-oxo-7,8-dihydroguanine by OG-Seq. *Journal of the American Chemical Society*, *139*, 2569–2572. http://dx.doi.org/10.1021/jacs.1026b12604.

Fleming, A. M., & Burrows, C. J. (2016). Formation and processing of DNA damage substrates for the hNEIL enzymes. *Free Radical Biology & Medicine*, http://dx.doi.org/10.1016/j.freeradbiomed.2016.1011.1030

Fleming, A. M., Ding, Y., & Burrows, C. J. (2016). Oxidative DNA damage is epigenetic by regulating gene transcription via base excision repair. *Proceedings of the National Academy of Sciences of the United States of America*, *114*, 2604–2609. http://dx.doi.org/10.1073/pnas.1619809114.

Gedik, C. M., & Collins, A. (2005). Establishing the background level of base oxidation in human lymphocyte DNA: Results of an interlaboratory validation study. *The FASEB Journal*, *19*, 82–84.

Goodwin, S., McPherson, J. D., & McCombie, W. R. (2016). Coming of age: Ten years of next-generation sequencing technologies. *Nature Reviews. Genetics*, *17*, 333–351.
Hailer-Morrison, M. K., Kotler, J. M., Martin, B. D., & Sugden, K. D. (2003). Oxidized guanine lesions as modulators of gene transcription. Altered p50 binding affinity and repair shielding by 7,8-dihydro-8-oxo-2′-deoxyguanosine lesions in the NF-kappaB promoter element. *Biochemistry*, *42*, 9761–9770.
Hosford, M. E., Muller, J. G., & Burrows, C. J. (2004). Spermine participates in oxidative damage of guanosine and 8-oxoguanosine leading to deoxyribosylurea formation. *Journal of the American Chemical Society*, *126*, 9540–9541.
Jain, M., Fiddes, I. T., Miga, K. H., Olsen, H. E., Paten, B., & Akeson, M. (2015). Improved data analysis for the MinION nanopore sequencer. *Nature Methods*, *12*, 351–356.
Malyshev, D. A., Dhami, K., Lavergne, T., Chen, T., Dai, N., Foster, J. M., et al. (2014). A semi-synthetic organism with an expanded genetic alphabet. *Nature*, *509*, 385–388.
Malyshev, D. A., Dhami, K., Quach, H. T., Lavergne, T., Ordoukhanian, P., Torkamani, A., et al. (2012). Efficient and sequence-independent replication of DNA containing a third base pair establishes a functional six-letter genetic alphabet. *Proceedings of the National Academy of Sciences of the United States of America*, *109*, 12005–12010.
Malyshev, D. A., Seo, Y. J., Ordoukhanian, P., & Romesberg, F. E. (2009). PCR with an expanded genetic alphabet. *Journal of the American Chemical Society*, *131*, 14620–14621.
Moore, S. P., Toomire, K. J., & Strauss, P. R. (2013). DNA modifications repaired by base excision repair are epigenetic. *DNA Repair*, *12*, 1152–1158.
Ohno, M., Miura, T., Furuichi, M., Tominaga, Y., Tsuchimoto, D., Sakumi, K., et al. (2006). A genome-wide distribution of 8-oxoguanine correlates with the preferred regions for recombination and single nucleotide polymorphism in the human genome. *Genome Research*, *16*, 567–575.
Pan, L., Zhu, B., Hao, W., Zeng, X., Vlahopoulos, S. A., Hazra, T. K., et al. (2016). Oxidized guanine base lesions function in 8-oxoguanine DNA glycosylase1-mediated epigenetic regulation of nuclear factor kappaB-driven gene expression. *The Journal of Biological Chemistry*, *291*, 25553–25566.
Pastukh, V., Roberts, J. T., Clark, D. W., Bardwell, G. C., Patel, M., Al-Mehdi, A. B., et al. (2015). An oxidative DNA "damage" and repair mechanism localized in the VEGF promoter is important for hypoxia-induced VEGF mRNA expression. *American Journal of Physiology. Lung Cellular and Molecular Physiology*, *309*, L1367–L1375.
Ramon, O., Sauvaigo, S., Gasparutto, D., Faure, P., Favier, A., & Cadet, J. (1999). Effects of 8-oxo-7,8-dihydro-2′-deoxyguanosine on the binding of the transcription factor Sp1 to its cognate target DNA sequence (GC box). *Free Radical Research*, *31*, 217–229.
Riedl, J., Ding, Y., Fleming, A. M., & Burrows, C. J. (2015). Identification of DNA lesions using a third base pair for amplification and nanopore sequencing. *Nature Communications*, *6*, 8807.
Riedl, J., Fleming, A. M., & Burrows, C. J. (2015). Sequencing of DNA lesions facilitated by site-specific excision via base excision repair DNA glycosylases yielding ligatable gaps. *Journal of the American Chemical Society*, *138*, 491–494.
Schibel, A. E. P., An, N., Jin, Q., Fleming, A. M., Burrows, C. J., & White, H. S. (2010). Nanopore detection of 8-oxo-7,8-dihydro-2′-deoxyguanosine in immobilized single-stranded DNA via adduct formation to the DNA damage site. *Journal of the American Chemical Society*, *132*, 17992–17995.
Seo, Y. J., Malyshev, D. A., Lavergne, T., Ordoukhanian, P., & Romesberg, F. E. (2011). Site-specific labeling of DNA and RNA using an efficiently replicated and transcribed class of unnatural base pairs. *Journal of the American Chemical Society*, *133*, 19878–19888.
Shibutani, S., Takeshita, M., & Grollman, A. P. (1991). Insertion of specific bases during DNA synthesis past the oxidation-damaged base 8-oxodG. *Nature*, *349*, 431–434.

Taghizadeh, K., McFaline, J. L., Pang, B., Sullivan, M., Dong, M., Plummer, E., et al. (2008). Quantification of DNA damage products resulting from deamination, oxidation and reaction with products of lipid peroxidation by liquid chromatography isotope dilution tandem mass spectrometry. *Nature Protocols*, *3*, 1287–1298.

Tornaletti, S., Maeda, L. S., Kolodner, R. D., & Hanawalt, P. C. (2004). Effect of 8-oxoguanine on transcription elongation by T7 RNA polymerase and mammalian RNA polymerase II. *DNA Repair*, *3*, 483–494.

Wood, M. L., Esteve, A., Morningstar, M. L., Kuziemko, G. M., & Essigmann, J. M. (1992). Genetic effects of oxidative DNA damage: Comparative mutagenesis of 7,8-dihydro-8-oxoguanine and 7,8-dihydro-8-oxoadenine in *Escherichia coli*. *Nucleic Acids Research*, *20*, 6023–6032.

Xue, L., & Greenberg, M. M. (2007). Facile quantification of lesions derived from 2′-deoxyguanosine in DNA. *Journal of the American Chemical Society*, *129*, 7010–7011.

Yoshihara, M., Jiang, L., Akatsuka, S., Suyama, M., & Toyokuni, S. (2014). Genome-wide profiling of 8-oxoguanine reveals its association with spatial positioning in nucleus. *DNA Research*, *21*, 603–612.

Zhang, Y., Liu, T., Meyer, C. A., Eeckhoute, J., Johnson, D. S., Bernstein, B. E., et al. (2008). Model-based analysis of ChIP-Seq (MACS). *Genome Biology*, *9*, R137.

CHAPTER NINE

Xenopus laevis as Model System to Study DNA Damage Response and Replication Fork Stability

Vincenzo Sannino, Federica Pezzimenti, Stefania Bertora, Vincenzo Costanzo[1]
DNA Metabolism Laboratory, IFOM—The FIRC Institute of Molecular Oncology, Milan, Italy
[1]Corresponding author: e-mail address: vincenzo.costanzo@ifom.eu

Contents

Methods in Enzymology, Volume 591
ISSN 0076-6879
http://dx.doi.org/10.1016/bs.mie.2017.03.018

Abstract

Although many players of the DNA damage response and DNA repair have been identified in several systems their biochemical role is still poorly understood. The use of the *Xenopus laevis* egg extract cell-free system allowed biochemical dissection of DNA replication and cell cycle events in a complex biological context. The possibility of manipulating the protein content by using protein depletion procedures makes egg extract a powerful system to study proteins whose inactivation results in cellular lethality. The egg extract has been increasingly used to study DNA damage response and the coordination of DNA replication with DNA repair. The recent development of advanced imaging techniques based on electron microscopy has allowed the characterization of replication intermediates formed in the absence of essential DNA repair proteins. These studies have been important to understand how cells maintain genome stability under unchallenged and stressful conditions. Here, we present a collection of protocols that have been developed to recapitulate DNA damage response activated by chromosome breakage in egg extract and to isolate replication intermediates for electron microscopy analysis using sperm nuclei or more defined genomic substrates.

1. INTRODUCTION

In the presence of DNA damage checkpoints delay cell cycle progression to gain time for DNA repair (Garner & Costanzo, 2009). DNA repair proteins promote specific processing of the different DNA lesions. Many of these repair reactions take place during DNA replication and must be coordinated with replication fork progression (Errico & Costanzo, 2010, 2012). Genetically tractable model systems have been instrumental to identify mutations in genes impaired in DNA damage cell cycle checkpoints and DNA repair. However, this approach has important limitations among which the fact that essential genes cannot be isolated in standard genetic screens.

The cell-free extract derived from *Xenopus* eggs is an ideal system to overcome some of these issues. We have used cell-free systems derived from *Xenopus* eggs to elucidate the biochemical basis of cell cycle transitions, activation of DNA damage checkpoints, and DNA replication control in the presence of DNA damage (Costanzo & Gautier, 2004; Costanzo, Paull, Gottesman, & Gautier, 2004; Costanzo, Robertson, & Gautier, 2004; Errico et al., 2009; Hashimoto, Puddu, & Costanzo, 2011; Jazayeri, Balestrini, Garner, Haber, & Costanzo, 2008; Smith et al., 2009; Trenz, Smith, Smith, & Costanzo, 2006). A major advantage of the *Xenopus* system is that biochemical analysis can be performed in the absence of transcription and protein synthesis due to the large amount of factors already present as maternal stockpile (Sannino, Kolinjivadi, Baldi, & Costanzo, 2016). Selective protein depletion from egg extract allows the characterization of the biochemical steps in which specific proteins operate (Sannino et al., 2016). The application of this procedure to essential DNA metabolism proteins facilitates the study of vertebrate gene products that would normally compromise viability when inactivated in other cellular systems, especially during DNA replication and repair (Sannino et al., 2016).

To study checkpoint signaling damaged DNA or DNA molecules reproducing damaged or stalled replication intermediates can be added to egg extract to activate ATM-, ATR-, and DNA-PK-dependent signal transduction (Costanzo, Paull, et al., 2004; Costanzo, Robertson, et al., 2004; Costanzo et al., 2000).

More recently, the development of novel techniques such as electron microscopy (EM) for single-molecule analysis has been helpful to understand the function of DNA repair and checkpoint proteins in DNA replication (Hashimoto, Ray Chaudhuri, Lopes, & Costanzo, 2010). The adoption of EM-mediated analysis allowed the expansion of the phenotypes that can be studied in egg extract (Hashimoto & Costanzo, 2011). In particular, the visualization of the native DNA configuration of replication intermediates isolated from egg extract depleted of specific factors led to discovery of new functions of essential DNA repair proteins such as RAD51 and Mre11 in DNA replication (Hashimoto et al., 2010). In addition, the use of large bacterial artificial chromosomes (BAC) templates, which are able to induce nuclear assembly and to support semiconservative DNA replication from multiple origins (Aze, Sannino, Soffientini, Bachi, & Costanzo, 2016) led to the discovery of how defined areas of the chromosomes replicate and respond to replication stress. To this end BACs containing DNA sequences derived from specific genomic loci such as centromeres were used to study the behavior of these regions (Aze et al., 2016). Here, we will review the basic protocols adopted in these studies.

2. PREPARATION OF INTERPHASE EGG EXTRACT

Several types of egg extracts (meiotic arrested, mitotic, cycling, and interphase) can be prepared by using different buffers and centrifugation steps. An example of the protocol used to study DNA replication, DNA damage checkpoint activation, and replication fork metabolism is described below. This protocol has the highest DNA replication efficiency among the protocols tested in the laboratory. Interphase extracts obtained through this procedure can be frozen and stored for long time before use without significant loss of activity. The protocols describing the preparation of egg extract and sperm nuclei have been adapted from other published protocols (Lohka & Masui, 1983; Murray, 1991).

2.1 Equipment

- Pasteur plastic pipettes.
- Ultracentrifuge tubes (Beckman 349622).
- TLA100.3 rotor (Beckman 349490).
- 2-mL Eppendorf tubes (Eppendorf 0030 120.094).

2.2 Buffers and Reagents

- PMSG (pregnant mare serum gonadotropin) (Sigma-Aldrich, G4877).
- HCG (human chorionic gonadotropin) (Sigma-Aldrich, CG10).
- Dejellying buffer: 20 m*M* Tris–HCl, pH 8.5; 110 m*M* NaCl (add 5 m*M* DTT just before use; prepare at least 30 mL for each egg set).
- S-buffer: 50 m*M* Hepes-KOH, pH 7.5; 50 m*M* KCl; 2.5 m*M* $MgCl_2$; 250 m*M* sucrose (add 2 m*M* β-mercaptoethanol and 15 μg/mL leupeptin just before use; prepare 40 mL for each egg set; keep at 4°C).
- MMR 5×: 100 m*M* Hepes-KOH, pH 7.5; 2 *M* NaCl; 10 m*M* KCl; 5 m*M* $MgSO_4$; 10 m*M* $CaCl_2$; 0.5 m*M* EDTA.
- ¼MMR: dilute 5× MMR to make ¼ (50 mL in 1 L; 1 L should be enough to process 2–3 egg sets).
- 10 mg/mL Leupeptin: dissolve in DMSO. Keep aliquots at −20°C.
- 10 m*M* Calcium ionophore A23187 (Sigma): dissolve in DMSO. Make 10 μL aliquots. Keep stocks at −20°C.
- 10 mg/mL Cytochalasin B: dissolve in DMSO. Keep stocks at −20°C.
- 10 mg/mL Cycloheximide: dissolve in MilliQ water. Keep stocks at −20°C.
- 1 *M* Creatine phosphate: dissolve in MilliQ water. Keep stocks at −20°C.
- 10 mg/mL Creatine phosphokinase: dissolve in MilliQ water. Keep stocks at −20°C.
- Cy3-dCTP working solution: dilute 1:50 in MilliQ water from the purchased batch (Cy3-dCTP, GE Healthcare, PA53021).

- Fixative solution: 15 m*M* PIPES, pH 7.2; 15 m*M* NaCl; 80 m*M* KCl; 10% formalin; 2 μg/mL Hoechst 33258; 50% Glycerol; 10 μg/mL DHCC (3,3′-dihexyloxacarbocyanine iodide).

2.3 Procedures

2.3.1 Female Frogs Injection

Seven or eight large female frogs are identified 5 days before injection and kept in isolation in a single tank. The day before egg collection frogs are injected with 250 U HCG. Six hours after the first injection, a second injection is performed using 650 U HCG. Each frog is placed in a single tank and in a 100 m*M* NaCl solution for 12 h in an incubator at 20°C. Eggs are collected in the morning and freshly used for extract preparation. Injected frogs can be reused after resting for 4 months.

2.3.2 Egg Extract Preparation

1. Use glass beakers to collect the eggs from each frog. We usually do not mix eggs from different frogs. Using a plastic Pasteur pipette remove imperfect eggs, which usually constitute less than 5% of the total eggs laid by each frog. Eggs that do not look homogeneously similar to the rest of the batch laid by the same frog are the ones to be discarded (Note 1).
2. Dejelly eggs by adding 15–30 mL Dejellying buffer containing fresh 5 m*M* DTT for 10 min. Refresh the Dejellying buffer several times. Eggs lose the jellycoat and become compact (Note 2).
3. Wash three times in ¼MMR (using about 50 mL buffer each time). Then discard as much buffer as possible.
4. Activate eggs by adding 2 μL calcium ionophore (from a 10 m*M* stock). Mix gently once and keep monitoring the eggs for 5–10 min. When animal pole becomes smaller (indicating egg activation) in most of the eggs go to next step (Note 3).
5. Wash three times with 50 mL ¼MMR then discard as much buffer as possible.
6. Wash twice with 20 mL ice-cold S-buffer supplemented with 2 m*M* β-mercaptoethanol and 15 μg/mL leupeptin.
7. Put the glass beaker on ice and collect the eggs in precooled 2-mL Eppendorf tubes. Remove excess buffer.
8. Pack the eggs by spinning for few seconds at 6000 rpm (just reach 6000 rpm and stop) in a refrigerated Eppendorf benchtop microcentrifuge.

9. Remove excess buffer with a small Pasteur plastic pipette from each tube while keeping the other tubes in the refrigerated rotor. Look out for liquid drops present under the cap of the tube and remove them.
10. Spin at 13,200 rpm for 10 min at 4°C using the same Eppendorf bench-top microcentrifuge to crush the eggs.
11. Collect cytoplasmic fraction (middle layer in Fig. 1A) while keeping the other tubes in a metal rack on ice: use a small plastic Pasteur pipette to make a hole across the top layer; clean the tip with Kimwipes; collect the cytoplasmic fraction present in the middle layer (Fig. 1B); clean the tip again and transfer the cytoplasm into a 15-mL Falcon tube on ice (Note 4). Alternatively, eggs can be crushed in a 15-mL snap-cap tube by centrifuging for 10 min at 10,000 rpm in a JS13.1 (Beckman, 346963) swinging-bucket rotor. In this case, after centrifugation collect cytoplasm with a syringe by punching the side of the tube. Collect just the cytoplasm above the dark layer and below the yellow plug on top of it (Fig. 1C).

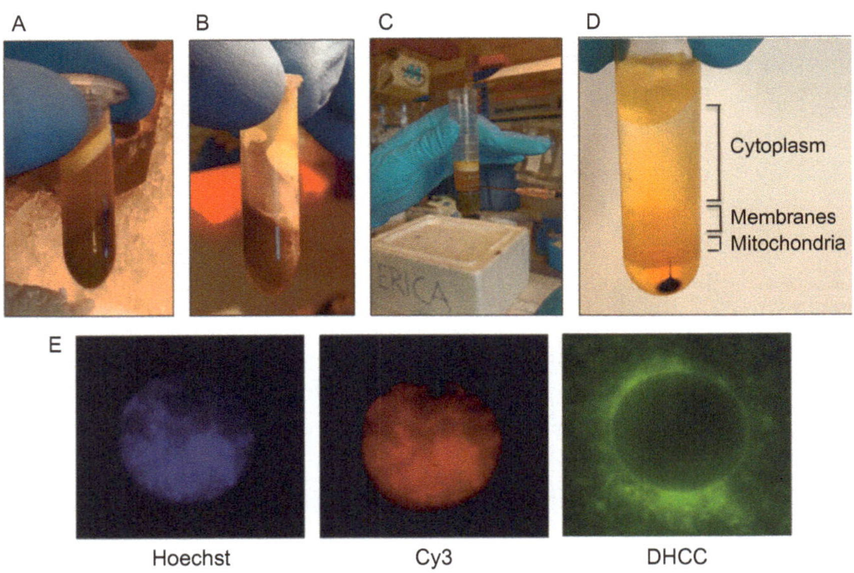

Fig. 1 *Xenopus* interphase egg extract preparation. (A) Extract stratification after the first crushing centrifugation step showing a top lipid layer, a central cytoplasmic layer and a lower pigment layer. (B) Residual material after the cytoplasmic fraction collection. (C) Collection of the cytoplasmic layer with the syringe method. (D) Layer stratification after the ultracentrifugation step: cytoplasmic, membranes, and mitochondria layers are indicated in the picture. (E) Replicating *Xenopus* nuclei monitored by fluorescence microscopy. Chromatin is visualized by Hoechst staining, nucleotide incorporation by Cy3 fluorescence, and membranes are visualized by fluorescent dye lipid stain DHCC.

12. Add 40 μg/mL cytochalasin B and mix gently by inverting the tubes.
13. Transfer this cytoplasmic extract to ultracentrifuge tubes (Beckman 349622).
14. Spin at 70,000 rpm for 18 min in TLA100.3 rotor at 4°C.
15. Collect the cytoplasmic fraction plus the membranes layer with a P-1000 pipette tip (Note 5). Make sure to avoid the brownish mitochondria layer underneath (Fig. 1D).
16. Add 40 μg/mL cycloheximide.
17. Mix gently and deeply with a cut P-1000 tip to homogenize the extract.
18. To freeze extracts add 3% glycerol, mix gently with a P-1000 pipette and freeze 20 μL drops directly in liquid nitrogen. Store the frozen pellets in cryovials in liquid nitrogen.
19. When ready to use thaw rapidly the pellets at room temperature in Eppendorf tubes. Thaw 3–4 pellets at the time separately in each tube and then collect the amount of extracts needed in the same tube.
20. After thawing supplemented extract with extra 40 μg/mL cycloheximide. Add 30 μ*M* creatine phosphate and 150 μg/mL creatine phosphokinase
21. To check the quality of the extract, nuclei formation and Cy3-dCTP incorporation can be rapidly tested. Add 0.5 μL of Cy3-dCTP working solution and sperm nuclei to a final concentration of 4000 nuclei/μL. Incubate at 23°C. After 30 min, take a 2 μL aliquot and mix it with an equal amount of fixative solution. Spot 2 μL of the mix on a microscope glass slide and cover with round coverslips. Quality of different batches of extract egg extract can be rapidly estimated by comparing nuclei formation and nucleotide incorporation at 30 min. Good quality extract should give nuclei as the ones shown in Fig. 1 E.

2.4 Notes

1. Eggs should be processed as fast as possible. To speed up the protocol, egg sorting can be directly performed in Dejellying buffer. Egg sorting should continue during all the procedure up to the crushing step.
2. Removal of jelly coat is crucial for full activation. The Dejellying step should last less than 10 min. To speed it up refresh the buffer every 2.5 min until eggs touch each other.
3. This step may depend on eggs quality. High quality eggs can be easily activated in 5 min by adding 2 μL calcium ionophore (usually an activation wave can be easily visualized). If this does not happen, we usually

add more calcium ionophore every 2.5 min. We proceed to interphase extract preparation only when we get at least 80%–90% of the eggs activated in 15 min.

4. Fast collection is important while recovering the middle layer.
5. It is important to collect the membranes layer avoiding the mitochondrial layer underneath. Membranes are essential for nuclei formation during the replication reactions.

3. *XENOPUS LAEVIS* DEMEMBRANATED SPERM NUCLEI PREPARATION

Preparing a proper stock of demembranated sperm nuclei from *X. laevis* male frogs is essential to carry out successful and reproducible experiments with the *Xenopus* egg extracts system. The procedure described here is aimed to get a high quality sperm preparation that can be efficiently used for routine assays to evaluate DNA replication activity and chromatin levels of DNA replication factors as well as more sensitive analyses based on fluorescence and EM.

A single sperm preparation from five male frogs will give a final amount of sperm nuclei that will be useful for several experiments. The entire procedure has been optimized to obtain a highly concentrated sperm preparation.

3.1 Equipment

- Dissection metal forceps.
- Razor blades.
- Dounce homogenizer (Sigma-Aldrich, P7984).
- Synthetic Nylon Mesh 25-μm (Plastock Group Ltd).
- Hemocytometer.

3.2 Buffers and Reagents

- MS-222 (Sigma-Aldrich, A5040).
- Spermidine (Sigma-Aldrich, S0266). Prepare a 0.5 *M* stock solution in H_2O.
- Spermine (Sigma-Aldrich, S4264). Prepare a 0.5 *M* stock solution in H_2O.
- Lysolecithin (Sigma-Aldrich, L4129). Prepare a 10 mg/mL stock solution in H_2O.

- EB buffer: 50 m*M* Hepes-KOH, pH 7.5; 50 m*M* KCl; 5 m*M* $MgCl_2$; 5 m*M* EGTA; 2 m*M* β-mercaptoethanol.
- SuNaSp buffer: 15 m*M* Hepes-KOH, pH 7.5; 250 m*M* sucrose; 75 m*M* NaCl; 0.5 m*M* spermidine; 0.15 m*M* spermine.
- SuNaSp-BSA: SuNaSp + 3% (w/v) bovine serum albumin.
- Storage buffer: EB buffer + 30% glycerol.

3.3 Procedures

3.3.1 Male Frogs Injection

To increase the final sperm yield each frog is usually primed with 50 U PMSG 7 days before and with 300 U HCG the day before the sperm preparation. During this time, frogs are maintained at 18°C in isolated tanks. The day of the sperm preparation, frogs are immersed in a solution containing an overdose of MS-222 (>3 g/L) and, after reaching a status of deep anesthetization, sacrificed by cutting through the heart. Testes are then easily recognized after removing the intestine, collected, and processed to isolate the sperm nuclei. All the passages of the sperm preparation are carried out as rapidly as possible, in order to minimize damage sperm nuclei.

3.3.2 Preparation of Demembranated Sperm DNA

1. Place the isolated testes in EB solution on ice. With the help of a pair of clean forceps and a razor blade carefully remove any residual tissue and blood vessel from the surface of the testes (Note 1). Rinse the testes by placing them in a glass Petri dish filled with EB placed on ice.
2. Transfer the clean testes in a second Petri dish containing 10 mL EB, placed on ice. Mince the testes by holding them with a pair of forceps and chopping them in fine pieces with the use of a razor blade.
3. Transfer all the content of the Petri dish in the Dounce homogenizer, placed on ice. Wash the Petri dish with further 5 mL EB and transfer them in the same homogenizer.
4. Homogenize the minced testes as finely as possible then let sediment any residual solid material and transfer the supernatant to a clean 50-mL Falcon tube on ice.
5. Wash again the glass Petri dish with 15 mL ice-cold EB and transfer them into the homogenizer, in order to harvest as much sperm as possible. Rehomogenize the mix, let sediment any residual material, and transfer the supernatant to the same 50-mL Falcon tube as before.
6. Cut a round hole in the screw cap of the Falcon tube and place a piece of 25-μm nylon membrane between the cap and the tube containing

the homogenate. Secure the membrane in place by closing the cap and pour the homogenate through the membrane into a clean 50-mL Falcon tube on ice. Help the passage of the mixture through the filter by piercing the wall of the falcon tube with a needle (Note 2).

7. Divide the filtered homogenate into two clean 15 mL snap-cap round-bottom tubes and centrifuge (without the cap) for 5 min in a swinging-bucket rotor (JS 13.1, Beckman) at 4250 × *g* at 4°C.
8. Save the supernatants. Resuspend the two pellets with 1.5 mL SuNaSp at room temperature and pool them together (Note 3). Collect any residue left in the tubes by rinsing with 1 mL fresh SuNaSp.
9. Spin again the supernatants saved at point 8 in new 15-mL tubes (Note 4). Pool together the two new pellets obtained at point 9 by resuspending them in a final volume of 1 mL of SuNaSp.
10. Pool all the pelleted fractions together (approximately 5 mL in total) and add 250 μL lysolecithin stock solution at room temperature.
11. After 5 min of incubation check the demembranation of the sperm by adding 1.5 μL of the mixture to 1.5 μL EB supplemented with 20 μg/mL Hoechst 33258. Monitor the demebranation by phase contrast and fluorescence microscopy. Demembranated sperm will appear as bright blue comma-shaped rods by fluorescence microscopy, while nondemebranated sperm will fail to incorporate Hoechst and will only be visible by phase contrast (Note 5). The monitoring is done every 5 min, until at least 95% of the sperm population appears positive to Hoechst staining. If after 15 min the demembranated sperm percentage is still low, add 100 μL lysolecithin solution, gently mix, and continue incubation at room temperature (Note 6).
12. To quench the action of the lysolecithin, divide the sperm mixture in two round-bottom 15-mL tubes and add to each of them 10 mL ice-cold SuNaSp supplemented with 3% BSA. From this step on, all the remaining ones will be performed on ice.
13. Centrifuge for 5 min in a swinging-bucket rotor at 4250 × *g*, at 4°C and discard the supernatants.
14. Resuspend the pellets in 1 mL of ice-cold SuNaSp-BSA. Add more SuNaSp-BSA to the tubes to reach a final volume of 12 mL and centrifuge again for 5 min at 4250 × *g*.
15. Resuspend the pellets in 1 mL ice-cold EB and then add some more EB to reach the final volume of 12 mL. Centrifuge again for 5 min at 4250 × *g*.
16. Repeat step 15 once.

17. Resuspend the pellets in 1 mL ice-cold EB and pool them together in one tube. Add more EB in order to reach the final volume of 12 mL and centrifuge for 5 min at 4250 × *g*.
18. Discard the supernatant and resuspend the pellet in 1 mL ice-cold Storage buffer.
19. To count the sperm, take a small aliquot from the preparation and prepare an appropriate dilution in EB plus 30% glycerol supplemented with 20 μg/mL Hoechst 33258. Count the sperm number using a hemocytometer and determine the sperm concentration of the stock preparation (Note 7).
20. Dilute the stock with Storage buffer in order to reach a final concentration of 200,000 nuclei/μL. Snap-freeze small aliquots (5–10 μL) in liquid nitrogen and store at −80°C (Note 8).

3.4 Notes

1. It is very important to remove blood vessels and tissues without damaging the testes. The testes should be white in color.
2. If the membrane clogs, it is possible to change it and continue the filtering procedure with a new one.
3. In order to disrupt any possible aggregate, resuspend sperm by using a 200 μL tip with a 10 μL tip inserted on it.
4. At this point, the sperm preparation might contain traces of somatic nuclei and erythrocytes, appearing as a red spots at the center of the white pellet. Such contamination will be washed off in the subsequent passages.
5. Usually sperm preparations contain a small percentage of somatic nuclei, which appear as round dots. If the percentage of somatic nuclei is above 5% of the total population, the preparation must be discarded.
6. It is important to follow the progress of demembranation at short intervals in order to stop it at the right moment, as a prolonged incubation with lysolecithin could damage the sperm.
7. The optimal dilution rate for the calculation of sperm concentration must be adjusted during the counting procedure itself. We usually obtain reliable counts by performing a 1:200–1:500 dilutions. Counting is done using a hemocytometer following manufacturer instructions. Perform at least four individual counts to ensure statistical accuracy.
8. A critical point is to keep resuspending the sperm suspension during the aliquots preparation, in order to avoid sedimentation and variability in the final concentration of sperm in the individual aliquots.

4. CHROMATIN BINDING OF DNA REPLICATION FACTORS

Xenopus egg extract can be used to investigate chromatin binding dynamics of DNA replication factors and other proteins involved in chromatin modeling and DNA transactions. We usually carry out DNA replication reactions with 4×10^3 nuclei/μL extract. In these conditions, complete replication of sperm nuclei can be achieved in about 2 h and chromatin-binding dynamics of individual factors can be assessed by recovering aliquots of the reaction at defined time points (e.g. 1, 30, 60, 90, and 120 min after sperm addition to the extract).

4.1 Equipment

1 mL syringe.

4.2 Buffers and Reagents

- EB (high stringency, HS): 50 m*M* Hepes-KOH, pH 7.5; 100 m*M* KCl; 2.5 m*M* $MgCl_2$.
- EB-HS-NP40: EB buffer + 0.25% NP40.
- EB-HS-NP40-Sucrose: EB buffer + 0.25% NP40 + 30% (w/v) sucrose.

4.3 Procedure

1. Thaw frozen *Xenopus* interphase extract. Prepare 50–200 μL extract (the amount of extract used can be optimized on the basis of the expected amount of protein to be detected in the chromatin fraction and the antibody sensitivity for the detection).
2. Add the appropriate amount of demembranated *Xenopus* sperm nuclei to reach a final concentration of 4×10^3 nuclei/μL.
3. Incubate at 23°C.
4. Collect 10–40 μL extract for each time-point in 2-mL Eppendorf tubes.
5. Stop the reaction of the corresponding sample by diluting the extract with 10 volumes of cold EB-HS-NP40 and resuspend the sample by pipetting.
6. Gently overlay the samples on 10 volumes of cold EB-HS-NP40-Sucrose.

7. Spin at 8300 × *g* at 4°C for 5 min in a swinging-bucket rotor (TLA 100.3, Beckman).
8. Carefully remove the supernatant and the dense sucrose layer without disturbing the pellet. Usually 30 μL of liquid are left at the bottom of the tube.
9. Resuspend the pellet in 10 volumes of cold EB-HS buffer.
10. Centrifuge the samples in a benchtop refrigerated microcentrifuge at max speed for 5 min. Use a fixed-angle rotor and properly orient the vials to recognize the position of the pellet.
11. Aspirate the supernatant with a 1-mL syringe (place the needle against the part of tube opposite to the expected pellet position not to disturb it).
12. Resuspend pellets with 20–30 μL Laemmli gel loading buffer (the chromatin pellet is barely visible).
13. Denature the samples for 3 min at 100°C and proceed with SDS-PAGE and WB analysis.

5. ACTIVATION OF ATM-, ATR-, AND DNA-PK-DEPENDENT DNA DAMAGE RESPONSE

5.1 Introduction

ATM, ATR, and DNA-PK kinases activation can be triggered by adding linear DNA to egg cytoplasm or by inducing DNA lesions directly on sperm before or after incubation in egg cytoplasm. Linear DNA can be obtained either by PCR amplification of defined sequences or by enzymatic digestion of intact plasmids.

5.2 Equipment

- P81 Phosphocellulose Squares (Merck, 20-134).

5.3 Buffers and Reagents

- 1 mg/mL linear DNA fragments (0.1–1 kb long linear double-stranded DNA) in water.
- Osmotic lysis solution (0.32 *M* sucrose; 1 m*M* $NaHCO_3$, pH 7.2).
- EB buffer: 50 m*M* Hepes, pH 7.5; 50 m*M* KCl; 2.5 m*M* $MgCl_2$.
- EB-Sucrose: EB buffer + 30% (w/v) sucrose.
- Histone H2AX C-terminal peptide (peptide sequence: KAPSGGKKATQASQEY).

- EB kinase buffer (20 m*M* Hepes-KOH, pH 7.5: 50 m*M* NaCl: 10 m*M* $MgCl_2$: 1 m*M* DTT: 1 m*M* NaF: 1 m*M* Na_3VO_4: 10 m*M* $MnCl_2$; 0.5 mg/mL Histone H2AX C-terminal peptide; 50 μ*M* ATP; 1 μL [γ-^{32}P]ATP 6000 Ci/mmol).
- ATR inhibitor VE-822 (Selleckchem, Cat No. S7102).
- ATM inhibitor KU-55933 (Sigma, Cat No. SML1109).
- DNA-PK inhibitor KU-57788 (Selleckchem, Cat No. S2638).
- EcoRI restriction enzyme (New England Biolabs).

5.4 Procedures

5.4.1 DNA Damage Response Activation in Cytoplasmic and Nuclear Extracts

1. 50 ng/μL linear DNA are added to 10–200 μL interphase egg extract prepared as described above. Control reactions are supplemented with an equivalent volume of buffer.
2. Reactions are incubated at 23°C for 30 min.
3. To activate ATM, ATR, and DNA-PK in nuclei, 50–300 μL egg extracts are incubated with 4×10^3 nuclei and 0.05–0.1 units/μL EcoRI restriction enzyme at 23°C for 60–120 min.

5.4.2 Monitoring ATM, ATR, and DNA-PK Kinase Activity

1. Prepare tubes containing 20 μL EB kinase buffer mix supplemented with 1 μL DMSO 10% or 50 μ*M* VE-822, KU-55933, or KU-57788.
2. Add 2.5 μL egg cytoplasm or 2.5 μL nuclei extract to these kinase mixes.
3. To obtain nuclei extracts, isolate nuclei by diluting of extract nuclei mix with 1 mL of EB buffer. Samples are then layered onto 0.4 mL EB-sucrose in 2-mL Eppendorf tubes and centrifuged at 3000 × *g* for 10 min at 4°C. Nuclei are then washed in EB, resuspended in three volumes of osmotic lysis solution, and then pelleted at 14,000 rpm in an Eppendorf benchtop refrigerated centrifuge for 20 min. The supernatant is collected and placed in a separate tube.
4. Kinase reactions are incubated at 30°C for 30 min.
5. Reactions are stopped by adding 45 μL 10% ice-cold trichloroacetic acid. After vortexing 35 μL samples are spotted on 2.1-cm diameter P81 phosphocellulose filter paper.
6. Filters are air-dried and washed three times with 500 mL cold 0.5% phosphoric acid (5–10 min per wash). The progress of the washing steps can be followed by removing a P81 filter circle and checking it with a Geiger counter.

7. Filters are washed once with 200 mL acetone.
8. Radioactivity is eventually quantified in a scintillation counter.
9. Activation status of ATM, ATR, and DNA-PK is determined by comparing the count per minutes (cpms) in the presence and absence of the respective inhibitors.

6. EM SAMPLE PREPARATION FOR VISUALIZATION OF DNA REPLICATION INTERMEDIATES ISOLATED FROM *X. LAEVIS* EGG EXTRACT

The analysis of replication forks can take advantage of the small replicon size and of the abundance of DNA replication intermediates that can be isolated from egg extracts. Here, we describe a detailed procedure to stabilize and isolate DNA replication intermediates prior to EM analysis. EM procedures and DNA structure analysis have been described elsewhere (Lopes, 2009).

In absence of any DNA lesion both parental and daughter DNA strands of replication intermediates will show the typical thickness associated to the presence of double-strand DNA (Fig. 2A). Genomic DNA isolated from extracts in which DNA replication has been challenged with exogenous agents added at different times after addition of sperm nuclei (e.g., 1.5 m*M* aphidicolin added 60 min after sperm nuclei) will instead show extended ssDNA accumulation on newly duplicated strands and reverse forks formation. ssDNA accumulating at the fork junction can be easily visualized because of the difference in the thickness of the DNA fiber (Fig. 2B). Reverse forks can usually be identified by the presence of a fourth arm (Fig. 2C) in a typical Y-shaped replication intermediate (Fig. 2D).

6.1 Equipment

- Stratalinker 2400 Bulbs, 365 nm (Agilent Technologies, 400079).
- QIAGEN Genomic-tip 20/G (Qiagen, 10223).
- Amicon Ultra-0.5 Centrifugal Filter Unit with Ultracel-100 (Millipore, UFC510096).

6.2 Buffers and Reagents

- Cy3-dCTP (GE Healthcare, PA53021).
- TMP (trimethylpsoralen) stock solution: 200 μg/mL in 100% ethanol (store at 4°C, avoid light exposure).

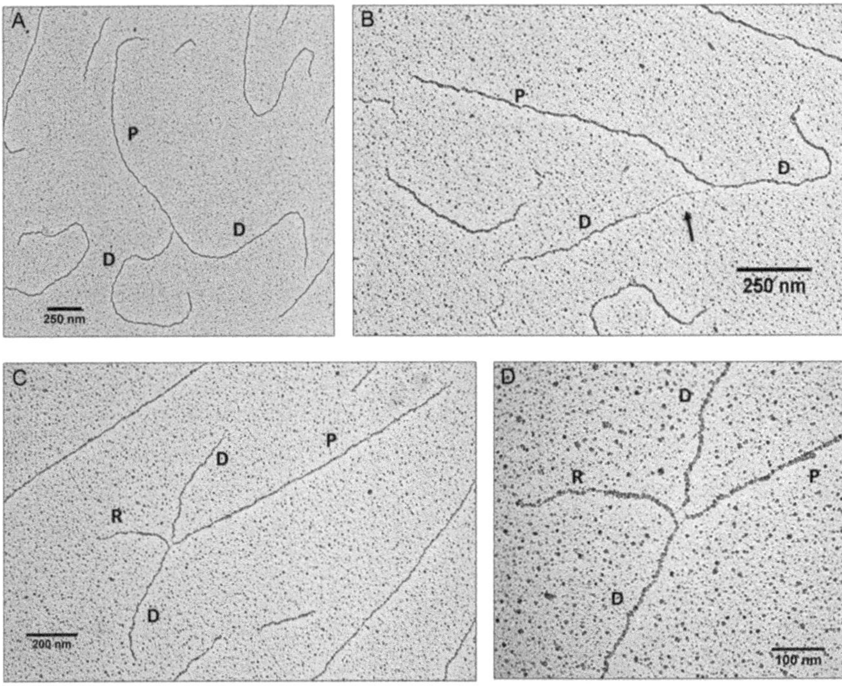

Fig. 2 EM analysis of DNA replication intermediates isolated from *Xenopus* egg extracts. (A) Typical Y-shaped DNA replication intermediate showing a parental strand (P) and two daughter strands (D) of equal length. (B) DNA replication intermediate showing ssDNA accumulating at the fork junction as indicated by the *black arrow*. (C) DNA replication intermediate showing a fourth regressed arm (R). (D) High magnification picture of the 4-way junction typical of the reverse fork structure.

- EB-EDTA buffer: 50 m*M* Hepes-KOH, pH 7.5; 100 m*M* KCl; 2,5 m*M* $MgCl_2$; 1 m*M* EDTA.
- EB-EDTA-Sucrose Buffer: EB-EDTA buffer + 30% (w/v) sucrose.
- 10% SDS stock.
- Chloroform:isoamylalcohol 24:1.
- Isopropanol.
- 70% ice-cold ethanol.
- 3 M sodium acetate pH 5.2.
- 18 mg/mL Proteinase K stock (Roche, 03115844001).
- 10 mg/mL RNase A stock (Sigma-Aldrich, R5503).
- TE buffer.
- 10 × NEBuffer 4 (New England Biolabs, B7004S).
- NdeI 20 U/μL (New England Biolabs, R0111).

- QBT buffer (Qiagen, 19054).
- 5 M NaCl.
- 10 m*M* Tris–HCl, pH 8; 300 m*M* NaCl.
- 10 m*M* Tris–HCl, pH 8; 500 m*M* NaCl.
- 10 m*M* Tris–HCl, pH 8; 1 *M* NaCl.
- 10 m*M* Tris–HCl, pH 8; 1 *M* NaCl; 1.8% (w/v) caffeine (Note 2).
- QBT buffer (Qiagen, 19054).

6.3 Procedures

6.3.1 DNA Replication

Replication reactions are carried out in large volumes to obtain sufficient amount of genomic DNA for the EM analysis.

1. Thaw frozen *Xenopus* interphase extract on ice. For each sample, prepare a tube containing 200 μL of extract.
2. Add the appropriate amount of demembranated *Xenopus* sperm nuclei to reach a final concentration of 4×10^3 nuclei/μL.
3. Prepare 2 × 1.5-mL tubes on ice for each sample.
4. Split each sample into two aliquots of 100 μL and incubate at 23°C.
5. As internal control, incubate 100 μL extract with Cy3-dCTP (5 μL of a 1:50 dilution from the stock) to monitor the efficiency of DNA replication at different time points.
6. When nuclei starts incorporating Cy3-dCTP (approximately between 45 and 75 min after the start of the incubation) reactions can be supplemented with agents interfering with DNA replication (eg., aphidicolin 10 μ*M* to 1.5 m*M*) for different lengths of time. Alternatively, sperm nuclei can be pretreated with DNA damaging agents UV or MMS as previously shown (Hashimoto et al., 2010).
7. Stop the reaction by diluting each 100 μL aliquot with 200 μL ice-cold EB-EDTA buffer and store on ice.
8. Gently lay the samples on 600 μL cold EB-EDTA-Sucrose Buffer.
9. Spin at 8300 × *g* at 4°C for 5 min in a swinging-bucket rotor (TLA 100.3, Beckman).
10. Carefully remove the supernatant and the dense sucrose layer without disturbing the pellet. Leave only a thin film of liquid. Usually about 30 μL liquid are left in the tube.
11. Resuspend the pellet in 100 μL (final volume) cold EB-EDTA buffer using a P-200 pipette with cut tips.

6.3.2 DNA Interstrand Cross-Linking

Psoralen cross-linking is required to stabilize replication intermediates prior to EM analysis.

1. Prechill a round-bottom 96-well microplate and a PCR metal rack on ice.
2. Cover the microplate lid with aluminum foil to protect from light.
3. Mount the 365 nm bulbs on the Stratalinker device.
4. Transfer the 100 μL DNA sample from the previous DNA replication procedure in the microplate in horizontal lanes.
5. Add 5 μL TMP stock solution to each 100 μL nuclei suspension and stir with the pipette tip. Incubate for 5 min in the dark (cover with foil) at 4°C or on the precooled metal rack.
6. Place the microplate on the metal rack and transfer everything on a height-adjustable support. Raise the support so that the microplate almost touches the bulbs. Try to align as much as possible the microplate lanes with the Stratalinker bulbs.
7. Irradiate at maximum power with UV-A for 7 min (this setting refers to Stratalinker UV 2400).
8. Repeat steps from 5 to 7 three more times (the four cross-linking cycles should take approximately 50 min in total).
9. Recover the nuclei suspension from the wells and rinse with 300 μL EB-EDTA buffer in order to recover eventual nuclei left in the wells. The final sample volume should be 400 μL.
10. Proceed with the DNA extraction procedure.

6.3.3 Genomic DNA Extraction From Replicated X. laevis Nuclei

1. Add 10 μL RNase A stock and 4 μL 10% SDS to the 400 μL nuclei suspension.
2. Mix and incubate for 1 h at 37°C.
3. Add 29 μL of Proteinase K stock and incubate at 50°C for 2 h (or alternatively overnight at 37°C).
4. Add an equal volume (400 μL) of chloroform:isoamyl alcohol 24:1 and deeply mix the samples.
5. Spin at 16,000 × *g* for 10 min.
6. Recover the supernatant (the aqueous phase) with a cut P-1000 tip and transfer in new 1.5-mL tubes.
7. Add one volume of isopropanol and 1:10 volume of 3 *M* sodium acetate to each sample. Mix by inversion. DNA precipitate should appear.

8. Spin the samples at 16,000 × *g* for 10 min at 4°C and remove the supernatant.
9. Wash the pellet with 500 μL 70% ethanol.
10. Spin at 16,000 × *g* for 5 min at 4°C. Remove the residual ethanol carefully with a P-20 tip.
11. If necessary, incubate the tubes at 37°C to evaporate the residual ethanol.
12. Pool together the two pellets corresponding to the same sample by resuspending them in a final volume of 100 μL of TE buffer.
13. Quantify the amount of extracted genomic DNA by Nanodrop (usually about 4–5 μg of DNA are expected starting from 200 μL of extract and 4000 nuclei/μL).

6.3.4 Gemomic DNA Digestion

Genomic DNA is digested to avoid aggregation of long DNA fibers and to help in the characterization of replication intermediates (restriction of duplicated sites will give rise to Y-shape intermediates with two replicated arms of equal length).

1. Prepare restriction enzyme digestion reactions by mixing (Note 1):
   - 100 μL DNA sample
   - 25 μL 10 × Buffer 4
   - 5 μL NdeI
   - 120 μL H_2O
2. Incubate at 37°C for 3–5 h.

6.3.5 Replication Intermediates Purification

Replication intermediates can be purified from other egg cytoplasm contaminants using Qiagen 20/G columns.

1. Equilibrate the QIAGEN Genomic-tips 20/G with 1 mL of QBT buffer. Let the liquid flow by gravity.
2. Wash the column three times with 1 mL of 10 m*M* Tris–HCl, pH 8; 1 *M* NaCl.
3. Equilibrate the column three times with 1 mL of 10 m*M* Tris–HCl, pH 8; 300 m*M* NaCl.
4. Adjust the final volume of the genomic DNA restriction reaction to 1 mL with 10 m*M* Tris–HCl, pH 8; 300 m*M* NaCl.
5. Load the 1 mL DNA mixtures on the 20/G tip column and let flow by gravity.

6. Wash the column two times with 1 mL of 10 m*M* Tris–HCl, pH 8; 500 m*M* NaCl.
7. Elute the DNA with 600 μL of 10 m*M* Tris–HCl, pH 8; 1 *M* NaCl; 1.8% caffeine. Repeat this step once again in order to increase DNA yield. Collect the two eluates in the same tube.

6.3.6 Final Purification and Quality Check

DNA obtained either from Qiagen 20/G columns purification is then further cleaned and concentrated, using Amicon size-exclusion devices.

1. Load the 600 μL DNA solution obtained from the elution step in the Amicon centrifugal filter and spin for 10 min at 2400 × *g* at room temperature (repeat this step twice for the 1.2 mL sample obtained from the Qiagen 20/G tips purification procedure).
2. Wash the membrane twice with 200 μL TE buffer (ensure the sample volume is down to about 30 μL before proceeding with each wash).
3. Spin the filters for 10 min at 10000 × *g* until 10–30 μL remain.
4. Recover the concentrated DNA mix by spinning the Amicon devices upside down.
5. Load 5 μL DNA samples on a 0.8% agarose gel to check for DNA quality and concentration.

6.4 Notes

1. As a standard reference 5 μg genomic DNA is usually digested with 100 U of NdeI for 3 h at 37°C. Units of enzyme and timing of digestion can be adjusted each time according to the purified DNA amount and the estimated efficiency of the restriction enzyme.

7. CONCLUSIONS

The procedures described here are intended to provide a guide through the most critical aspects of the protocols concerning the use of *Xenopus* egg extract to study DNA damage response and DNA replication. However, researchers using this system need to be aware that one of the most important factor for successful experiments is the quality of the biological material. Therefore, extreme care should be dedicated to the selection and maintenance of the animals providing eggs and sperm nuclei.

ACKNOWLEDGMENTS

We thank the members of the Costanzo lab for helpful discussion. Work in V.C. laboratory is funded by the Associazione Italiana per Ricerca sul Cancro (AIRC), the European Research Council (ERC) Consolidator Grant (614541), the Giovanni Armenise-Harvard foundation award, the Epigen Progetto Bandiera (4.7), the AICR-Worldwide Cancer Research (13-0026), and the Fondazione Telethon (GGP13-071) grant awarded to V.C.

REFERENCES

Aze, A., Sannino, V., Soffientini, P., Bachi, A., & Costanzo, V. (2016). Centromeric DNA replication reconstitution reveals DNA loops and ATR checkpoint suppression. *Nature Cell Biology*, *18*(6), 684–691.

Costanzo, V., & Gautier, J. (2004). Xenopus cell-free extracts to study DNA damage checkpoints. *Methods in Molecular Biology*, *241*, 255–267.

Costanzo, V., Paull, T., Gottesman, M., & Gautier, J. (2004a). Mre11 assembles linear DNA fragments into DNA damage signaling complexes. *PLoS Biology*, *2*(5), E110.

Costanzo, V., Robertson, K., & Gautier, J. (2004b). Xenopus cell-free extracts to study the DNA damage response. *Methods in Molecular Biology*, *280*, 213–227.

Costanzo, V., Robertson, K., Ying, C. Y., Kim, E., Avvedimento, E., Gottesman, M., et al. (2000). Reconstitution of an ATM-dependent checkpoint that inhibits chromosomal DNA replication following DNA damage. *Molecular Cell*, *6*(3), 649–659.

Errico, A., Cosentino, C., Rivera, T., Losada, A., Schwob, E., Hunt, T., et al. (2009). Tipin/Tim1/And1 protein complex promotes pol alpha chromatin binding and sister chromatid cohesion. *The EMBO Journal*, *28*(23), 3681–3692.

Errico, A., & Costanzo, V. (2010). Differences in the DNA replication of unicellular eukaryotes and metazoans: Known unknowns. *EMBO Reports*, *11*(4), 270–278.

Errico, A., & Costanzo, V. (2012). Mechanisms of replication fork protection: A safeguard for genome stability. *Critical Reviews in Biochemistry and Molecular Biology*, *47*(3), 222–235.

Garner, E., & Costanzo, V. (2009). Studying the DNA damage response using in vitro model systems. *DNA Repair*, *8*(9), 1025–1037.

Hashimoto, Y., & Costanzo, V. (2011). Studying DNA replication fork stability in Xenopus egg extract. *Methods in Molecular Biology*, *745*, 437–445.

Hashimoto, Y., Puddu, F., & Costanzo, V. (2011). RAD51- and MRE11-dependent reassembly of uncoupled CMG helicase complex at collapsed replication forks. *Nature Structural & Molecular Biology*, *19*(1), 17–24.

Hashimoto, Y., Ray Chaudhuri, A., Lopes, M., & Costanzo, V. (2010). Rad51 protects nascent DNA from Mre11-dependent degradation and promotes continuous DNA synthesis. *Nature Structural & Molecular Biology*, *17*(11), 1305–1311.

Jazayeri, A., Balestrini, A., Garner, E., Haber, J. E., & Costanzo, V. (2008). Mre11-Rad50-Nbs1-dependent processing of DNA breaks generates oligonucleotides that stimulate ATM activity. *The EMBO Journal*, *27*(14), 1953–1962.

Lohka, M. J., & Masui, Y. (1983). Formation in vitro of sperm pronuclei and mitotic chromosomes induced by amphibian ooplasmic components. *Science*, *220*(4598), 719–721.

Lopes, M. (2009). Electron microscopy methods for studying in vivo DNA replication intermediates. *Methods in Molecular Biology*, *521*, 605–631.

Murray, A. W. (1991). Cell cycle extracts. *Methods in Cell Biology*, *36*, 581–605.

Sannino, V., Kolinjivadi, A. M., Baldi, G., & Costanzo, V. (2016). Studying essential DNA metabolism proteins in Xenopus egg extract. *The International Journal of Developmental Biology*, *60*(7-8-9), 221–227.

Smith, E., Dejsuphong, D., Balestrini, A., Hampel, M., Lenz, C., Takeda, S., et al. (2009). An ATM- and ATR-dependent checkpoint inactivates spindle assembly by targeting CEP63. *Nature Cell Biology*, *11*(3), 278–285.

Trenz, K., Smith, E., Smith, S., & Costanzo, V. (2006). ATM and ATR promote Mre11 dependent restart of collapsed replication forks and prevent accumulation of DNA breaks. *The EMBO Journal*, *25*(8), 1764–1774.

CHAPTER TEN

Ensemble and Single-Molecule Analysis of Non-Homologous End Joining in Frog Egg Extracts

Thomas G.W. Graham[*,1], **Johannes C. Walter**[*,†,2], **Joseph J. Loparo**[*,2]
[*]Harvard Medical School, Boston, MA, United States
[†]Howard Hughes Medical Institute, Harvard Medical School, Boston, MA, United States
[2]Corresponding authors: e-mail address: johannes_walter@hms.harvard.edu; joseph_loparo@hms.harvard.edu

Contents

Abstract

Non-homologous end joining (NHEJ) repairs the majority of DNA double-strand breaks in human cells, yet the detailed order of events in this process has remained obscure. Here, we describe how to employ *Xenopus laevis* egg extract for the study of NHEJ. The egg extract is easy to prepare in large quantities, and it performs efficient end joining that requires the core end joining proteins Ku, DNA-PKcs, XLF, XRCC4, and DNA ligase IV.

[1] Current address: Laboratory of Neurophysiology and Behavior, The Rockefeller University, New York, NY, United States.

Methods in Enzymology, Volume 591
ISSN 0076-6879
http://dx.doi.org/10.1016/bs.mie.2017.03.020

These factors, along with the rest of the soluble proteome, are present at endogenous concentrations, allowing mechanistic analysis in a system that begins to approximate the complexity of cellular end joining. We describe an ensemble assay that monitors covalent joining of DNA ends and fluorescence assays that detect joining of single pairs of DNA ends. The latter assay discerns at least two discrete intermediates in the bridging of DNA ends.

1. INTRODUCTION

DNA double-strand breaks (DSBs) are common and extremely toxic DNA lesions that must be repaired to maintain genomic integrity. The major pathway for DSB repair in human cells is nonhomologous end joining (NHEJ), which rejoins DNA ends by direct ligation. NHEJ is a versatile mechanism that can ligate incompatible and chemically damaged ends. It does this by using a variety of different enzymes, including polymerases and exonucleases, to make DNA ends compatible for joining. How end processing is regulated to minimize mutations is poorly understood.

During NHEJ, ends are first bound by a heterodimer of the Ku70 and Ku80 proteins (Ku), which encircles the broken DNA end like a ring (Walker, Corpina, & Goldberg, 2001). Ku recruits the DNA-dependent protein kinase catalytic subunit (DNA-PKcs), creating the DNA-PK holoenzyme (Carter, Vancurová, Sun, Lou, & DeLeon, 1990; Dvir, Peterson, Knuth, Lu, & Dynan, 1992; Dvir, Stein, Calore, & Dynan, 1993; Gottlieb & Jackson, 1993; Lees-Miller, Chen, & Anderson, 1990). DNA-PK phosphorylates a number of proteins, including itself, and DNA-PKcs autophosphorylation is important for its function (Dobbs, Tainer, & Lees-Miller, 2010; Jette & Lees-Miller, 2015; Jiang et al., 2015). A variety of different enzymes, including DNA polymerases λ and μ, polynucleotide kinase/phosphatase (PNKP), aprataxin, aprataxin and PNKP-like factor (APLF), and Artemis, are used to process damaged or mismatched ends and make them compatible for joining (Menon & Povirk, 2016; Waters, Strande, Wyatt, Pryor, & Ramsden, 2014). Ends are ligated by DNA ligase IV (LIG4), which resides in a complex with its essential accessory factor XRCC4 (Critchlow, Bowater, & Jackson, 1997; Grawunder et al., 1997; Li et al., 1995). XRCC4 in turn binds to its homolog XRCC4-like factor (XLF), which stimulates the activity of the LIG4:XRCC4 complex (Ahnesorg, Smith, & Jackson, 2006; Buck et al., 2006; Gu, Lu, Tsai, Schwarz, & Lieber, 2007; Hentges et al., 2006; Lu, Pannicke, Schwarz, & Lieber, 2007; Tsai, Kim, & Chu, 2007). XLF also interacts with the Ku-DNA complex (Yano, Morotomi-Yano, Lee, & Chen, 2011; Yano et al., 2008).

Recently, a new paralog of XRCC4 and XLF (PAXX) has been discovered, which is redundant with XLF in some but not all contexts (Balmus et al., 2016; Craxton et al., 2015; Kumar, Alt, & Frock, 2016; Lescale et al., 2016; Liu, Shao, Jiang, Lee, & Zha, 2017; Ochi et al., 2015; Roy et al., 2015; Tadi et al., 2016; Xing et al., 2015). Unlike XLF, PAXX does not interact with XRCC4, although it does interact with Ku.

Cell-free systems provide a powerful tool to study biochemical mechanisms. Cell-free extracts derived from human cells recapitulate NHEJ, although the efficiency of end joining is generally low (Akopiants et al., 2009; Baumann & West, 1998; Chappell, Hanakahi, Karimi-Busheri, Weinfeld, & West, 2002; Cortes et al., 1996; Feldmann, Schmiemann, Goedecke, Reichenberger, & Pfeiffer, 2000; Hanakahi, Bartlet-Jones, Chappell, Pappin, & West, 2000; Jayaram, Ketner, Adachi, & Hanakahi, 2008; Labhart, 1999a; Lee et al., 2004; Lee, Yannone, Chen, & Povirk, 2003; Pfeiffer, Feldmann, Odersky, Kuhfittig-Kulle, & Goedecke, 2005; Pfeiffer, Odersky, Goedecke, & Kuhfittig-Kulle, 2014; Smeaton, Miller, Ketner, & Hanakahi, 2007; Weis-Garcia et al., 1997). Similarly, defined systems using purified NHEJ proteins join DNA ends less efficiently than in vivo NHEJ, possibly because they do not contain all of the proteins that participate in and regulate NHEJ in the cell (Chang et al., 2016; Ma & Lieber, 2006; Ma et al., 2004; Nick McElhinny et al., 2005; Tsai et al., 2007). For these reasons, a whole egg lysate from the frog *Xenopus laevis* is an attractive alternative. These extracts contain the entire proteome of the egg and can be produced in large quantities. Early experiments showed that egg extracts join DNA ends rapidly at room temperature dependent on Ku and DNA-PK (Chen et al., 2001; Di Virgilio & Gautier, 2005; Labhart, 1999b; Postow et al., 2008; Taylor et al., 2010). More recently, we showed that end joining additionally requires XLF, XRCC4-LIG4, and LIG4 catalytic activity (Graham, Walter, & Loparo, 2016).[a] In addition to joining compatible DNA ends, egg extract joins incompatible ends by filling in or degrading incompatible overhangs (Chen et al., 2001; Daza et al., 1996; Di Virgilio & Gautier, 2005; Gu, Bennett, & Povirk, 1996; Labhart, 1999b; Pfeiffer & Vielmetter, 1988; Sandoval & Labhart, 2002; Thode, Schäfer,

[a] End joining independent of Ku and DNA-PKcs may also occur in egg extract (Di Virgilio & Gautier, 2005; Graham et al., 2016; Labhart, 1999b; Sandoval & Labhart, 2002). The relative contributions of classical Ku/DNA-PKcs-dependent and alternative Ku/DNA-PKcs-independent end joining pathways seem to depend on the exact reaction conditions and DNA substrates employed. Under the conditions described here, end joining occurs primarily through the classical pathway (Graham et al., 2016).

Pfeiffer, & Vielmetter, 1990; Zhu & Peng, 2016). These processing events depend on the factors mentioned earlier (our unpublished results), indicating that egg extract will be useful to elucidate how end processing is regulated during NHEJ.

Here, we describe ensemble and single-molecule NHEJ assays in egg extract. Ensemble assays report on the processing and ligation of DNA ends by monitoring DNA products, while single-molecule assays reveal transient intermediates in end joining. We describe two variations of a single-molecule fluorescence assay in which DNA substrates are labeled near their ends with the fluorescent dyes Cy3 and Cy5 and bridging of ends is indicated by colocalization and Förster resonance energy transfer (FRET) between these two fluorophores.

2. ENSEMBLE END JOINING ASSAYS

In this section, we describe ensemble biochemical assays for monitoring end joining in egg extract. We typically use a high-speed supernatant (HSS) of unfertilized eggs arrested in interphase (Lebofsky, Takahashi, & Walter, 2009). However, end joining also occurs in low-speed egg lysates that contain membranes and in "CSF-arrested" extract that is arrested in metaphase II of meiosis (Di Virgilio & Gautier, 2005; Postow et al., 2008). Highly concentrated nucleoplasmic extracts derived from either in vitro-reconstituted nuclei or germinal vesicles can join DNA ends under certain conditions, although they primarily tend to resect DNA ends and may join ends through alternative end joining (Alt-EJ) mechanisms (Budzowska, Graham, Sobeck, Waga, & Walter, 2015; Lehman & Carroll, 1991; Lehman, Clemens, Worthylake, Trautman, & Carroll, 1993; Liao, Toczylowski, & Yan, 2008; Toczylowski & Yan, 2006; Yan, McCane, Toczylowski, & Chen, 2005; B. Stinson & R. Amunugama, personal communication).

Linear DNA fragments introduced into egg extract are joined end to end by the NHEJ machinery. This yields circular or linear end joining products (Fig. 1A), the ratio of which varies depending on substrate length and concentration. Egg extracts including HSS are also capable of performing the 5′–3′ end resection step that initiates homologous recombination or Alt-EJ (Liao, Guay, Toczylowski, & Yan, 2012; Liao, Toczylowski, & Yan, 2011). We find that the overall DNA concentration added to extract influences the types of products formed and the balance between end joining and resection. We include 10–100 ng/μL

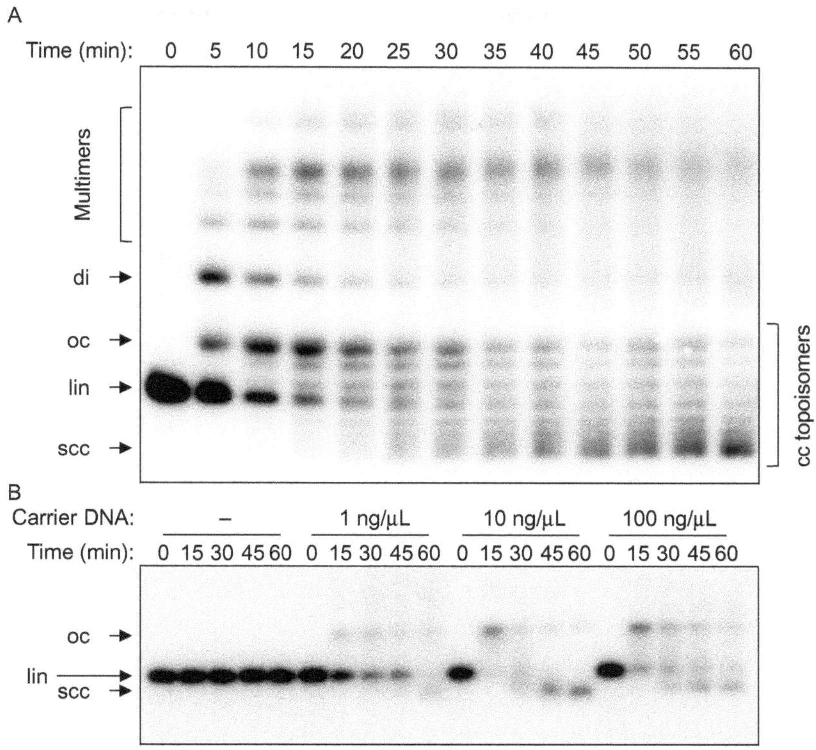

Fig. 1 Ensemble end joining assay. (A) Time course of an end joining reaction, showing conversion of linear substrate (lin) into open circular (oc), supercoiled closed-circular (scc), dimeric (di), and multimeric products. Supercoiling of closed-circular DNA in extract arises from nucleosome assembly. Closed-circular (cc) topoisomers are visible between the oc and scc bands. The substrate DNA band is slightly overexposed at the 0-min timepoint. (B) Carrier DNA dependence of end joining. A very low concentration of substrate DNA (~0.05 ng/μL) was incubated with an extract containing different amounts of closed-circular carrier DNA. End joining does not occur at all if the overall DNA concentration is too low.

of closed-circular plasmid DNA as a "carrier" in our reactions, as this suppresses resection and promotes end joining (through an unknown mechanism; see Fig. 1B). The efficiency of DNA replication in egg extract has similarly been shown to depend on the total DNA concentration (Lebofsky, van Oijen, & Walter, 2011). The optimal concentration of carrier DNA may vary between different preparations of extract. Importantly, the effect of carrier DNA does not depend on the presence of homology between the substrate and the carrier DNA, as both heterologous and homologous carrier plasmids support efficient end joining.

We typically incubate end joining reactions at room temperature; however, joining at 13–15°C has also been reported (Gu et al., 1996; Labhart, 1999b).

End joining reactions may be performed with unlabeled DNA, which is then separated on an agarose gel and stained with an intercalating dye such as SYBR Gold. Alternatively, substrate DNA may be site-specifically labeled with a fluorescent dye or a radioactive nucleotide. Below, we describe how to perform end joining reactions using radiolabeled substrate DNA, which provides high sensitivity and good signal to noise for reactions with small amounts of substrate. At low substrate DNA concentrations (~1 ng/μL), the products formed are predominantly circles, which result from joining of the two ends of the same molecule of DNA. We commonly use linearized 3–5 kb plasmid DNA to permit efficient circularization.

2.1 Substrate Preparation

2.1.1 Materials

- Maxiprepped plasmid DNA containing a single *Eco*RI site
- EcoRI-HF® (New England Biolabs, Cat. #R3101). Any restriction enzyme may be used, provided that it cuts the plasmid to leave 5′ overhangs that can be filled in with a suitable radiolabeled dNTP
- Molecular biology grade agarose
- Horizontal gel electrophoresis rig, casting tray (~10 cm × 10 cm), and comb
- Heavy-duty plastic packing tape
- DNA ladder (e.g., 2-log ladder from New England Biolabs, Cat. #N3200)
- Klenow fragment of DNA polymerase I, 5000 U/mL (New England Biolabs, Cat. #M0210L). Note: Do not use the 3′–5′ exonuclease-deficient version of the Klenow fragment, as this enzyme can add non-templated nucleotides to blunt-ended DNA
- [α-^{32}P]-dATP (EasyTides, Perkin Elmer)
- Unlabeled dCTP, dGTP, and dTTP
- Ethidium bromide, 10 mg/mL solution
- 6 × gel loading dye (New England Biolabs, Cat. #B7024S)
- Tris–borate–EDTA (TBE) buffer: Dissolve the following per liter to make a 10 × stock: 108 g of Tris base, 55 g of boric acid, and 40 mL of 0.5 M EDTA solution, pH 8.0. Use EDTA, not EDTA disodium salt, and adjust the pH of the EDTA stock solution to 8.0 with sodium hydroxide.
- Long-wavelength (365 nm) UV light box

- Dialysis tubing (SpectraPor #132650, 23 mm, 6–8 kDa MWCO) and clips
- Cooled microcentrifuge
- Isobutanol, equilibrated with aqueous buffer at pH 8
- 3 *M* sodium acetate, pH 5.2
- 100% and 70% ethanol
- Clean razor blade or scalpel
- Broad-tipped forceps
- 1 × TE buffer: 10 m*M* Tris–HCl, pH 8, 1 m*M* EDTA
- 10 m*M* Tris–HCl, pH 7.5
- PCR purification kit (Qiagen)
- Nanodrop spectrophotometer

2.1.2 Protocol

1. Digest 100 μg of plasmid DNA with 5 μL of 20 U/μL *Eco*RI in a 100 μL volume of CutSmart buffer at 37°C for 3 h.
2. Tape all wells except one of a gel loading comb together with packing tape, and cast a 1 × TBE agarose gel containing 1 μg/mL ethidium bromide. Add 20 μL of loading dye to the DNA digestion reaction and load in the large well. Load a DNA ladder in the small well. Separate by electrophoresis at 7.5 V/cm.
3. On a UV light box, cut out the band containing the DNA with a sharp razor blade or a scalpel. To prevent photodamage, avoid excessive exposure of the DNA to UV light. Alternatively, the band may be visible under room light when viewed against a white background.
4. Electroelute the DNA from the gel slice:
   a. Cut a piece of dialysis tubing several centimeters longer than the gel slice. Rinse the tubing with water and open the ends by rubbing them between your thumb and forefinger. Clamp one end with a plastic clip.
   b. Submerge the dialysis tubing in a gel box containing 1 × TBE to fill it with buffer, pushing out any bubbles.
   c. Slide the product band (preferably as a single piece of agarose) into the dialysis bag, squeeze out any bubbles and excess buffer, and clamp the other end.
   d. Add ethidium bromide to the 1 × TBE running buffer to a final concentration of ~1 μg/mL.
   e. Orient the tubing perpendicular to the direction of the electric field and push the gel slice to the edge of the tubing nearest the cathode.

- **f.** Apply voltage at 7.5 V/cm. The DNA will migrate out of the gel slice and accumulate on the opposite edge of the dialysis tubing. The stained DNA is often visible in room light, and it can also be visualized more clearly under ultraviolet light.
- **g.** After all of the DNA has exited the gel (about 30 min to 1 h), remove one of the dialysis clips, and pull out the gel slice with a pair of forceps, being careful not to disturb the DNA that has accumulated on the opposite side of the dialysis tubing.
- **h.** Carefully rub the dialysis tubing between your thumb and forefinger to free DNA bound to the sides, being careful not to spill the solution from the tubing. Pipette out the DNA.
- **i.** To remove ethidium bromide, extract the DNA solution twice with 1 volume of isobutanol. Mix well at each extraction step by vortexing, and centrifuge briefly to separate the phases. Isobutanol forms the top phase.
- **j.** Precipitate DNA by adding 0.1 volumes of 3 *M* sodium acetate, pH 5.2 and 2.5 volumes of 100% ethanol. Incubate at −20°C for ≥10 min and spin at 16,000 × *g* for 30 min at 4°C in a microcentrifuge. Large volumes can be divided between microcentrifuge tubes or spun at ≥4000 × *g* in 15 mL conical tubes in a swinging bucket rotor. Wash the pellet briefly with 70% ethanol and centrifuge again for 1 min. For samples divided between multiple tubes or large samples in 15 mL conical tubes, disrupt all pellets by pipetting up and down with 70% ethanol, transfer to a single microcentrifuge tube, and centrifuge for 10 min at 16,000 × *g*. Thoroughly aspirate the 70% ethanol and dissolve the pellet in 100 μL of 10 m*M* Tris–HCl pH 7.5. Determine the concentration by 280 nm absorbance on a low-volume UV–Vis spectrophotometer (e.g., Nanodrop).

5. Set up the following radiolabeling reaction:

- **a.** 1.0 μg of gel purified linear DNA fragment
- **b.** 2 μL 10 × NEBuffer 2 (New England Biolabs; supplied with Klenow fragment)
- **c.** 1 μL [α-^{32}P]-dATP
- **d.** 0.67 μL each 1 m*M* dTTP, dCTP, and dGTP. All dNTPs are included in the reaction to prevent resection by the 3′–5′ exonuclease activity of the Klenow fragment
- **e.** 0.5 μL of 5000 U/mL Klenow fragment
- **f.** Water to 20 μL

6. Incubate at 25°C for 15 min.

7. Stop the reaction by adding 100 μL of PCR purification kit buffer PB (Qiagen). Purify the DNA using a PCR purification kit (Qiagen) following the manufacturer's instructions. Elute the DNA from the spin column with 50 μL of 10 m*M* Tris–HCl, pH 7.5. Full recovery of the DNA would yield a final concentration of 20 ng/μL.

2.2 End Joining Reactions

2.2.1 Materials

- HSS of unfertilized *X. laevis* eggs (Lebofsky et al., 2009). We typically flash-freeze 33 μL aliquots in liquid nitrogen and store them at −80°C.
- Nocodazole (0.5 mg/mL solution in dimethylsulfoxide)
- Creatine phosphokinase (CPK; Sigma, #C-3755, 35,000 U; Type I from rabbit muscle), 5 mg/mL solution in 10 m*M* HEPES, pH 7.5, 50% glycerol, 50 m*M* NaCl. Store at −20°C.
- Adenosine triphosphate (ATP; Sigma, #A-5394), 200 m*M* solution. Adjust the pH to approximately 7 with NaOH using pH indicator paper, and store aliquots at −20°C.
- Phosphocreatine (PC; Sigma, #P-6502), 1 *M* solution in 10 m*M* KH_2PO_4, pH 7.0. Store at −20°C.
- Closed circular plasmid DNA, 1 μg/μL in 1 × TE buffer
- Stop solution/loading dye: 8 m*M* EDTA, 0.13% phosphoric acid, 10% Ficoll, 5% SDS, 0.2% bromophenol blue, 80 m*M* Tris, pH 8
- Proteinase K, 1 mg/mL solution
- Hybond-XL membrane (GE Healthcare Life Sciences)
- Heavy Whatman filter paper
- C-fold or multifold paper towels
- Plastic wrap
- Large book or other flat, heavy object
- Vacuum gel dryer
- Storage phosphorscreen and phosphorimager

2.2.2 Protocol

1. Prepare an ATP regeneration system by combining the following:
   - 0.5 μL of 5 mg/mL CPK
   - 5 μL of 200 m*M* ATP
   - 10 μL of 1 *M* PC
2. Add 0.5 μL of 0.5 mg/mL nocodazole to a single 33 μL aliquot of egg extract, and mix thoroughly by pipetting, being careful to avoid bubbles.

Also add 1 μL of ATP regeneration system and 3.3 μL of 1 μg/μL closed-circular carrier plasmid. Mix thoroughly.

3. For each end joining reaction, combine 10 μL of this extract mixture with 0.5 μL of radiolabeled substrate DNA on ice.
4. Withdraw an initial 2 μL sample into 5 μL of stop solution/loading dye. Transfer the reactions to room temperature and withdraw additional 2 μL samples at desired time points, e.g., 20, 40, and 60 min.
5. Treat the stopped reactions with proteinase K (1 μg per sample) for 30 min at 37°C or overnight at room temperature.
6. Load the samples on a 0.8% 1 × TBE agarose gel and separate by electrophoresis for 1.5–2 h at 7.5 V/cm.
7. Sandwich the gel between two pieces of the Hybond-XL membrane to trap the DNA and then sandwich this between two pieces of a heavy Whatman filter paper. Place the gel sandwich on a stack of paper towels, cover with a piece of plastic wrap, and compress under a large book or other heavy, flat object for at least 15 min to remove most of the buffer from the gel.
8. Remove the paper towels and dry the gel for at least 1 h on a vacuum gel dryer set to 80°C.
9. Remove the Whatman filter paper, cover the dried gel/Hybond-XL sandwich with a single layer of plastic wrap, and expose to a storage phosphorscreen for several hours to overnight. Image exposed screen on a phosphorimager.

2.2.3 Notes

1. For imaging nonradiolabeled DNA by intercalating dye staining, treat the samples with RNase A (2 μg per sample for 30 min at 37°C) prior to PK treatment to degrade endogenous egg RNA in the extract. Separate DNA on a gel without intercalating dye and stain the gel after electrophoresis.
2. The amount of DNA in different bands may be quantified using image analysis software such as ImageJ. The intensity of radiolabeled DNA summed over the entire lane usually decreases over the course of the reaction, likely due to some fraction of the substrate DNA being resected. End joining efficiency varies between batches of extract, but the majority of labeled DNA remaining at 30 min is typically found in joined products (Fig. 1).

2.3 Immunodepletion Analysis

Immunodepletion of proteins from egg extract can be used to study their functions in end joining, both in ensemble and in single-molecule assays. Custom antibodies are raised against the desired factor by immunizing rabbits with the purified full-length protein, a fragment of the protein, or a synthetic peptide. We have had good success with antibodies directed against 12- to 16-amino acid synthetic peptides derived from the C-termini of proteins of interest. The antibody is typically affinity-purified from serum by coupling the antigen to a solid support (e.g., SulfoLink resin [Thermo Fisher]), passing crude serum over the resin, eluting bound antibody with low pH, and rapidly neutralizing the eluate with concentrated high-pH Tris buffer (Harlow & Lane, 1988). Commercial services are available for immunization and bleeding of rabbits and affinity purification of antibody.

2.3.1 *Materials*

- Affinity-purified antibody (1 mg/mL)
- Nonspecific IgG antibody (see Note 1)
- Phosphate-buffered saline (PBS; Teknova): 135 m*M* NaCl, 2.7 m*M* KCl, 4.3 m*M* Na_2HPO_4, 1.4 m*M* KH_2PO_4
- PBS + 0.02% sodium azide (Warning: Sodium azide is toxic!)
- Protein A sepharose CL-4B beads (GE Healthcare Life Sciences), stored as a 33% slurry in PBS + 0.02% sodium azide
- Low-binding 0.65 mL and 1.7 mL microcentrifuge tubes (Corning Costar® #3206 and Sigma Aldrich #T3406)
- Swinging bucket or horizontal microcentrifuge
- Vacuum aspirator
- Gel loading pipette tips
- Ultrafine gel loading pipette tips (USA Scientific, 0.25 mm OD orifice, #1022-8950)
- Egg lysis buffer: 10 m*M* HEPES, pH 7.7, 50 m*M* KCl, 2.5 m*M* $MgCl_2$, 250 m*M* sucrose. Filter sterilize and store at 4°C.
- Clean scissors

2.3.2 *Protocol*

1. Bead preparation:
   a. To prepare enough beads for several depletion experiments, dispense 300 μL of a 33% protein A sepharose bead slurry (i.e., 100 μL bed

volume) to a 1.7-mL low-binding microcentrifuge tube. Centrifuge for 30 s at 2000 × *g* to pellet the beads. Aspirate most of the supernatant with a vacuum aspirator fitted with a gel loading tip.

b. Add 300 μg of 1 mg/mL affinity-purified antibody and incubate on a rotary mixer for 1 h at room temperature. A larger quantity of antibody is necessary in some cases for proteins that are difficult to deplete. Crude serum can sometimes be used without affinity purification.

c. Wash beads three times with 1 mL of PBS, centrifuging 30 s at 2000 × *g* between washes. Aspirate thoroughly with a gel loading tip followed by an ultrafine gel loading tip, and resuspend with 2 bed volumes (200 μL) of PBS + 0.02% sodium azide to make a 33% slurry.

2. Immunodepletion:

a. Dispense 0.1 bed volumes of beads for each volume of extract you wish to deplete in two 0.65 mL low-binding microcentrifuge tubes (see Note 2). For instance, to deplete 33 μL of extract, aliquot 10 μL of 33% bead slurry to each tube.

b. Wash beads twice with 500 μL of 1 × ELB, centrifuging for 1 min at 2000 × *g* after each wash.

c. Supplement extract with 0.5 mg/mL nocodazole in DMSO to a final concentration of 8 ng/μL to prevent microtubule formation. Mix well.

d. Thoroughly aspirate buffer from the first tube of beads using an ultrafine gel loading tip and add extract to the beads. Inject a small (~1 μL) bubble to facilitate mixing.

e. Place the tube on a rotary mixer and rotate at a moderate rate (~0.1–0.2 revolutions per second) for 20 min at room temperature.

f. Pellet the beads by centrifuging for 1 min at 2000 × *g*. Transfer the supernatant to the second tube of beads. To avoid transferring beads from one tube to another, it helps to first remove the bulk of the extract using a regular pipette tip and then remove the remainder with an ultrafine gel loading tip immersed in the beads. Cut off the end of the tip with a pair of clean scissors to avoid transferring beads and to make it easier to dispense the extract from the tip.

g. Repeat steps e and f. After the final round of immunodepletion, the extract may be stored on ice for over an hour without loss of activity.

2.3.3 Notes

1. A mock depletion control with nonspecific IgG bound to protein A sepharose beads is typically included. Nonspecific IgG can be purchased or purified from the serum of nonimmunized rabbits (normal rabbit serum) using protein A sepharose (Harlow & Lane, 1988). The most important control is to rescue any defects resulting from immunodepletion with the addition of recombinant protein added at or near endogenous concentration.
2. Immunodepletion conditions are optimized for each protein. For each round, we use 0.1–0.2 bed volumes of beads per 1 volume of extract. A 20-min incubation time at room temperature has been effective for all NHEJ factors tested so far; however, some proteins may require a longer incubation at 4°C.

3. SINGLE-MOLECULE END JOINING ASSAY

While the ensemble assay described earlier allows covalently joined NHEJ products to be observed, it does not permit the detection of transient, noncovalent reaction intermediates. Single-molecule fluorescence colocalization and FRET can be used to monitor bridging of DNA ends. We have developed two assay configurations (Fig. 2): In the "intermolecular" version of the assay, a short DNA duplex is tethered to a glass surface at one end and labeled at the other end with Cy3 (Fig. 2, upper panel). This tethered DNA duplex is incubated with egg extract containing a second DNA duplex that is labeled with Cy5 near both ends. This assay revealed two stages in DNA bridging (Graham et al., 2016): the Cy3- and Cy5-labeled DNA duplexes are first tethered in a "long-range" complex in which Cy3 and Cy5 colocalize, but the dyes are not sufficiently close for detection of FRET. The long-range synaptic complex is subsequently converted into a "short-range" complex in which DNA ends are aligned, as indicated by an increase in FRET signal. The short-range complex can also be studied using an "intramolecular" assay in which the two ends of a longer, internally tethered DNA are labeled with Cy3 and Cy5 (Fig. 2, lower panel). Though formation of the long-range complex cannot be detected using this substrate, it is useful for characterizing the high-FRET short-range complex, which forms at a high rate due to the increased local concentration of DNA ends. Using these two assays, we previously showed that formation of the long-range synaptic complex requires Ku and DNA-PKcs but not DNA-PK catalytic activity, XLF, or XRCC4-LIG4, while formation of the short-range

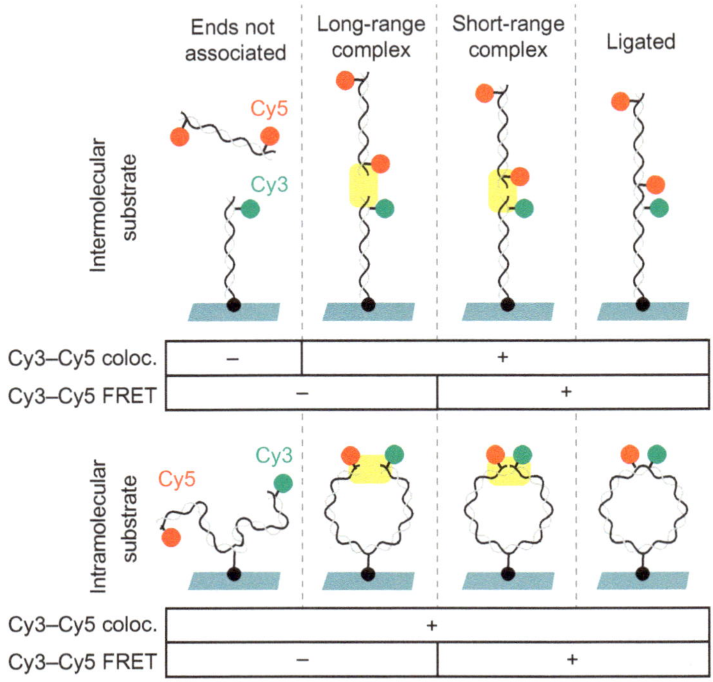

Fig. 2 Schematic of intramolecular and intermolecular single-molecule DNA bridging assays. The intermolecular assay (*upper panel*) relies on a 100-bp DNA substrate labeled near one end with Cy3 (*cyan circle*) and tethered to the coverslip at the other end by a biotin–streptavidin attachment (*black circle*). This is incubated with egg extract containing a second 100-bp DNA labeled near both ends with Cy5 (*red circles*). Formation of the long-range synaptic complex (Graham et al., 2016) is detected in the intermolecular assay based on the appearance of a discrete Cy5 spot that colocalizes with a Cy3-labeled DNA on the surface. Subsequent formation of the short-range synaptic complex is indicated by the appearance of FRET between Cy3 and Cy5. The intramolecular DNA substrate (*lower panel*) consists of a single, 2-kb DNA labeled near one end with Cy3 and near the other end with Cy5 and tethered to a streptavidin-coated coverslip via an internal biotin. Formation of the short-range synaptic complex (Graham et al., 2016) is indicated by appearance of FRET between Cy3 and Cy5.

synaptic complex requires DNA-PK catalytic activity, XLF, and XRCC4-LIG4, but not LIG4 catalytic activity.

3.1 Substrate Preparation

3.1.1 Short Duplexes for Intermolecular Assay

This protocol describes the preparation of labeled 100 bp DNA duplexes suitable for the intermolecular single-molecule NHEJ assay (Figs. 2, upper

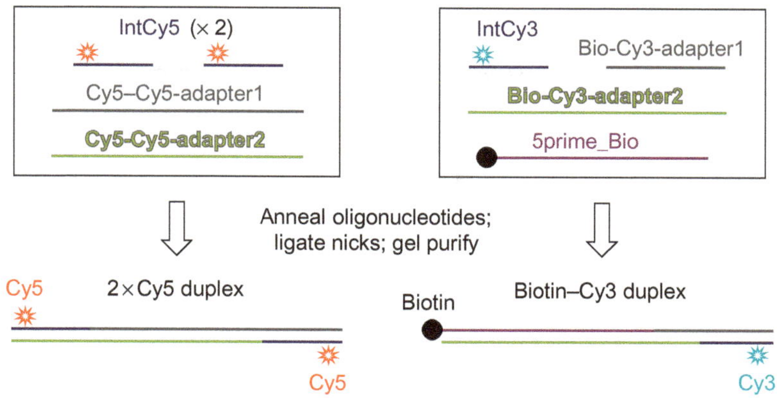

Fig. 3 Preparation of substrates for intermolecular single-molecule assay.

panel, and 3). Similar results have been obtained for longer (~1 kb) DNA fragments; however, these longer DNAs are susceptible to aggregation in the extract, probably as a result of chromatinization.

3.1.1.1 Materials

- Fluorescently labeled oligonucleotides. Oligonucleotides labeled internally 7 nt from the 5′ end permit end joining while allowing FRET to be observed. We used the following sequences:

  `IntCy3` 5′ GGATCT/iCy3N/ACCGCTGTTGAGATC 3′.

  `IntCy5` 5′ AACTCT/iCy5N/TTTCCGAAGGTAACTGG 3′.

  `/iCy3N/` and `/iCy5N/` represent C6-amino-deoxyuridine residues labeled with Cy3 and Cy5 NHS esters (off-catalog modification from Integrated DNA Technologies).
- The following long oligonucleotides:

2xCy5 substrate

`Cy5-Cy5-adapter1`

5′/5Phos/acatttactctctaacatcacgcctagatagaaacagatagcttgaacagatCC AGTTACCTTCGGAAAAAGAGTT 3′

`Cy5-Cy5-adapter2`

5′/5Phos/atctgttcaagctatctgtttctatctaggcgtgatgttagagagtaaatgtCCA GTTACCTTCGGAAAAAGAGTT 3′

Biotin-Cy3 substrate

`Bio-Cy3-adapter1`

5′/5Phos/GTTACATCGAACTGGATCTCAACAGCGGTAAGATCC 3′

Bio-Cy3-adapter2

5′/5Phos/CAGTTCGATGTAACTTAGAACCGCATCTTT
CACAGGTTCTTTTTGCTCTCATTGTTAGTCATTTGT
CAGATTCAACTG 3′

5prime_Bio

5′/5Biosg/CAGTTGAATCTGACAAATGACTAACAATGA
GAGCAAAAAGAACCTGTGAAAGATGCGGTTCTAA 3′

/5Phos/ represents 5′ phosphorylation and /5Biosg/ represents a 5′ biotin modification (Integrated DNA Technologies)

- 5% 19:1 bis-acrylamide 0.5 × TBE-PAGE gel
- Low-binding 0.65 and 1.7 mL microcentrifuge tubes (Corning Costar® #3206 and Sigma Aldrich #T3406)
- 27-gauge needle
- Isopropanol
- NEBuffer 3 (New England Biolabs; 100 m*M* NaCl, 50 m*M* Tris–HCl, 10 m*M* $MgCl_2$, 1 m*M* DTT, pH 7.9)
- 20 m*M* ATP
- T4 DNA ligase (New England Biolabs, #M0202M)
- Cellulose acetate spin filters (Corning Costar Spin-X)
- 10-bp DNA ladder (Invitrogen, #10821-015)

3.1.1.2 Procedure

1. Mix equimolar quantities of the following oligonucleotides in 25 μL of 1 × NEBuffer 3 (New England Biolabs; 100 m*M* NaCl, 50 m*M* Tris–HCl, 10 m*M* $MgCl_2$, 1 m*M* DTT, pH 7.9):
   a. Cy5 duplex: Cy5–Cy5-adapter1, Cy5–Cy5-adapter2, and IntCy5 (add two molar equivalents of IntCy5, which anneals on both sides of the duplex)
   b. Biotin-Cy3 duplex: Bio-Cy3-adapter1, Bio-Cy3-adapter2, 5prime_Bio, and IntCy3
2. Heat to 95°C for 1 min in a PCR machine and then cool at 0.1°C/s to 25°C.
3. Add 1.3 μL of 20 m*M* ATP and 1 μL of T4 DNA ligase and incubate for 1 h at 37°C to seal nicks in the DNA.
4. Separate products on a 5% 19:1 bis-acrylamide 0.5 × TBE-PAGE gel, dividing the reaction between several lanes. Run a DNA ladder and

the original oligonucleotides in separate lanes to verify the size of the final product and to confirm that it is well separated from the reactants.

5. Place the gel against a light background. Using a sharp scalpel or a razor blade, excise the product, which should be visible without staining as a bright pink (Cy3) or blue (Cy5) band. Excise as narrow a slice of polyacrylamide as possible while recovering the bulk of the product. Stain the remainder of the gel for about 2–5 min with 1 μg/mL ethidium bromide in 0.5 × TBE to visualize the molecular weight standards. Image promptly to minimize diffusion of the DNA out of the gel.
6. Place the gel band in a low-binding 0.65 mL microcentrifuge tube that is clean on the outside. Regular tubes may also be used. Use a 27-gauge needle to poke a hole in the bottom of the 0.65 mL tube and place it in a 1.7-mL low-binding microcentrifuge tube. Spin at maximum speed for 1 min in a microcentrifuge, which will force the gel through the needle hole, grinding it into small pieces. Use a pipette tip or clean, fine-tipped forceps to transfer residual gel pieces from the small tube to the large tube. Add 500 μL of 1 × TE buffer (10 m*M* Tris–HCl, pH 8, 1 m*M* EDTA) to the gel fragments in the large tube and rotate for several hours to overnight on a rotary mixer at room temperature.
7. Using a P1000 pipette tip with the end cut off, transfer the polyacrylamide slurry to a cellulose acetate spin filter and spin at 16,000 × *g* in a microcentrifuge. Add 50 μL of 3 *M* sodium acetate, pH 5.2 and 1 mL of isopropanol to the filtrate. Incubate at −20°C for ≥10 min and then spin at maximum speed in a 4°C microcentrifuge for 30 min. A brightly colored pellet should be visible. Aspirate the supernatant and wash the pellet with 70% ethanol. Respin for 1 min and dissolve in 1 × ELB salts (10 m*M* HEPES, pH 7.7, 50 m*M* KCl, 2.5 m*M* $MgCl_2$). Determine the concentration of the DNA and the fluorescent dyes using a Nanodrop spectrophotometer.

3.1.2 Blunt-Ended Intramolecular (Circularization) Substrate

This section describes how to prepare a single-molecule FRET reporter for monitoring bridging of the two ends of a single linear DNA molecule (Figs. 2, lower panel, and 4). This reporter is a 2-kbp long, internally biotinylated DNA fragment labeled 7 nt from one end with Cy3 and 7 nt from the other end with Cy5. A PCR product is first generated with internally

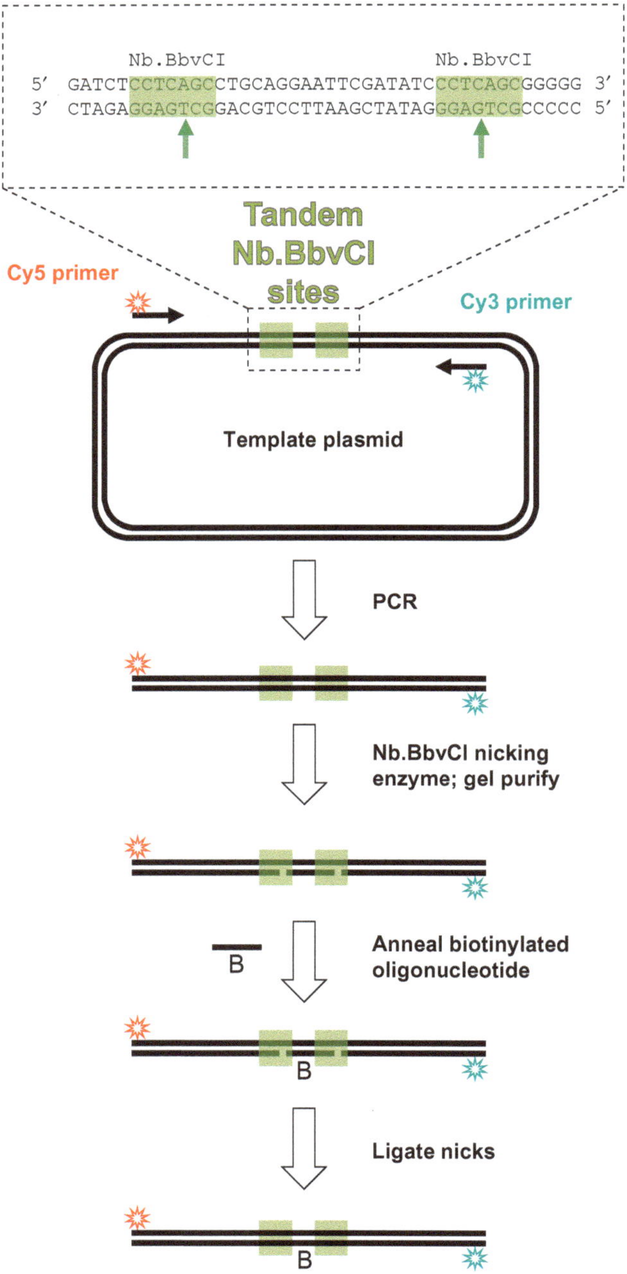

Fig. 4 Intramolecular FRET substrate preparation.

Cy3- and Cy5-labeled primers using a template that contains tandem sites for the nicking restriction endonuclease Nb.BbvCI (see Note 1). After digestion of the PCR product with the nicking enzyme, the short oligonucleotide between these restriction sites is replaced with an excess of internally biotinylated, phosphorylated oligonucleotide by heat denaturation followed by annealing. T4 DNA ligase is used to seal the nicks in the DNA, and the final product is gel purified, yielding an internally biotinylated, double-stranded DNA molecule with Cy3 and Cy5 labels near each end.

3.1.2.1 Materials

- The same fluorescently labeled oligonucleotides used for making the 100 bp duplex substrates (see above).
- 2 × Q5 polymerase master mix (NEB) or similar PCR master mix. Taq polymerase should not be used as it will add a nontemplated dA nucleotide to the 3′ end of the PCR product.
- Template DNA. The template contains tandem recognition sites for the nicking restriction endonuclease Nb.BbvCI (Fig. 4). Our full-length template plasmid is available upon request.
- Internally biotinylated oligonucleotide; identical to the intervening sequence (lower strand) between the nicking sites of the PCR template:

  `5′ /5Phos/TGAGGGATATCGAA/iBiodUK/TCCTGCAGGC 3′`

  `/5Phos/` represents a 5′ phosphate modification and `/iBiodUK/` represents an internal biotinylated deoxyuridine (Integrated DNA Technologies).
- Dialysis tubing (SpectraPor #132650, 23 mm, 6–8 kDa MWCO) and clips
- 4°C microcentrifuge
- Nb.BbvCI restriction enzyme (New England Biolabs, Cat. #R0631S)
- T4 DNA ligase (New England Biolabs, #M0202M)
- Molecular biology grade agarose
- TBE buffer (see above)
- 1 × TE buffer (see above)
- Phenol:chloroform:isoamyl alcohol (25:24:1) mixture, equilibrated with aqueous buffer at pH 8
- Isobutanol, equilibrated with aqueous buffer at pH 8
- 3 *M* sodium acetate, pH 5.2
- 100% and 70% ethanol
- Clean razor blade or scalpel
- Ethidium bromide 10 mg/mL solution
- Nanodrop spectrophotometer

3.1.2.2 Protocol

1. In a 400 μL total reaction volume, PCR amplify a 2-kbp fragment using the above primer and template combination. We use Q5 polymerase 2 × master mix (New England Biolabs). Note: Prior to ordering expensive fluorescently labeled primers, verify that the PCR works with unlabeled primers.
2. Extract the PCR product once with 1 volume of 25:24:1 phenol:chloroform:isoamyl alcohol and once with 1 volume of chloroform. Centrifuge briefly at maximum speed in a microcentrifuge after each extraction step. Retain the aqueous fraction, which will be on top.
3. Precipitate the DNA by adding 40 μL of 3 *M* sodium acetate, pH 5.2 and 800 μL of 100% ethanol. Incubate on ice for ≥10 min and then spin in a 4°C microcentrifuge for 30 min at maximum speed. Wash the pellet with 70% ethanol, spin briefly, and aspirate the ethanol thoroughly with an ultrafine gel loading tip.
4. Redissolve the pellet in 200 μL of 1 × CutSmart buffer (New England Biolabs; 50 m*M* potassium acetate, 20 m*M* Tris–acetate, 10 m*M* magnesium acetate, 100 μg/mL bovine serum albumin, pH 7.9).
5. Digest for 1 h at 37°C with 4 μL (40 U) of Nb.BbvCI.
6. Separate the reaction products on a 1 × TBE 0.8% agarose gel containing 1 μg/mL ethidium bromide.
7. Cut out the product band with a clean razor blade or a scalpel.
8. Extract and purify DNA from the gel slice using the electroelution protocol given earlier for preparation of the ensemble end joining substrate. DNA may also be purified by splitting the product between several spin columns of a commercial gel purification kit; however, this typically gives a lower yield.
9. Determine the concentration of the purified DNA using a spectrophotometer that can accommodate small volumes (e.g., Nanodrop). Add 10 molar equivalents of an internally biotinylated oligonucleotide and 0.1 volumes of 10 × T4 DNA ligase buffer (New England Biolabs). Heat to 80°C for 5 min and slowly cool to room temperature.
10. Add 1 μL of T4 DNA ligase and incubate either at 37°C for 1 h or overnight at room temperature to seal the nicks in the DNA. (Note: Because the DNA ends are not phosphorylated, they will not be ligated in this step.)
11. Heat-inactivate T4 DNA ligase at 65°C for 10 min and store the ligated product at −20°C.

3.1.2.3 Notes

1. Nb.BbvCI retains residual activity toward the top strand, meaning that this enzyme can generate some double-strand breaks in DNA as well as nicks. Nt.BbvCI, the BbvCI mutant that nicks the top strand rather than the bottom strand, has higher fidelity (New England Biolabs, personal communication). Using Nt.BbvCI, in combination with a top-strand biotinylated oligonucleotide, may help to limit unwanted double-strand cleavage of the PCR product.
2. Our protocol does not remove free biotinylated oligonucleotide, which does not seem to interfere with the single-molecule assay.

3.2 Preparation of Flowcells

Glass coverslips are passivated and functionalized with a mixture of polyethylene glycol (PEG) and biotin-PEG, as described previously (Tanner & van Oijen, 2010). A simple flowcell is constructed by cutting a small channel in a piece of double-sided tape and sandwiching this between the functionalized coverslip and a quartz top containing two drilled holes. Plastic tubing is inserted into the holes, and the assembly is sealed with epoxy (Fig. 5).

3.2.1 Materials

For coverslip functionalization (see Tanner & van Oijen, 2010 for protocol):

- 1 *M* potassium hydroxide
- 100% ethanol
- Acetone, ACS reagent grade

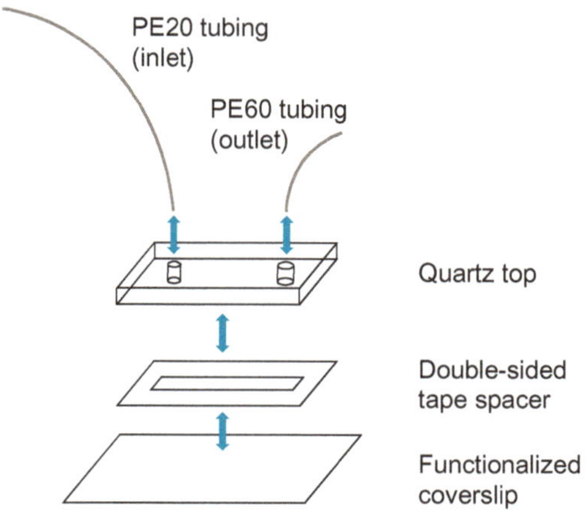

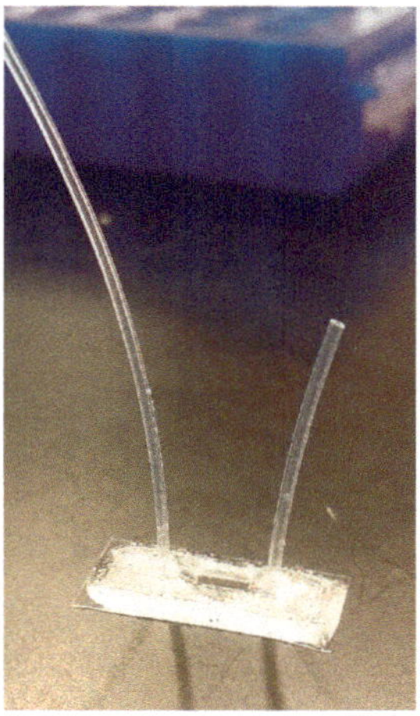

Fig. 5 Schematic of flowcell assembly (*left*) and completed flowcell (*right*).

- Aminopropyltriethoxysilane
- Sodium bicarbonate
- mPEG-SVA and Biotin-mPEG-SVA (Laysan)
- Water bath sonicator
- Coverslips (VWR Micro Cover Glasses, No. 1.5)

For flowcell construction:

- PE20 and PE60 polyethylene tubing (BD Intramedic, #427416 and 427406)
- Quartz tops (OZ grade clear fused quartz plate, 20 × 7 × 1 mm custom size)
- Epoxy (Devcon 5 Minute Epoxy #14250)
- Diamond-tipped scribe
- Double-sided tape sheet (7″ × 10″, 0.12-mm thick, double-sided SecureSeal™ Adhesive Sheet, Grace Biolabs #620001)
- Flat plastic coverslip forceps
- Flat metal forceps
- No. 5 scalpel handle and stainless steel blades
- Scissors
- Plastic cutting board
- Dremel tool with a diamond-tipped drill bit (A&M Instruments; 1.2 mm Flame Medium 3/32″ Shank, #HP863-012)

3.2.2 Protocol

1. Prepare functionalized coverslips essentially as described before (Tanner & van Oijen, 2010), with the following modifications (step numbers refer to the protocol provided in the reference):
   a. Do not quench the silanization reaction by flooding staining jars with a large volume of water (step 3). Rather, discard the silanization mixture in a hazardous waste bottle and rinse the coverslips five to six times with ultrapure water.
   b. In step 5, use mPEG-SVA and Biotin-mPEG-SVA from Laysan Bio, Inc.
   c. Place silanized coverslips directly on top of each other during PEGylation in step 6 instead of using spacer coverslips.
   d. Do not use compressed gas to dry the coverslips in step 7. Instead, extensively wash the coverslip with ultrapure water until it forms tight beads on the surface. Blot remaining beads of water from the edges using a lint-free wipe.
2. Prepare quartz tops:

a. Drill two holes in a 20 × 6 × 1 -mm piece of quartz using a Dremel rotary tool with a diamond-tipped bit. Support the piece of quartz on a stack of paper towels submerged in water. A relatively small amount of force is required to drill through the quartz. Pressing too hard can easily break the quartz. Continue drilling each hole until it is just wide enough to accommodate a piece of PE20 (inlet) or PE60 (outlet) tubing.
b. If you are recycling an already-used quartz top, wash it thoroughly with acetone, scrape off remnants of double-sided tape and epoxy with a razor blade and remove debris from the holes by pushing through a piece of PE20/PE60 tubing and cutting it off on the other side with a sharp razor blade.

3. Cut a 1-mm wide, 1-cm long channel out of double-sided tape using a razor blade or a sharp scalpel.
4. Peel off the transparent side of the double-sided tape, and stick it to a quartz top, making sure that the holes are aligned with the channel.
5. Using the end of a pair of blunt forceps or the bottom of a microcentrifuge tube, press against the double-sided tape to eliminate any air-filled spaces between it and the quartz.
6. Use a diamond-tipped scribe to cut a piece of a functionalized coverslip slightly larger than the quartz top. When cutting coverslips, hold the scribe vertically and do not press very hard. Practice with non-functionalized coverslips before using functionalized coverslips.
7. Peel off the other side of the double-sided tape on the quartz top and place the quartz top in the center of the coverslip.
8. With the quartz top flat on the benchtop, carefully press on the coverslip with the tip of a pair of blunt plastic forceps or the bottom of a microcentrifuge tube to eliminate any air-filled spaces between the coverslip and the tape. It is important to press gently, as the coverslip can easily crack.
9. Cut a 2-cm piece of PE60 tubing for the outlet and a 10-cm piece of PE20 tubing for the inlet. To avoid clogging, angle each cut slightly so that the end of the tubing will not be pressed flat against the coverslip.
10. Turn the flowcell over and insert the PE20 and PE60 tubing into the appropriate holes. It helps to grasp the tubing near the end using a pair of flat metal forceps.
11. Use epoxy to glue the tubing in place. Touch the tubing on each side with a drop of epoxy using a gel loading pipette tip and then drag the tip to connect the two drops. Then, use epoxy to seal the edges of the

quartz top. The epoxy should extend from the top surface of the quartz down to the coverslip.

3.3 Single-Molecule NHEJ Assay

This assay monitors the bridging of single pairs of DNA ends by smFRET. The intramolecular and intermolecular assays are performed in essentially the same way, except that in the latter case, a second labeled DNA fragment is included in the extract mixture. Alternating laser excitation (ALEX) is used to visualize both the donor (Cy3) and the acceptor (Cy5) dyes. This is important, as it allows loss of FRET signal due to dye separation to be distinguished from loss of FRET signal due to bleaching or blinking of Cy5.

3.3.1 Materials

- Labeled DNA substrate(s) (see above)
- 1 × egg lysis buffer (ELB) salts—10 m*M* HEPES, pH 7.7, 50 m*M* KCl, 2.5 m*M* $MgCl_2$
- Bell jar connected to a vacuum line
- Streptavidin—1 mg/mL in PBS
- Laser power meter (LaserMate Q, Coherent)
- Control sample for aligning channels (see Section 4.3). We use one of the two samples:
  - A "nanogrid" of 150 nm holes in a thin metal film, generated by electron beam lithography, which is imaged with transillumination to provide an array of calibration points (Baday et al., 2012).
  - TetraSpeck fluorescent microspheres (Thermo Fisher) deposited on a glass coverslip, which are imaged using total internal reflection fluorescence (TIRF).
- Protocatechuic acid (PCA), 250 m*M* solution in 1 × ELB salts. Adjust the stock solution to pH 7.7 with NaOH. Store large aliquots at −80°C and a small working aliquot at −20°C. Make a new working aliquot when the old one begins yellowing, which indicates oxidation.
- Protocatechuate 3,4-dioxygenase (PCD), 5 μ*M* solution in 10 m*M* HEPES, pH 7.5, 50 m*M* KCl, 1.25 m*M* $MgCl_2$, 50% glycerol. Store aliquots at −20°C.
- Creatine phosphokinase (CPK), adenosine triphosphate (ATP), and phosphocreatine (PC) stock solutions; see ensemble end joining reaction protocol earlier.
- Maxiprepped plasmid DNA to use as carrier, 1 μg/μL in 1 × TE (10 m*M* Tris–HCl, pH 8, 1 m*M* EDTA)

3.3.2 Microscope Setup

For our single-molecule experiments, we use a home-built, through-objective TIRF microscope with laser illumination. A home-built dual view device (Fig. 6) permits simultaneous wide-field imaging of Cy3 and Cy5 emission. Our instrument is constructed around an Olympus IX-71 upright microscope fitted with a ZT532/638rpc filter cube (Chroma). Beams from a Coherent Sapphire 532-nm laser and a Coherent CUBE 641-nm laser are expanded with telescopes made of pairs of converging lenses, combined using dichroic mirrors, expanded again with another two-lens telescope, and directed into the back port of the microscope. The laser beam is focused at the back focal plane of the objective using a lens on a vertical translation mount, whose position can be adjusted to set the TIRF angle. Fluorescence emission is directed through the side port of the microscope into a home-made dual view (Fig. 6), which images Cy3 and Cy5 emission on the two halves of an EMCCD camera (ImageEM, Hamamatsu). Lasers are switched on and off with Uniblitz VS14 shutters controlled by a Uniblitz VMM-D3 3-channel shutter controller. Shutter timing is synchronized with the output trigger signal of the camera using an NI USB-6009 DAQ card (National Instruments) controlled by custom software written in LabView. A motorized microstage (Mad City Labs) is used to position the sample in the x–y plane and is controlled with LabView software provided by

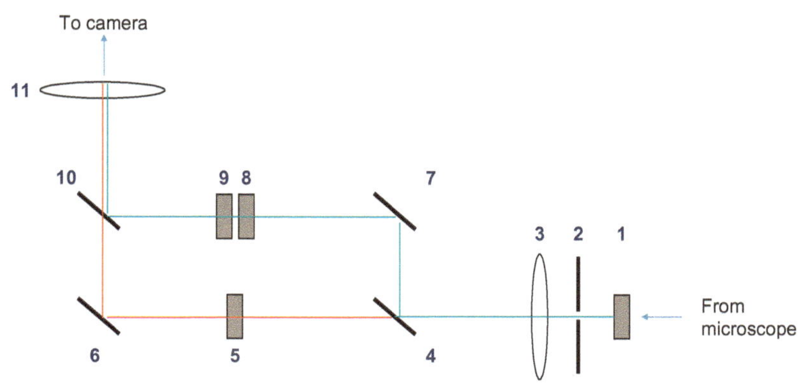

(1) StopLine 488/532/635 notch filter
(2) Thorlabs VA100 adjustable mechanical slit
(3) Thorlabs AC508-100-A-ML $f = 100$ mm, Ø2″ Achromatic Doublet
(4) Chroma T640lpxr dichroic mirror
(5) Chroma ET700/75m bandpass filter
(6) Mirror on 2-axis adjustable mount (Thorlabs)
(7) Mirror on 2-axis adjustable mount (Thorlabs)
(8) Chroma ET650sp shortpass filter
(9) Chroma ZET 488/532 m emission filter
(10) Chroma T640lpxr dichroic mirror
(11) Thorlabs AC508-150-A $f = 150.0$ mm, Ø2″ Achromatic Doublet

Fig. 6 Schematic of home-built dual view for separating Cy3 and Cy5 emission signals.

the manufacturer. A Cordless Rumble Pad 2 game controller (Logitech) is used for manual stage control, and custom LabView software (available upon request) is used to coordinate automated movement sequences with shutter timing.

3.3.3 Protocol

1. Turn on the EMCCD camera of the microscope to allow the camera chip to cool down, and turn on the lasers.
2. Place 10 mL of 1 × ELB salts in a conical tube with its lid loosened; allow to sit under vacuum in the bell jar for at least 10 min to degas. Degassing will reduce the amount of oxygen in the solution and prevent the formation of bubbles in the flowcell.
3. While the 1 × ELB salts is degassing, use a gel loading tip to inject 15 μL of 1 mg/mL streptavidin in PBS into the outlet PE60 tubing of a microfluidic flowcell. Ensure that streptavidin gets all the way through the channel and comes out into the PE20 tubing on the other side. Allow the streptavidin to sit in the flowcell for $\geq$5 min.
4. Set the power of each laser to the desired value using a laser power meter. For simplicity, we typically measure the laser power at one of its focal points on the table and later determine the corresponding power density at the sample.
5. Place a drop of immersion oil on the objective lens, being very careful to avoid air bubbles, which can skew the angle of the TIRF beam. Moving fringes of light and dark intensity or a large dark patch that creeps into the image are symptomatic of immersion oil bubbles.
6. Mount the flowcell securely on the microscope and raise the objective until a circle of immersion oil touches the coverslip.
7. Using brightfield illumination (by eye or with EM gain set to 0 on the camera), focus on the double-sided tape on the edge of the channel. This will bring the objective near the correct focal plane.
8. Wash the streptavidin out of the flowcell with ELB salts. To this end, place the inlet tubing in the bottom of a 0.65-mL microcentrifuge tube containing 0.5 mL of ELB salts. Insert a gel loading tip attached to a P200 pipette into the outlet and draw 200 μL of the buffer through the flowcell. When removing the gel loading tip from the tubing, hold on to the tubing so as not to accidentally rip it out of the flowcell. Alternatively, a syringe pump connected to the outlet may be used to draw solutions into the flowcell.

9. Draw ~30–50 μL of a dilution of the appropriate biotinylated DNA substrate into the flowcell. We do not generally determine the absolute concentration of the substrate, but instead titrate the concentration empirically for each new substrate preparation and batch of coverslips until the desired density of substrates on the surface is achieved. Substrates should be tethered at a low enough density that they are mostly nonoverlapping (~300 substrates per field of view (FOV) on our microscope).
10. Open the shutter on the 532-nm laser, and focus on the substrate DNA, imaging at about 100 ms per frame. Focus slowly using the fine focus knob. Near the correct plane of focus, the relative background intensity in the two channels changes in a characteristic way that becomes recognizable with experience. One common mistake is focusing on the wrong (bottom) side of the coverslip. For the intramolecular substrate, check that the microscope is focused on the top side by switching to the 641-nm laser; a similar density of spots in the Cy5 channel should be visible. If there is any ambiguity, focus downward to the bottom surface of the coverslip (where surface-bound fluorescent contaminants should be visible) and then focus back up to the top surface.
11. Wash the channel again with ELB salts.
12. Supplement a 33-μL aliquot of HSS with 0.5 μL of 0.5 mg/mL nocodazole in DMSO. Mix thoroughly.
13. Prepare an ATP regeneration system by combining the following:
    - 0.5 μL of 5 mg/mL creatine phosphokinase
    - 5 μL of 200 m*M* ATP
    - 10 μL of 1 *M* phosphocreatine
14. Make a reaction mix with HSS of egg cytosol:
    - 25 μL HSS supplemented with nocodazole
    - 2.5 μL of 1 μg/μL closed-circular plasmid DNA ("carrier DNA"). A lower concentration of carrier DNA may be optimal for some preparations of extract. This can be optimized in ensemble end joining experiments (see above)
    - 0.8 μL ATP regeneration system
    - 0.6 μL of 250 m*M* PCA
    - 0.6 μL of 5 μ*M* PCD
    - 0.6 μL of 50 m*M* Trolox. Mix well after adding the Trolox.
    - 1 μL of 100 n*M* 2xCy5-labeled 100 bp duplex (for intermolecular assay only)

15. Pop any bubbles by holding the tube firmly against one index finger and flicking sharply with the nail of the other index finger. Spin 30 s at maximum speed in a microcentrifuge.
16. Prepare shutter control software (we use custom LabView code, which is available upon request) and image acquisition software to begin recording as soon as extract is pulled into the flowcell. We typically excite Cy3 with one or two frames of 532 nm light for every one frame of Cy5 excitation with 641 nm light. Illumination may be continuous or stroboscopic. Continuous illumination at a lower laser power is generally preferable. Typical laser powers through the objective for a 500-ms exposure are 0.1 mW for the 647 nm laser and 0.5 mW for the 532 nm laser.
17. Pull the extract into the flowcell, turn off the room lights, and begin recording a movie. Manual refocusing may be necessary if a single FOV is imaged for an extended period of time. Switch to fresh FOVs as desired. To collect long trajectories, we sometimes record in a single FOV for 30 min with intermittent illumination. To obtain overall rates of long- and short-range synaptic complex formation, each FOV may be sampled for a shorter period of time—typically 18–30 s for the intramolecular assay or 3 min for the intermolecular assay.
18. At the end of the experiment, acquire several images of a nanogrid of subdiffraction-limited holes under transillumination (with camera EM gain turned off) or a slide with TetraSpeck fluorescent beads under TIRF illumination to serve as a standard for registering the two channels. A new set of images should be acquired for every day of experiments, as the optics may shift (or be deliberately realigned) from day to day. For bead samples, data can be combined from several FOVs to achieve uniform coverage of the visible area.
19. Quartz tops can be recycled by plucking the tubing out of the flowcell and submerging the flowcell in acetone overnight. This will soften the epoxy and double-sided tape, allowing the coverslip and quartz top to be peeled apart.

4. ANALYSIS OF SINGLE-MOLECULE DATA

4.1 Opening Data in MATLAB

We analyze our imaging data using custom MATLAB scripts, which are available upon request. Data are recorded using HCImage software and are saved as ".cxd" files. This file format, and a variety of other imaging file formats, can be imported into MATLAB using the BioFormats package. To use this plugin, download the MATLAB Toolbox from the BioFormats website (http://downloads.openmicroscopy.org/bio-formats/5.2.4/), extract the downloaded archive, and add the extracted folder to the MATLAB search path using the `addpath` command.

4.2 Field of View Segmentation

A single movie file may contain data from many different FOVs. Movie segments from different FOVs are divided using an automated function that calculates the correlation between pixel intensities in successive frames. A change of FOV is indicated by a drop in this correlation below a user-defined threshold, which may need to be manually adjusted between movies, depending on the number of tethered substrates and the amount of background noise in the images.

4.3 Channel Alignment

The Cy3 and Cy5 channels of the microscope are realigned to within a few pixels prior to each experiment. However, because of slight imperfections in the alignment and distortions in the optics, it is necessary to register the channels more precisely during data analysis. We do this in one of the two ways.

For the intramolecular FRET reporter, for which the same molecules are reliably labeled with Cy3 and Cy5, the substrates themselves may be used for channel alignment. Each channel is divided into eight sectors, and the optimal translation for aligning each sector is determined by image correlation between the Cy3 and Cy5 channels across several FOVs.

Alternatively, a reference sample containing point sources that appear in both the Cy3 and Cy5 channels may be used for alignment. A coverslip with nonspecifically adsorbed TetraSpeck fluorescent microspheres (Thermo Fisher) works for this purpose. However, we typically image a trans-illuminated "nanogrid" consisting of a thin film of aluminum containing 150 nm holes generated by electron beam lithography. The channels are first roughly aligned using image correlation to find an approximate translation over the entire visible area. Small defects in the array facilitate unambiguous alignment. Points are detected in one channel using a peak-finding function, and approximate positions in the other channel are found by applying the translation. The points in the two channels are then localized more precisely by two-dimensional Gaussian fitting. The MATLAB function `cp2tform` is used to generate a spatial transformation function between the fitted points in the two channels.

4.4 Drift Correction

Within each single-FOV movie segment, drift is corrected by calculating the spatial cross-correlation function between the initial frame in the segment and each subsequent frame. This, and other operations involving image correlation, can be done efficiently by Fourier transforming each image (MATLAB function `fft2`) multiplying in the frequency domain, and inverse transforming (MATLAB function `ifft2`). The peak of the cross-correlation function, rounded to the nearest integer, is taken to be the offset of the frame in pixels from the initial frame in the segment.

4.5 Extraction of Integrated Intensities

For locating single DNA substrates, the first several frames in each FOV are averaged and processed with a spatial bandpass filter to extract features the size of a diffraction-limited spot. Points of interest are identified in one of the two channels using a peak-finding function that identifies local maxima that are not closer together than some minimum distance (typically 7 pixels). For the intermolecular substrate, statically bound DNAs are located in the Cy3 channel. For the intramolecular FRET reporter, which is labeled with both Cy3 and Cy5, points are located in the Cy5 channel to avoid bias against high-FRET spots. A transformation function between the two channels (see above) is used to find corresponding points in the other channel.

The fluorescence intensity of each substrate is determined by summing the pixel intensities in a disc of radius 4 pixels centered on the peak. Local background is determined as the mean intensity of a circle of pixels surrounding this disc. The background-subtracted intensity is given by: (summed intensity of the disc) − (average background intensity) * (area of the disc).

4.6 Analysis and Interpretation of Single-Molecule Traces

The earlier analysis yields single-substrate traces of Cy3 and Cy5 emission with 532 nm excitation as well as Cy5 emission with 641 nm excitation. FRET efficiency (E_{FRET}) is calculated after applying a correction for Cy5 direct excitation by 532 nm light, Cy3 bleedthrough into the Cy5 channel, and differences in quantum yield and detection efficiency of Cy3 and Cy5 emission, as described previously (Lee et al., 2005).

4.6.1 Analysis of Traces From the Intermolecular Assay

For the intermolecular assay, we manually inspect traces using a homemade browser function written in MATLAB. The times of Cy5-DNA binding (long-range complex formation) and transition to FRET (short-range complex formation) are manually annotated within the browser function. An example of a Cy5-DNA-binding event followed by a transition to high FRET is shown in Fig. 7A. A waiting time histogram can be compiled for the transition between long- and short-range complexes.

A survival curve for the long-range complex and the overall rate of long-range complex formation are determined by detecting Cy5-DNA-binding events in an automated fashion: A histogram of background-subtracted Cy5 intensity is first compiled for all particles over all frames of the movie. Because most Cy3-DNAs are not bound by a Cy5-DNA, the highest peak of this histogram represents zero Cy5 molecules. This peak is fit locally to a Gaussian function, and a threshold is chosen two standard deviations above the mean. Cy5-binding events are assigned whenever the Cy5 intensity of a particular spot crosses this threshold. A survival time distribution is then calculated from the durations of these binding events.

Calculating the survival time distribution of the long-range complex from Cy5 traces is complicated by two considerations: (1) some Cy5-positive spots may transition to a short-range complex, and (2) some Cy5-binding events may be cut off by the end of a movie. To deal with this issue, we calculate survival curves using the Kaplan–Meier estimator, which takes into account the fact that survival times may be known incompletely if individuals (in this case, molecules) are lost to follow-up measurements, a condition known as "right censorship." We consider Cy5-binding events to be right-censored by transitions to high FRET or by the end of a movie. Events are also considered to be right-censored when Cy3 photobleaches, which makes it impossible (by measuring FRET) to determine whether the complex is in a long-range or short-range/ligated state. Our criterion for transitions to high FRET for the purpose of censoring is that the calculated FRET efficiency exceeds 0.4 for four consecutive frames (see Note 1). Our criterion for Cy3 photobleaching is that the Cy3 intensity drops two standard deviations below the center of the single-Cy3 peak in the Cy3 intensity histogram.

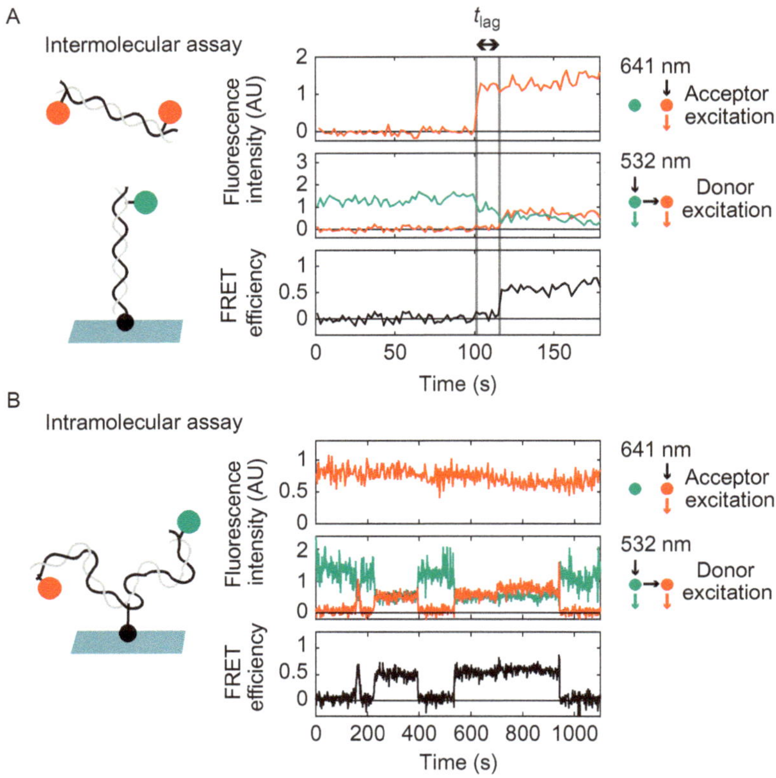

Fig. 7 Example single-molecule traces from the intermolecular (A) and intramolecular (B) NHEJ assays. The *upper panel* in (A) and (B) shows Cy5 emission with direct excitation by 641 nm light. The *middle panel* shows Cy3 and Cy5 emission with excitation of Cy3 by 532 nm light. The *lower panel* shows calculated FRET efficiency. Substrate schematics are shown to the *left*. Note that 0 s corresponds to the first frame in the particular field of view being imaged, not the time of extract addition. (A) In the intermolecular assay, binding of a Cy5-DNA to a Cy3-DNA on the surface is detected by appearance of Cy5 signal with 641 nm excitation (*first dashed gray line*). Formation of the short-range synaptic complex is indicated by the appearance of a high-FRET signal (*second dashed gray line*) after a time delay ("t_{lag}"). (B) A sample trace from the intramolecular assay showing three rounds of short-range complex formation and dissolution, as indicated by increases and decreases in FRET. This trace was acquired in extract immunodepleted of XRCC4-LIG4 and supplemented with catalytically inactive XRCC4-LIG4^{K278R} complex, which supports short-range complex formation but not ligation.

The rate of long-range complex formation is calculated by dividing the number of observed binding events by the number of frames in which Cy3-DNAs are "available" to be bound (i.e., not already bound and not photobleached). Cy5-DNA-binding events are not counted after Cy3 has photobleached, as this makes it impossible to determine whether the complex is in a low-FRET or a high-FRET state.

4.6.2 Analysis of Traces From the Intramolecular Assay

Similar to the intermolecular assay, traces from the intramolecular assay are inspected using a homemade browser function written in MATLAB. Photobleaching events and transitions between low- and high-FRET states are manually annotated. Events are considered right-censored for Kaplan–Meier survival curve analysis if they are terminated by the end of the movie or by photobleaching of Cy3 or Cy5.

To obtain the kinetics of high-FRET complex formation (including both short-range complex and ligated product), we use an automated stage control program to switch to a new FOV every 15–30 s. This provides more statistically independent data points than imaging a single FOV, and it avoids loss of signal due to photobleaching. The FRET efficiency histogram can be obtained for the substrates within each FOV, and the fraction of FRET-positive spots ($E_{FRET} > 0.25$) can be plotted as a function of time (e.g., Fig. 4 in Graham et al., 2016).

4.7 Notes

1. The requirement that $E_{FRET} > 0.4$ over four consecutive frames helps to avoid spurious detection of FRET transitions due to measurement noise. This requirement is reasonable, given that high-FRET short-range complexes generally persist much longer than this.

5. CONCLUSIONS AND OUTLOOK

X. laevis egg extract provides a powerful approach for detailed mechanistic studies of nonhomologous end joining. The system is unique because it supports highly efficient, cell-free end joining that depends on the core end joining factors (Ku, DNA-PKcs, DNA ligase IV-XRCC4, and XLF). This property, together with the presence of other NHEJ regulators in extract, both known (e.g., DNA polymerase λ, DNA polymerase μ, PNKP, Artemis) and unknown, makes it likely that any mechanistic insights gained reflect the complexity of NHEJ as it occurs in the cell. The use of single-molecule approaches within this system makes it possible to probe transient end joining intermediates. Measuring smFRET between labeled DNA ends, while a useful readout of end proximity, is only one possible configuration of the assay. Many variations can be envisioned, including multiwavelength imaging of fluorescently labeled proteins together with FRET between DNA ends. The extract-based approach could potentially be extended to study the competition between NHEJ and other repair pathways, such as homologous recombination (in particular, the initial 5′–3′ resection step) or Alt-EJ. Furthermore, replication-competent extracts could be used to investigate how NHEJ is suppressed at double-strand breaks arising from replication fork collapse or at DNA interstrand cross-links (Long, Räschle, Joukov, & Walter, 2011; Räschle et al., 2008; Zhang & Walter, 2014).

REFERENCES

Ahnesorg, P., Smith, P., & Jackson, S. P. (2006). XLF interacts with the XRCC4-DNA ligase IV complex to promote DNA nonhomologous end-joining. *Cell*, *124*, 301–313.

Akopiants, K., Zhou, R.-Z., Mohapatra, S., Valerie, K., Lees-Miller, S. P., Lee, K.-J., et al. (2009). Requirement for XLF/Cernunnos in alignment-based gap filling by DNA polymerases lambda and mu for nonhomologous end joining in human whole-cell extracts. *Nucleic Acids Research*, *37*, 4055–4062.

Baday, M., Cravens, A., Hastie, A., Kim, H., Kudeki, D. E., Kwok, P.-Y., et al. (2012). Multicolor super-resolution DNA imaging for genetic analysis. *Nano Letters*, *12*, 3861–3866.

Balmus, G., Barros, A. C., Wijnhoven, P. W. G., Lescale, C., Hasse, H. L., Boroviak, K., et al. (2016). Synthetic lethality between PAXX and XLF in mammalian development. *Genes & Development*, *30*, 2152–2157.

Baumann, P., & West, S. C. (1998). DNA end-joining catalyzed by human cell-free extracts. *Proceedings of the National Academy of Sciences of the United States of America*, *95*, 14066–14070.

Buck, D., Malivert, L., de Chasseval, R., Barraud, A., Fondanèche, M.-C., Sanal, O., et al. (2006). Cernunnos, a novel nonhomologous end-joining factor, is mutated in human immunodeficiency with microcephaly. *Cell*, *124*, 287–299.

Budzowska, M., Graham, T. G. W., Sobeck, A., Waga, S., & Walter, J. C. (2015). Regulation of the Rev1-pol ζ complex during bypass of a DNA interstrand cross-link. *The EMBO Journal*, *34*, 1971–1985.

Carter, T., Vancurová, I., Sun, I., Lou, W., & DeLeon, S. (1990). A DNA-activated protein kinase from HeLa cell nuclei. *Molecular and Cellular Biology*, *10*, 6460–6471.

Chang, H. H. Y., Watanabe, G., Gerodimos, C. A., Ochi, T., Blundell, T. L., Jackson, S. P., et al. (2016). Different DNA end configurations dictate which NHEJ components are most important for joining efficiency. *The Journal of Biological Chemistry*, *291*, 24377–24389.

Chappell, C., Hanakahi, L. A., Karimi-Busheri, F., Weinfeld, M., & West, S. C. (2002). Involvement of human polynucleotide kinase in double-strand break repair by nonhomologous end joining. *The EMBO Journal*, *21*, 2827–2832.

Chen, S., Inamdar, K. V., Pfeiffer, P., Feldmann, E., Hannah, M. F., Yu, Y., et al. (2001). Accurate in vitro end joining of a DNA double strand break with partially cohesive 3′-overhangs and 3′-phosphoglycolate termini: Effect of Ku on repair fidelity. *The Journal of Biological Chemistry*, *276*, 24323–24330.

Cortes, P., Weis-Garcia, F., Misulovin, Z., Nussenzweig, A., Lai, J. S., Li, G., et al. (1996). In vitro V(D)J recombination: Signal joint formation. *Proceedings of the National Academy of Sciences of the United States of America*, *93*, 14008–14013.

Craxton, A., Somers, J., Munnur, D., Jukes-Jones, R., Cain, K., & Malewicz, M. (2015). XLS (c9orf142) is a new component of mammalian DNA double-stranded break repair. *Cell Death and Differentiation*, *22*, 890–897.

Critchlow, S. E., Bowater, R. P., & Jackson, S. P. (1997). Mammalian DNA double-strand break repair protein XRCC4 interacts with DNA ligase IV. *Current Biology*, 7, 588–598.

Daza, P., Reichenberger, S., Göttlich, B., Hagmann, M., Feldmann, E., & Pfeiffer, P. (1996). Mechanisms of nonhomologous DNA end-joining in frogs, mice and men. *Biological Chemistry*, *377*, 775–786.

Di Virgilio, M., & Gautier, J. (2005). Repair of double-strand breaks by nonhomologous end joining in the absence of Mre11. *The Journal of Cell Biology*, *171*, 765–771.

Dobbs, T. A., Tainer, J. A., & Lees-Miller, S. P. (2010). A structural model for regulation of NHEJ by DNA-PKcs autophosphorylation. *DNA Repair (Amst)*, *9*, 1307–1314.

Dvir, A., Peterson, S. R., Knuth, M. W., Lu, H., & Dynan, W. S. (1992). Ku autoantigen is the regulatory component of a template-associated protein kinase that phosphorylates RNA polymerase II. *Proceedings of the National Academy of Sciences of the United States of America*, *89*, 11920–11924.

Dvir, A., Stein, L. Y., Calore, B. L., & Dynan, W. S. (1993). Purification and characterization of a template-associated protein kinase that phosphorylates RNA polymerase II. *The Journal of Biological Chemistry*, *268*, 10440–10447.

Feldmann, E., Schmiemann, V., Goedecke, W., Reichenberger, S., & Pfeiffer, P. (2000). DNA double-strand break repair in cell-free extracts from Ku80-deficient cells: Implications for Ku serving as an alignment factor in non-homologous DNA end joining. *Nucleic Acids Research*, *28*, 2585–2596.

Gottlieb, T. M., & Jackson, S. P. (1993). The DNA-dependent protein kinase: Requirement for DNA ends and association with Ku antigen. *Cell*, *72*, 131–142.

Graham, T. G. W., Walter, J. C., & Loparo, J. J. (2016). Two-stage synapsis of DNA ends during non-homologous end joining. *Molecular Cell*, *61*, 850–858.

Grawunder, U., Wilm, M., Wu, X., Kulesza, P., Wilson, T. E., Mann, M., et al. (1997). Activity of DNA ligase IV stimulated by complex formation with XRCC4 protein in mammalian cells. *Nature*, *388*, 492–495.

Gu, X. Y., Bennett, R. A., & Povirk, L. F. (1996). End-joining of free radical-mediated DNA double-strand breaks in vitro is blocked by the kinase inhibitor wortmannin at a step preceding removal of damaged 3′ termini. *The Journal of Biological Chemistry*, *271*, 19660–19663.

Gu, J., Lu, H., Tsai, A. G., Schwarz, K., & Lieber, M. R. (2007). Single-stranded DNA ligation and XLF-stimulated incompatible DNA end ligation by the XRCC4-DNA ligase IV complex: Influence of terminal DNA sequence. *Nucleic Acids Research*, *35*, 5755–5762.

Hanakahi, L. A., Bartlet-Jones, M., Chappell, C., Pappin, D., & West, S. C. (2000). Binding of inositol phosphate to DNA-PK and stimulation of double-strand break repair. *Cell*, *102*, 721–729.

Harlow, E., & Lane, D. (1988). *Antibodies: A laboratory manual*. Woodbury, NY: Cold Spring Harbor Laboratory Press.

Hentges, P., Ahnesorg, P., Pitcher, R. S., Bruce, C. K., Kysela, B., Green, A. J., et al. (2006). Evolutionary and functional conservation of the DNA non-homologous end-joining protein, XLF/Cernunnos. *The Journal of Biological Chemistry*, *281*, 37517–37526.

Jayaram, S., Ketner, G., Adachi, N., & Hanakahi, L. A. (2008). Loss of DNA ligase IV prevents recognition of DNA by double-strand break repair proteins XRCC4 and XLF. *Nucleic Acids Research*, *36*, 5773–5786.

Jette, N., & Lees-Miller, S. P. (2015). The DNA-dependent protein kinase: A multifunctional protein kinase with roles in DNA double strand break repair and mitosis. *Progress in Biophysics and Molecular Biology*, *117*, 194–205.

Jiang, W., Crowe, J. L., Liu, X., Nakajima, S., Wang, Y., Li, C., et al. (2015). Differential phosphorylation of DNA-PKcs regulates the interplay between end-processing and end-ligation during nonhomologous end-joining. *Molecular Cell*, *58*, 172–185.

Kumar, V., Alt, F. W., & Frock, R. L. (2016). PAXX and XLF DNA repair factors are functionally redundant in joining DNA breaks in a G1-arrested progenitor B-cell line. *Proceedings of the National Academy of Sciences of the United States of America*, *113*, 10619–10624.

Labhart, P. (1999a). Nonhomologous DNA end joining in cell-free systems. *European Journal of Biochemistry*, *265*, 849–861.

Labhart, P. (1999b). Ku-dependent nonhomologous DNA end joining in Xenopus egg extracts. *Molecular and Cellular Biology*, *19*, 2585–2593.

Lebofsky, R., Takahashi, T., & Walter, J. C. (2009). DNA replication in nucleus-free Xenopus egg extracts. *Methods in Molecular Biology*, *521*, 229–252.

Lebofsky, R., van Oijen, A. M., & Walter, J. C. (2011). DNA is a co-factor for its own replication in Xenopus egg extracts. *Nucleic Acids Research*, *39*, 545–555.

Lee, J. W., Blanco, L., Zhou, T., Garcia-Diaz, M., Bebenek, K., Kunkel, T. A., et al. (2004). Implication of DNA polymerase lambda in alignment-based gap filling for Nonhomologous DNA end joining in human nuclear extracts. *The Journal of Biological Chemistry*, *279*, 805–811.

Lee, N. K., Kapanidis, A. N., Wang, Y., Michalet, X., Mukhopadhyay, J., Ebright, R. H., et al. (2005). Accurate FRET measurements within single diffusing biomolecules using alternating-laser excitation. *Biophysical Journal*, *88*, 2939–2953.

Lee, J. W., Yannone, S. M., Chen, D. J., & Povirk, L. F. (2003). Requirement for XRCC4 and DNA ligase IV in alignment-based gap filling for nonhomologous DNA end joining in vitro. *Cancer Research*, *63*, 22–24.

Lees-Miller, S. P., Chen, Y. R., & Anderson, C. W. (1990). Human cells contain a DNA-activated protein kinase that phosphorylates simian virus 40 T antigen, mouse p53, and the human Ku autoantigen. *Molecular and Cellular Biology*, *10*, 6472–6481.

Lehman, C. W., & Carroll, D. (1991). Homologous recombination catalyzed by a nuclear extract from Xenopus oocytes. *Proceedings of the National Academy of Sciences of the United States of America*, *88*, 10840–10844.

Lehman, C. W., Clemens, M., Worthylake, D. K., Trautman, J. K., & Carroll, D. (1993). Homologous and illegitimate recombination in developing Xenopus oocytes and eggs. *Molecular and Cellular Biology*, *13*, 6897–6906.

Lescale, C., Lenden Hasse, H., Blackford, A. N., Balmus, G., Bianchi, J. J., Yu, W., et al. (2016). Specific roles of XRCC4 paralogs PAXX and XLF during V(D)J recombination. *Cell Reports*, *16*, 2967–2979.

Li, Z., Otevrel, T., Gao, Y., Cheng, H. L., Seed, B., Stamato, T. D., et al. (1995). The XRCC4 gene encodes a novel protein involved in DNA double-strand break repair and V(D)J recombination. *Cell*, *83*, 1079–1089.

Liao, S., Guay, C., Toczylowski, T., & Yan, H. (2012). Analysis of MRE11's function in the $5' \rightarrow 3'$ processing of DNA double-strand breaks. *Nucleic Acids Research*, *40*, 4496–4506.

Liao, S., Toczylowski, T., & Yan, H. (2008). Identification of the Xenopus DNA2 protein as a major nuclease for the $5' \rightarrow 3'$ strand-specific processing of DNA ends. *Nucleic Acids Research*, *36*, 6091–6100.

Liao, S., Toczylowski, T., & Yan, H. (2011). Mechanistic analysis of Xenopus EXO1's function in 5′-strand resection at DNA double-strand breaks. *Nucleic Acids Research*, *39*, 5967–5977.

Liu, X., Shao, Z., Jiang, W., Lee, B. J., & Zha, S. (2017). PAXX promotes KU accumulation at DNA breaks and is essential for end-joining in XLF-deficient mice. *Nature Communications*, *8*, 13816.

Long, D. T., Räschle, M., Joukov, V., & Walter, J. C. (2011). Mechanism of RAD51-dependent DNA interstrand cross-link repair. *Science*, *333*, 84–87.

Lu, H., Pannicke, U., Schwarz, K., & Lieber, M. R. (2007). Length-dependent binding of human XLF to DNA and stimulation of XRCC4.DNA ligase IV activity. *The Journal of Biological Chemistry*, *282*, 11155–11162.

Ma, Y., & Lieber, M. R. (2006). In vitro nonhomologous DNA end joining system. *Methods in Enzymology*, *408*, 502–510.

Ma, Y., Lu, H., Tippin, B., Goodman, M. F., Shimazaki, N., Koiwai, O., et al. (2004). A biochemically defined system for mammalian nonhomologous DNA end joining. *Molecular Cell*, *16*, 701–713.

Menon, V., & Povirk, L. F. (2016). End-processing nucleases and phosphodiesterases: An elite supporting cast for the non-homologous end joining pathway of DNA double-strand break repair. *DNA Repair (Amst)*, *43*, 57–68.

Nick McElhinny, S. A., Havener, J. M., Garcia-Diaz, M., Juárez, R., Bebenek, K., Kee, B. L., et al. (2005). A gradient of template dependence defines distinct biological roles for family X polymerases in nonhomologous end joining. *Molecular Cell*, *19*, 357–366.

Ochi, T., Blackford, A. N., Coates, J., Jhujh, S., Mehmood, S., Tamura, N., et al. (2015). DNA repair. PAXX, a paralog of XRCC4 and XLF, interacts with Ku to promote DNA double-strand break repair. *Science*, *347*, 185–188.

Pfeiffer, P., Feldmann, E., Odersky, A., Kuhfittig-Kulle, S., & Goedecke, W. (2005). Analysis of DNA double-strand break repair by nonhomologous end joining in cell-free extracts from mammalian cells. *Methods in Molecular Biology*, *291*, 351–371.

Pfeiffer, P., Odersky, A., Goedecke, W., & Kuhfittig-Kulle, S. (2014). Analysis of double-strand break repair by nonhomologous DNA end joining in cell-free extracts from mammalian cells. *Methods in Molecular Biology*, *1105*, 565–585.

Pfeiffer, P., & Vielmetter, W. (1988). Joining of nonhomologous DNA double strand breaks in vitro. *Nucleic Acids Research*, *16*, 907–924.

Postow, L., Ghenoiu, C., Woo, E.M., Krutchinsky, A.N., Chait, B.T., and Funabiki, H. (2008). Ku80 removal from DNA through double strand break-induced ubiquitylation. The Journal of Cell Biology 182, 467–479.

Räschle, M., Knipscheer, P., Knipsheer, P., Enoiu, M., Angelov, T., Sun, J., et al. (2008). Mechanism of replication-coupled DNA interstrand crosslink repair. *Cell*, *134*, 969–980.

Roy, S., de Melo, A. J., Xu, Y., Tadi, S. K., Négrel, A., Hendrickson, E., et al. (2015). XRCC4/XLF interaction is variably required for DNA repair, and is not required for ligase IV stimulation. *Molecular and Cellular Biology*, *35*, 3017–3028.

Sandoval, A., & Labhart, P. (2002). Joining of DNA ends bearing non-matching 3′-overhangs. *DNA Repair (Amst)*, *1*, 397–410.

Smeaton, M. B., Miller, P. S., Ketner, G., & Hanakahi, L. A. (2007). Small-scale extracts for the study of nucleotide excision repair and non-homologous end joining. *Nucleic Acids Research*, *35*, e152.

Tadi, S. K., Tellier-Lebègue, C., Nemoz, C., Drevet, P., Audebert, S., Roy, S., et al. (2016). PAXX is an accessory c-NHEJ factor that associates with Ku70 and has overlapping functions with XLF. *Cell Reports*, *17*, 541–555.

Tanner, N. A., & van Oijen, A. M. (2010). Visualizing DNA replication at the single-molecule level. *Methods in Enzymology*, *475*, 259–278.

Taylor, E. M., Cecillon, S. M., Bonis, A., Chapman, J. R., Povirk, L. F., & Lindsay, H. D. (2010). The Mre11/Rad50/Nbs1 complex functions in resection-based DNA end joining in Xenopus laevis. *Nucleic Acids Research*, *38*, 441–454.

Thode, S., Schäfer, A., Pfeiffer, P., & Vielmetter, W. (1990). A novel pathway of DNA end-to-end joining. *Cell*, *60*, 921–928.

Toczylowski, T., & Yan, H. (2006). Mechanistic analysis of a DNA end processing pathway mediated by the Xenopus Werner syndrome protein. *The Journal of Biological Chemistry*, *281*, 33198–33205.

Tsai, C. J., Kim, S. A., & Chu, G. (2007). Cernunnos/XLF promotes the ligation of mismatched and noncohesive DNA ends. *Proceedings of the National Academy of Sciences of the United States of America*, *104*, 7851–7856.

Walker, J. R., Corpina, R. A., & Goldberg, J. (2001). Structure of the Ku heterodimer bound to DNA and its implications for double-strand break repair. *Nature*, *412*, 607–614.

Waters, C. A., Strande, N. T., Wyatt, D. W., Pryor, J. M., & Ramsden, D. A. (2014). Nonhomologous end joining: A good solution for bad ends. *DNA Repair (Amst)*, *17*, 39–51.

Weis-Garcia, F., Besmer, E., Sawchuk, D. J., Yu, W., Hu, Y., Cassard, S., et al. (1997). V(D)J recombination: In vitro coding joint formation. *Molecular and Cellular Biology*, *17*, 6379–6385.

Xing, M., Yang, M., Huo, W., Feng, F., Wei, L., Jiang, W., et al. (2015). Interactome analysis identifies a new paralogue of XRCC4 in non-homologous end joining DNA repair pathway. *Nature Communications*, *6*, 6233.

Yan, H., McCane, J., Toczylowski, T., & Chen, C. (2005). Analysis of the Xenopus Werner syndrome protein in DNA double-strand break repair. *The Journal of Cell Biology*, *171*, 217–227.

Yano, K., Morotomi-Yano, K., Lee, K.-J., & Chen, D. J. (2011). Functional significance of the interaction with Ku in DNA double-strand break recognition of XLF. *FEBS Letters*, *585*, 841–846.

Yano, K., Morotomi-Yano, K., Wang, S.-Y., Uematsu, N., Lee, K.-J., Asaithamby, A., et al. (2008). Ku recruits XLF to DNA double-strand breaks. *EMBO Reports*, *9*, 91–96.

Zhang, J., & Walter, J. C. (2014). Mechanism and regulation of incisions during DNA interstrand cross-link repair. *DNA Repair (Amst)*, *19*, 135–142.

Zhu, S., & Peng, A. (2016). Non-homologous end joining repair in Xenopus egg extract. *Scientific Reports*, *6*, 27797.

CHAPTER ELEVEN

Analysis of Structure-Selective Endonuclease Activities From Yeast and Human Extracts

Joao Matos*,[1], Stephen C. West†,[1]
*Institute of Biochemistry, ETH Zürich, Zürich, Switzerland
†The Francis Crick Institute, London, United Kingdom
[1]Corresponding authors: e-mail address: joao.matos@bc.biol.ethz.ch; stephen.west@crick.ac.uk

Contents

Abstract

The efficient separation of two equal DNA masses to the daughter cells is an essential step in mitosis. This process is dependent upon the removal of any remaining recombination or replication intermediates that link sister chromatids, and a failure to resolve these intermediates leads to genome instability. Similarly, a failure to resolve meiotic recombination intermediates that link homologous chromosomes can cause chromosome nondisjunction and aneuploidy. Cleavage of these potentially toxic replication/recombination intermediates requires the Mus81 endonuclease, which is active upon flaps, forks, and more complex secondary structures in DNA such as Holliday junctions. Recent studies of Mus81 revealed that it is regulated throughout the cell cycle: Mus81 activity is controlled in S-phase to limit the cleavage of replication fork structures, whereas it is activated at G2/M to ensure the cleavage of recombination and late replication intermediates. In this chapter,

Methods in Enzymology, Volume 591
ISSN 0076-6879
http://dx.doi.org/10.1016/bs.mie.2017.03.005

we describe a simple method that can monitor the activity of Mus81, which involves the immunoprecipitation of epitope-tagged Mus81 and use of an on-bead assay for nuclease activity.

1. INTRODUCTION

MUS81 is a member of the MUS81/XPF family of structure-selective endonucleases (Ciccia, McDonald, & West, 2008). The heterodimeric proteins, Mus81–Mms4 in *Saccharomyces cerevisiae* and MUS81–EME1 in *Homo sapiens*, play important roles in the resolution of recombination intermediates (Boddy et al., 2001; Castor et al., 2013; Doe, Ahn, Dixon, & Whitby, 2002; Garner, Kim, Lach, Kottemann, & Smogorzewska, 2013; Jessop & Lichten, 2008; Oh, Lao, Taylor, Smith, & Hunter, 2008; Wechsler, Newman, & West, 2011; Wyatt, Sarbajna, Matos, & West, 2013) and the cleavage of late replication intermediates at common fragile sites (Naim, Wilhelm, Debatisse, & Rosselli, 2013; Ying et al., 2013). Cells lacking MUS81 show defects in chromosome segregation and exhibit genome instability (Mayle et al., 2015; Minocherhomji et al., 2015; Sarbajna, Davies, & West, 2014).

Recent research has shown that the activities of MUS81 are regulated throughout the cell cycle. In meiotic and mitotic yeast cells, Mus81–Mms4 is activated by Cdk/Cdc5-dependent phosphorylation (Gallo-Fernandez, Sauger, Ortiz-Bazan, Vazquez, & Tercero, 2012; Matos, Blanco, Maslen, Skehel, & West, 2011; Matos, Blanco, & West, 2013; Szakal & Branzei, 2013), and related regulatory networks have been shown to occur in humans (Wyatt et al., 2013). Previously, we developed an assay for MUS81 activity in which epitope-tagged MUS81 was immunoprecipitated and on-bead nuclease assays were carried out, providing a relatively quick and easy tool capable of monitoring its activity at different stages of the cell cycle. Here, we describe these assays in more detail. The general schemes for the analysis of yeast Mus81–Mms4 or human MUS81–EME1 activities are similar. As shown in Fig. 1 for *S. cerevisiae*, cells carrying myc13-tagged Mus81 are lysed and the Myc tag is used to immunoprecipitate Mus81 from the whole cell extracts. The immunoprecipitated beads are then washed extensively and incubated directly with the DNA substrate. In the experiments described here, we use nicked Holliday junctions (nHJs) as they are a particularly good substrate for Mus81–Mms4, but the method can also be used to analyze the cleavage of 3′-flaps or replication

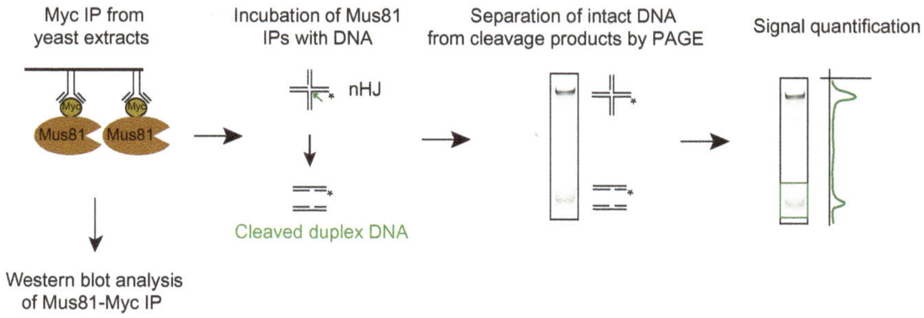

Fig. 1 On-bead resolution of nicked HJs by yeast Mus81–Mms4 endonuclease. Schematic diagram illustrating the assay for cleavage of a nHJ. Mus81-myc13 is immunoaffinity purified from mitotically proliferating budding yeast cells using anti-Myc affinity beads. After extensive washing, the beads are divided and either analyzed by Western blotting for Mus81 or incubated with fluorescently labeled ATTO 550-nHJ DNA. The uncut DNA is separated from the nicked duplex cleavage products by neutral PAGE. The DNA species are visualized using a Typhoon scanner in the Cy3 channel. The relative activity of Mus81 is calculated by determining the efficiency of conversion of the DNA substrate into nicked duplex DNA products. * indicates 5′-labels with ATTO 550.

fork structures. The nHJs are produced by annealing partially complementary oligonucleotides, one of which is fluorescently labeled with ATTO 550. Our previous work utilized ^{32}P-labeled DNA, so the use of fluorescent labels may represent an advantage for those laboratories that do not wish to use radioactivity. The input DNA and products of the cleavage reactions may then be assayed by polyacrylamide gel electrophoresis (PAGE) and quantified by imaging the fluorescent label. The advantages of these on-bead cleavage assays are many-fold: (i) the assay is relatively quick and does not require extensive protein purification, (ii) the proteins are extracted from their endogenous species, (iii) posttranslational modifications are preserved, (iv) protein complexes and interaction partners are maintained, and (v) assays are carried out in conditions that are as close as possible to the physiological environment for enzymatic reaction. The assays are not without their limitations, however, and these include (i) the sensitivity of the assay (low abundance nucleases may be difficult to detect); (ii) the potential presence of contaminating nucleases which requires extensive washing, stringent controls, and nuclease-dead variants that can validate the assay conditions; and (iii) attachment to the beads prevents free diffusion of the nuclease, which may interfere with the mechanism of cleavage and reduce cleavage efficiency. We also recommend that care is taken with regard to the use of N- or C-terminal tags, as tag location can affect nuclease activity.

2. MONITORING THE ACTIVITY OF SEMIPURIFIED Mus81/MUS81 FROM *S. CEREVISIAE* AND HUMAN CELL EXTRACTS

2.1 Immunoprecipitation of Epitope-Tagged Mus81 From Yeast Lysates

This procedure is derived from that described previously for the analysis of epitope-tagged Mus81 endonuclease from *S. cerevisiae* (Matos et al., 2011, 2013). The DNA used for these nuclease assays could be a 3′-flap or replication fork structure, or a nHJ, as all serve as excellent substrates for Mus81–Mms4. DNA substrates are prepared by annealing partially complementary oligonucleotides, essentially as described previously (Rass & West, 2006). However, in the methods described here, we use fluorescently labeled, rather than ^{32}P-labeled, DNA, as this may be more useful for those laboratories that do not carry out radioactive work. The protocol is optimized for the immunoprecipitation (IP) and assay of the activity of myc13-tagged Mus81 from *S. cerevisiae*, expressed from its endogenous promoter. As negative controls, we use the parental wild type strain and, in addition, a strain expressing a nuclease-dead version of Mus81 (Mus81ND-myc13). Identical conditions can be used to successfully immunoprecipitate Mus81-ha3, Mus81-myc9, Mus81-FLAG3, Mus81-FLAG6, Mms4-myc9, and Mms4-FLAG3 fusions. However, due to slightly different IP efficiencies, the total amount of protein used in each case needs to be optimized for each Mus81 or Mms4-tag fusion.

Notes

- The entire procedure is performed at 4°C unless specified otherwise
- All buffers are chilled on ice before use
- Anti-myc agarose beads were prepared by conjugating purified anti-myc antibody (clone 9E10, Sigma-Aldrich) using the AminoLink Plus immobilization kit (Thermo Scientific) according to the manufacturer's manual.
- In preparation for the IP, 10 μL of packed anti-myc beads/nHJ resolution reaction should be extensively washed and incubated for at least 30 min in buffer R + bovine serum albumin (BSA). The BSA reduces the level of nonspecific binding of proteins in the extract to the beads.

Solutions and reagents

- Yeast strains (Matos et al., 2011)
  - YWL1041; YWL1042; YWL2776 (see Table 1 for details)
- Glass beads, 0.5 mm diameter (Carl Roth GmbH)

Table 1 *S. cerevisiae* Strains Used in This Study

Strain Number	Relevant Genotype (SK1 Background, Homozygous Diploid)
YML1041	ho::LYS2 ura3 leu2::hisG trp1::hisG his3::hisG *MUS81*
YML1042	ho::LYS2 ura3 leu2::hisG trp1::hisG his3::hisG *mus81Δ::KanMX::MUS81*^{WT}*-myc13::LEU2*
YML2776	ho::LYS2 ura3 leu2::hisG trp1::hisG his3::hisG *mus81Δ::KanMX::MUS81*^{DD}*-myc13::LEU2*

- 50% slurry of anti-myc agarose beads (10 mg antibody/mL of packed agarose beads)
- Protease inhibitors
  - EDTA-free Protease Inhibitor Cocktail (Roche). Grind tablet to a powder and dissolve the powder in buffer shortly before use
  - 0.2 *M* phenylmethylsulfonyl fluoride (PMSF) in DMSO. Because PMSF is relatively unstable in DMSO, this needs to be prepared immediately prior to use
- Water + 1 m*M* PMSF/DMSO
- Buffer W
  - 40 m*M* Tris–HCl, pH 7.5
  - 150 m*M* NaCl
  - 0.1% NP40
  - 10% glycerol
- Buffer R
  - 40 m*M* Tris–HCl, pH 7.5
  - 150 m*M* NaCl
  - 0.1% NP40
  - 10% glycerol
  - 1 m*M* NaF
  - 20 m*M* beta-glycerophosphate
  - 1 m*M* dithiothreitol (DTT), added freshly from a 1 *M* stock in water
- Buffer R + BSA (1 mg/mL)
- Buffer R + protease inhibitors
  - 1 tablet of EDTA-free Protease Inhibitor Cocktail/15 mL buffer
  - 1 m*M* PMSF
- Bio-Rad Protein Assay kit (Bio-Rad)

- Protein sample buffer (2 ×)
  - NuPAGE LDS-sample buffer (Invitrogen), 2 ×
  - 200 m*M* DTT, added fresh
- 21G hypodermic needles (BD Microlance 3)

2.1.1 Harvesting *S. cerevisiae* Cells

Procedure

1. Grow 200 mL of yeast culture, proliferating exponentially, to $OD_{600} \sim 0.8$.
2. Transfer the culture to a precooled 250 mL-centrifuge bottle; add 2 mL of PMSF/DMSO and mix.
3. Spin at 1000 rcf for 3 min. Resuspend cells in 25 mL of ice-cold water containing PMSF and transfer the suspension into a 50-mL Falcon tube.
4. Spin at 1000 rcf for 3 min and carefully remove supernatant. Resuspend cells in 1 mL of ice-cold water containing PMSF and transfer the suspension into a 1.5-mL screw-cap microcentrifuge tube.
5. Spin at 1000 rcf for 3 min and carefully remove supernatant. At this point, the cell pellet may be frozen in liquid nitrogen and stored at −80°C.

2.1.2 Preparation of Cell Extracts

1. Thaw the cell pellet on ice and adjust its packed volume to ~0.5 mL. Resuspend the cells in 0.5 mL of buffer R + protease inhibitors and add 0.5 mL of glass-beads. Mix well to ensure that the cell pellet is completely resuspended.
2. To break cells, homogenize the suspension three times for 45 s (Fast-Prep 24 5G, MP Bio, highest settings). Cool for 2 min in ice water between runs. Verify cell lysis on the microscope. >90% of cells should be disrupted with this procedure.
3. Punch a hole into the bottom of the 1.5-mL tube using a strong needle and carefully put the tube into a 2-mL microcentrifuge tube. Centrifuge the assembly at 400 rcf, 1 min, to transfer the extract but not the glass beads into the 2-mL tube.
4. Centrifuge the whole cell lysate at >13,000 rpm for 30 min in a cooled benchtop centrifuge and carefully transfer the cleared lysate to a new 1.5-mL microcentrifuge tube.
5. Measure protein concentration using the Bio-Rad Protein Assay kit. The protein concentration is typically 10–20 mg/mL.

2.1.3 Immunoprecipitation

1. Adjust all samples to the same protein concentration and volume with buffer R + protease inhibitors. For the experiments in Fig. 2A and B, we used 6 mg of total protein in a volume of 0.6 mL. This was sufficient

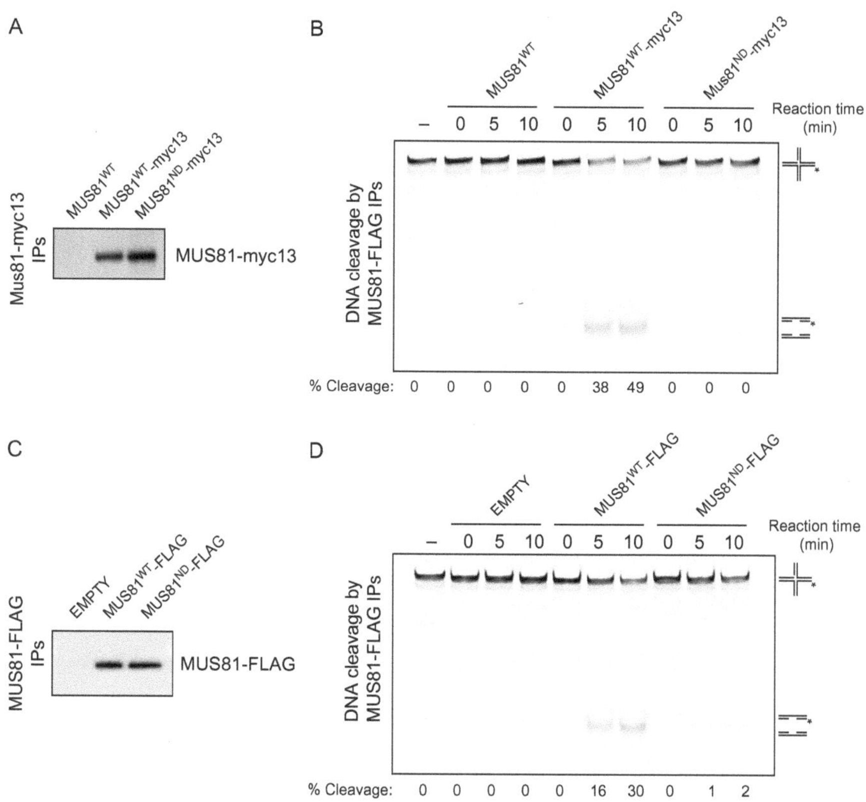

Fig. 2 Mus81/MUS81-dependent cleavage of nHJs by immunoprecipitates from yeast and human cells. (A) *S. cerevisiae* cells expressing Mus81^{WT}, Mus81^{WT}-myc13, or Mus81ND-myc13 were harvested during exponential proliferation. Extracts were prepared and affinity-purified Mus81-myc13 was analyzed by Western blotting. (B) Mus81-myc13 immunoprecipitates from (A) were assayed for the cleavage of fluorescently labeled nHJ DNA. DNA substrates and products were separated by neutral PAGE and imaged using a Typhoon scanner. The intensity of each band was determined using ImageQuant software. Mus81ND contains the D414A and D415A mutations, which impair its catalytic activity. * indicates 5′-labels with ATTO 550. (C) Subconfluent HeLa FRT/TO cells conditionally expressing MUS81^{WT}-FLAG or MUS81ND-FLAG were induced for 16 h. MUS81-FLAG was affinity-purified using anti-FLAG-beads and the immunoprecipitates analyzed by Western blotting. (D) MUS81-FLAG immunoprecipitates generated as in (C) were analyzed for nuclease activity using nHJ DNA as in (B). MUS81ND contains the D307A mutation, which impairs its catalytic activity. * indicates 5′-labels with ATTO 550.

to obtain enough Mus81 for three DNA cleavage assays and for Western blot analysis of the IPs.

2. For Western blotting, transfer 25 μL of the normalized extract into a microcentrifuge tube containing 25 μL of 2 × protein sample buffer. Boil the sample for 3 min and freeze in liquid nitrogen. This sample can be used to verify the expression levels of Mus81-myc13 prior to the IP.
3. To the remaining 575 μL of cleared lysate add 80 μL of a 50% slurry of BSA-blocked anti-myc agarose beads. Incubate on a rotating wheel for 90 min, 14 rpm, with an angle of 45 degree.
4. Sediment the beads by centrifugation at 100 rcf, for 1 min, on a benchtop centrifuge. Carefully remove the supernatant, using a hypodermic needle connected to a vacuum pump, without drying the beads.
5. Wash beads sequentially with 1 mL aliquots of the following buffers:
   - 2 × 1 mL of buffer R + protease inhibitors
   - 2 × 1 mL of buffer R + BSA
   - 2 × 1 mL of buffer R
   - 1 × 1 mL of buffer W
6. After the final wash, centrifuge again and remove excess washing buffer.
7. Resuspend the beads in 800 μL of buffer W and split them equally into four precooled 1.5-mL microcentrifuge tubes.
8. Sediment the beads by centrifugation at 100 rcf and gently remove the buffer without completely drying the beads.
9. Add 20 μL of protein sample buffer to 1 of the 4 sets of beads and boil for 5 min before freezing in liquid nitrogen. This sample is used for Western blotting to determine whether Mus81-myc13 was successfully immunoprecipitated and to normalize the relative amounts of Mus81 in each sample. The remaining 3 sets of beads can be used for the cleavage assays described in Section 3.1.

2.2 IP of Epitope-Tagged MUS81 From Extracts of Human Cells

This method is derived from previous studies in which MUS81–EME1 activity was analyzed from extracts prepared from human cells carrying FLAG-tagged MUS81 (Duda et al., 2016; Wyatt et al., 2013). In the experiments described here, we used Flp-In T-Rex HeLa cells in which MUS81 expression was under the control of a doxycycline-inducible promoter.

Notes

- HeLa FRT/TO cells were maintained in Dulbecco's Modified Eagle Media (DMEM), supplemented with 10% fetal bovine serum (FBS)

and appropriate antibiotic selection. All cultures were grown at 37°C in a humidified atmosphere containing 5% CO_2
- The induction of MUS81-FLAG expression was initiated 24h prior to harvesting by addition of 1 μg/mL doxycycline
- Unless specified otherwise, all steps are performed at 4°C
- All buffers are chilled on ice before use

Solutions and reagents
- Cell lines (Duda et al., 2016)
  - CLML182; CLML155; CLML156 (see Table 2 for details)
- Anti-FLAG agarose beads (Anti-FLAG® M2 Affinity Gel Agarose Beads, Sigma-Aldrich)
- 21G needles (BD Microlance 3)
- 5-mL syringes (BD Luer)
- Buffers are the same as described in Section 2.1

2.2.1 Harvesting HeLa Cells

Procedure

1. Grow each cell line in four tissue culture plates (144 × 21 mm, Nunc) to a confluency of ~70%.
2. Scrape cells into their medium (25 mL) on the culture dishes, transfer into a 50-mL centrifuge tube, and pellet the cells (150 rcf, 5 min).
3. Remove the supernatant and wash the combined pellet from the four plates with 20 mL of ice-cold phosphate-buffered saline (PBS), spin again, and aspirate the supernatant. Resuspend the cells in 1 mL ice-cold PBS and transfer into a 1.5-mL microcentrifuge tube.
4. Centrifuge the cells, remove the supernatant, and snap-freeze the cell pellet in liquid nitrogen.

2.2.2 Preparation of HeLa Cell-Free Extracts

1. Thaw the cell pellet on ice and adjust its packed volume to ~0.3 mL. Resuspend the cells with 1.2 mL of buffer R + protease inhibitors.

Table 2 HeLa Cell Lines Used in This Study

Cell Line Number	Relevant Genotype (Flp-In T-Rex)
CLML182	—
CLML155	$MUS81^{WT}$-SF[a]
YML156	$MUS81^{ND}$-SF[a]

[a]SF is StrepII-FLAG.

Mix well to ensure that the cell pellet is completely resuspended before the next step.

2. To break the cells, force the cell suspension through a 21G needle 20 times, using a 5-mL syringe. Keep the suspension on ice for 5 min and then repeat the procedure. Verify cell lysis using a microscope.
3. Centrifuge the whole cell lysate at >20,000 rcf for 30 min in a benchtop centrifuge and carefully transfer the cleared lysate to a new 1.5-mL microcentrifuge tube.
4. Measure protein concentration with the Bio-Rad Protein Assay kit. The protein concentration is typically ~10 mg/mL.

2.2.3 Immunoprecipitation

For the experiment in Fig. 2C and D, we used 6 mg of total protein in a volume of 0.6 mL. This was sufficient to obtain enough MUS81-FLAG for three cleavage assays and for Western blot analysis of the IPs. The IP was performed essentially as described for yeast Mus81-myc13 in Section 2.1.

3. NUCLEASE ACTIVITY ASSAY ON AGAROSE BEADS

3.1 Preparation and Labeling of nHJ DNA

Notes

- Oligonucleotide X01 was fluorescently labeled at the 5′-terminus with ATTO 550 and purified by PAGE (Mycrosynth). The oligo was then annealed with four other partially complementary oligonucleotides (Wyatt et al., 2013) (Table 3) to generate a nHJ which was purified as described (Rass & West, 2006). The amounts described later are sufficient for >200 cleavage reactions
- When light protected and stored at 4°C, PAGE-purified ATTO 550-labeled nHJ is stable for at least 6 months
- Unless specified otherwise, all steps are performed at room temperature

Solutions and reagents

- Oligos at 100 μ*M* concentration in water (see Table 3): X01, X02, X03 1/2, X03 2/2, X04
- PAGE-purified ATTO 550-labeled nHJ (30 ng/μL)
- 10 × buffer 2 (NEB)
  - 100 m*M* Tris–HCl, pH 7.9
  - 0.5 *M* NaCl

Table 3 Oligonucleotides Used to Prepare the nHJ Substrate

Oligonucleotide	Sequence (5′–3′)
X01[a]	ACGCTGCCGAATTCTACCAGTGCCTTGCTA GGACATCTTTGCCCACCTGCAGGTTCACCC
X02	GGGTGAACCTGCAGGTGGGCAAAGATGTCC ATCTGTTGTAATCGTCAAGCTTTATGCCGT
X03 1/2[b]	ATGGAGCTGTCTAGAGGATCCGACTATCG
X03 2/2	ACGGCATAAAGCTTGACGATTACAACAGATC
X04	CGATAGTCGGATCCTCTAGACAGCTCCATG TAGCAAGGCACTGGTAGAATTCGGCAGCGT

[a]This oligo was 5′-labeled with ATTO 550.
[b]This oligo carried a 5′-phosphate group.

  - 100 m*M* $MgCl_2$
  - 10 m*M* DTT
- DNA storage buffer
  - 10 m*M* Tris–HCl, pH 8.0
  - 1 m*M* $MgCl_2$
  - 50 m*M* NaCl
- DNA loading dye
  - 0.03% bromophenol blue
  - 60% glycerol
  - 60 m*M* EDTA, pH 8.0
  - 10 m*M* Tris–HCl, pH 8.0
- TBE buffer
  - 89 m*M* Tris
  - 89 m*M* boric acid
  - 2 m*M* EDTA

Procedure

1. In a 1.5-mL screw-cap microcentrifuge tube, mix 48 μL of PAGE-purified ATTO 550-labeled oligonucleotide X01 with 15 μL of each of the other oligonucleotides (X02, X03 1/2, X03 2/2, and X04). Add 12 μL of 10 × buffer 2 (NEB).
2. Place in a boiling water bath. After 2 min switch off the water bath and allow to cool slowly overnight to room temperature. Cover the water bath with aluminium foil to protect the fluorescently labeled DNA from light.

3. Prepare a 12% polyacrylamide gel in TBE (PROTEAN II xi Cell system, 20 × 20 cm gel).
4. Add 1/6th volume of loading dye to the annealed substrate and separate DNA species by PAGE.
5. Run for 4.5 h at 200 V at 4°C in TBE. Protect from light by covering the gel chamber with aluminium foil.
6. Remove one of the gel plates and excise the substrate band using a scalpel. The ATTO 550-labeled DNA substrate is visible without the need for special illumination.
7. Transfer the gel slice into a screw-cap microcentrifuge tube and break it into small pieces using the back of a plastic inoculation loop.
8. Add 250 μL of DNA storage buffer and elute overnight on a rotating wheel at 4°C.
9. Spin at max speed on a benchtop centrifuge for 1 min and transfer the eluted DNA into a new tube.
10. Determine the DNA concentration using a NanoDrop system (Thermo Fisher Scientific).
11. Dilute the DNA in storage buffer to 30 ng/μL. Store at 4°C, protected from light.

3.2 Resolution Assay on Agarose Beads

Notes

- Resolution assays should be performed immediately following the IP step (2.1.3 or 2.2.3). Storage of the immunoprecipitates reduces nuclease activity and prevents appropriate comparison of activity from cells collected after different treatments.

Solutions and reagents

- ATTO 550-labeled nHJ (30 ng/μL)
- 1 *M* Tris–HCl, pH 7.5
- 10 m*M* $MgCl_2$
- Stop solution
  - 10 mg/mL proteinase K
  - 2.5% SDS
  - 50 m*M* EDTA
  - 100 m*M* Tris–HCl, pH 7.5
- DNA loading dye

Procedure

1. Prepare a master mix of nHJ in cleavage buffer (50 m*M* Tris–HCl, pH 7.5, 3 m*M* $MgCl_2$) according to Table 4.

Table 4 Pipetting Scheme for the Master Mix of DNA Cleavage Reactions

	Volume in μL for 1 10 μL Reaction
1 *M* Tris–HCl, pH 7.5	0.5
10 m*M* $MgCl_2$	3
ATTO 550 nHJ	0.5
H_2O	6

2. Add 10 μL of the reaction mix to each tube containing the IPs generated in Section 2.1.3 or 2.2.3.
3. Incubate for the desired times at 37°C in a thermomixer with gentle agitation (800 rpm). In Fig. 2B and D, incubation was for 0, 5, or 10 min.
4. Reactions are stopped by addition of 5 μL of Stop solution, followed by incubation for 45 min at 37°C.
5. Add 5 μL of DNA loading dye. Samples can be stored at 4°C overnight, if necessary.

3.3 PAGE Separation of the Cleavage Products and Data Analysis

Solutions and reagents

- 30% acrylamide/bis-acrylamide solution 29:1 (Bio-Rad)
- 10% ammonium persulphate
- 10 × TBE
- TEMED
- DNA loading dye
  - 0.03% bromophenol blue
  - 60% glycerol
  - 60 m*M* EDTA
  - 10 m*M* Tris–HCl, pH 8.0

Procedure

1. Prepare a 10% polyacrylamide gel in TBE (10 × 20 cm gel).
2. Load sample onto the gel and run for 75 min at 150 V, room temperature, in TBE. Protect from light by covering the gel chamber with aluminium foil.
3. Carefully remove the gel from the glass plates and image in a Typhoon Scanner (GE Healthcare) using the Cy3 channel and the highest sensitivity.

3.4 Data Analysis

Fig. 2B and D show the expected outcome of a reaction in which nHJs are cleaved by a Mus81-myc13 immunoprecipitate from yeast and a MUS81-FLAG IP from human cells. The extent of cleavage of the nHJ is dependent on the incubation time and on the nuclease activity of Mus81/MUS81.

To quantify the portion of cleaved substrate (Fig. 2B and D), the percent nHJ cleaved was calculated by determining the fraction of nicked duplex DNA product relative to the sum of the intact substrate and resolution product. Band intensities were determined using ImageQuant software (GE Healthcare).

4. ADAPTATION TO OTHER STRUCTURE-SELECTIVE ENDONUCLEASES AND DNA SUBSTRATES

The protocols outlined here can be easily modified to analyze the activity of Mus81–Mms4 or MUS81–EME1 immunoprecipitates with different DNA structures. The specific conditions described can be directly used for DNA substrates that serve as models for 3′-flaps or replication forks. To observe the cleavage of nHJs, which are resolved with low efficiency, greater amounts of immunoprecipitated Mus81/MUS81 is required and longer incubation times are recommended.

The activities of other structure-selective endonucleases such as Yen1/GEN1 and SLX1–SLX4 can be monitored by adapting the protocols described here. Indeed, these protocols may be useful for almost any structure-selective endonuclease. Key modifications to take into consideration are (1) the amount of starting material both in terms of total protein and the expression levels of the nuclease to be assayed; (2) the optimal buffer, and in particular the concentration of $MgCl_2$ in the cleavage reaction; (3) the incubation time.

ACKNOWLEDGMENTS

We thank Miguel G. Blanco for help in establishing the conditions for the resolution assays described here, and Heike Duda and Philipp Wild for help in generating the cell lines for inducible MUS81 expression. Work in the laboratory of J.M. is supported by ETH Zurich and the Swiss National Science Foundation (Grants 31003A_153058 and 155823), while that in S.C.W's laboratory is supported by the Francis Crick Institute (FC10212), the European Research Council (ERC-ADG-66640), and the Louis-Jeantet Foundation. The Francis Crick Institute receives core funding from Cancer Research UK, the Medical Research Council, and the Wellcome Trust.

REFERENCES

Boddy, M. N., Gaillard, P. H. L., McDonald, W. H., Shanahan, P., Yates, J. R., & Russell, P. (2001). Mus81-Eme1 are essential components of a Holliday junction resolvase. *Cell*, *107*, 537–548.

Castor, D., Nair, N., Déclais, A. C., Lachaud, C., Toth, R., Macartney, T. J., et al. (2013). Cooperative control of Holliday junction resolution and DNA repair by the SLX1 and MUS81-EME1 nucleases. *Molecular Cell*, *52*, 221–233.

Ciccia, A., McDonald, N., & West, S. C. (2008). Structural and functional relationships of the XPF/MUS81 family of proteins. *Annual Review of Biochemistry*, 77, 259–287.

Doe, C. L., Ahn, J. S., Dixon, J., & Whitby, M. C. (2002). Mus81-Eme1 and Rqh1 involvement in processing stalled and collapsed replication forks. *The Journal of Biological Chemistry*, *277*, 32753–32759.

Duda, H., Arter, M., Gologgnitzer, J., Teloni, F., Wild, P., Blanco, M. G., et al. (2016). A mechanism for controlled breakage of under-replicated chromosomes during mitosis. *Developmental Cell*, *39*, 740–755.

Gallo-Fernandez, M., Sauger, I., Ortiz-Bazan, M. A., Vazquez, M. V., & Tercero, J. A. (2012). Cell cycle-dependent regulation of the nuclease activity of Mus81-Eme1/Mms4. *Nucleic Acids Research*, *40*, 8325–8335.

Garner, E., Kim, Y., Lach, F. P., Kottemann, M. C., & Smogorzewska, A. (2013). Human GEN1 and the SLX4-associated nucleases MUS81 and SLX1 are essential for the resolution of replication-induced Holliday junctions. *Cell Reports*, *5*, 207–215.

Jessop, L., & Lichten, M. (2008). Mus81/Mms4 endonuclease and Sgs1 helicase collborate to ensure proper recombination intermediate metabolism during meiosis. *Molecular Cell*, *31*, 313–323.

Matos, J., Blanco, M. G., Maslen, S. L., Skehel, J. M., & West, S. C. (2011). Regulatory control of the resolution of DNA recombination intermediates during meiosis and mitosis. *Cell*, *147*, 158–172.

Matos, J., Blanco, M. G., & West, S. C. (2013). Cell cycle kinases coordinate the resolution of recombination intermediates with chromosome segregation. *Cell Reports*, *4*, 76–86.

Mayle, R., Campbell, I. M., Beck, C. R., Yu, Y., Wilson, M., Shaw, C. A., et al. (2015). Mus81 and converging forks limit the mutagenicity of replication fork breakage. *Science*, *349*, 742–747.

Minocherhomji, S., Ying, S., Bjerregaard, V. A., Bursomanno, S., Aleliunaite, A., Wu, W., et al. (2015). Replication stress activates DNA repair synthesis in mitosis. *Nature*, *528*, 286–290.

Naim, V., Wilhelm, T., Debatisse, M., & Rosselli, F. (2013). ERCC1 and MUS81-EME1 promote sister chromatid separation by processing late replication intermediates at common fragile sites during mitosis. *Nature Cell Biology*, *15*, 1008–1015.

Oh, S. D., Lao, J. P., Taylor, A. F., Smith, G. R., & Hunter, N. (2008). RecQ helicase, Sgs1, and XPF family endonuclease, Mus81-Mms4, resolve aberrant joint molecules during meiotic recombination. *Molecular Cell*, *31*, 324–336.

Rass, U., & West, S. C. (2006). Synthetic junctions as tools to identify and characterise Holliday junction resolvases. In J. L. Campbell & P. Modrich (Eds.), *Methods in Enzymology* (pp. 485–501). San Diego: Elsevier.

Sarbajna, S., Davies, D., & West, S. C. (2014). Roles of SLX1-SLX4, MUS81-EME1 and GEN1 in avoiding genome instability and mitotic catastrophe. *Genes & Development*, *28*, 1124–1136.

Szakal, B., & Branzei, D. (2013). Premature Cdk1/Cdc5/Mus81 pathway activation induces aberrant replication and deleterious crossover. *The EMBO Journal*, *32*, 1155–1167.

Wechsler, T., Newman, S., & West, S. C. (2011). Aberrant chromosome morphology in human cells defective for Holliday junction resolution. *Nature*, *471*, 642–646.

Wyatt, H. D. M., Sarbajna, S., Matos, J., & West, S. C. (2013). Coordinated actions of SLX1-SLX4 and MUS81-EME1 for Holliday junction resolution in human cells. *Molecular Cell, 52*, 234–247.
Ying, S. M., Minocherhomji, S., Chan, K. L., Palmai-Pallag, T., Chu, W. K., Wass, T., et al. (2013). MUS81 promotes common fragile site expression. *Nature Cell Biology, 15*, 1001–1007.

CHAPTER TWELVE

Strategies and Methods of Transcription-Coupled Repair Studies In Vitro and In Vivo

Vitaly Epshtein*, Venu Kamarthapu*,†, Evgeny Nudler*,†,1
*New York University School of Medicine, New York, NY, United States
†Howard Hughes Medical Institute, New York University School of Medicine, New York, NY, United States
[1]Corresponding author: e-mail address: evgeny.nudler@nyumc.org

Contents

Abstract

Transcription-coupled repair (TCR) serves an important role in preserving genome integrity and maintaining fidelity of replication. Coupling transcription to DNA repair requires a coordinated action of several factors, including transcribing RNA polymerase and various transcription modulators and repair proteins. To study TCR in molecular detail, it is important to employ defined protein complexes in vitro and defined genetic backgrounds in vivo. In this chapter, we present methods to interrogate various aspects of TCR at different stages of repair. We describe promoter-initiated and nucleic acid scaffold-initiated transcription as valid approaches to recapitulate various stages of TCR, and discuss their strengths and weaknesses. We also outline an approach to study TCR in its cellular context using *Escherichia coli* as a model system.

Methods in Enzymology, Volume 591
ISSN 0076-6879
http://dx.doi.org/10.1016/bs.mie.2017.03.014

1. INTRODUCTION

Transcription-coupled repair (TCR) is a subpathway of the nucleotide excision repair (NER) process that allows for rapid repair of DNA lesions on actively transcribed template strands (Ganesan, Spivak, & Hanawalt, 2012; Kamarthapu & Nudler, 2015). TCR was first discovered in eukaryotes (Bohr, Smith, Okumoto, & Hanawalt, 1985; Mellon, Bohr, Smith, & Hanawalt, 1986) and subsequently in prokaryotes (Mellon & Hanawalt, 1989), indicating it to be an evolutionarily conserved mechanism of genome maintenance. TCR operates in a variety of organisms from bacteria to humans and utilizes the ubiquitous nature of RNA transcription to survey and repair DNA lesions throughout the genome. In TCR, RNA polymerase functions as an impromptu scanner that halts at the sites of bulky DNA damage, which is followed by recruitment of the NER components to the lesion. Two distinct molecular pathways couple transcription to repair (Epshtein, 2015; Kamarthapu & Nudler, 2015; Rasouly, Pani, & Nudler, 2017). Initially, the transcription coupling repair factor, also known as Mfd protein, binds to an elongation complex arrested at the site of DNA damage and "pushes" it forward using its ATPase activity to dislodge RNA polymerase from the lesion (Park, Marr, & Roberts, 2002; Selby & Sancar, 1993). After removing transcription complex, Mfd then recruits UvrA to the lesion site and repair proceeds via a common NER mechanism (Kisker, Kuper, & Van Houten, 2013; Reardon & Sancar, 2005; Truglio, Croteau, Van Houten, & Kisker, 2006). This type of TCR may function particularly well during nonstress conditions when NER protein expression is not activated (Backendorf, Brandsma, Kartasova, & van de Putte, 1983; Siegel, 1983). Recently, our laboratory demonstrated that during genotoxic stress cells employ a different TCR strategy based on UvrD protein (also known as helicase II). In this process, UvrD dimerizes during the SOS response and actively pulls the stalled RNA polymerases backward from the lesion sites, essentially working as a pro-backtracking factor. The transcription elongation factor NusA and the bacterial alarmone ppGpp render RNA polymerase backtracking prone to facilitate UvrD-mediated backtracking (Epshtein et al., 2014; Kamarthapu et al., 2016). Removing RNA polymerase from the damaged site helps to unmask the site for repair and allows UvrD together with NusA to recruit NER proteins to the site of damage (Cohen & Walker, 2010; Manelyte et al., 2009).

To study the TCR process in molecular detail, it is important to reconstruct its main features in vitro and in vivo in a defined system. We employed several techniques that allowed us to reconstitute different stages of UvrD-dependent TCR pathway with various degrees of complexity (Epshtein et al., 2014; Kamarthapu et al., 2016). To study the initial recognition of DNA damage and the lateral dynamics of the transcribing complexes, we employed linear DNA fragments with natural promoters and well-studied transcription units. Such an approach simplifies data interpretation and, at the same time, allows monitoring of the behavior of RNA polymerase in response to different protein and nonprotein factors that participate in TCR. It is easy to substitute linear DNA fragments by circular plasmid DNA if the effects of DNA-binding proteins and supercoiling have to be taken into consideration or if free DNA ends interfere with the experimental design.

For an in-depth study of the damage repair process, we used a nucleic acid scaffold assembly method (Daube & von Hippel, 1992; Korzheva, Mustaev, Nudler, Nikiforov, & Goldfarb, 1998) that allows the incorporation of any commercially available DNA modification at defined positions in the template. This approach circumvents issues of ambiguity that arise when DNA damage is randomly placed in DNA by chemical treatment or UV irradiation. In general, this method could be applied to studies of DNA repair pathways other than NER, if a suitable modification is incorporated into a specific DNA strand. The same scaffold method is also useful for mapping contacts within the repair complexes using photocross-linkable nucleotides incorporated in defined positions of assembled oligonucleotides (Epshtein et al., 2014). This approach can be even more powerful when it is coupled with mass spectrometry analysis of the resulting adducts.

Last but not least, in vivo analysis of TCR can be used to validate in vitro findings and also to evaluate the natural context of a particular repair mechanism. We use the primer extension technique that relies on the inability of Taq DNA polymerase to proceed past bulky lesions (Chandrasekhar & Van Houten, 1994; Epshtein et al., 2014; Wellinger & Thoma, 1996). We monitor the disappearance of the intermediate DNA bands corresponding to the DNA lesions undergoing repair, allowing us to deduce the TCR factors and their roles in the process by manipulating the genetic background of cells subjected to genotoxic stress.

2. GENERAL CONSIDERATIONS FOR STUDYING BACTERIAL TCR IN VITRO

Reconstruction of TCR in vitro requires a DNA template that has two features: a defined DNA lesion that can be recognized by NER factors and a promoter to initiate transcription on the desired DNA strand. There are two general methods to introduce lesions into DNA. The first method relies on using DNA-damaging agents such as UV light or certain chemicals to inflict DNA damage on single-stranded (ss) or double-stranded (ds) DNA fragments (for example, see Selby, Witkin, & Sancar, 1991). Such an approach is inexpensive and rapid, but nonspecific and of variable efficiency. Strand-, but not site-, specificity can be achieved by treatment of ssDNA followed by its annealing to a cognate nontemplate strand. The second method is based on the use of specific DNA oligos with a lesion positioned at a precise location. This method allows for 100% efficiency in the placement and specificity of the DNA damage site, but it is more expensive and typically requires modified oligonucleotides obtained from an outside source. Here, we concentrate on this oligo-based method, which makes interpretation of the experimental results much more straightforward and rigorous.

Historically, the method of assembly of a DNA template with a defined lesion site utilized a DNA oligo bearing a modified nucleotide annealed to the complementary nonmodified ssDNA (Selby & Sancar, 1993; Shi, Gamper, & Hearst, 1987; Shi, Gamper, Van Houten, & Hearst, 1988). By designing the entire DNA template, one can place the promoter, terminator, and other desired sequences in any combination, creating a group of various templates with different transcriptional properties. In addition, the annealing can be performed on a plasmid, followed by ligation, thereby generating a circular template for transcription (Donahue, Yin, Taylor, Reines, & Hanawalt, 1994). The advantages of this method are the use of natural dsDNA substrates that have native promoters and the relative ease with which the templates are produced for various experiments. Disadvantages of this method are the lower than 100% occupancy of the promoter by the RNA polymerase and the inability to use preassembled complexes of RNA polymerase with accessory proteins because such complexes could be compromised in their ability to initiate transcription (e.g., when components of the complex compete with the promoter-specific sigma subunit). We (Epshtein et al., 2014) recently adopted and modified another method that uses artificially assembled nucleic acid scaffolds (Daube & von Hippel, 1992; Korzheva et al., 1998). In this method, a modified DNA oligo serves as the template and an RNA oligo is annealed to the desired position, followed by RNA polymerase binding to DNA–RNA hybrid. The resulting scaffold is then completed by the addition of the nontemplate strand, immobilizing the complex on beads by an affinity tag and washing away the excess nonassembled oligos using appropriate buffers. This method allows complexes to be assembled with 100% efficiency and permits manipulations of the nucleic acid structure, such as the introduction of a noncomplementary transcription bubble or intrastrand and interstrand cross-links. It also allows the precise placement of cross-linkable moieties and fluorophores in the DNA, RNA:DNA hybrid, or nontemplated RNA. Furthermore, the assembly of the elongation complex bypasses the transcription initiation step, thereby allowing the use of various multiprotein complexes and/or RNA polymerase mutants that otherwise would be incapable of initiating RNA synthesis.

2.1 Measuring Pro-Backtracking Activity of UvrD During Transcription In Vitro

UvrD functions as a strong $3' \rightarrow 5'$ helicase with a relatively moderate processivity of ~30 nt per binding event (Hickson, Arthur, Bramhill, & Emmerson, 1983; Matson, 1986; Matson & George, 1987). Interacting with the transcribing RNA polymerase, UvrD binds at the rear end of the transcription bubble and unwinds the DNA duplex in the upstream direction, forcing RNA polymerase to backtrack. To demonstrate this in vitro, we monitored the appearance of new RNA products corresponding to UvrD-derived backtracked complexes during a single-round transcription of a linear dsDNA fragment (Nudler, Gusarov & Bar-Nahum 2003). Our templates contained the T7A1 promoter from position −70 to +20 fused to different transcribed fragments (Fig. 1A). We recommend the use of strong promoters, such as T7A1, to achieve a higher signal-to-noise ratio. The general strategy of the experimental design is shown in Fig. 1B.

2.1.1 Equipment

- Thermal (PCR) cycler
- Agarose and acrylamide gel apparatus
- Table-top mini centrifuge
- Standard Urea-PAGE equipment (OWL Scientific); 6% urea gels (acrylamide:bis-acrylamide 19:1)
- Gel dryer (Model 583 Bio-Rad)
- Typhoon scanner (GE) with a phosphor-imaging screen
- ImageQuant TL software (GE) for data visualization and analysis

2.1.2 Buffers and Reagents

- 5′ Biotin-tagged DNA primers
- Phusion High-Fidelity DNA polymerase (NEB)
- Phusion PCR buffer (NEB)

A

tccagatcccgaaaatttatcaaaaagagtattgacttaaagtctaacctataggatacttacagcc

ATCGAGAGGGACACGGCGAAGGCGGCAttttaactttcttgaattcagccggaaaaatcctaaattcatttaatatttatctttttaccgtttcg
cttaccccggtcgaacgtcaacttacgtcattttccgcccaacagtaatataatcaaacaaattaatcccgcaacataacaccagtaaaatcaa
taattttctctaagtcacttattcctcaggtaattgttaatatatccagaatgttcctcaaaatatatttccctctatcttctcgttgcgctta
atttgactaattctcattagcgactaattttaatgagtgtcgacacacaacactcatattaatgaaacaatgcaacgcaacgggagaaataacat
ggccgaacatcgtggtggttcaggaaatttcgccgaagaccgtgagaaggcatccgacgcaggccgtaaaggcggtcagcatagcggcggtaatt
ttaaaaatgatccgcaacgcgcatctgaagcgggtaaaaaaggcggtcaacaaagcggtggtaataaatcaggcaaatcctg

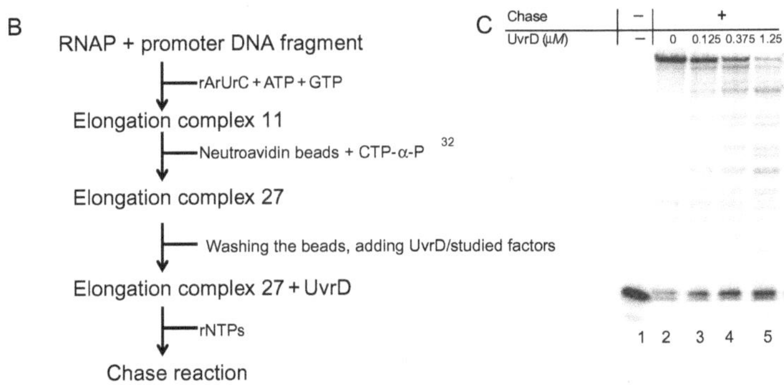

Fig. 1 Measuring the pro-backtracking activity of UvrD in vitro. (A) Sequence of the DNA template. The nontranscribed region of the fragment is shown in *italic font*; the initially transcribed region of the transcription unit up to position 27 is shown in *capital letters*. (B) Flowchart of the experimental design. (C) Titration of the UvrD protein in the chase reaction. Lower migrating bands corresponding to backtracked elongation complexes begin to appear.

- 10 m*M* dNTPs stock solutions (Amerasham)
- Gel extraction kit (Qiagen)
- RNA polymerase holoenzyme (purified according to Svetlov & Artsimovitch, 2015)
- TB50: 40 m*M* Tris–HCl, pH 8.0; 50 m*M* NaCl; 10 m*M* $MgCl_2$; 0.003% Igepal-60
- TB1000: 40 m*M* Tris–HCl, pH 8.0; 1 *M* NaCl; 10 m*M* $MgCl_2$; 0.003% Igepal-60
- TB100: 40 m*M* Tris–HCl, pH 8.0; 100 m*M* NaCl; 10 m*M* $MgCl_2$; 0.003% Igepal-60
- NeutrAvidin beads (Pierce)
- 1 m*M* set of ribonucleotides (NTPs) (Amersham)
- Ultrapure 250 μ*M* ATP and GTP (Amersham)
- rArUrC primer (IDT) 100 μ*M*
- [α-^{32}P] CTP (3000 Ci/mmol) (MP Biomedicals)
- Heparin solution in DNase/RNase-free water, 15 mg/mL
- UvrD protein (purified according to Epshtein et al., 2014)
- Stop buffer: 1 × TBE buffer, 8 *M* urea, 20 m*M* EDTA, 0.025% xylene cyanol, 0.025% bromophenol blue

- Whatman blotting paper (Thermo Scientific)
- Saran Wrap film (FisherBrand)

2.1.3 Procedure

1. Assemble a PCR reaction using a pair of primers, one of which is labeled with biotin at its 5′-end and separate the products on a 1%–2% agarose gel.
2. Purify the resulting DNA using a commercially available kit.
3. Mix 2pmol of nontagged RNA polymerase holoenzyme with 2pmol of DNA template in 20μL of TB50 buffer; incubate for 5min at 37°C.
4. Add to the sample 2μL of 100μ*M* rArUrC RNA primer, 2μL of 250μ*M* ATP, and 2μL of 250μ*M* GTP (final concentration 25μ*M* for ATP and GTP, 10μ*M* rArUrC RNA primer); continue the incubation for 5min at 37°C. An elongation complex with an 11nt RNA forms (EC11, Fig. 1B).
5. Transfer the reaction mixture to a new LowBind 1.7-mL Eppendorf tube with ~10μL of NeutrAvidin beads (Pierce) supplemented with 1.5mg/mL heparin (3μL of 15mg/mL stock solution) and suspended in TB50 buffer. Shake the mixture at room temperature for 5min.
6. Add 2μCu [α-^{32}P] CTP (3000 Ci/mmol) (MP Biomedicals) for 5min at room temperature to form EC27 (Fig. 1B).
7. Wash the beads two times in 1mL of TB1000 and two times in 1mL of TB100 using a table-top mini centrifuge; retain ~10–100μL of the reaction mixture as needed.
8. Distribute the resulting complex into 10μL aliquots in fresh Eppendorf tubes and add an appropriate amount of UvrD protein (final concentration 125n*M* to 1.25μ*M*); incubate for 5min at room temperature.
9. Add 2μL of 1m*M* NTPs to a final concentration of 100μ*M* and incubate the reaction mixture for 5min at 37°C to chase the EC27.
10. Stop the reaction by adding an equal volume (10μL) of stop buffer.
11. Heat the reaction mixture at 98°C for 5min.
12. Load samples of the stopped reaction mixture on a 6% 20cm × 20cm 1 × TBE-8 *M* urea gel and apply voltage for 20min at a constant output of 50W.
13. Transfer the gel to Whatman paper and cover it with a Saran Wrap film on one side.
14. Dry the gel in a gel dryer for 30min at +80°C.

15. Expose the gel to a phosphor-imaging screen for 1–6 h.
16. Scan the screen with a Typhoon scanner and quantify the signal with ImageQuant TL software (see example at Fig. 1C).

2.1.4 Notes

1. To test the effects of various factors on the pro-backtracking activity of UvrD, we added the factors together with UvrD protein at step 8.
2. The template DNA PCR fragment could be additionally purified by ethanol precipitation followed by dissolving the resulting pellet in RNAse-free deionized water.

2.2 Assembly of the Nucleic Acid Scaffold and TCR

Studies of TCR benefit greatly from the assembly of defined transcription complexes with lesions placed at specific positions. We used T7A1 promoter sequences as a base for scaffold assembly, but it is not a prerequisite. In general, any DNA fragment can be used as an assembly platform, restricted only by its size. The minimal DNA size is determined by the binding of RNA polymerase (footprint) and by the annealing of the nontemplate strand, which is generally about 45 nucleotides long, whereas the RNA should be at least 9–10 nucleotides long to form a stable elongation complex (Nudler, Avetissova, Markovtsov, & Goldfarb, 1996). The upper limit of the DNA fragment is restricted only by its cost and the synthetic yield of the desired oligos. We employed ssDNA oligos of between 70 and 200 nucleotides for assembly of the scaffold (Fig. 2A). It is important to immobilize the resulting elongation complex on affinity beads using a tag attached to the complex. We usually use biotin-tagged RNA polymerase and avidin-coated beads; however, biotinylated DNA or RNA oligos can be used as well. As an alternative, hexahistidine-tagged RNA polymerase can be used in conjunction with a Metal Affinity Resin (Clontech). Immobilization of the complexes is essential for quality control because it allows washing away nonbound components that could otherwise interfere with the later steps of the procedure, for example, by binding to various components of the NER complex. The principle scheme of an experiment is shown in Fig. 2B.

We reconstruct only the part of the TC-NER required for the recognition/excision of the damaged site, omitting the subsequent general NER steps that lead to filling the gap and ligating the resulting nicks. Those steps can be included by the addition of DNA polymerase and DNA ligase to the reaction mixture (Selby et al., 1991) with dNTPs.

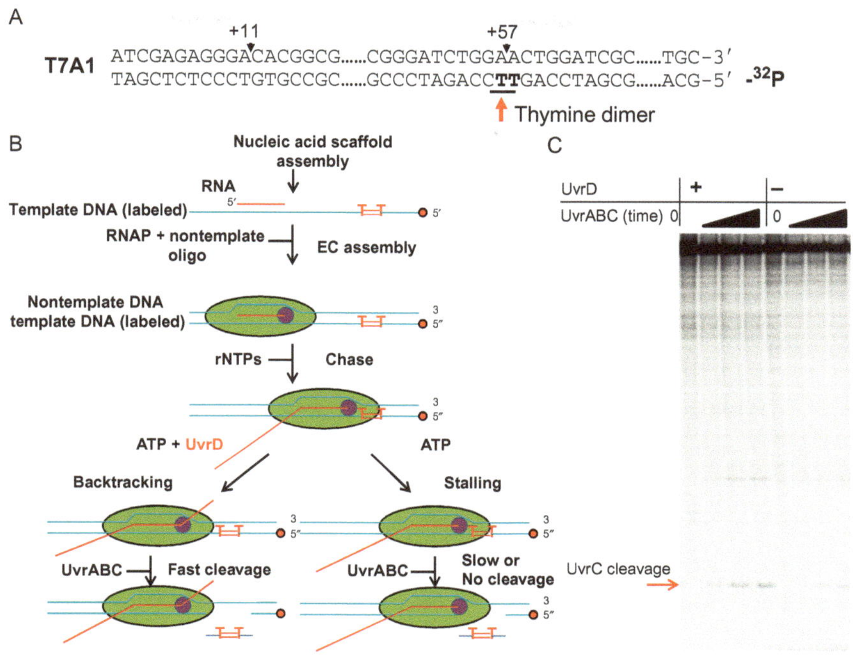

Fig. 2 The nucleic acid scaffold assembly and the TCR assay. (A) Template (*lower*) and nontemplate (*upper*) oligos. The positions of the radioactive label and the thymidine dimer are indicated. (B) Principal scheme of the experimental design. *Green oval* represents RNA polymerase, and the *purple circle* indicates the catalytic site. RNA is represented in *red*, and DNA is represented in *blue*. The *red dot* at the end of DNA template strand corresponds to the radioactive phosphate. T=T indicates the position of the thymidine dimer. (C) UvrABC-induced cleavage of the assembled and chased scaffold. The radioactive band corresponding to the cleavage reaction is indicated by a *red arrow*.

2.2.1 Equipment

- Thermal (PCR) cycler
- Table-top mini centrifuge
- Standard Urea-PAGE equipment (OWL Scientific); 10% urea gels (acrylamide:bis-acrylamide 19:1)
- Typhoon scanner (GE) with phosphor-imaging screen
- ImageQuant TL software (GE) for visualization and analysis

2.2.2 Buffers and Reagents

- Annealing buffer: 20 m*M* Tris–HCl, pH 8.0; 12% glycerol; 40 m*M* KCl; 5 m*M* $MgCl_2$
- TB50: 40 m*M* Tris–HCl, pH 8.0; 50 m*M* NaCl; 10 m*M* $MgCl_2$; 0.003% Igepal-60

- TB0: 20 m*M* Tris–HCl, pH 8.0; 10 m*M* $MgCl_2$
- TB1000: 40 m*M* Tris–HCl, pH 8.0; 1 *M* NaCl; 10 m*M* $MgCl_2$; 0.003% Igepal-60
- TB100: 40 m*M* Tris–HCl, pH 8.0; 100 m*M* NaCl; 10 m*M* $MgCl_2$; 0.003% Igepal-60
- RNA polymerase holoenzyme (purified according to Svetlov & Artsimovitch, 2015)
- NeutrAvidin beads (Pierce)
- 10 × PNK buffer, T4 Polynucleotide kinase (both from New England Biolabs)
- 1 m*M* set of ribonucleotides (NTPs)
- 1 m*M* ATP ultra pure (Amersham)
- UvrD protein (purified according to Epshtein et al., 2014)
- Stop buffer: 1 × TBE buffer, 8 *M* urea, 20 m*M* EDTA, 0.025% xylene cyanol, 0.025% bromophenol blue
- 150 pmol/μL stocks of assembly RNA and DNA oligos (IDT)
- [γ-^{32}P] ATP (~6000 Ci/mmol, e.g., 0135020, MP Biomedicals)
- Saran Wrap film (FisherBrand)
- Used X-ray films

2.2.3 Procedure

1. Add 5 μL (75 pmol, 15 pmol/μL) of the lesion-containing template DNA oligo and 5 μL (75 pmol, 15 pmol/μL) of scaffolding RNA to 40 μL of annealing buffer and anneal the RNA to the DNA in a PCR cycler by cycling from 98°C to 25°C with an initial incubation at 98°C for 30 s followed by a 1-s incubation and subsequently increasing the annealing time by 1 s for each 1°C drop. Store the resulting scaffold (15 pmol/μL) at −20°C, defrosting it as needed.
2. Mix 1 μL (15 pmol) of the scaffold with an equal amount of biotinylated RNA polymerase and add 20 μL of the annealing buffer. Incubate the mixture for 10 min at room temperature, add 1 μL of nontemplate DNA oligo (150 pmol/μL, 10 × excess), and continue the incubation for another 10 min at the same temperature.
3. Transfer the reaction mixture to a fresh Eppendorf tube with ~10 μL of NeutrAvidin beads (Pierce) suspended in TB50 buffer and shake it at room temperature for 5 min.
4. Wash the beads two times with 1 mL of TB1000, two times with 1 mL of TB0, and once with 1 mL of DNase/RNase-free water using a tabletop mini centrifuge. Retain about 20 μL of the beads in the buffer.

5. Add 2.5 μL of 10× PNK buffer (NEB), 2.5 μL of [γ-^{32}P] ATP (~6000 Ci/mmol, e.g., 0135020, MP Biomedicals), and 1 μL of T4 Polynucleotide kinase (20 U, NEB); incubate for 30 min at 37°C.
6. Wash the beads two times with 1 mL of TB1000 and two times with 1 mL of TB100 using a table-top mini centrifuge; retain about 20 μL of the reaction mixture.
7. Add 2 μL of 1 m*M* NTPs to a final concentration of 100 μ*M* and incubate the reaction mixture for 10 min at 37°C to chase the complex to the lesion site.
8. Add UvrD protein to a final concentration of 125 n*M* to 1.25 μ*M* plus 2 μL of 10 m*M* ATP to a final concentration of 1 m*M* and incubate the reaction mixture at room temperature for 5 min.
9. Wash the beads two times with 1 mL of TB100 using a table-top mini centrifuge; retain about 20 μL of the bead slurry.
10. To initiate the cleavage reaction, add 2 μL of premixed 25 n*M* UvrA, 1 μ*M* UvrB, and 250 n*M* UvrC proteins plus 2 μL of 10 m*M* ATP for the desired time (10–30 min) at 37°C. The final concentration should be 2.5 n*M* UvrA, 100 n*M* UvrB, 25 n*M* UvrC, and 1 m*M* ATP.
11. Stop the reaction by adding an equal volume (20 μL) of stop buffer.
12. Heat reaction mixture at 98°C for 5 min.
13. Load samples of the stopped reaction mixture on a 10% 20 cm × 40 cm 1 × TBE–8 *M* urea gel and apply voltage for 60 min at a constant output of 50 W.
14. Transfer the gel to used X-ray film and wrap it with a Saran Wrap film.
15. Expose the gel to a phosphor-imaging screen 10–16 h.
16. Scan the screen using a Typhoon scanner and quantify signal with ImageQuant TL software (see example in Fig. 2C).

2.2.4 Notes

1. Proposed method does not involve labeling of RNA to avoid obscuring DNA cleavage products with RNA bands. However, for control purposes, RNA can be labeled with [α-^{32}P] NTP (0.02 μCu, 3000 Ci/mmol) for 5 min after step 6 followed by washing four times this 1 mL of TB100.
2. We use biotin-tagged RNA polymerase because in our case UvrABC proteins were purified using a hexahistidine tag and would bind to the resin instead of their DNA target. On the other hand, if those proteins are purified without a hexahistidine tag, then, as alternative, RNA polymerase can be used with such a tag instead.

3. It is very important to wash out UvrD after treating complex with it. If not removed, UvrD will displace UvrC from the lesion site, using its translocase activity, thus completely suppressing cleavage reaction.

2.3 Mapping the Components of the TCR Complex Using Protein–DNA Photocross-Linking In Vitro

2-Thiouridine is the most commonly used photoactivated cross-linking reagent that is suitable for mapping contacts between nucleic acids and proteins in various multicomponent complexes (Mustaev et al., 2003; Nudler et al., 2003). It is similar to native uridine and thymidine and can be incorporated into both DNA and RNA to preserve the geometry and biochemical properties of the complexes. Photoactivation of the 2-thiouridine requires irradiation with long wavelength UV (308 nm), which does not damage nucleic acids and minimizes off-target nonspecific cross-linking. The efficiency of the cross-linking between 2-thiouridine and a protein depends on the local amino acid composition of the target and requires a close (so-called zero-length) contact between the cross-linker and the protein.

2.3.1 Equipment

- Thermal (PCR) cycler
- Table-top mini centrifuge
- Hand-held 308 nm UV lamp (Cole-Palmer)
- Standard SDS-PAGE equipment (Novex); 4%–12% gradient acrylamide Bis-Tris SDS gels (Novex)
- Gel dryer (Model 583 Bio-Rad)
- Typhoon scanner (GE) with a phosphor-imaging screen
- ImageQuant TL software (GE) for data visualization and analysis

2.3.2 Buffers and Reagents:

- Annealing buffer: 20 m*M* Tris–HCl, pH 8.0; 12% glycerol; 40 m*M* KCl; 5 m*M* $MgCl_2$
- TB50: 40 m*M* Tris–HCl, pH 8.0; 50 m*M* NaCl; 10 m*M* $MgCl_2$; 0.003% Igepal-60
- TB0: 20 m*M* Tris–HCl, pH 8.0; 10 m*M* $MgCl_2$
- TB1000: 40 m*M* Tris–HCl, pH 8.0; 1 *M* NaCl; 10 m*M* $MgCl_2$; 0.003% Igepal-60
- TB100: 40 m*M* Tris–HCl, pH 8.0; 100 m*M* NaCl; 10 m*M* $MgCl_2$; 0.003% Igepal-60

- RNA polymerase holoenzyme (purified according to Svetlov & Artsimovitch, 2015)
- NeutrAvidin beads (Pierce)
- 10 × PNK buffer, T4 Polynucleotide kinase (both from New England Biolabs)
- UvrD protein (purified according to Epshtein et al., 2014)
- NuPAGE LDS sample buffer (Novex)
- MOPS running buffer (Novex)
- 150 pmol/μL stocks of assembly RNA and DNA oligos (IDT)
- [γ-^{32}P] ATP (~6000 Ci/mmol, e.g., 0135020, MP Biomedicals)
- Whatman blotting paper (Thermo Scientific)

2.3.3 Procedure

1. Add 5 μL (75 pmol, 15 μ*M*) of the template DNA oligo and 5 μL (75 pmol, 15 μ*M*) of scaffolding RNA to 40 μL of annealing buffer and anneal the RNA to the DNA in a PCR cycler by cycling from 98°C to 25°C with an initial incubation at 98°C is 30 s followed by a 1-s incubation and subsequently increasing the annealing time by 1 s for each 1°C degree drop. Store the resulting scaffold (15 pmol/μ L) at −20°C, defrosting it as needed.
2. Mix 1 μL (15 pmol) of the scaffold with an equal amount of RNA polymerase and add 20 μL of the annealing buffer. Incubate the mixture for 10 min at room temperature, add 1 μL of nontemplate DNA oligo (150 pmol/μL, 10 × excess), and continue the incubation for another 10 min at the same temperature.
3. Transfer the reaction mixture into a fresh 1.7-mL Eppendorf tube with 10 μL of NeutrAvidin beads (Pierce) suspended in TB50 buffer and shake it at room temperature for 5 min.
4. Wash the beads two times with 1 mL of TB1000, two times with 1 mL of TB0, and once with 1 mL of water using a table-top mini centrifuge. Retain about 20 μL of the beads in the buffer.
5. Add 2.5 μL of 10 × PNK buffer (NEB), 2.5 μL of [γ-^{32}P] ATP (~6000 Ci/mmol, e.g., 0135020, MP Biomedicals), and 1 μL of T4 Polynucleotide kinase (20 U, NEB), incubate for 30 min at 37°C.
6. Wash beads two times with 1 mL of TB1000 and two times with 1 mL of TB100 using a table-top mini centrifuge; retain about 20–200 μL of the reaction mixture as needed.
7. Distribute the resulting complex to 10 μL aliquots in new 1.7-mL Eppendorf tubes and add an appropriate amount of UvrD protein (125 n*M* to 1.25 μ*M*), incubate for 5 min at room temperature.

8. Place the sample tubes in ice and position a UV lamp directly above the uncapped tubes. To fix the tubes in ice use standard floaters. Irradiate the samples 10 min at a wavelength of 308 nm.
9. Stop the reaction by adding 5 μL of NuPAGE LDS sample buffer.
10. Heat the reaction mixture at 37°C for 10 min.
11. Load the samples onto a 4%–12% gradient acrylamide Bis-Tris SDS gel in MOPS running buffer (Novex) and apply a constant output of 200 V for 45 min.
12. Transfer the gel to Whatman paper and cover it on one side with a Saran Wrap film.
13. Dry the gel in a gel dryer for 30 min at +80°C.
14. Expose the gel to a phosphor-imaging screen.
15. Scan the screen using a Typhoon scanner and quantify the signal with ImageQuant TL software.

2.3.4 Notes

1. For mapping various additional components in the UvrD–TCR complex, such as NusA, we add them together with UvrD in equimolar amounts at step 7.
2. This method allows probing of the contacts between different parts of the nucleic acid scaffold and TCR proteins, but it can be coupled with more precise mapping such as partial protein cleavage, which is described elsewhere (Grachev et al., 1989; Mustaev et al., 2003), or XLMS (cross-linking coupled to mass spectrometry analysis) allowing the more precise mapping of areas of contact to smaller regions on the surface of interacting proteins.

3. ASSAY OF TCR IN VIVO

The first assay of in vivo TCR was developed in the Hanawalt lab (Bohr et al., 1985; Mellon et al., 1986) utilizing the Southern hybridization method. This method remains very useful for monitoring gene-specific repair. However, it relies on a time-consuming and complex procedure and is limited by the specific restriction sites flanking the region of interest. To overcome these limitations, other assays were developed, including ligation-mediated PCR (Spivak, Pfeifer, & Hanawalt, 2006), quantitative PCR (qPCR) (Chandrasekhar & Van Houten, 1994), and primer extension (Chandrasekhar & Van Houten, 1994; Epshtein et al., 2014; Wellinger &

Thoma, 1996). qPCR and primer extension depend on the blockage of Taq DNA polymerase at DNA lesions sites, allowing the detection of damage and/or repair at single-nucleotide resolution. In this chapter, we describe the plasmid-based primer extension assay used in our laboratory.

3.1 Plasmid-Based Primer Extension Method

Escherichia coli cells harboring reporter plasmid are grown until log phase. The resulting culture is exposed to UV light (254 nm) to generate cyclobutane pyrimidine dimers (CPDs) in DNA. Plasmid DNA is purified from the treated cells and used as a template for primer extension. CPDs block primer extension and are identified as aberrant-sized fragments on sequencing gels. Depending on the repair efficiency of the *E. coli* strain under study, CPD's either disappear or persist, causing a change in the yield of the primer extension product. This method can be applied to any bacterial strain that supports plasmids with an inducible gene system such as *LacZ*. It can also be adapted to any plasmid containing an inducible or constitutively expressed gene.

3.1.1 Equipment

- Shaker
- Incubator
- Standard Urea-PAGE equipment (acrylamide:bis-acrylamide 19:1).
- Thermal (PCR) cycler
- UV lamp primarily emitting at 254 nm (Fisher Scientific)
- Centrifuge (Thermo Scientific)
- Gel dryer (Model 583 Bio-Rad)
- Typhoon scanner (GE) with a phosphor-imaging screen
- ImageQuant TL software (GE) for visualization and analysis

3.1.2 Buffers and Reagents

- LB medium
- M9 minimal medium
- Minimal medium supplements: 10% casamino acids and 20% glucose
- 1 *M* IPTG
- Hydroxyurea
- NET buffer: 20 m*M* NaCl, 25 m*M* EDTA, and 20 m*M* Tris–Cl, pH 8.0
- 10 × PNK buffer, T4 Polynucleotide kinase (New England Biolabs)
- Formamide stop buffer: 80% formamide, 1 × TBE, 0.05% xylene cyanol, 0.05% bromophenol blue

- [γ-^{32}P] ATP (~6000 Ci/mmol, e.g., 0135020, MP Biomedicals)
- Taq buffer and Taq DNA polymerase enzyme from New England Biolabs (M0267L)

3.1.3 Procedure

1. Transform an *E. coli* strain with a desired plasmid containing either an inducible or a constitutively expressed gene to measure the progression of DNA repair. Inoculate a single colony into 5 mL of LB medium with the appropriate antibiotic and grow it overnight at 37°C with shaking.
2. The next day, dilute the overnight culture 1:100 with LB medium supplemented with the appropriate antibiotic and grow the cells in a 37°C shaker with shaking at 250 rpm, monitoring cell growth using absorbance at 600 nm. At an OD_{600} of 0.3, collect the cells by centrifugation at 5000 × *g* for 10 min.
3. Discard the supernatant and wash the pellet with M9 minimal medium. Resuspend the pellet in prewarmed M9 minimal medium to an OD_{600} of 0.4. Induce the *LacZ* gene by adding 1 m*M* IPTG.
4. Aliquot 15 mL of NET buffer (20 m*M* NaCl, 25 m*M* EDTA, and 20 m*M* Tris–Cl, pH 8.0) into a 50-mL tube on ice for each timed cell sample to be assayed. For the untreated control sample, collect 15 mL of the culture into one of the prechilled ice cold NET buffer tubes.
5. Place the remaining cell suspension into a Pyrex glass drying dish (Corning, Inc). The depth of the solution should be between 0.1 and 0.2 cm. Irradiate the cells with 40–60 J/m^2 of UV light (254 nm) generated by a germicidal lamp. Work under yellow light to avoid photoreactivation of CPDs.
6. Collect 15 mL of culture immediately after UV treatment into one of the prechilled NET buffer-containing tubes. Supplement the remaining culture with 0.25% casamino acids, 0.4% glucose, and 1 m*M* IPTG and incubate at 37°C in the dark to avoid photoreactivation of CPDs. If required, add 50 m*M* hydroxyurea to prevent replicative DNA synthesis.
7. Place the remaining culture in a 250 mL culture flask and incubate it in a 37°C shaker in the dark. At each sample time (at 10, 20, 30, 40, 50, and 60 min) collect 15 mL of culture and transfer it into a prechilled NET buffer-containing tube.
8. After the last sample is taken, pellet the cells by centrifuging at 5000 × *g* for 10 min.

9. Plasmid isolation: Isolate plasmid DNA from the cell pellet using the Qiagen plasmid mini prep kit following the manufacturer's instructions.
10. Primer design: The primer should be designed to anneal to the template strand of the gene of interest. Primers are designed to be 20–30 nucleotides long and contain a G + C content of 40%–60%. For example, a suitable primer for the *LacZ* gene template strand is 5′-CCATGATTACGGATTCACTGGCCGTCG-3′. Calculate the T_m of the oligonucleotide primer from the IDT website http://www.idtdna.com/calc/analyzer.
11. Primer labeling: Combine 10–20 pmol of primer with 5 μL of [γ-^{32}P] ATP (~6000 Ci/mmol, e.g., 0135020, MP Biomedicals), 5 μL of 10× buffer, and 20 U of polynucleotide kinase enzyme in a total reaction volume of 50 μL. Incubate the reaction at 37°C for 60 min and then inactivate the enzyme by incubating at 65°C for 20 min. After labeling, remove unincorporated [γ-^{32}P] ATP by passing the sample through a gel filtration column (e.g., Micro Bio-Spin from Bio-Rad).
12. Primer extension: In a 0.2-mL PCR tube (Axygen), mix 10 μL of plasmid DNA (250 ng), 1 pmol of [γ-^{32}P]-labeled primer, 2.5 μL of 10× Taq buffer, 400 μ*M* of dNTPs, 1 U of Taq polymerase, and DNase/RNase-free water in a total reaction volume of 25 μL. In a thermal cycler, the sample is denatured at 98°C for 5 min, annealed at 55°C for 10 min, and primer extension performed at 68°C for 5 min. Add formamide stop buffer to the reaction mixture and load aliquots onto a 6% sequencing gel.

3.1.4 Note

1. If the sequencing gel bands appear weak, concentrate the primer extension products or perform multiple cycles of primer extension before loading the products on the gel.

4. SUMMARY AND TECHNICAL NOTES

The study of bacteria TCR requires complex approaches to verify and validate the data produced by different methods. Despite a decades-long search, only recently have TCR factors other than Mfd been identified. Due to the complex nature of the DNA repair process, the redundancy of different pathways, and the pleiotropic functions of the participating proteins, it is essential to use highly defined in vitro and in vivo systems to avoid ambiguous interpretation of the results and/or omission of the crucial factors. High purity of components and precise stoichiometry also play an important role in studies of TCR. For example, UvrD functions as a monomer when for its translocase activity, but requires dimerization for its helicase activity (Lee, Balci, Jia, Lohman, & Ha, 2013; Maluf, Fischer, & Lohman, 2003). On the other hand, the dimerization of UvrD is a crucial requirement in the control of this repair pathway, making it possible to rapidly recover from extensive backtracking, and allowing additional factors such as NusA and ppGpp to function in fine-tuning the process (Kamarthapu et al., 2016).

We also emphasize the importance of selection the target lesion for study because it can dramatically affect the experimental outcome. We chose cyclobutane adducts not only because it is the most common UV-induced DNA lesion but also because it completely blocks transcription (Donahue et al., 1994; Selby, Drapkin, Reinberg, & Sancar, 1997) and is large enough to require an extensive RNA polymerase backtracking for efficient repair. Smaller DNA adducts could be either inefficient in stalling transcription or could be quickly repaired independent of coupling if RNA polymerase moves backward far enough on its own, without additional factors, to expose the lesion. We tested both the natural lesions, such as thymine dimers, and artificial fluorescein adducts and found that both performed well in the experimental system described here.

The local DNA sequence is another important consideration, as it can affect the results of the experiment. UvrD is a powerful helicase that pulls RNA polymerase backward with enough force to terminate transcription, if a stable RNA hairpin is formed immediately behind the lesion site. Those termination events could complicate data analysis because terminated complexes will be removed from the reaction mixture during sample washing steps. Alternatively, if the lesion site is placed too close to the transcription start site, the pro-backtracking activity of UvrD could move RNA polymerase backward far enough to separate it completely from the RNA, again resulting in premature termination. We recommend a transcription unit design that places the lesion site at least 50 nucleotides from the transcription start site and omits palindromic sequences liable to produce stable RNA hairpins reminiscent of intrinsic terminators.

ACKNOWLEDGMENTS

The authors would like to thank Dr. Svetlov for critical reading of the manuscript. This work was supported by the NIH Grant R01 GM107329 and by the Howard Hughes Medical Institute.

REFERENCES

Backendorf, C., Brandsma, J. A., Kartasova, T., & van de Putte, P. (1983). In vivo regulation of the uvrA gene: Role of the "-10" and "-35" promoter regions. *Nucleic Acids Research*, *11*, 5795–5810.

Bohr, V. A., Smith, C. A., Okumoto, D. S., & Hanawalt, P. C. (1985). DNA repair in an active gene: Removal of pyrimidine dimers from the DHFR gene of CHO cells is much more efficient than in the genome overall. *Cell*, *40*, 359–369.

Chandrasekhar, D., & Van Houten, B. (1994). High resolution mapping of UV-induced photoproducts in the Escherichia coli lacI gene: Inefficient repair of the non-transcribed strand correlates with high mutation frequency. *Journal of Molecular Biology*, *238*, 319–332.

Cohen, S. E., & Walker, G. C. (2010). The transcription elongation factor NusA is required for stress-induced mutagenesis in Escherichia coli. *Current Biology*, *20*, 80–85.

Daube, S. S., & von Hippel, P. H. (1992). Functional transcription elongation complexes from synthetic RNA-DNA bubble duplexes. *Science*, *258*, 1320–1324.

Donahue, B. A., Yin, S., Taylor, J. S., Reines, D., & Hanawalt, P. C. (1994). Transcript cleavage by RNA polymerase II arrested by a cyclobutane pyrimidine dimer in the DNA template. *Proceedings of the National Academy of Sciences of the United States of America*, *91*, 8502–8506.

Epshtein, V. (2015). UvrD helicase: An old dog with a new trick: How one step backward leads to many steps forward. *BioEssays*, *37*, 12–19.

Epshtein, V., Kamarthapu, V., McGary, K., Svetlov, V., Ueberheide, B., Proshkin, S., et al. (2014). UvrD facilitates DNA repair by pulling RNA polymerase backwards. *Nature*, *505*, 372.

Ganesan, A., Spivak, G., & Hanawalt, P. C. (2012). Transcription-coupled DNA repair in prokaryotes. *Progress in Molecular Biology and Translational Science*, *110*, 25–40.

Grachev, M. A., Lukhtanov, E. A., Mustaev, A. A., Zaychikov, E. F., Abdukayumov, M. N., Rabinov, I. V., et al. (1989). Studies of the functional topography of Escherichia coli RNA polymerase. A method for localization of the sites of affinity labelling. *European Journal of Biochemistry*, *180*, 577–585.

Hickson, I. D., Arthur, H. M., Bramhill, D., & Emmerson, P. T. (1983). The E. coli uvrD gene product is DNA helicase II. *Molecular & General Genetics*, *190*, 265–270.

Kamarthapu, V., Epshtein, V., Benjamin, B., Proshkin, S., Mironov, A., Cashel, M., et al. (2016). ppGpp couples transcription to DNA repair in E. coli. *Science*, *352*, 993–996.

Kamarthapu, V., & Nudler, E. (2015). Rethinking transcription coupled DNA repair. *Current Opinion in Microbiology*, *24*, 15–20.

Kisker, C., Kuper, J., & Van Houten, B. (2013). Prokaryotic nucleotide excision repair. *Cold Spring Harbor Perspectives in Biology*, *5*(3), 1–18.

Korzheva, N., Mustaev, A., Nudler, E., Nikiforov, V., & Goldfarb, A. (1998). Mechanistic model of the elongation complex of Escherichia coli RNA polymerase. *Cold Spring Harbor Symposia on Quantitative Biology*, *63*, 337–345.

Lee, K. S., Balci, H., Jia, H., Lohman, T. M., & Ha, T. (2013). Direct imaging of single UvrD helicase dynamics on long single-stranded DNA. *Nature Communications*, *4*, 1878.

Maluf, N. K., Fischer, C. J., & Lohman, T. M. (2003). A dimer of Escherichia coli UvrD is the active form of the helicase in vitro. *Journal of Molecular Biology*, *325*, 913–935.

Manelyte, L., Guy, C. P., Smith, R. M., Dillingham, M. S., McGlynn, P., & Savery, N. J. (2009). The unstructured C-terminal extension of UvrD interacts with UvrB, but is dispensable for nucleotide excision repair. *DNA Repair*, *8*, 1300–1310.

Matson, S. W. (1986). Escherichia coli helicase II (urvD gene product) translocates unidirectionally in a 3′ to 5′ direction. *The Journal of Biological Chemistry*, *261*, 10169–10175.

Matson, S. W., & George, J. W. (1987). DNA helicase II of Escherichia coli. Characterization of the single-stranded DNA-dependent NTPase and helicase activities. *The Journal of Biological Chemistry*, *262*, 2066–2076.

Mellon, I., Bohr, V. A., Smith, C. A., & Hanawalt, P. C. (1986). Preferential DNA repair of an active gene in human cells. *Proceedings of the National Academy of Sciences of the United States of America*, *83*, 8878–8882.

Mellon, I., & Hanawalt, P. C. (1989). Induction of the *Escherichia coli* lactose operon selectively increases repair of its transcribed DNA strand. *Nature*, *342*, 95–98.

Mustaev, A., Zaychikov, E., Grachev, M., Kozlov, M., Severinov, K., Epshtein, V., et al. (2003). Strategies and methods of cross-linking of RNA polymerase active center. *Methods in Enzymology*, *371*, 191–206.

Nudler, E., Avetissova, E., Korzheva, N., & Mustaev, A. (2003a). Characterization of protein-nucleic acid interactions that are required for transcription processivity. *Methods in Enzymology*, *371*, 179–190.

Nudler, E., Avetissova, E., Markovtsov, V., & Goldfarb, A. (1996). Transcription processivity: Protein-DNA interactions holding together the elongation complex. *Science*, *273*, 211–217.

Nudler, E., Gusarov, I., & Bar-Nahum, G. (2003b). Methods of walking with the RNA polymerase. *Methods in Enzymology*, *371*, 160–169.

Park, J. S., Marr, M. T., & Roberts, J. W. (2002). E. coli transcription repair coupling factor (Mfd protein) rescues arrested complexes by promoting forward translocation. *Cell*, *109*, 757–767.

Rasouly, A., Pani, B., & Nudler, E. (2017). A magic spot in genome maintenance. *Trends in Genetics*, *33*, 58–67.

Reardon, J. T., & Sancar, A. (2005). Nucleotide excision repair. *Progress in Nucleic Acid Research and Molecular Biology*, *79*, 183–235.

Selby, C. P., Drapkin, R., Reinberg, D., & Sancar, A. (1997). RNA polymerase II stalled at a thymine dimer: Footprint and effect on excision repair. *Nucleic Acids Research*, *25*, 787–793.

Selby, C. P., & Sancar, A. (1993). Molecular mechanism of transcription-repair coupling. *Science*, *260*, 53–58.

Selby, C. P., Witkin, E. M., & Sancar, A. (1991). Escherichia coli mfd mutant deficient in "mutation frequency decline" lacks strand-specific repair: In vitro complementation with purified coupling factor. *Proceedings of the National Academy of Sciences of the United States of America*, *88*, 11574–11578.

Shi, Y. B., Gamper, H., & Hearst, J. E. (1987). The effects of covalent additions of a psoralen on transcription by E. coli RNA polymerase. *Nucleic Acids Research*, *15*, 6843–6854.

Shi, Y. B., Gamper, H., Van Houten, B., & Hearst, J. E. (1988). Interaction of Escherichia coli RNA polymerase with DNA in an elongation complex arrested at a specific psoralen crosslink site. *Journal of Molecular Biology*, *199*, 277–293.

Siegel, E. C. (1983). The Escherichia coli uvrD gene is inducible by DNA damage. *Molecular & General Genetics*, *191*, 397–400.

Spivak, G., Pfeifer, G. P., & Hanawalt, P. (2006). In vivo assays for transcription-coupled repair. *Methods in Enzymology*, *408*, 223–246.

Svetlov, V., & Artsimovitch, I. (2015). Purification of bacterial RNA polymerase: Tools and protocols. *Methods in Molecular Biology*, *1276*, 13–29.

Truglio, J. J., Croteau, D. L., Van Houten, B., & Kisker, C. (2006). Prokaryotic nucleotide excision repair: The UvrABC system. *Chemical Reviews*, *106*, 233–252.

Wellinger, R.-E., & Thoma, F. (1996). Taq DNA polymerase blockage at pyrimidine dimers. *Nucleic Acid Research*, *24*, 1578–1579.

CHAPTER THIRTEEN

Reconstituted System for the Examination of Repair DNA Synthesis in Homologous Recombination

Youngho Kwon, James M. Daley, Patrick Sung[1]
Yale University School of Medicine, New Haven, CT, United States
[1]Corresponding author: e-mail address: patrick.sung@yale.edu

Contents

Methods in Enzymology, Volume 591
ISSN 0076-6879
http://dx.doi.org/10.1016/bs.mie.2017.03.021

Abstract

In homologous recombination (HR), DNA polymerase δ-mediated DNA synthesis occurs within the displacement loop (D-loop) that is made by the recombinase Rad51 in conjunction with accessory factors. We describe in this chapter the reconstitution of the D-loop and repair DNA synthesis reactions using purified *Saccharomyces cerevisiae* HR (Rad51, RPA, and Rad54) and DNA replication (PCNA, RFC, and DNA polymerase δ) proteins and document the role of the Pif1 helicase in DNA synthesis via a migrating DNA bubble intermediate. These reconstituted systems are particularly valuable for understanding the conserved mechanism of repair DNA synthesis dependent on DNA polymerase δ and its cognate helicase in eukaryotic organisms.

1. INTRODUCTION

HR is a conserved pathway for the removal of DNA double strand breaks (DSBs) and the repair of injured DNA replication forks. Herein, a homologous DNA sequence is engaged by the processed lesion and serves as the template for DNA synthesis to initiate a usually error-free repair process (Jasin & Rothstein, 2013; Symington, Rothstein, & Lisby, 2014). Lesion processing entails nucleolytic resection of the 5′ DNA strand associated with a break end to generate a 3′-ended DNA tail of a considerable length (Daley, Niu, Miller, & Sung, 2015). The ssDNA tail is bound by the conserved recombinase enzyme Rad51 and its accessory factors, and the resulting nucleoprotein ensemble then searches for and engages the homologous locale either in the sister chromatid or in the homologous chromosome, followed by invasion of the latter to form a hybrid DNA joint called the displacement loop, or D-loop (Kowalczykowski, 2015; San Filippo, Sung, & Klein, 2008). It should be noted that in the vast majority of eukaryotes, an additional, meiosis-specific recombinase called Dmc1 is needed for optimal interhomolog recombination (Brown & Bishop, 2014; Hunter, 2015), but here we focus on Rad51 only.

Rad51 catalyzes D-loop formation within the context of a right-handed nucleoprotein filament known as the presynaptic filament, whose assembly requires ATP binding by Rad51, although ATP hydrolysis prompts the dissociation of Rad51 protomers from the DNA ligand (Sung, 1994). D-loop formation by the presynaptic filament is enhanced by associated factors including the Swi2/Snf2 family proteins Rad54 and Rdh54 and the ssDNA-binding protein RPA (Daley, Gaines, Kwon, & Sung, 2014;

Heyer, 2007). Following strand invasion, de novo DNA synthesis occurs within the D-loop. Several DNA polymerases (Pols), namely, Pol δ and Pol ε in yeast (Deem et al., 2011; Hicks, Kim, & Haber, 2010; Maloisel, Fabre, & Gangloff, 2008) and Pol η in vertebrates (Buisson et al., 2014; Kawamoto et al., 2005; McIlwraith et al., 2005), have been proposed to contribute to DNA synthesis, and among them, Pol δ has emerged as the major player (Lydeard, Jain, Yamaguchi, & Haber, 2007; Maloisel et al., 2008; Wilson et al., 2013). Following repair DNA synthesis, the extended D-loop is resolved via one of the several mechanistically distinct pathways to yield mature recombinants of different classes (Sung & Klein, 2006; Symington et al., 2014). The involvement of Pol δ in repair DNA synthesis has been studied most extensively within the context of a HR pathway called break-induced DNA replication (BIR), in which an invading DNA strand primes DNA synthesis capable of copying an entire arm of the donor chromatid (Fig. 1) (Costantino et al., 2014; Dilley et al., 2016; Donnianni & Symington, 2013; Lydeard et al., 2007; Saini et al., 2013; Wilson et al., 2013).

Importantly, in genetic studies, *Saccharomyces cerevisiae* Pif1, a member of the SF1 family of helicases, has been implicated in BIR in conjunction with Pol δ (Chung, Zhu, Papusha, Malkova, & Ira, 2010; Saini et al., 2013; Wilson et al., 2013). Using highly purified *S. cerevisiae* proteins, a system that permits dissection of the mechanistic underpinnings of D-loop-primed DNA synthesis reaction has been developed (Li, Stith, Burgers, & Heyer, 2009; Sebesta, Burkovics, Haracska, & Krejci, 2011; Wilson et al., 2013). Using this system, we have shown that Pif1 greatly stimulates Pol δ-mediated DNA extension within the context of the Rad51-made D-loop. Importantly, we have furnished evidence that Pif1 fulfills two distinct functions in the DNA synthesis reaction; namely, (i) it enhances the ability of the polymerase ensemble to catalyze DNA strand displacement synthesis via an interaction with the proliferating cell nuclear antigen (PCNA), the polymerase processivity clamp, and (ii) concomitant with DNA synthesis, Pif1 dissociates the invading strand to establish a migrating DNA bubble structure (Fig. 1) (Wilson et al., 2013). This chapter describes the materials and experimental procedures for reconstituting the repair DNA synthesis reaction using a ssDNA oligonucleotide as the invading strand and supercoiled dsDNA as the information donor. Our method utilizes Rad51, RPA, and Rad54 to generate D-loops and PCNA, the multi-subunit PCNA loader replication factor C (RFC), the trimeric Pol δ and

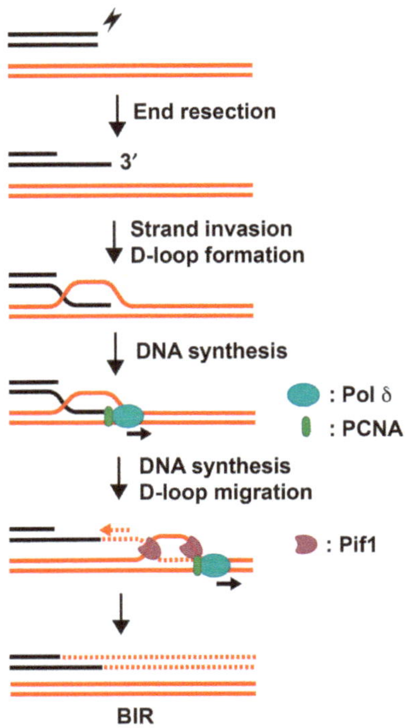

Fig. 1 DNA intermediates and the role of Pif1 in repair DNA synthesis during HR. The 3′ ssDNA tail resulting from DNA end resection is engaged by the Rad51 recombinase, which then invades a homologous sequence to form a D-loop. DNA synthesis within the D-loop is carried out by DNA Pol δ in conjunction with PCNA. In *S. cerevisiae*, the Pif1 helicase greatly stimulates the extent of DNA synthesis via a migrating D-loop. Note that PCNA is loaded onto the 3′ terminus of the invading DNA strand by RFC, which is not depicted in the figure.

Pif1 helicase in the DNA synthesis phase of the reaction. The methods for product analyses are also described.

2. ASSEMBLING AND ANALYSIS OF THE D-LOOP REACTION

This protocol describes the procedures for forming the Rad51 presynaptic filament on ^{32}P-labeled 90-mer ssDNA (Fig. 2A) and the generation of D-loops using supercoiled pBluescript plasmid DNA as recipient and the ssDNA-binding protein RPA and the dsDNA translocase Rad54 as accessory factors (Petukhova, Stratton, & Sung, 1998; Raschle, Van Komen, Chi,

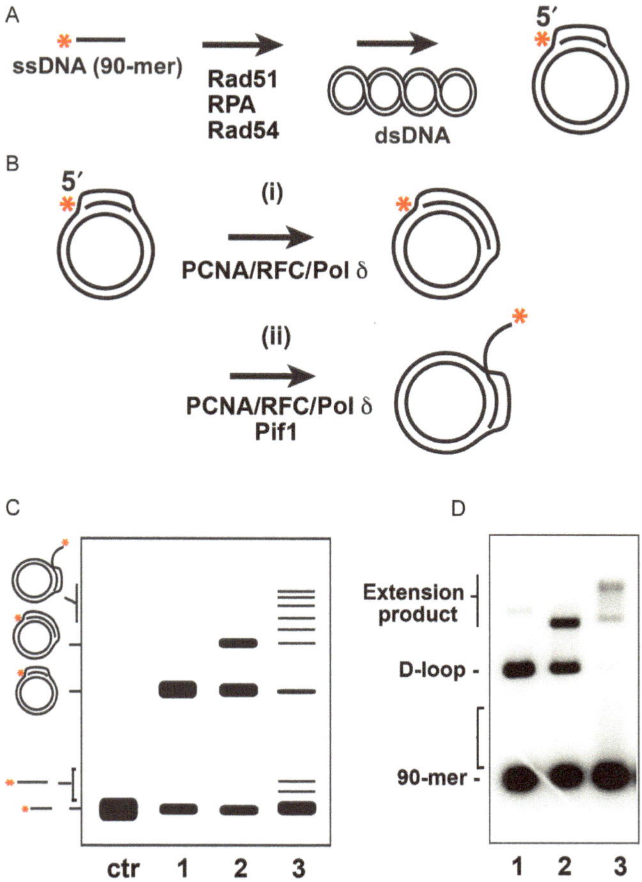

Fig. 2 Schematics for the D-loop reaction (A) and DNA synthesis reaction primed from a D-loop (B). The reaction products can be separated by native gel electrophoresis as shown in the cartoon in (C) and as the gel image of an actual experiment (D). (C) and (D) *Lane 1*, D-loop product; *lane 2*, D-loop that has been extended by Pol δ–PCNA–RFC; and *lane 3*, D-loop that has been extended by Pol δ–PCNA–RFC in conjunction with Pif1. The *bracket* identifies the extended invading ssDNA strand that has been dissociated from the D-loop by Pif1. The asterisk identifies the ^{32}P label.

Ellenberger, & Sung, 2004; Sugiyama, Zaitseva, & Kowalczykowski, 1997). Since Rad54 hydrolyzes a large quantity of ATP, an ATP-regenerating system consisting of creatine phosphate (CP) and creatine kinase (CK) should be included to avoid ATP depletion (Petukhova et al., 1998). Note that RPA, but not Rad54, is also needed for the efficiency of the subsequent DNA synthesis reaction (Yuzhakov, Kelman, Hurwitz, & O'Donnell, 1999).

2.1 Purification of ssDNA by Denaturing Polyacrylamide Gel Electrophoresis

2.1.1 Equipment

- Standard polyacrylamide gel electrophoresis (PAGE) equipment (e.g., PROTEAN II xi cell (Bio-Rad))
- Standard agarose gel electrophoresis equipment
- Circulating water bath with cooling and heating functions
- Dialysis membrane (e.g., FlexTube (IBI Scientific))
- Handheld UV lamp
- Amicon Ultra-4 microconcentrators (Millipore)
- Standard centrifuge (e.g., Sorvall Evolution RC (Thermo Scientific)), rotor (e.g., SLA 600TC (Thermo Scientific)), and conical tube adaptor

2.1.2 Reagents and Buffers

- Oligo-D (90-mer): synthetic DNA oligonucleotide that is homologous to residues 1932–2022 of pBluescript SK plasmid DNA (5′-AAATCAATCTAAAGTATATATGAGTAAACT TGGTCTGACAGTTACCAATGCT TAATCAGTGAGGCACCTATCTCAGCGATCT GTCTATTT-3′) (Raschle et al., 2004; Raynard & Sung, 2009)
- TE buffer (10 m*M* Tris–HCl, pH 7.5, 1 m*M* EDTA)
- Gel-loading buffer (94% formamide, 10 m*M* Tris–HCl, pH 7.5, 1 m*M* EDTA, 0.05% bromophenol blue)
- TAE buffer (40 m*M* Tris–acetate, pH 8.0, 1 m*M* EDTA)
- 10% polyacrylamide gel (20 cm × 16 cm × 1 mm) made in TAE buffer containing 7 *M* urea

2.1.3 Procedure

- Dissolve Oligo-D in TE buffer to 10 μg/μL.
- Mix 500 μg Oligo-D and 100 μL gel-loading buffer and heat the sample at 55°C for 5 min.
- With the use of the circulating bath, preheat the urea-containing polyacrylamide gel to 55°C (Sambrook, Fritsch, & Maniatis, 1989).
- Load the oligonucleotide solution onto the polyacrylamide gel and carry out electrophoresis at 150 V for 2 h (or until the bromophenol blue dye migrates through ~70% of the gel).
- Remove the top glass plate from the gel and identify the major DNA band with the UV lamp.

- Excise the gel slice containing the major DNA species and recover the DNA by electroelution at 100 V for 2 h in a dialysis membrane in the gel box normally used for agarose gel electrophoresis.
- Concentrate the eluted DNA in the microconcentrator to ~0.9 μg/μL (3 m*M* nucleotides) and store it at −20°C.

2.2 5′ ^{32}P-End Labeling of ssDNA

2.2.1 Equipment

- Standard PAGE equipment (e.g., Mini-PROTEAN vertical electrophoresis cell (Bio-Rad))
- Micro Bio-Spin 6 column (Bio-Rad)
- Vacuum gel dryer
- Phosphorimager (e.g., Personal Molecular Imager System (Bio-Rad))
- Quantity One software (Bio-Rad) for phosphorimage analysis
- Heating block
- Microcentrifuge
- Amersham Hybond-N+ membrane (GE Healthcare)
- Whatman 3MM chromatography paper (GE Healthcare)

2.2.2 Reagents and Buffers

- Gel-purified Oligo-D (3 m*M* nucleotides)
- T4 Polynucleotide kinase (PNK, 10 U/μL) and 10 × buffer (NEB)
- [γ-^{32}P]ATP (6000 Ci/mmol, 10 μCi/μL) (Perkin Elmer Life Science)
- TAE buffer (40 m*M* Tris–acetate, pH 8.0, 1 m*M* EDTA)
- TE buffer (10 m*M* Tris–HCl, pH 7.5, 1 m*M* EDTA)
- Gel-loading buffer (15 m*M* Tris–HCl, pH 7.5, 25% glycerol, 0.05% Orange G)
- 8% polyacrylamide gel made in TAE buffer

2.2.3 Procedure

- Mix the following and incubate at 37°C for 60 min.
  - Oligo-D, 3 μL
  - PNK buffer (10 ×), 5 μL
  - PNK, 2 μL
  - [γ-^{32}P]ATP (10 μCi/μL), 5 μL
  - H_2O, to 50 μL
- Heat the reaction mixture at 65°C for 20 min to inactivate PNK.
- Set aside 0.5 μL of the reaction mixture (Sample A) for PAGE analysis later.

- Remove the unincorporated [γ-^{32}P]ATP using the Micro Bio-Spin 6 column (Bio-Rad) at 1000 × *g* for 4 min in a microcentrifuge and then adjust the volume to 100 μL with H_2O (Sample B).
- Mix separately Sample A and 1 μL Sample B with 10 μL of gel-loading buffer and resolve them in the polyacrylamide gel at 100 V for 30 min with TAE as the running buffer (Sambrook et al., 1989).
- Dry the gel onto Amersham Hybond-N+ membrane layered on Whatman 3MM Chromatography paper in the vacuum drier at 80°C.
- Expose the dried gel to a phosphor screen and quantify the signals in the phosphorimager using the Quantity One software; this serves to calculate the yield of ^{32}P-labeled Oligo-D.
- Store the labeled DNA in small aliquots at −20°C.

2.3 D-Loop Reaction

2.3.1 Equipment

- Heating block
- Microcentrifuge

2.3.2 Reagents and Buffers

- ^{32}P-labeled Oligo-D (90 μ*M* nucleotides)
- pBluescript SK supercoiled DNA[1] (440 μ*M* base pairs, prepared as described in Raynard & Sung, 2009)
- Buffer A (5 × stock) (135 m*M* Tris–HCl, pH 7.5, 5 m*M* DTT)
- Protein storage buffer (25 m*M* Tris–HCl, pH 7.5, 150 m*M* KCl, 10% glycerol, 0.5 m*M* EDTA, 0.01% Igepal CA-630 (Sigma), 1 m*M* DTT)
- $MgCl_2$ (100 m*M*)
- BSA (10 μg/μL in TE buffer)
- ATP[2] (100 m*M*, pH 7.5)
- CP[2] (purchased from Roche Diagnostics, 1 *M* in H_2O)
- CK[2] (purchased from Roche Diagnostics, 1.5 mg/mL in H_2O)
- Solution of dATP, dCTP, dGTP, and TTP[3] (25 m*M* each in H_2O)
- Rad51 (23 μ*M*, expressed and purified as described in Van Komen, Macris, Sehorn, & Sung, 2006)
- Rad54 (2.5 μ*M*, expressed and purified as described in Raschle et al., 2004)
- RPA (5 μ*M*, expressed and purified as described in Van Komen et al., 2006)
- SDS (10%)
- Proteinase K (PK, purchased from Roche Diagnostics, 10 mg/mL in H_2O)

Note 1: D-loop formation is enhanced by supercoiling in the dsDNA (Van Komen, Petukhova, Sigurdsson, Stratton, & Sung, 2000). Analyze the pBluescript plasmid DNA by native gel electrophoresis in a 0.9% agarose gel and by ethidium bromide staining (Sambrook et al., 1989). The majority ($\geq$90%) of the DNA should be in the supercoiled form.

Note 2: Keep the stocks of ATP, CP, and CK in small aliquots at −80°C and use only freshly thawed reagents for the reaction.

Note 3: dNTPs are not needed for D-loop formation, but are essential for the DNA synthesis reaction to be described later.

2.3.3 Procedure

- For each 20 μL reaction, mix the following reagents in a 1.5-mL microcentrifuge tube:
  4 μL Buffer A (5 × stock)
  0.4 μL ATP
  0.4 μL CP
  0.4 μL CK
  1.4 μL $MgCl_2$
  0.53 μL ^{32}P-labeled Oligo-D (2 μ*M* nucleotides final concentration)
  0.08 μL dNTP mixture (100 μ*M* final concentration)
  An appropriate amount of H_2O to adjust the final reaction volume (see later) to 20 μL after protein additions.
- Add Rad51[1] (0.6 μ*M* final concentration) to the above and incubate at 37°C for 10 min.
- Add RPA[1] (0.4 μ*M* final concentration) to the above and incubate at 30°C for 5 min.
- Add Rad54[1] (0.3 μ*M* final concentration) to the above and incubate at 23°C for 2 min.
- Add pBluescript (35 μ*M* base pairs final concentration) and incubate at 30°C for 2 min.
- Leave the tube on ice and then proceed to the steps in Section 3 or terminate and deproteinize the reaction by adding 1 μL SDS and 1 μL PK followed by a 5-min incubation at 37°C.

Note 1: The optimal amount of each protein can vary depending on the specific activity of protein preparations. In our experience, when stored on ice, Rad51 and RPA remain stable for at least 1 week, while Rad54 begins to lose its enzymatic activity after 2 days. Once thawed from –80°C storage, the protein preparations should be used immediately and not be refrozen.

3. REPAIR DNA SYNTHESIS REACTION

This protocol is for assembling the DNA synthesis reaction using D-loops generated as described in Section 2.3.3. The protein species needed for this reaction are the three-subunit Pol δ, the five-subunit RFC, homotrimeric PCNA, and the Pif1 helicase. As shown in Fig. 2B, Pol δ together with RFC and PCNA can efficiently extend the ^{32}P-labeled invading strand, while the addition of Pif1 leads to a strong stimulation of the DNA synthesis track length and the formation of a migrating DNA bubble structure with a growing 5′ ssDNA tail produced as a result of Pif1-mediated dissociation of the extended invading strand. Methods for DNA synthesis product analysis are described in Sections 4 and 5.

3.1 Equipment

- Heating block
- Microcentrifuge
- Water bath

3.2 Reagents and Buffers

- Protein storage buffer (25 m*M* Tris–HCl, pH 7.5, 150 m*M* KCl, 10% glycerol, 0.5 m*M* EDTA, 0.01% Igepal CA-630 (Sigma), 1 m*M* DTT)
- Pol δ (2 μ*M*[1], purified as described in Wilson et al., 2013)
- RFC (4 μ*M*[1], expressed and purified as described in Wilson et al., 2013)
- PCNA (4 μ*M*[1], expressed and purified as described in Sebesta et al., 2011)
- Pif1 (800 n*M* supplemented with 100 ng/μL BSA, expressed and purified as described in Wilson et al., 2013)
- SDS (10%)
- PK (10 mg/mL)

Note 1: The optimal amount of each protein can vary depending on the specific activity of protein preparations. In our experience, when stored on ice, PCNA is stable for at least 1 week, while RFC, Pol δ, and Pif1 lose activity after 2 days. Once thawed from −80°C storage, the protein preparations should be used immediately and not be refrozen.

3.3 Procedure

- Mix RFC and PCNA (final concentration of 200 n*M* each) and leave on ice.

- Add 16 μL of the D-loop reaction from Section 2.3.3 to the PCNA–RFC mixture and incubate on ice for 2 min.
- Add Pol δ (100 n*M* final concentration) and Pif1 (40 n*M* final concentration) and incubate at 15°C up to 8 min.
- Terminate and deproteinize the synthesis reaction by adding 1 μL SDS and 1 μL PK and a 10-min incubation at 37°C.

4. ANALYSIS OF D-LOOPS AND EXTENDED D-LOOPS.

The products (D-loops, extended D-loops, and extended D-loops with a migrating bubble) from the D-loop and DNA synthesis reactions can be resolved by native gel electrophoresis, while the length of the DNA synthesis track can be more accurately determined by denaturing gel electrophoresis and 2D gel electrophoresis.

4.1 Equipment

- Standard PAGE equipment (e.g., PROTEAN II xi cell (Bio-Rad)).
- Standard agarose gel electrophoresis equipment and setup for phosphorimaging analysis as described earlier.

4.2 Reagents and Buffers

- Native gel-loading buffer (15 m*M* Tris–HCl, pH 7.5, 25% glycerol, 0.05% Orange G)
- Denaturing gel-loading buffer (94% formamide, 10 m*M* Tris–HCl, pH 7.5, 1 m*M* EDTA, 0.05% bromophenol blue)
- TAE buffer (40 m*M* Tris–acetate, pH 8.0, 1 m*M* EDTA)
- TBE buffer (90 m*M* Tris–HCl, pH 8.3, 90 m*M* boric acid, 2 m*M* EDTA)
- Alkaline solution 1 (0.5 *M* NaOH, 10 m*M* EDTA)
- Alkaline solution 2 (50 m*M* NaOH, 1 m*M* EDTA)
- Neutralization buffer (0.5 *M* Tris–HCl, pH 7.5)
- 0.8% agarose gel made in TAE buffer
- 4% polyacrylamide gel (20 cm × 16 cm × 1 mm) made in TBE buffer containing 7 *M* urea
- 0.9% agarose gel (11 cm × 15 cm × 6 mm) made in alkaline solution 2

4.3 Native Gel Electrophoresis

- Mix 10 μL of the reaction mixture described in Section 3.3 with 10 μL of native gel-loading buffer.

- Resolve the mixture in the 0.8% agarose gel at 90 mA for 120 min at room temperature.
- Dry the gel as described in Section 2.2.3.
- Subject the dried gel to phosphorimaging analysis as described in Section 2.2.3. A schematic of the expected results and actual results from a typical experiment are shown in Fig. 2C and D, respectively.

4.4 Denaturing Gel Electrophoresis

4.4.1. This protocol is appropriate for characterizing DNA synthesis products up to 1.5 kb in length.

- Mix 10 μL of the reaction from Section 3.3 with 10 μL of denaturing gel-loading buffer and incubate at 95°C for 5 min.
- Resolve the mixture in the denaturing polyacrylamide gel at 200 mA at 55°C for 90 min (Sambrook et al., 1989).
- Dry the gel and analyze the dried gel in the phosphorimager as described in Section 2.2.3.

4.4.2. This protocol is for characterizing DNA synthesis products up to several kilobases in length.

- Mix 10 μL of the reaction from Section 3.3 with 2 μL alkaline solution 1 and 8 μL native gel-loading buffer.
- Resolve the mixture in the agarose gel in alkaline solution 2 at 50 mA for 300 min (Sambrook et al., 1989).
- Soak the gel in neutralization buffer for 20 min.
- Dry the gel and analyze the dried gel in the phosphorimager as described in Section 2.2.3.

4.5 2D Gel Electrophoresis

The 2D gel analytical protocol provides an independent means for determining the length of DNA synthesis products and also of the population of extended invading strand that has been dissociated from the D-loop by Pif1. Herein, the reaction products from Section 3.3 are first separated by native agarose gel electrophoresis (Fig. 3A, (i)) as described in Section 4.3, followed by a second electrophoretic step in a denaturing agarose gel (Fig. 3A, (ii)).

- Carry out native gel electrophoresis of the DNA synthesis products in duplicate following the procedure described in Section 4.3.
- Dry down one of the two lanes as in Section 2.2.3.

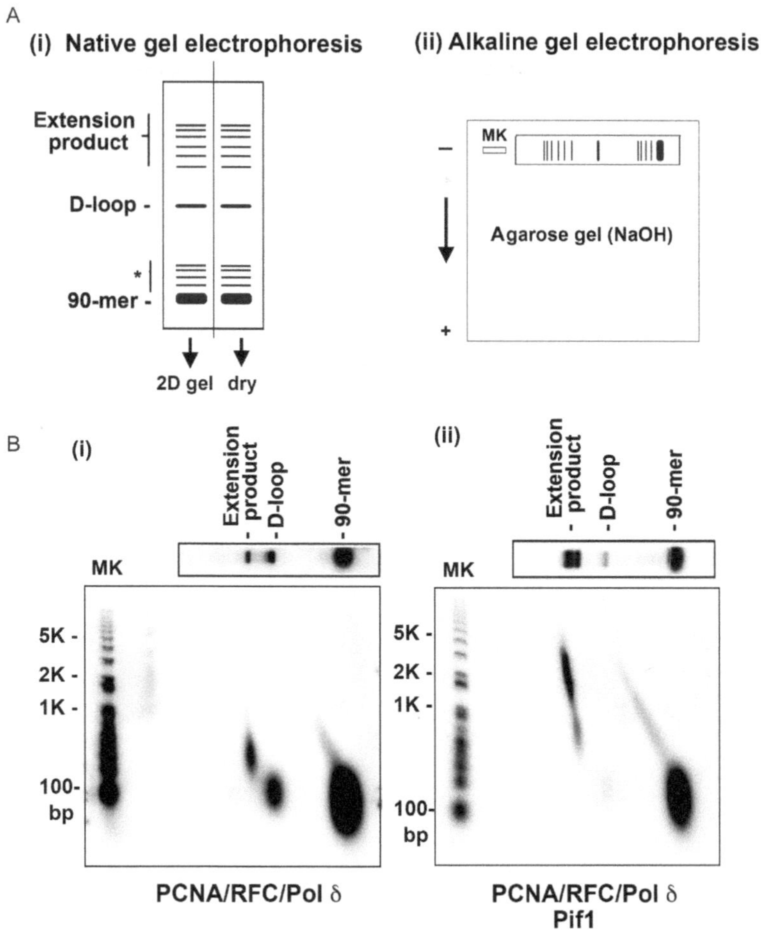

Fig. 3 2D gel electrophoresis of DNA synthesis products. (A) *Cartoon representation* of the DNA species first resolved by native gel electrophoresis (i) and then by a second electrophoretic step in an alkaline gel together with DNA size markers (MK) (ii). The asterisk identifies the extended invading ssDNA strand that has been dissociated from the D-loop. (B) 2D gel analysis of DNA synthesis reactions mediated by Pol δ–PCNA–RFC (i) and Pol δ–PCNA–RFC–Pif1 (ii). *The gel images are taken from our published study Wilson, M. A., Kwon, Y., Xu, Y., Chung, W. H., Chi, P., Niu, H., et al. (2013). Pif1 helicase and Poldelta promote recombination-coupled DNA synthesis via bubble migration.* Nature, 502*(7471), 393–396.*

- Place the other lane on the leading edge of a gel tray (11 cm × 15 cm × 6 mm) and pour 100 mL of molten 0.9% agarose in alkaline solution 2 to imbed the lane.
- Develop the gel in alkaline solution 2 at 50 mA for 8 h.
- Soak the gel in neutralization buffer for 20 min.

– Dry the gel and analyze the dried gel in the phosphorimager as described in Section 2.2.3 (Fig. 3B).

5. ANALYSIS OF DNA SYNTHESIS WITHIN A MIGRATING D-LOOP

5.1 Equipment

- Standard PAGE equipment (e.g., PROTEAN II xi cell (Bio-Rad))
- Standard agarose gel electrophoresis equipment and setup for phosphorimaging analysis as described earlier
- Heating block
- Microcentrifuge
- Gel documentation station fitted with a UV light source

5.2 Reagents and Buffers

- Restriction enzymes *Ahd*I and *Xmn*I (NEB)
- Calf thymus topoisomerase I (Invitrogen)
- SDS (10%)
- PK (10 mg/mL)
- BSA (10 μg/μL in TE buffer)
- Phenol–chloroform–isoamyl alcohol or PCI mix (25:24:1)
- NaOAc (3 *M*, pH 5.2)
- Ethanol
- Chloroform
- Ethidium bromide (Sigma-Aldrich)
- Native gel-loading buffer (15 m*M* Tris–HCl, pH 7.5, 25% glycerol, 0.05% Orange G)
- Denaturing gel-loading buffer (94% formamide, 10 m*M* Tris–HCl, pH 7.5, 1 m*M* EDTA, 0.05% bromophenol blue)
- TAE buffer (40 m*M* Tris–acetate, pH 8.0, 1 m*M* EDTA)
- TBE buffer (90 m*M* Tris–HCl, pH 8.3, 90 m*M* boric acid, 2 m*M* EDTA)
- Alkaline solution 1 (0.5 *M* NaOH, 10 m*M* EDTA)
- Alkaline solution 2 (50 m*M* NaOH, 1 m*M* EDTA)
- Neutralization buffer (0.5 *M* Tris–HCl, pH 7.5)
- TE buffer (10 m*M* Tris–HCl, pH 7.5, 1 m*M* EDTA)
- NEBuffer 4 (50 m*M* K–acetate, 20 m*M* Tris–acetate, pH 7.9, 10 m*M* Mg–acetate, 1 m*M* DTT)
- 0.8% agarose gel made in TAE buffer
- 4% polyacrylamide gel (20 cm × 16 cm × 1 mm) made in TBE buffer containing 7 *M* urea

5.3 Verification of the Supercoiled State of Duplex DNA

It is crucial to verify that the pBluescript plasmid DNA retains its supercoiled state during DNA synthesis. The analysis provides assurance that extensive DNA synthesis is mediated via D-loop migration and not through a DNA relaxation or rolling circle mechanism that could result from a contaminating topoisomerase or endonuclease activity, respectively. The method described here helps verify that the plasmid DNA template remains intact.

- Mix 10 μL of the reaction from Section 3.3 with 10 μL native gel-loading buffer and incubate at 95°C for 2 min to dissociate the extended invading DNA strand from the supercoiled pBluescript plasmid DNA.
- Resolve the mixture along with supercoiled pBluescript marker DNA in the 0.8% agarose gel at 90 mA for 120 min using TAE as the electrophoresis buffer.
- Soak the gel in H_2O with ethidium bromide (1 μL/mL final concentration) for 10 min to stain DNA and then in a large volume of H_2O for 1 h to reduce background staining.
- Record the stained DNA species in the gel documentation station.

5.4 Testing for Dependence of DNA Synthesis Track Length on Topoisomerase

Topological stress that accumulates during extension of the invading strand would inhibit the movement of the polymerase ensemble within the D-loop. In this case, the addition of a topoisomerase can lead to an increase of the DNA synthesis track length (Li et al., 2009; Wilson et al., 2013). However, in the bubble migration mode of DNA synthesis, topological stress in the plasmid DNA is relieved via dissociation of the extended invading strand mediated by Pif1 (Wilson et al., 2013). Therefore, the lack of any stimulatory effect of topoisomerase addition provides experimental support for the bubble migration mechanism (Wilson et al., 2013). Here, we describe the method to test whether DNA synthesis is responsive to topoisomerase addition:

- Perform the DNA synthesis reaction with RFC, PCNA, and Pol δ as described in Section 3.3.
- After a 4-min incubation, add Pif1 (40 n*M* final concentration) and 8 U of calf thymus topoisomerase I to the reaction mixture (20 μL final volume) and incubate for 8 min.
- As controls, carry out the reaction in the same fashion but omit Pif1 and/or topoisomerase I.

- Terminate and deproteinize the reaction by adding 0.5 μL SDS and 1 μL PK and incubating it at 37°C for 10 min.
- Resolve the reaction mixtures in a denaturing polyacrylamide gel and analyze the DNA products in the phosphorimager as described in Section 2.2.3.

5.5 Analysis of the DNA Synthesis Products by Restriction Enzyme Digests

This analysis is to provide definitive biochemical evidence for a migrating DNA bubble established during DNA synthesis. The conceptual basis for the method, which relies on monitoring the susceptibility of sites in the pBluescript plasmid molecule to restriction enzymes, is explained in Fig. 4A:

- Perform the DNA synthesis reaction (20 μL) with RFC, PCNA, and Pol δ as described in Section 3.3.
- After a 4-min incubation, add Pif1 (40 n*M* final concentration) and incubate for 8 min.
- As control, carry out the reaction in the same fashion but omit Pif1.
- Mix the reaction with 80 μL TE and extract with 100 μL PCI and subsequently with 100 μL chloroform twice to deproteinize as described in Sambrook et al. (1989).
- Add 10 μL NaOAc (3 *M*, pH 5.2) and 300 μL absolute ethanol followed by a 1-h incubation on ice to precipitate DNA (Sambrook et al., 1989).
- Collect DNA precipitate by centrifugation and wash the pellet with 300 μL ice-cold 70% ethanol twice and then air dry the pellet.
- Dissolve the pellet with 10 μL of NEBuffer 4 supplemented with 0.1 mg/mL BSA.
- Add 2.5 U of *Ahd*I or 10 U of *Xmn*I (20 U/μL) and incubate at 37°C for 10 min.
- Terminate and deproteinize the reaction by adding 0.5 μL SDS and 1 μL PK and incubating it at 37°C for 10 min.
- Add 5 μL of denaturing gel-loading buffer and incubate at 95°C for 5 min.
- Resolve the DNA species by electrophoresis in the denaturing polyacrylamide gel at 55°C and 200 mA for 90 min using TBE buffer as the electrophoresis buffer.
- Dry the gel and analyze the dried gel by phosphorimaging analysis as described in Section 2.2.3 (Fig. 4B).

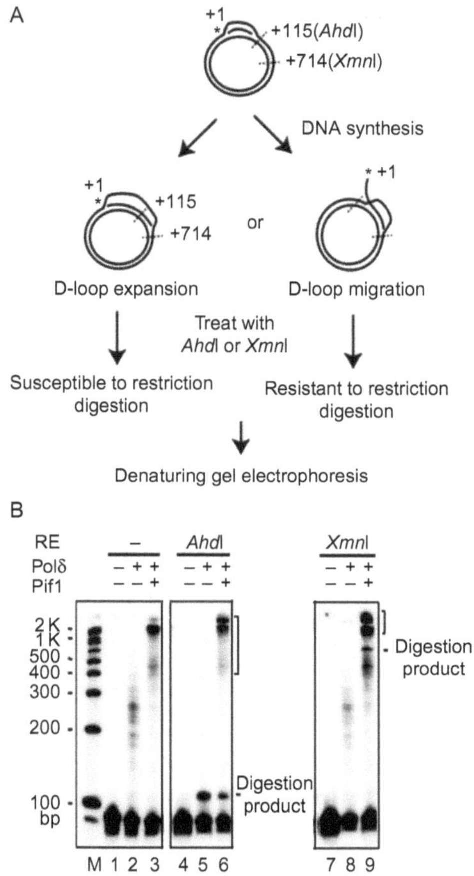

Fig. 4 Analysis of DNA extension products by restriction digests. (A) *Ahd*I and *Xmn*I incise dsDNA at 115 and 714 nucleotides from the 5′ end of the nonextended ^{32}P-labeled invading strand, respectively. The extended DNA strand is normally susceptible to the action of either restriction enzyme (RE). However, upon its release from the D-loop by Pif1, the extended invading strand becomes resistant to the restriction enzymes. The asterisk identifies the ^{32}P label. (B) Gel images from a typical experiment are shown. The *bracket* identifies extended DNA species resistant to the restriction enzymes. *Panel B: Modified from our published study Wilson, M. A., Kwon, Y., Xu, Y., Chung, W. H., Chi, P., Niu, H., et al. (2013). Pif1 helicase and Poldelta promote recombination-coupled DNA synthesis via bubble migration.* Nature, 502*(7471), 393–396.*

5.6 Special Note

The presence of a migrating DNA bubble in the DNA synthesis reaction can also be verified by electron microscopy (Wilson et al., 2013). However, this requires highly specialized equipment and training, and is therefore not covered in this chapter.

ACKNOWLEDGMENTS

Work in our laboratory is supported by research grants from the National Institutes of Health including the SBDR program project grant from the National Cancer Institute.

REFERENCES

Brown, M. S., & Bishop, D. K. (2014). DNA strand exchange and RecA homologs in meiosis. *Cold Spring Harbor Perspectives in Biology*, 7(1), a016659.

Buisson, R., Niraj, J., Pauty, J., Maity, R., Zhao, W., Coulombe, Y., et al. (2014). Breast cancer proteins PALB2 and BRCA2 stimulate polymerase η in recombination-associated DNA synthesis at blocked replication forks. *Cell Reports*, *6*(3), 553–564.

Chung, W. H., Zhu, Z., Papusha, A., Malkova, A., & Ira, G. (2010). Defective resection at DNA double-strand breaks leads to de novo telomere formation and enhances gene targeting. *PLoS Genetics*, *6*(5), e1000948.

Costantino, L., Sotiriou, S. K., Rantala, J. K., Magin, S., Mladenov, E., Helleday, T., et al. (2014). Break-induced replication repair of damaged forks induces genomic duplications in human cells. *Science*, *343*(6166), 88–91.

Daley, J. M., Gaines, W. A., Kwon, Y., & Sung, P. (2014). Regulation of DNA pairing in homologous recombination. *Cold Spring Harbor Perspectives in Biology*, *6*(11), a017954.

Daley, J. M., Niu, H., Miller, A. S., & Sung, P. (2015). Biochemical mechanism of DSB end resection and its regulation. *DNA Repair (Amst)*, *32*, 66–74.

Deem, A., Keszthelyi, A., Blackgrove, T., Vayl, A., Coffey, B., Mathur, R., et al. (2011). Break-induced replication is highly inaccurate. *PLoS Biology*, *9*(2), e1000594.

Dilley, R. L., Verma, P., Cho, N. W., Winters, H. D., Wondisford, A. R., & Greenberg, R. A. (2016). Break-induced telomere synthesis underlies alternative telomere maintenance. *Nature*, *539*(7627), 54–58.

Donnianni, R. A., & Symington, L. S. (2013). Break-induced replication occurs by conservative DNA synthesis. *Proceedings of the National Academy of Sciences of the United States of America*, *110*(33), 13475–13480.

Heyer, W. D. (2007). Biochemistry of eukaryotic homologous recombination. *Topics in Current Genetics*, *17*, 95–133.

Hicks, W. M., Kim, M., & Haber, J. E. (2010). Increased mutagenesis and unique mutation signature associated with mitotic gene conversion. *Science*, *329*(5987), 82–85.

Hunter, N. (2015). Meiotic recombination: The essence of heredity. *Cold Spring Harbor Perspectives in Biology*, 7(12), a016618.

Jasin, M., & Rothstein, R. (2013). Repair of strand breaks by homologous recombination. *Cold Spring Harbor Perspectives in Biology*, *5*(11), a012740.

Kawamoto, T., Araki, K., Sonoda, E., Yamashita, Y. M., Harada, K., Kikuchi, K., et al. (2005). Dual roles for DNA polymerase eta in homologous DNA recombination and translesion DNA synthesis. *Molecular Cell*, *20*(5), 793–799.

Kowalczykowski, S. C. (2015). An overview of the molecular mechanisms of recombinational DNA repair. *Cold Spring Harbor Perspectives in Biology*, 7(11), a016410.

Li, X., Stith, C. M., Burgers, P. M., & Heyer, W. D. (2009). PCNA is required for initiation of recombination-associated DNA synthesis by DNA polymerase delta. *Molecular Cell*, *36*(4), 704–713.

Lydeard, J. R., Jain, S., Yamaguchi, M., & Haber, J. E. (2007). Break-induced replication and telomerase-independent telomere maintenance require Pol32. *Nature*, *448*(7155), 820–823.

Maloisel, L., Fabre, F., & Gangloff, S. (2008). DNA polymerase delta is preferentially recruited during homologous recombination to promote heteroduplex DNA extension. *Molecular and Cellular Biology*, *28*(4), 1373–1382.

McIlwraith, M. J., Vaisman, A., Liu, Y., Fanning, E., Woodgate, R., & West, S. C. (2005). Human DNA polymerase eta promotes DNA synthesis from strand invasion intermediates of homologous recombination. *Molecular Cell*, *20*(5), 783–792.

Petukhova, G., Stratton, S., & Sung, P. (1998). Catalysis of homologous DNA pairing by yeast Rad51 and Rad54 proteins. *Nature*, *393*(6680), 91–94.

Raschle, M., Van Komen, S., Chi, P., Ellenberger, T., & Sung, P. (2004). Multiple interactions with the Rad51 recombinase govern the homologous recombination function of Rad54. *The Journal of Biological Chemistry*, *279*(50), 51973–51980.

Raynard, S., & Sung, P. (2009). Assay for human Rad51-mediated DNA displacement loop formation. *Cold Spring Harbor Protocols*, *2009*(1), prot5120.

Saini, N., Ramakrishnan, S., Elango, R., Ayyar, S., Zhang, Y., Deem, A., et al. (2013). Migrating bubble during break-induced replication drives conservative DNA synthesis. *Nature*, *502*(7471), 389–392.

Sambrook, J., Fritsch, E. F., & Maniatis, T. (1989). *Molecular cloning—A laboratory manual* (2nd ed.). Plainview, NY: Cold Spring Harbor Laboratory Press.

San Filippo, J., Sung, P., & Klein, H. (2008). Mechanism of eukaryotic homologous recombination. *Annual Review of Biochemistry*, *77*, 229–257.

Sebesta, M., Burkovics, P., Haracska, L., & Krejci, L. (2011). Reconstitution of DNA repair synthesis in vitro and the role of polymerase and helicase activities. *DNA Repair (Amst)*, *10*(6), 567–576.

Sugiyama, T., Zaitseva, E. M., & Kowalczykowski, S. C. (1997). A single-stranded DNA-binding protein is needed for efficient presynaptic complex formation by the Saccharomyces cerevisiae Rad51 protein. *The Journal of Biological Chemistry*, *272*(12), 7940–7945.

Sung, P. (1994). Catalysis of ATP-dependent homologous DNA pairing and strand exchange by yeast RAD51 protein. *Science*, *265*(5176), 1241–1243.

Sung, P., & Klein, H. (2006). Mechanism of homologous recombination: Mediators and helicases take on regulatory functions. *Nature Reviews. Molecular Cell Biology*, 7(10), 739–750.

Symington, L. S., Rothstein, R., & Lisby, M. (2014). Mechanisms and regulation of mitotic recombination in Saccharomyces cerevisiae. *Genetics*, *198*(3), 795–835.

Van Komen, S., Macris, M., Sehorn, M. G., & Sung, P. (2006). Purification and assays of Saccharomyces cerevisiae homologous recombination proteins. *Methods in Enzymology*, *408*, 445–463.

Van Komen, S., Petukhova, G., Sigurdsson, S., Stratton, S., & Sung, P. (2000). Superhelicity-driven homologous DNA pairing by yeast recombination factors Rad51 and Rad54. *Molecular Cell*, *6*(3), 563–572.

Wilson, M. A., Kwon, Y., Xu, Y., Chung, W. H., Chi, P., Niu, H., et al. (2013). Pif1 helicase and Poldelta promote recombination-coupled DNA synthesis via bubble migration. *Nature*, *502*(7471), 393–396.

Yuzhakov, A., Kelman, Z., Hurwitz, J., & O'Donnell, M. (1999). Multiple competition reactions for RPA order the assembly of the DNA polymerase delta holoenzyme. *The EMBO Journal*, *18*(21), 6189–6199.

CHAPTER FOURTEEN

Current and Emerging Assays for Studying the Primer Synthesis Activities of DNA Primases

Thomas A. Guilliam, Aidan J. Doherty[1]
Genome Damage and Stability Centre, School of Life Sciences, University of Sussex, Brighton, United Kingdom
[1]Corresponding author: e-mail address: ajd21@sussex.ac.uk

Contents

Abstract

Primases play a crucial role in the initiation of DNA synthesis during replication by de novo synthesis of short RNA or DNA "primers." In recent years, evidence has accumulated which expands the essential roles of primases to include, not only the initiation of replication but also other critical roles in DNA metabolism, including damage tolerance and repair. Despite the broadening roles for these enzymes, the methods used to identify and characterize primase activities are limited. Historically, biochemical analysis of primases has been based on the synthesis of radioactively labeled primers and their detection on denaturing polyacrylamide gels. In the last two decades, a number of alternative primase assays have been developed in an effort to supersede radioactive

Methods in Enzymology, Volume 591
ISSN 0076-6879
http://dx.doi.org/10.1016/bs.mie.2017.03.006

methods. However, the radioactive gel-based assay, which has not significantly changed since its conception in the late 1970s, remains the most widely used and favored method. In this chapter, we discuss the background to, and the advantages and disadvantages of, the current techniques used to characterize primase activity in vitro. Finally, we describe an alternative, gel-based, fluorescent primase assay, which we have successfully used in the characterization of a recently identified primase-polymerase, PrimPol.

1. INTRODUCTION

Primases possess the unique ability to utilize single-stranded (ss) DNA for the initiation of de novo RNA/DNA synthesis. The short RNA or DNA chains produced from this synthesis are termed primers and provide the 3′ hydroxyl required for further extension by DNA polymerases during the initiation of replication. Due to the semidiscontinuous nature of DNA replication, primase activity is not only essential during initiation but also to continuously prime Okazaki fragment synthesis on the lagging strand. All domains of life employ primases, however, two distinct primase superfamilies, DnaG primases and archaeo-eukaryotic primases (AEPs), facilitate bacterial and archaeal/eukaryotic DNA replication, respectively. Recently, evidence has accumulated suggesting that primase-polymerases of the AEP superfamily also play key roles in DNA damage tolerance and repair, where their primase activity is essential for replication restart mechanisms including, repriming of replication downstream of lesions and secondary structures (Guilliam & Doherty, 2017; Guilliam, Keen, Brissett, & Doherty, 2015).

Despite a move away from radioactivity and toward fluorescence in primer extension-based polymerase assays, gel-based primase assays still routinely make use of radiolabeled nucleotides. In these assays, primase activity is determined by the quantification of radiolabeled nucleotide containing primers, visualized on denaturing polyacrylamide gels. In this chapter, we discuss the advantages and limitations of existing methods used to study primases in vitro, including both traditional radioactive gel-based assays and more recently developed nonradioactive high-throughput screening (HTS) approaches. Finally, we describe a gel-based primase assay of particular use in the analysis of primase-polymerases. This assay, which utilizes fluorescently labeled nucleotides, removes the need for potentially hazardous radioactivity, allowing the assay to be performed in any laboratory without requiring training in handling radioactivity. Furthermore, this assay is used in the same way as traditional radioactive primase assays to study primase activity, as well as evaluating the effect of binding partners, reaction conditions, sequence preference, and the location of priming. To demonstrate the effectiveness of this assay, we make use of purified human PrimPol, a primase-polymerase involved in DNA damage tolerance in eukaryotes.

2. RADIOACTIVE-BASED PRIMASE ASSAYS

2.1 Traditional Radioactive Primase Assays

In the 1970s, in vivo studies of T7 DNA replication implicated phage gene 4 protein in priming DNA synthesis (Strätling & Knippers, 1973; Wolfson & Dressler, 1972). Subsequent in vitro studies developed a primase complementation assay to analyze T7 gene 4 protein. This assay confirmed that it synthesizes primers required for the initiation of DNA synthesis by T7 DNA polymerase (Hinkle & Richardson, 1975; Romano & Richardson, 1979; Scherzinger, Lanka, Morelli, Seiffert, & Yuki, 1977; Scherzinger & Litfin, 1974). Consequently, the T7 gene 4 product became the first designated DNA primase (Scherzinger et al., 1977).

The assay used in these early studies measured the ability of purified T7 primase to stimulate DNA synthesis in extracts prepared from *Escherichia coli* infected with T7 lacking the gene 4 protein (Hinkle & Richardson, 1975). Reactions were assembled containing rNTPs and dNTPs, one of which was ^{3}H or ^{32}P labeled, extracts from T7 infected *E. coli*, purified T7 primase and T7 linear duplex DNA. Following incubation, the reaction products were precipitated and washed on filter paper and radioactivity was measured by a liquid scintillation counter (Hinkle & Richardson, 1975). Thus, the ability of T7 primase to stimulate DNA synthesis in the extracts could be determined by the increase in acid insoluble radioactivity produced following incubation. In addition to analyzing the effect of T7 primase on DNA synthesis in extracts, these reports also examined the effect of the enzyme on T7 polymerase activity using a similar approach. Reactions were assembled in the same manner; however, extract was omitted and replaced with purified T7 DNA polymerase. Again, it was identified that T7 primase markedly stimulated DNA synthesis by the polymerase on duplex T7 DNA (Hinkle & Richardson, 1975).

Despite indicating that the stimulation observed in these early studies was due to de novo primer synthesis by T7 primase, interpretation of results was somewhat limited due to the duplex linear DNA template used. To gain further clarity, pycnographic analysis of products was required. Template T7 DNA was ^{3}H-, ^{13}C-, and ^{15}N-labeled and α-^{32}P dATP was provided for primer synthesis. Subsequent CsCl density gradient centrifugation confirmed that ^{32}P-labeled reaction products separated from the heavy template DNA, thereby indicating that the products and template DNA were not covalently linked, allowing inference that de novo initiation of DNA synthesis occurred (Hinkle & Richardson, 1975). Follow-up studies of T7 primase avoided this issue by using circular ss phage ΦX174 DNA as a template, removing the possibility of synthesis being initiated by a loop mechanism (Scherzinger et al., 1977). Interestingly, it was later found that the T7 gene 4 protein is also a helicase, thus explaining the initial observation of stimulation of DNA synthesis on duplex linear DNA (Bernstein & Richardson, 1988).

A similar complementation-based primase assay was used in the initial characterization of *E. coli* primase DnaG (Bouché, Zechel, & Kornberg, 1975; Rowen & Kornberg, 1978). It was determined that DnaG is required for stimulation of DNA synthesis in extracts prepared from *E. coli* expressing a temperature-sensitive DnaG mutant (Rowen & Kornberg, 1978). Furthermore, in the absence of extracts, DnaG was able to initiate DNA synthesis on phage G4 DNA in the presence of *E. coli* single-strand DNA-binding protein (SSB) and DNA polymerase III holoenzyme (Rowen & Kornberg, 1978). Additionally, priming by DnaG was also observed on G4 and M13 DNA, as well as poly(dT) templates, in the presence of the DnaB helicase but in the absence of SSB or DNA polymerase III (Arai & Kornberg, 1979).

Similar assays were also used in the basic characterization of other prokaryotic and phage primases (Krevolin & Calendar, 1985; Lanka, Scherzinger, Günther, & Schuster, 1979; Morris, Sinha, & Alberts, 1975). These techniques were very useful in the early analysis and identification of primase activity. However, the information gleaned from the assays was limited. Notably, primer synthesis could often only be detected in the presence of additional replisome components and the results could be obscured by the DNA template. Furthermore, analysis of reaction products by liquid scintillation counting was unable to provide qualitative information about the length and sequence of the synthesized primers. Consequently, follow-up studies of phage and prokaryotic primases, as well as early analyses of the replicative eukaryotic primase Prim1/2 from yeast and *Drosophila*, further resolved primase products by polyacrylamide gel electrophoresis (PAGE) (Biswas, Joseph, & Biswas, 1987; Bouché,

Rowen, & Kornberg, 1978; Conaway & Lehman, 1982a, 1982b; Romano & Richardson, 1979; Scherzinger et al., 1977; Wu, Zechner, & Marians, 1992). Reactions were assembled in essentially the same manner as described previously using ss phage or poly(dT) templates, however other replisome components such as polymerases and helicases were omitted, allowing direct synthesis by the primase of interest to be analyzed (Biswas et al., 1987). Following incubation, reactions were stopped with EDTA and ionic surfactants. Additionally, reactions were incubated with proteinase K for 1–2 h before phenol chloroform extraction or ethanol precipitation. Pellets were resuspended in buffer containing formamide and dyes, denatured by heating before loading onto gels. Electrophoresis was performed on urea–polyacrylamide gels which were autoradiographed using X-ray film (Biswas et al., 1987; Conaway & Lehman, 1982a).

By resolving reaction products on polyacrylamide gels, the synthesized primers could be analyzed to single nucleotide resolution, providing more direct information about primase activity than liquid scintillation counting alone. Many studies have utilized this approach in the characterization of primases in order to determine the length of synthesized primers (Frick & Richardson, 2001). Further processing of RNA primase reaction products by limited alkali or ribonuclease digestion, prior to resolution on urea–polyacrylamide gels, has also been used to determine the sequence of these primers (Romano & Richardson, 1979).

Since the development of gel-based radioactive primase assays in the late 1970s–1980s, the basic outline of this technique has largely remained unchanged. However, clean-up steps, including proteinase K digestion and phenol chloroform extraction, are now generally omitted. This greatly increases the speed and ease of the technique, without significantly affecting the quality of results. In addition, advances in oligonucleotide synthesis technology have allowed specific DNA templates of known sequence to be generated and analyzed in the assay, increasing the applicability of the technique. One example of this is in the analysis of T7 primase recognition sites. Numerous oligonucleotides containing different modified primase recognition sites were synthesized and tested for their ability to act as primer synthesis templates (Frick & Richardson, 1999). This allowed identification of the requirements for sequence-specific DNA binding and primer synthesis by T7 primase. Synthetic oligonucleotides were also used in the early characterization of human Prim1/2 to investigate the stimulatory effect of manganese on primase activity (Kirk & Kuchta, 1999). The use of synthetic oligonucleotides in primase assays has a number of advantages over traditional phage templates. These include the ability to use shorter oligonucleotides of known sequence, allowing initiation sites and product sequences to be determined, and the ease with which they can be modified, permitting analysis of different sequences and DNA secondary structures on primase activity. Nevertheless, phage DNA, e.g., ss M13, is still often used as a template to identify primase activity due to its affordability and circular nature, which prevents snap-back and extension of the 3′ end of the template.

Developments in other accessory components of the radioactive primase assay have also increased the speed and ease with which it can be performed. Traditionally, autoradiography was used to detect radioactivity by exposing X-ray films to polyacrylamide gels following resolution. The resulting images were quantified by densitometry, permitting analysis of the reaction products. These techniques can be time-consuming and require long exposure times. Film is also easily over exposed, resulting in signal lying outside the linear dynamic range, preventing accurate quantification. Generally, radioactive gels are now analyzed by phosphorimaging. Storage-phosphor screens replace film, subsequent scanning of these with a helium–neon laser causes emission of luminescence proportional to the level of radiation which is quantified using a photomultiplier, consequently producing a digital image. The resulting image and reaction products can then be quantified using image analysis software. In comparison to X-ray film detection, phosphorimaging has greatly increased, sensitivity, linear dynamic range, and speed of image development, despite increased expense and lower resolution (Van Kirk, Feinberg, Robertson, Freeman, & Vrana, 2010).

In summary, the radioactive gel-based primase assay has remained largely unchanged since its initial development in the late 1970s; however, refinements of the technique and advances in accessory components have greatly increased the speed and ease with which it can be performed. An outline of a typical radioactive gel-based primase assay used today is shown in Fig. 1.

Despite relatively few changes to the original gel-based primase assay, the technique is still the "go-to" option for analyzing primase activity due to its excellent sensitivity and ability to provide information on the size and yield of reaction products. This is highlighted by the use of the assay in the characterization of recently discovered archaeal and viral primases (De Silva, Paran, & Moss, 2009; Galal, Pan, Kelman, & Hurwitz, 2012; Lipps, Weinzierl, von Scheven, Buchen, & Cramer, 2004; Silva, Lewis, Berglund, Koonin, & Moss, 2007; Zuo, Rodgers, Mikheikin, & Trakselis, 2010). However, the assay also has a number of limitations and

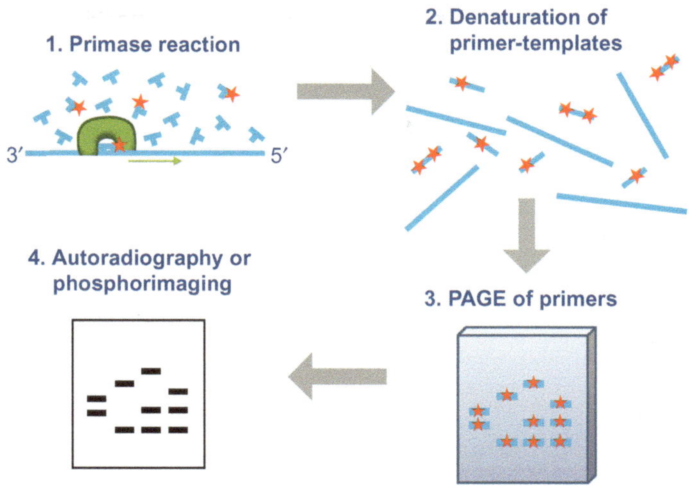

Fig. 1 Outline of the radioactive gel-based primase assay. 1. Primase reactions are assembled containing the primase of interest (shown in *green*), a ssDNA template, native rNTPs or dNTPs, and radiolabeled rNTPs or dNTPs (indicated by *red star*) in an appropriate buffer. Incorporation of radiolabeled nucleotides during synthesis and extension generates radiolabeled primers. 2. Following primer synthesis, reactions are quenched through addition of EDTA and primer-template duplexes are denatured by heating in buffer containing formamide. 3. Reaction products are loaded onto a denaturing urea–polyacrylamide gel and electrophoresis is performed to resolve the radiolabeled primers. 4. Synthesized primers are visualized by autoradiography using X-ray film or phosphorimaging.

disadvantages. Perhaps the most significant of these is the reliance on potentially hazardous radioactivity. This requires training in its handling and disposure, in addition to the implementation of rigorous safety measures before the assay can be performed. Radiolabeled dNTPs have a short half-life, limiting the time that it can be used and increasing assay costs. Radioactivity imaging methods can also suffer from poor linear dynamic range, with overexposure resulting in a signal outside of the linear range thereby preventing accurate quantification. Additionally, in spite of refinements of the original assay, the technique is still relatively time consuming thus making it unsuitable for certain applications, such as HTS of primase inhibitors.

2.2 A High-Throughput Radioactive-Based Primase Assay

Due to the limited applicability of traditional primase assays in HTS approaches, a modified 96-well plate scintillation proximity assay (SPA) was developed to screen *E. coli* DnaG inhibitors (Fig. 2) (Zhang et al.,

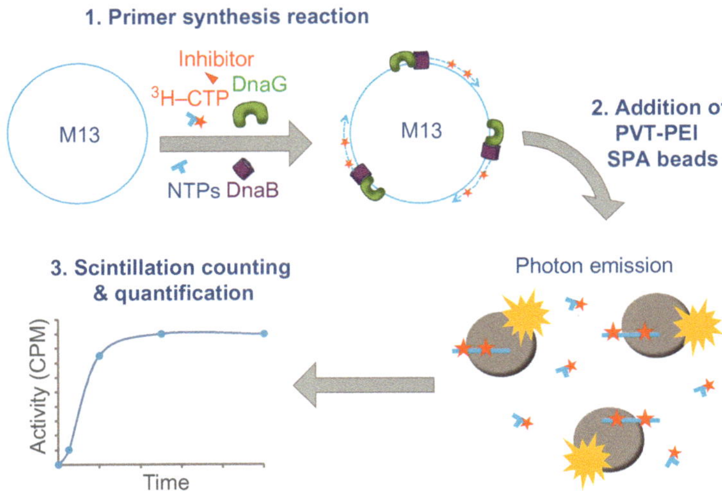

Fig. 2 Overview of the high-throughput radioactive-based primase assay. 1. Primer synthesis reactions are assembled in 96-well plates containing *E. coli* DnaG primase (shown in *green*), DnaB helicase (shown in *purple*), native rNTPs, ^{3}H–CTP (indicated by *red star*), ss M13 template DNA, and the test compound (shown as *red triangle*) or DMSO, in an appropriate buffer. Incubation of reactions permits synthesis of ^{3}H–CTP-labeled primers on the M13 template. 2. Polyvinyl toluene-polyethyleneimine (PVT-PEI) scintillation proximity assay (SPA) beads are added to the reaction. Capture of the radiolabeled primers on the PVT-PEI SPA beads stimulates photon emission from the scintillant in the beads. Free ^{3}H–CTP does not bind the beads, and therefore does not stimulate photon emission. 3. Photon emission is detected and quantified using a photomultiplier tube based scintillation counter.

2002). In this assay, reactions are assembled in a 96-well plate containing buffer, metal ions, unlabeled dNTPs, [^{3}H]CTP, a ssDNA template, DnaG, DnaB, and the test compound or DMSO. Following incubation, a suspension of polyvinyl toluene-polyethyleneimine (PVT-PEI)-coated SPA beads are added and plates are read after ~1 h on a Topcount instrument (Packard) (Zhang et al., 2002). The assay is based upon the capture of primase products on PVT-PEI SPA beads. This capture brings ^{3}H-labeled products in close proximity to the SPA beads. Consequently, decay of the ^{3}H releases β-particles, which stimulate the scintillant in the beads to emit photons. The emitted photons can then be detected and quantified using a photomultiplier tube-based scintillation counter. Free ^{3}H[CTP], not bound to the SPA beads, does not stimulate photon emission due to the insufficient energy of the β-particles to reach the beads. Thus, SPA has the advantage that no separation or washing step is required to remove the free ^{3}H[CTP].

This assay overcomes the time-consuming nature of the traditional gel-based primase assay and extends the application of the technique to HTS.

The assay provides a sensitive and efficient method to quantify primase activity and screen DnaG inhibitors that, when coupled with a DnaB helicase assay, can provide insights into the mechanism of action of those compounds (Zhang et al., 2002). However, the technique is still reliant on the use of radioactivity and, due the HTS approach, is costly and may generate large amounts of liquid waste. Additionally, unlike gel-based approaches, the assay does not provide qualitative information about the size or sequence of the synthesized primers.

3. NONRADIOACTIVE PRIMASE ASSAYS

Given the drawbacks of radioactive-based primase assays, a number of alternative nonradioactive assays have been developed, some of which are applicable to HTS. In this section, the background to these methods and their advantages and disadvantages will be discussed.

3.1 Thermally Denaturing High-Performance Liquid Chromatography Primase Assay

To avoid the cost and safety issues associated with radioactive assays, an alternative primase assay was developed, based upon high-performance liquid chromatography (HPLC) analysis of products (Fig. 3) (Koepsell, Bastola, Hinrichs, & Griep, 2004). *E. coli* DnaG and enzymatic reactions were assembled in appropriate buffer conditions containing the primase, ssDNA template (<30 nt in length) blocked at the 3′ end, and native rNTPs. Importantly, by blocking the 3′ end of the template, the user can directly examine de novo synthesis by the primase, rather than elongation from a 3′ end hairpin on the synthetic ssDNA template, which can be produced by template snap-back. Notably, other assays, including the radioactive HTS method, did not control for this phenomenon, potentially generating misleading data and interpretation of results. Koepsell et al. found that when the 3′ end of the template was not blocked, 10-fold more primase was required for de novo primer synthesis and the rate constant of primer synthesis was three times greater than that reported when using the radioactive HTS method (Koepsell et al., 2004; Zhang et al., 2002). Following incubation, reactions were stopped by heat denaturation, desalted, and dried, before resuspension in water and analysis by HPLC on an alkylated nonporous polystyrene–divinylbenzene copolymer microsphere bead column under thermally denaturing conditions. UV detection of eluted oligonucleotides at 260 nm produced chromatograms with peaks corresponding to the template and various smaller products (Koepsell et al.,

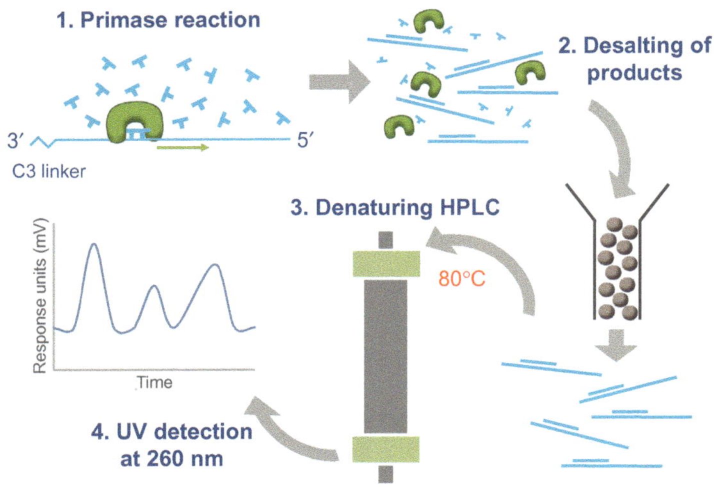

Fig. 3 Summary of the thermally denaturing HPLC primase assay. 1. Primase reactions are assembled containing the primase of interest (shown in *green*), native dNTPs or rNTPs, and a ssDNA template blocked at the 3′ end by a C3 linker, in an appropriate buffer. Assembled reactions are incubated allowing synthesis of unlabeled primers. 2. Reactions are stopped by heating, before being desalted, dried, and resuspended in water. 3. Samples are then loaded onto an alkylated nonporous polystyrene–divinylbenzene copolymer microsphere bead column and analyzed by HPLC under thermally denaturing conditions (80°C). 4. Primase reaction products and template DNA are detected upon elution by monitoring UV absorbance at 260 nm. The resulting chromatogram can be used for quantification of primer synthesis by comparison to a standard curve, taking into account the extinction coefficient of the oligonucleotide.

2004). Reaction products were quantified by analyzing the area under each peak, taking into account variations in extinction coefficients between oligonucleotides, and compared to a standard curve. Importantly, analysis by denaturing HPLC allows products to be separated by both size and hydrophobicity, thus producing qualitative, as well as quantitative, information about the synthesized primers. This assay was used to determine the kinetics of de novo primer synthesis by DnaG, as well as to identify the IC50 for dNTP inhibition of primase activity (Koepsell et al., 2004).

The denaturing HPLC primase assay has a number of advantages over traditional gel-based radioactive assays. Perhaps the greatest of these is that it can be performed with native rNTPs/dNTPs, removing the hazards associated with radioactivity. Additionally, removal of radioactivity also decreases the cost of the assay, discounting initial costs for equipment. Like traditional gel-based assays, this method provides sensitive qualitative and quantitative information about primer synthesis. Furthermore, the HPLC analysis is automated and scalable to a degree, with each run taking ~20 min (Koepsell et al., 2004). This makes the assay much quicker for individual experiments requiring only a short number of runs, compared to gel-based assays. However, with larger experiments requiring multiple runs, the analysis time can be much greater and it is here that gel-based assays have the advantage of resolving multiple samples at the same time. Likewise, this method lacks sufficient throughput for HTS approaches, making it unsuitable for screening large chemical libraries to identify inhibitors. Quantification of reaction products using chromatogram peaks can also be more difficult than analyzing gels. Notably, variations in extinction coefficients between products require knowledge of the nucleotide content of the peak, in addition to the generation of a standard curve. Coupled with this, HPLC analysis requires optimization for each specific ssDNA template used, prior to performing experiments.

3.2 A Fluorometric High-Throughput Primase Assay

Given the limitations of the HPLC primase assay in HTS, Koepsell et al. developed a high-throughput microplate-based fluorescent primase assay, adaptable to robotic screening methods (Fig. 4) (Koepsell, Hanson, Hinrichs, & Griep, 2005). This assay is based on PicoGreen nucleic acid dye, a fluorochrome that binds specifically to double-stranded (ds) DNA. When bound, PicoGreen fluoresces at an excitation maximum of 480 nm, with an emission peak at 520 nm. The dye was previously found to offer an effective and sensitive way to quantify dsDNA due to its high level of fluorescence enhancement upon DNA binding, thus making it suitable for the detection of primer-template duplexes (Ahn, Costa, & Emanuel, 1996). In this high-throughput fluorescent primase assay, reactions are assembled in a 96-well microplate and incubated for the desired time to allow primer synthesis. Following incubation, PicoGreen dye is added, which binds to the RNA–DNA duplexes generated from primer synthesis and fluoresces upon excitation, allowing detection and quantification of primase products using a spectrofluorometer. Additionally, PicoGreen dye quenches the primase reaction, removing the need for quenching with EDTA (Koepsell et al., 2005).

This high-throughput fluorometric primase assay therefore offers a non-radioactive alternative for HTS studies of potential primase inhibitors. The assay is able to provide quantitative information on primer synthesis but does not generate qualitative information, such as primer length or sequence.

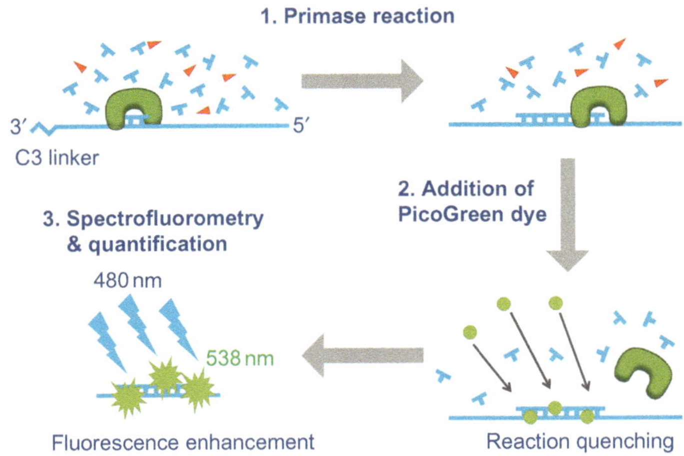

Fig. 4 Overview of the fluorometric high-throughput primase assay. 1. Reactions are assembled in a 96-well plate containing the primase of interest (shown in *green*), a ssDNA template with a blocked 3′ end, unlabeled dNTPs, and the chemical compound to be screened (indicated by *red triangle*) or DMSO, in a suitable buffer. Incubation of reactions facilitates primer synthesis and extension, generating primer-template duplexes. 2. PicoGreen dye (shown as a *green circle*) is added to the reaction which binds to the primer-template duplexes, producing fluorescence enhancement. Addition of PicoGreen dye also quenches the reaction, removing the need to add EDTA. 3. The 96-well plate is scanned using a spectrofluorometer with an excitation at 485 nm and an emission at 538 nm. Thus, the level of fluorescence is determined by the amount of PicoGreen dye bound to dsDNA, which is dependent upon the amount of primer synthesis.

Although the microplate format, fast analysis time, and ability to function in the presence of DMSO makes the method an attractive option for HTS, there are a number of potential drawbacks to the technique. First, the assay is effective in detecting primers longer than ~6 nt due to their stable association with the template DNA. However, shorter primers may not provide the stable duplex required for PicoGreen binding and fluorescence enhancement. Second, a potential issue in using fluorometric assays for HTS is the interaction of nonpolar and aromatic compounds with the fluorescent label, which may interfere with the signal and obscure results (Biswas, Resto-Roldán, Sawyer, Artsimovitch, & Tsodikov, 2013).

3.3 A High-Throughput Primase-Pyrophosphatase Activity Assay

A more recently developed alternative to the fluorometric primase assay, with similar HTS applications, is the primase-pyrophosphatase assay

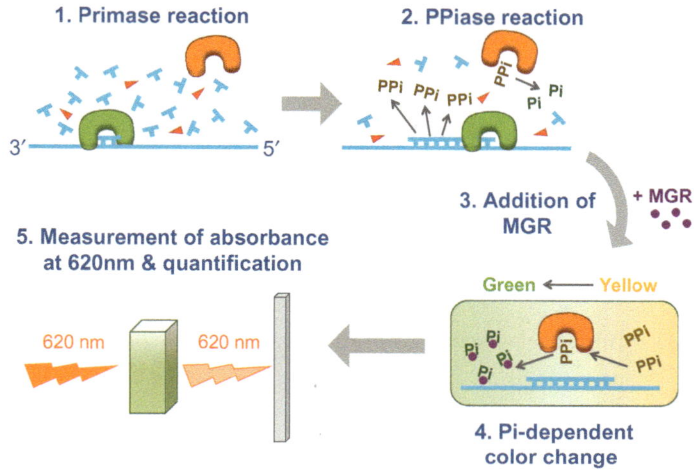

Fig. 5 Outline of the high-throughput primase-pyrophosphatase activity assay. 1. For HTS, primase reactions are assembled in 384-well plates containing *Mtb* DnaG (shown in *green*), *Mtb* pyrophosphatase (PPase) (shown in *orange*), unlabeled rNTPs, a ssDNA template, and the inhibitors to be tested (shown as a *red triangle*) or DMSO, in a suitable buffer. Incubation of the reactions permits primer synthesis and extension. 2. Incorporation of nucleotides by the primase during synthesis releases pyrophosphate (PPi). The released pyrophosphate is then cleaved by PPase into phosphate (Pi). 3. Malachite green reagent (MGR) (indicated by *purple circle*) and sodium citrate are added to the reaction. MGR forms a complex with the phosphate which produces a color change from *yellow* to *green*. 4. The color change is dependent on the level of free Pi, which is consequently dependent upon PPi release, and ultimately the level of primase activity. 5. The color change is detected and quantified by measuring absorbance at 620 nm in a plate reader.

(Fig. 5) (Biswas et al., 2013). In this assay, primase activity is coupled to inorganic pyrophosphatase (PPase), which cleaves the pyrophosphate (PPi) released during the priming reaction into phosphate (Pi). The Pi concentration can then be measured using malachite green reagent (MGR), which displays increasing absorbance at 620 nm as the concentration of Pi increases, producing a color change from yellow to green that is quantifiable using a plate reader. Importantly, PPase does not exhibit any cleavage activity on NTPs, consequently making the enzyme's activity dependent upon PPi released during primer synthesis (Biswas et al., 2013). Additionally, ssDNA templates lacking thymidine nucleotides in their 5′ half are used, allowing ATP to be omitted and thus preventing any background signal from being generated by ATPase activity present in the protein preparations.

Biswas et al. successfully used this assay to screen 2560 small molecules for *Mycobacterium tuberculosis* DnaG (*Mtb* DnaG) inhibition (Biswas et al., 2013).

For HTS, reactions were assembled in 384-well plates containing the inhibitors, rNTPs, ssDNA template, *Mtb* DnaG, and *Mtb* PPase, in the appropriate buffer. After incubation, MGR and sodium citrate were added and absorbance at 620 nm was measured in a plate reader (Biswas et al., 2013). Thus, PPi release can be used to provide a quantitative measure of primer synthesis, permitting kinetic analyses of *Mtb* DnaG. Using this method for HTS, the authors identified a number of "hits" for DnaG inhibitors, including suramin and doxorubicin. Further kinetic analysis of primase activity, in the presence of these inhibitors, measured the release of PPi as a function of inhibitor concentration under various DNA and NTP concentrations. This analysis provided insights into the mode of inhibition, which suggested that the inhibitors may act by blocking the binding of DnaG to DNA (Biswas et al., 2013).

The primase-pyrophosphate assay is thus a proven method for HTS of primase inhibitors. Like the previously discussed assays, the method benefits from being nonradioactive, significantly reducing potential hazards and making it more suitable for HTS in an academic setting. Additionally, the assay is quick to perform and directly measures NTP incorporation, allowing any primase activity to be detected regardless of primer length. Importantly, however, the use of PPase in the method requires further analysis to confirm that the screened inhibitors are acting upon the primase itself and not the PPase. Indeed, the authors used the traditional gel-based radioactive primase assay to confirm that DnaG was inhibited by the identified compounds (Biswas et al., 2013).

4. A FLUORESCENCE GEL-BASED PRIMASE ASSAY

Given the disadvantages of radioactive assays, and the limited ability of nonradioactive HTS methods to generate qualitative information (e.g. primer length or sequence), we aimed to develop a gel-based nonradioactive primase assay with the same benefits and basic set-up as the classic radioactive assay but without the hazards and time-consuming nature of radioactive work. We have developed a fluorescent gel-based primase assay, which utilizes 6-carboxyfluoroscein (6-FAM) labeled nucleotides instead of radioactivity. Here, the background and outline of the method is described followed by a detailed method for this assay, using PrimPol as an example primase and, finally, a discussion of the advantages and disadvantages of this technique.

4.1 Theory and Overview of the Fluorescence-Based Primase Assay

The fluorescence-based primase assay described here has the same basic layout as the traditional radioactive-based primase assay (Fig. 6). Reactions are assembled in appropriate buffer conditions containing the purified primase, ssDNA template, dNTPs or rNTPs, 6-FAM dNTPs or rNTPs (~100-fold lower concentration than unmodified nucleotides), and divalent metal ions. Incubation of the assembled reactions permits primer synthesis and extension on the ssDNA template. The majority of synthesis occurs using unlabeled nucleotides; however, incorporation of 6-FAM dNTPs or rNTPs during extension of the primers allows the reaction products to be visualized.

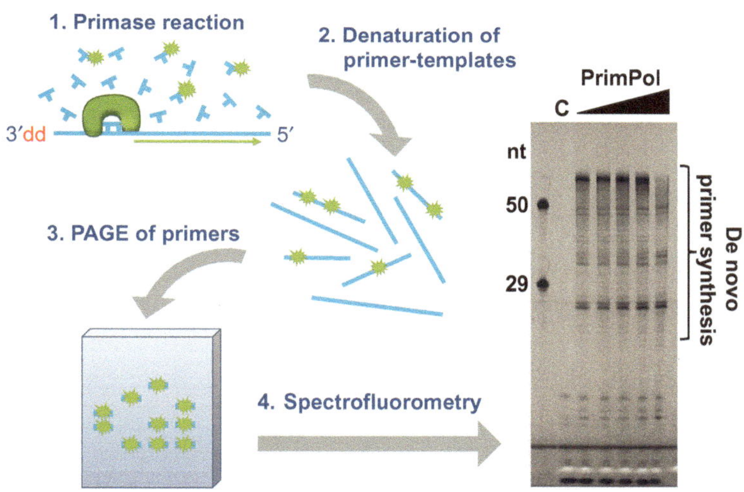

Fig. 6 Summary of the fluorescence gel-based primase assay. 1. Primase reactions are assembled containing the primase of interest (shown in *green*), a ssDNA template blocked at the 3′ end with a dideoxynucleotide, unlabeled dNTPs or rNTPs, and 6-FAM labeled dNTPs or rNTPs (indicated by *green star*), in an appropriate buffer. Incubation of the reaction facilitates primer synthesis and extension with incorporation of 6-FAM dNTPs/rNTPs during synthesis by the primase, generating fluorescently labeled primers. 2. Reactions are quenched in buffer containing EDTA and formamide, and primer-template duplexes are denatured by heating. 3. Reaction products are resolved on a denaturing urea-polyacrylamide gel. 4. Following electrophoresis, gels are scanned with a fluorescent image reader. An example gel image is shown. Here, the fluorescence gel-based primase assay was performed as described in Section 4.3.3 using increasing concentrations of PrimPol (0.25, 0.5, 1, 2, and 4 μ*M*) (indicated by *black triangle*) on a 66 nt template (sequence shown in Section 4.3.1), blocked at the 3′ end with a dideoxynucleotide. Reactions were incubated for a single 15-min time point. C indicates the "no enzyme control." Nucleotide size markers of 50 and 29 nt are shown to the *left* of the image.

Following quenching of the reactions, background given by the fluorescent nucleotides can be reduced using a number of optional DNA precipitation and clean-up approaches. The resulting reaction products are subsequently resolved on a denaturing urea–polyacrylamide gel and visualized on a fluorescence image reader.

4.2 Preparation of Primase Assay Reagents

Prior to performing the assay, the primase of interest must be expressed and isolated to a high level of purity. Importantly, the purified primase must be free from contaminating primases, polymerases, and nucleases, which can interfere with the interpretation of results. Additionally, suitable purified ssDNA template must be obtained. Both synthetic linear and circular phage templates are suitable for this assay, although reaction products smaller than ~10 nucleotides (nt) can be difficult to distinguish from background without further extension, therefore templates >10 nt are recommended.

4.3 Primer Synthesis Reaction

4.3.1 Buffers and Reagents

- 10× TBE: 1 *M* Tris (pH 7.6), 1 *M* boric acid, 20 m*M* EDTA
- 7 *M* urea, 15% polyacrylamide gel mix 60 mL: 28.8 g urea, 22.5 mL acrylamide:bisacrylamide (19:1), 199.2 μL APS, 24 μL TEMED, 6 mL 10× TBE (add APS and TEMED immediately before pouring gel) (see tip 1 and 2)
- 10× reaction buffer: 100 m*M* Bis-Tris-Propane-HCl (pH 7.0), 100 m*M* $MgCl_2$, 10 m*M* DTT (buffer available from NEB as NEBuffer 1)
- 10 μ*M* ssDNA template (5′-Biot-GTCTTCTATCTCGTCTATAT TCTATTGTCTCTATGAATACCTTCATCAGTCTCACATAGA-TGCAT-dideoxyC-3′ or another suitable ssDNA template)
- 2.5 m*M* dNTP stock solution: 2.5 m*M* of each dNTP (NEB), diluted in ddH_2O (dNTPs can be replaced with rNTPs if required)
- 25 μ*M* FAM dNTP stock solution: 25 μ*M* N^6-(6-amino)hexyl-dATP-6-FAM, 25 μ*M* 5-propargylamino-dCTP-6-FAM, 25 μ*M* aminoallyl-dUTP-6-FAM (JenaBioscience), diluted in ddH_2O (can also be replaced with 6-FAM rNTPs)
- Purified Primase: PrimPol (or another primase of interest)
- 2× stop buffer: 95% formamide, 20 m*M* EDTA, 0.25% bromophenol blue and xylene cyanol (see tip 3)
- FAM-labeled oligonucleotide size marker

4.3.2 Equipment

- Vertical nucleic acid PAGE setup (adjustable gel slab system, 165 mM × 280 mm glass plates, 0.75 mm spacers, 20 well combs) (C.B.S.·Scientific) or equivalent (see tip 4)
- Incubator, water bath, or dry heating block
- Standard microcentrifuge
- FLA-5100 fluorescent image analyzer (Fujifilm)
- ImageQuant TL for image analysis

4.3.3 Experimental Procedure

1. Before assembling the primase assay reactions, urea–polyacrylamide gels should be prepared and poured according to the manufacturer's instructions. This will allow sufficient time for setting. Note that the gel should be prerun in 1 × TBE for 0.5–1 h before loading (see tip 5).
2. On ice, assemble 10 μL reactions for each variable in the following order: 5 μL ddH_2O, 1 μL 10 × reaction buffer (1 x final), 1 μL ssDNA template (1 μM final), 1 μL dNTP stock solution (250 μM final), and 1 μL FAM dNTP stock solution (2.5 μM final) (see tip 6). If taking multiple time points, make one stock reaction with 10 μL per time point and an additional 10 μL to account for pipetting errors, e.g., if taking five-time points make a 60 μL reaction. A "no enzyme" control should also be prepared. Exposure of the FAM dNTP stock, and reactions containing FAM dNTPs, to light should be kept to a bare minimum.
3. On ice, make a 10 × stock of PrimPol using 1 × reaction buffer for dilution. For each primase, the concentration required to give the desired level of activity should be determined by testing a range of concentrations. In this case, the amount of ddH_2O added to the reaction can be adjusted to account for the changing volume.
4. Preincubate the assembled reactions at 37°C for 5 min (see tip 7).
5. Initiate the reaction by adding 1 μL of 10 × PrimPol stock and mix by pipetting.
6. Incubate the reactions for the desired time point or time course at 37°C (see tip 7).
7. Stop the reaction by adding 10 μL 2 × stop buffer. If using a stock reaction and taking multiple time points, 10 μL of the reaction should be removed and added to 10 μL of 2 × stop buffer.
8. Incubate the quenched reactions at 90°C for 3 min and spin briefly in a microcentrifuge.

9. Load each 20 μL sample onto the pre-run urea–polyacrylamide gel and resolve according to the manufacturers recommendations (see tips 5 and 8). Observe the migration of the bromophenol blue and xylene cyanol dyes to monitor progression of the samples.
10. A FAM-labeled oligonucleotide size marker should be run alongside the samples to allow determination of product sizes.
11. Before imaging, the gel system should be disassembled and the plates thoroughly cleaned with dH_2O and ethanol. Failure to do this will affect the image quality. Note that it is not necessary to remove the gel from the glass plates before scanning.
12. Visualize the gel using an FLA-5100 image reader, or an equivalent imager.
13. The resulting digital image can be analyzed using image analysis software, such as ImageQuant TL, if quantification of primase reaction products is desired.

4.3.4 Tips

1. For optimal resolution, the concentration of acrylamide should be adjusted according to the expected product size. A higher polyacrylamide concentration will resolve smaller oligonucleotide products.
2. Following addition of APS, and again after addition of TEMED, the gel mix should be mixed gently by inverting to avoid aeration and the generation of bubbles.
3. The inclusion of marker dyes in the stop buffer can interfere with the fluorescent signal, if migrating at the same size as the reaction product. In this case, the dyes can be omitted from the stop buffer and run in an empty well to still allow monitoring of sample progression.
4. Using a large sequencing PAGE setup allows much great resolution of reaction products compared to smaller gels. Addition of an aluminium heat-dispersion plate in the set-up can help prevent "smiling" of the samples.
5. Prior to prerunning the gel, and again before loading samples, the wells should be washed 2× using a 1× TBE filled syringe and needle to remove any gel pieces and urea.
6. If assembling a large number of reactions, it is beneficial to make a single reaction stock and aliquot this for each sample, taking into account the different amount of ddH_2O which might be required for each reaction depending on the variable being analyzed. This also minimizes variation between reactions caused by pipetting error.

7. Although a bench-top dry heating block is sufficient for short time courses, an incubator should be used for long time points to prevent evaporation and condensation on the inside of the Eppendorf tube lid.
8. If all of the wells of the gel are not being used, loading of "blank samples," containing 1 × TBE and 1 × stop buffer, into the empty wells can help prevent gel "smiling."

4.3.5 Methods to Reduce Background

Although the primase assay can be used reliably without any clean-up steps to remove unincorporated 6-FAM dNTPs, background signal from the dNTPs is visible lower down the gel. To remove this background and improve the quality of results, a number of DNA precipitation techniques can be implemented. These clean-up steps should be performed following incubation of the reactions, but before the addition of stop buffer (between steps 6 and 7).

We have reliably removed free 6-FAM dNTPs using magnetic streptavidin beads (Roche) (note that the ssDNA template must be biotin labeled to use this method), Oligo Clean & Concentrator columns (Zymo Research), and ethanol precipitation (Fig. 7). Of these techniques, ethanol precipitation, following a standard protocol is the most cost-effective and efficient method. If using this method, note that the pellet will likely not be visible. Use of the Oligo Clean & Concentrator columns (Zymo Research) or equivalent, following the manufacturer's protocol, may be desirable in the interest of time when analyzing a large number of samples. Following precipitation, the DNA should be suspended in 1 × stop buffer and the protocol continued as described earlier.

4.4 Considerations When Performing the Fluorescence-Based Primase Assay

The fluorescent gel-based primase assay described earlier has been successfully used in both published and unpublished primase studies from our group (Kobayashi et al., 2016; Schiavone et al., 2016). However, there are a number of points that must be considered before performing the assay. First, we have used this assay to study both human PrimPol and archaeal replicative primases with success. These primases are able to incorporate the 6-FAM-labeled dNTPs during primer synthesis and extension (Fig. 8). However, it must be noted that 6-FAM-labeled dNTPs are significantly modified in comparison to native or radiolabeled dNTPs and thus care must be taken to ensure that the primase of interest is able to efficiently

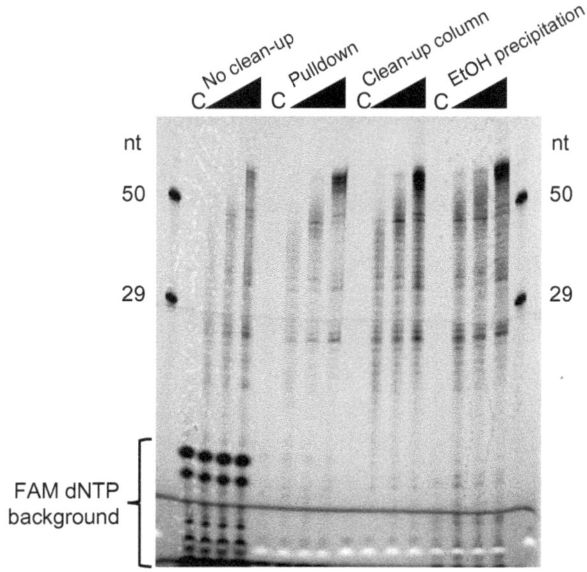

Fig. 7 DNA precipitation methods used to reduce FAM dNTP background. The assay was performed as described in Section 4.3.3 using 250 n*M* PrimPol. The *black triangle* indicates increasing time points of 1, 5, and 10 min. Following incubation, reactions were either immediately quenched with stop buffer (as indicated in Section 4.3.1) (shown on the *left* as "no clean-up") or were subject to DNA precipitation clean-up techniques. "Pull down" reaction samples were quenched with binding-washing buffer (10 m*M* Tris–HCl, pH 7.5, 500 m*M* NaCl, 10 m*M* EDTA) and supplemented with 20 μL streptavidin-coated beads (Roche). Binding was performed for 1 h at 4°C and samples were subsequently washed 3 × 1 mL with binding-washing buffer, before resuspension in 1 × stop buffer. "Clean-up column" reaction samples were bound, washed, and eluted from Oligo Clean & Concentrator columns (Zymo Research) according to the manufacturer's instructions, before resuspension in 1 × stop buffer. "EtOH precipitation" samples were supplemented with 1/10 vol. 3 *M* NaOAc and 3 vol. of 100% EtOH, before incubation for 15 min on ice. Following incubation, samples were spun in a microcentrifuge at top speed for 30 min at 4°C, washed with 70% EtOH, centrifuged again for 15 min, dried, and resuspended in 1 × stop buffer. FAM dNTP background can be seen at bottom of the gel for the "no clean-up" samples, but not in any of the DNA precipitation sample lanes. "C" indicates the no enzyme control. Nucleotide (nt) size markers of 50 and 29 nt are shown on either side of the image.

incorporate these modified nucleotides. Note that PrimPol and replicative archaeal primases preferentially synthesize primers using dNTPs over rNTPs, hence the use of 6-FAM dNTPs here. In the case that the primase of interest synthesizes RNA primers, 6-FAM dNTPs can be substituted for 6-FAM rNTPs, which are also commercially available.

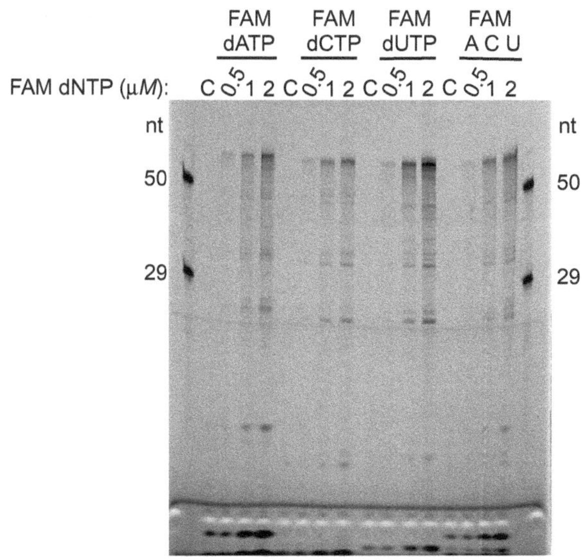

Fig. 8 PrimPol can incorporate 6-FAM dATP, dCTP, and dUTP. Reactions were assembled containing 250 n*M* PrimPol and either 6-FAM dATP, dCTP, dUTP, or all three, at increasing concentrations (0.5, 1, and 2 μ*M*), and performed as outlined in Section 4.3.3 for a single 10 min time point. "C" indicates the no enzyme control. Nucleotide (nt) size markers of 50 and 29 nt are shown on either side of the image.

Second, when using a linear ssDNA template with a free 3′ end, products larger than the template may be observed (Fig. 9). We have determined that this is likely due to formation of a hairpin at the 3′ end of the template, produced by snap-back. Extension of this 3′ end hairpin by the primase consequently produces reaction products much larger than expected and may additionally sequester the enzyme away from performing de novo primer synthesis. This can be avoided by blocking the 3′ end of the template with a 3′ C3 spacer, dideoxynucleotide, or other suitable modification. Alternatively, a circular ssDNA template may be used.

Lastly, this assay is most applicable for the analysis of primase polymerases, such as PrimPol, which are able to synthesize and extend their own primers. When using linear ssDNA templates <100 nt, these enzymes can perform extension up to the end of the template. By analyzing these reaction products, taking into account the length of the oligonucleotide product and template, the primase initiation site can be determined. This is particularly useful when assessing activities such as repriming (Kobayashi et al., 2016; Schiavone et al., 2016). Very short primase reaction products, which are not extended after the initial synthesis, may be difficult to distinguish from

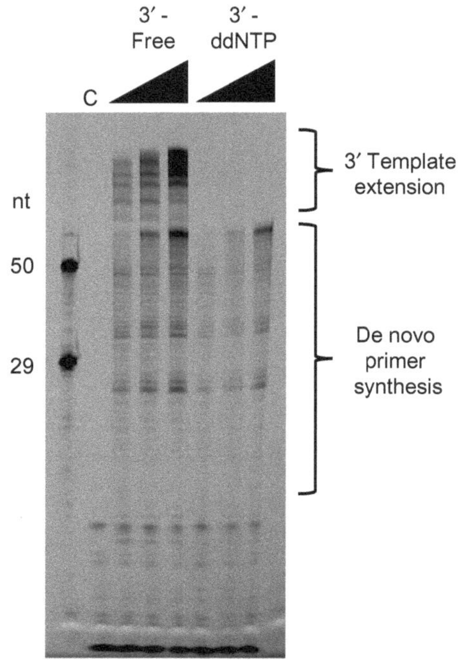

Fig. 9 Comparison of the fluorescence gel-based primase assay on a linear ssDNA template with either a free 3′ end, or a 3′ end blocked with a dideoxynucleotide. Assays were performed as detailed in Section 4.3.3 containing 250 n*M* PrimPol, and a 66 nt template with a free 3′ end or containing a dideoxynucleotide at the 3′ end (sequence in Section 4.3.1). When the 3′ end is free, reaction products larger than the template are observed ("3′-free" sample lanes); however, addition of a 3′ dideoxynucleotide removes these products ("3′-ddNTP" sample lanes), suggesting they are produced by extension of the 3′ end of the template due to template snap-back. The *black triangle* indicates increasing time points of 5, 10, and 15 min. "C" indicates the no enzyme control. Nucleotide (nt) size markers of 50 and 29 nt are shown on the *left* side of the image.

the background given by 6-FAM dNTPs without any additional clean-up steps. However, the short nature of these primers may make them more liable to being washed away during clean-up. A higher concentration of 6-FAM dNTPs may also be needed in assays where the product size is very small, in order to increase the probability of the primase incorporating the 6-FAM dNTP into its primer. Alternatively, a single native dNTP may be omitted and replaced with the equivalent 6-FAM dNTP. This problem can also be overcome by coupling the primase with a processive polymerase, as has been described previously in both fluorescent and radioactive assays (Bianchi et al., 2013; Galal et al., 2012; Keen, Bailey, Jozwiakowski, & Doherty, 2014; Keen, Jozwiakowski, Bailey, Bianchi, & Doherty, 2014).

4.5 Advantages and Limitations

This assay is intended to be used in the place of the classic radioactive gel-based primase assay. By utilizing fluorescence, rather than radioactivity, the assay has a number of major advantages. First, the potential hazards, rigorous safety measures, training, and cost of waste disposal, associated with handling radioactivity are avoided. Second, substitution of radioactivity with fluorescence permits more accurate quantification of reaction products due to the improved linear dynamic range. Radioactive dNTPs also have a short half-life of days or weeks. In contrast, 6-FAM dNTPs can be stored at −20°C for up to 1 year before performance decreases. Consequently, despite their initial cost, fluorescent dNTPs can be a more affordable option if assays are performed over a long period of time. Additionally, fluorescent gels can be immediately and rapidly scanned (~10 min scanning time) following electrophoresis. This avoids the lengthy phosphor screen exposure times required for radioactivity detection, which can often take up to 12 h. Coupled with this, the gel-based nature of the assay allows a large number of samples (up to 20 per gel) to be resolved and imaged at the same time, taking only 2–3 h in total. This potentially makes the technique faster than alternatives, such as the denaturing HPLC primase assay (20 min run time per sample), if a large number of samples are to be analyzed.

Another major advantage of the assay is its similar setup and readout to the traditional radioactive method. This allows the technique to be easily adopted by laboratories used to performing gel-based radioactive primase assays without extensive alterations to the method and equipment, or additional training. Indeed, despite alternatives, such as denaturing HPLC, gel-based primase assays are still most commonly used due to the ease of interpretation and lack of requirement for extensive optimization when changing templates and enzymes.

However, the fluorescence-based primase assay also shares some drawbacks with other qualitative assays. Most notably, similar to the gel-based radioactive and HPLC primase assays, the fluorescence primase assay is not yet amenable to HTS. However, gel-based radioactive assays have previously been used to confirm inhibitors identified from large HTS methods, such as the primase-pyrophosphatase activity assay (Biswas et al., 2013). In these instances, fluorescence could be used to replace radioactivity, in order to confirm hits from HTS, and thus make the approach completely non-radioactive. Nevertheless, it must be noted that the fluorescence does not provide the same level of sensitivity as radioactivity, requiring micromolar, in comparison to nanomolar, concentrations of labeled nucleotides. Lastly, it is possible that some primases may not tolerate the FAM-labeled nucleotides and we have yet to test the assay with FAM-rNTPs.

5. SUMMARY AND CONCLUSION

Since the identification of the first primases in the 1970s, characterization of these enzymes has largely relied upon radioactive gel-based methods. Despite possessing excellent sensitivity and generating valuable qualitative primer synthesis information, these assays have major disadvantages, primarily due to their use of radioactivity. Consequently, in the last two decades, a number of alternative nonradioactive primase assays have been developed. In most cases, these techniques have focused on enabling HTS of potential primase inhibitor compounds. Although these assays suit this purpose well, they generally lack the capability to provide qualitative information about reaction products and often require large-scale optimization prior to being performed. As a consequence, in spite of its time-consuming and hazardous nature, the gel-based radioactive primase assay remains the go-to option for the identification and basic characterization of primases.

In this chapter, we have described how fluorescence provides a reliable alternative to radioactivity in the traditional gel-based primase assay, without requiring significant changes to the procedure or setup. By replacing radioactive dNTPs with 6-FAM labeled dNTPs, all the disadvantages associated with radioactive work are eliminated. Furthermore, this substitution also offers clear advantages in speed over the traditional technique. We have used this fluorescent primase assay in published studies of eukaryotic primase-polymerase, PrimPol, thereby highlighting the general applicability of this technique in primase characterization. This assay can be used in place of radioactive techniques to characterize basic primase activity, identify initiation sites, assess the impact of binding partners and accessory proteins, determine the effect of different reaction conditions, and to confirm primase inhibitor compounds identified through HTS. In summary, the fluorescent gel-based primase assay described here offers a safer and faster alternative to the classic, but still widely used, radioactive assay.

ACKNOWLEDGMENTS

A.J.D.'s laboratory is supported by grants from the Biotechnology and Biological Sciences Research Council (BBSRC: BB/H019723/1, BB/M008800/1, and BB/M004236/1). T.A.G. was supported by a University of Sussex Ph.D. studentship.

Conflict of interest statement. None declared.

REFERENCES

Ahn, S. J., Costa, J., & Emanuel, J. R. (1996). PicoGreen quantitation of DNA: Effective evaluation of samples pre-or Psost-PCR. *Nucleic Acids Research*, *24*(13), 2623–2625. https://doi.org/10.1093/nar/24.13.2623.

Arai, K., & Kornberg, A. (1979). A general priming system employing only dnaB protein and primase for DNA replication. *Proceedings of the National Academy of Sciences*, *76*(9), 4308–4312.

Bernstein, J. A., & Richardson, C. C. (1988). Purification of the 56-kDa component of the bacteriophage T7 primase/helicase and characterization of its nucleoside 5′-triphosphatase activity. *Journal of Biological Chemistry*, *263*(29), 14891–14899.

Bianchi, J., Rudd, S. G., Jozwiakowski, S. K., Bailey, L. J., Soura, V., Taylor, E., … Doherty, A. J. (2013). PrimPol bypasses UV photoproducts during eukaryotic chromosomal DNA replication. *Molecular Cell*, *52*(4), 566–573. https://doi.org/10.1016/j.molcel.2013.10.035.

Biswas, E. E., Joseph, P. E., & Biswas, S. B. (1987). Yeast DNA primase is encoded by a 59-kilodalton polypeptide: Purification and immunochemical characterization. *Biochemistry*, *26*(17), 5377–5382. https://doi.org/10.1021/bi00391a024.

Biswas, T., Resto-Roldán, E., Sawyer, S. K., Artsimovitch, I., & Tsodikov, O. V. (2013). A novel non-radioactive primase–pyrophosphatase activity assay and its application to the discovery of inhibitors of Mycobacterium tuberculosis primase DnaG. *Nucleic Acids Research*, *41*, e56. https://doi.org/10.1093/nar/gks1292 gks1292.

Bouché, J. P., Rowen, L., & Kornberg, A. (1978). The RNA primer synthesized by primase to initiate phage G4 DNA replication. *Journal of Biological Chemistry*, *253*(3), 765–769.

Bouché, J. P., Zechel, K., & Kornberg, A. (1975). dnaG gene product, a rifampicin-resistant RNA polymerase, initiates the conversion of a single-stranded coliphage DNA to its duplex replicative form. *Journal of Biological Chemistry*, *250*(15), 5995–6001.

Conaway, R. C., & Lehman, I. R. (1982a). A DNA primase activity associated with DNA polymerase alpha from *Drosophila melanogaster* embryos. *Proceedings of the National Academy of Sciences*, *79*(8), 2523–2527.

Conaway, R. C., & Lehman, I. R. (1982b). Synthesis by the DNA primase of *Drosophila melanogaster* of a primer with a unique chain length. *Proceedings of the National Academy of Sciences*, *79*(15), 4585–4588.

De Silva, F. S., Paran, N., & Moss, B. (2009). Products and substrate/template usage of vaccinia virus DNA primase. *Virology*, *383*(1), 136–141. https://doi.org/10.1016/j.virol.2008.10.008.

Frick, D. N., & Richardson, C. C. (1999). Interaction of bacteriophage T7 Gene 4 primase with its template recognition site. *Journal of Biological Chemistry*, *274*(50), 35889–35898. https://doi.org/10.1074/jbc.274.50.35889.

Frick, D. N., & Richardson, C. C. (2001). DNA Primases. *Annual Review of Biochemistry*, *70*(1), 39–80. https://doi.org/10.1146/annurev.biochem.70.1.39.

Galal, W. C., Pan, M., Kelman, Z., & Hurwitz, J. (2012). Characterization of DNA primase complex isolated from the archaeon, Thermococcus kodakaraensis. *Journal of Biological Chemistry*, *287*(20), 16209–16219. https://doi.org/10.1074/jbc.M111.338145.

Guilliam, T. A., & Doherty, A. J. (2017). PrimPol—Prime time to reprime. *Genes*, *8*(1), 20. https://doi.org/10.3390/genes8010020.

Guilliam, T. A., Keen, B. A., Brissett, N. C., & Doherty, A. J. (2015). Primase-polymerases are a functionally diverse superfamily of replication and repair enzymes. *Nucleic Acids Research*, *43*(14), 6651–6664. https://doi.org/10.1093/nar/gkv625.

Hinkle, D. C., & Richardson, C. C. (1975). Bacteriophage T7 deoxyribonucleic acid replication in vitro. Purification and properties of the gene 4 protein of bacteriophage T7. *Journal of Biological Chemistry*, *250*(14), 5523–5529.

Keen, B. A., Bailey, L. J., Jozwiakowski, S. K., & Doherty, A. J. (2014b). Human PrimPol mutation associated with high myopia has a DNA replication defect. *Nucleic Acids Research*, *42*(19), 12102–12111. https://doi.org/10.1093/nar/gku879.

Keen, B. A., Jozwiakowski, S. K., Bailey, L. J., Bianchi, J., & Doherty, A. J. (2014a). Molecular dissection of the domain architecture and catalytic activities of human PrimPol. *Nucleic Acids Research*, *42*(9), 5830–5845. https://doi.org/10.1093/nar/gku214.

Kirk, B. W., & Kuchta, R. D. (1999). Human DNA primase: Anion inhibition, manganese stimulation, and their effects on in vitro start-site selection. *Biochemistry*, *38*(31), 10126–10134. https://doi.org/10.1021/bi990351u.

Kobayashi, K., Guilliam, T. A., Tsuda, M., Yamamoto, J., Bailey, L. J., Iwai, S., … Hirota, K. (2016). Repriming by PrimPol is critical for DNA replication restart downstream of lesions and chain-terminating nucleosides. *Cell Cycle (Georgetown, Texas)*, *15*(15), 1997–2008. https://doi.org/10.1080/15384101.2016.1191711.

Koepsell, S., Bastola, D., Hinrichs, S. H., & Griep, M. A. (2004). Thermally denaturing high-performance liquid chromatography analysis of primase activity. *Analytical Biochemistry*, *332*(2), 330–336. https://doi.org/10.1016/j.ab.2004.06.019.

Koepsell, S. A., Hanson, S., Hinrichs, S. H., & Griep, M. A. (2005). Fluorometric assay for bacterial primases. *Analytical Biochemistry*, *339*(2), 353–355. https://doi.org/10.1016/j.ab.2004.12.004.

Krevolin, M. D., & Calendar, R. (1985). The replication of bacteriophage P4 DNA in vitro. *Journal of Molecular Biology*, *182*(4), 509–517. https://doi.org/10.1016/0022-2836(85)90237-2.

Lanka, E., Scherzinger, E., Günther, E., & Schuster, H. (1979). A DNA primase specified by I-like plasmids. *Proceedings of the National Academy of Sciences*, *76*(8), 3632–3636.

Lipps, G., Weinzierl, A. O., von Scheven, G., Buchen, C., & Cramer, P. (2004). Structure of a bifunctional DNA primase-polymerase. *Nature Structural & Molecular Biology*, *11*(2), 157–162. https://doi.org/10.1038/nsmb723.

Morris, C. F., Sinha, N. K., & Alberts, B. M. (1975). Reconstruction of bacteriophage T4 DNA replication apparatus from purified components: Rolling circle replication following de novo chain initiation on a single-stranded circular DNA template. *Proceedings of the National Academy of Sciences*, *72*(12), 4800–4804.

Romano, L. J., & Richardson, C. C. (1979). Characterization of the ribonucleic acid primers and the deoxyribonucleic acid product synthesized by the DNA polymerase and gene 4 protein of bacteriophage T7. *Journal of Biological Chemistry*, *254*(20), 10483–10489.

Rowen, L., & Kornberg, A. (1978). Primase, the dnaG protein of *Escherichia coli*. An enzyme which starts DNA chains. *Journal of Biological Chemistry*, *253*(3), 758–764.

Scherzinger, E., Lanka, E., Morelli, G., Seiffert, D., & Yuki, A. (1977). Bacteriophage-T7-induced DNA-priming protein. *European Journal of Biochemistry*, *72*(3), 543–558. https://doi.org/10.1111/j.1432-1033.1977.tb11278.x.

Scherzinger, E., & Litfin, F. (1974). Initiation of the replication of single-stranded DNA by concerted action of phage T7 RNA and DNA polymerases. *European Journal of Biochemistry*, *46*(1), 179–180. https://doi.org/10.1111/j.1432-1033.1974.tb03610.x.

Schiavone, D., Jozwiakowski, S. K., Romanello, M., Guilbaud, G., Guilliam, T. A., Bailey, L. J., … Doherty, A. J. (2016). PrimPol is required for replicative tolerance of G Quadruplexes in vertebrate cells. *Molecular Cell*, *61*(1), 161–169. https://doi.org/10.1016/j.molcel.2015.10.038.

Silva, F. S. D., Lewis, W., Berglund, P., Koonin, E. V., & Moss, B. (2007). Poxvirus DNA primase. *Proceedings of the National Academy of Sciences*, *104*(47), 18724–18729. https://doi.org/10.1073/pnas.0709276104.

Strätling, W., & Knippers, R. (1973). Function and purification of Gene 4 protein of phage T7. *Nature*, *245*(5422), 195–197. https://doi.org/10.1038/245195a0.

Van Kirk, C., Feinberg, L., Robertson, D., Freeman, W., & Vrana, K. (2010). *Phosphorimager. In John Wiley & Sons, Ltd (Ed.), Encyclopedia of life sciences*. Chichester, UK: John Wiley & Sons, Ltd. Retrieved from http://doi.wiley.com/10.1002/9780470015902.a0002973.pub2.

Wolfson, J., & Dressler, D. (1972). Regions of single-stranded DNA in the growing points of replicating bacteriophage T7 chromosomes. *Proceedings of the National Academy of Sciences*, *69*(9), 2682–2686.

Wu, C. A., Zechner, E. L., & Marians, K. J. (1992). Coordinated leading- and lagging-strand synthesis at the Escherichia coli DNA replication fork. I. Multiple effectors act to modulate Okazaki fragment size. *Journal of Biological Chemistry*, *267*(6), 4030–4044.

Zhang, Y., Yang, F., Kao, Y.-C., Kurilla, M. G., Pompliano, D. L., & Dicker, I. B. (2002). Homogenous assays for *Escherichia coli* DnaB-stimulated DnaG primase and DnaB helicase and their use in screening for chemical inhibitors. *Analytical Biochemistry*, *304*(2), 174–179. https://doi.org/10.1006/abio.2002.5627.

Zuo, Z., Rodgers, C. J., Mikheikin, A. L., & Trakselis, M. A. (2010). Characterization of a functional DnaG-type primase in archaea: Implications for a dual-primase system. *Journal of Molecular Biology*, *397*(3), 664–676. https://doi.org/10.1016/j.jmb.2010.01.057.

CHAPTER FIFTEEN

Electrical Probes of DNA-Binding Proteins

Jacqueline K. Barton[1], Phillip L. Bartels, Yingxin Deng, Elizabeth O'Brien
California Institute of Technology, Pasadena, CA, United States
[1]Corresponding author: e-mail address: jkbarton@caltech.edu

Contents

Abstract

A DNA electrochemistry platform has been developed to probe proteins bound to DNA electrically. Here gold electrodes are modified with thiol-modified DNA, and DNA charge transport chemistry is used to probe DNA binding and enzymatic reaction both with redox-silent and redox-active proteins. For redox-active proteins, the electrochemistry permits the determination of redox potentials in the DNA-bound form, where

Methods in Enzymology, Volume 591
ISSN 0076-6879
http://dx.doi.org/10.1016/bs.mie.2017.03.024

comparisons to DNA-free potentials can be made using graphite electrodes without DNA modification. Importantly, electrochemistry on the DNA-modified electrodes facilitates reaction under aqueous, physiological conditions with a sensitive electrical measurement of binding and activity.

1. INTRODUCTION

The fundamental properties of DNA charge transport (DNA CT), particularly the picosecond timescale over which charge migrates (O'Neill, Becker, Wan, Barton, & Zewail, 2003) and the exquisite sensitivity of DNA CT to perturbations in the base pair π-stacking interactions (Arnold, Grodick, & Barton, 2016), facilitate the use of DNA electrochemistry in detecting the activity of many different DNA-binding proteins, as well as in sensing DNA damage (Fig. 1) (Boal et al., 2009, 2005; Boon et al., 2002; DeRosa et al., 2005; Gorodetsky, Ebrahim, et al., 2008; Grodick, Segal, Zwang, & Barton, 2014; Mui, Fuss, Ishida, Tainer, & Barton, 2011; Slinker et al., 2011). Here, we describe the characteristics, protocols, and platforms, we have used to detect and monitor these DNA-binding proteins electrically. This detection sensitively depends on an electrochemical signal readout from either a redox-active moiety in the DNA-binding protein (Boal et al., 2009, 2005; DeRosa et al., 2005; Grodick et al., 2014; Mui et al., 2011) or from a DNA-intercalating redox probe bound to the DNA electrode (Boon et al., 2002; Gorodetsky, Ebrahim, et al., 2008; Slinker et al., 2011). The DNA-mediated electrode platform can sense a DNA-binding protein because the protein kinks the DNA, interfering with DNA CT (Gorodetsky, Ebrahim, et al., 2008), or perhaps because of the protein binding and cutting DNA attached to the electrode (Boon et al., 2002; Slinker et al., 2011). The platform can even detect electrically the unwinding of a duplex substrate by a helicase enzyme (Grodick et al., 2014; Mui et al., 2011). DNA-modified electrodes thus serve as substrates and templates for a wide variety of DNA-binding proteins. Indeed, the limits of protein detection can be nanomolar concentrations (Boon et al., 2002; Gorodetsky, Ebrahim, et al., 2008; Slinker et al., 2011) and depend on protein binding affinity more so than any property of DNA CT.

As an illustration, consider the transcription factor TATA-binding protein (TBP), which is responsible for activation of several different eukaryotic genes (Kornberg, 2007). TBP kinks duplex DNA approximately

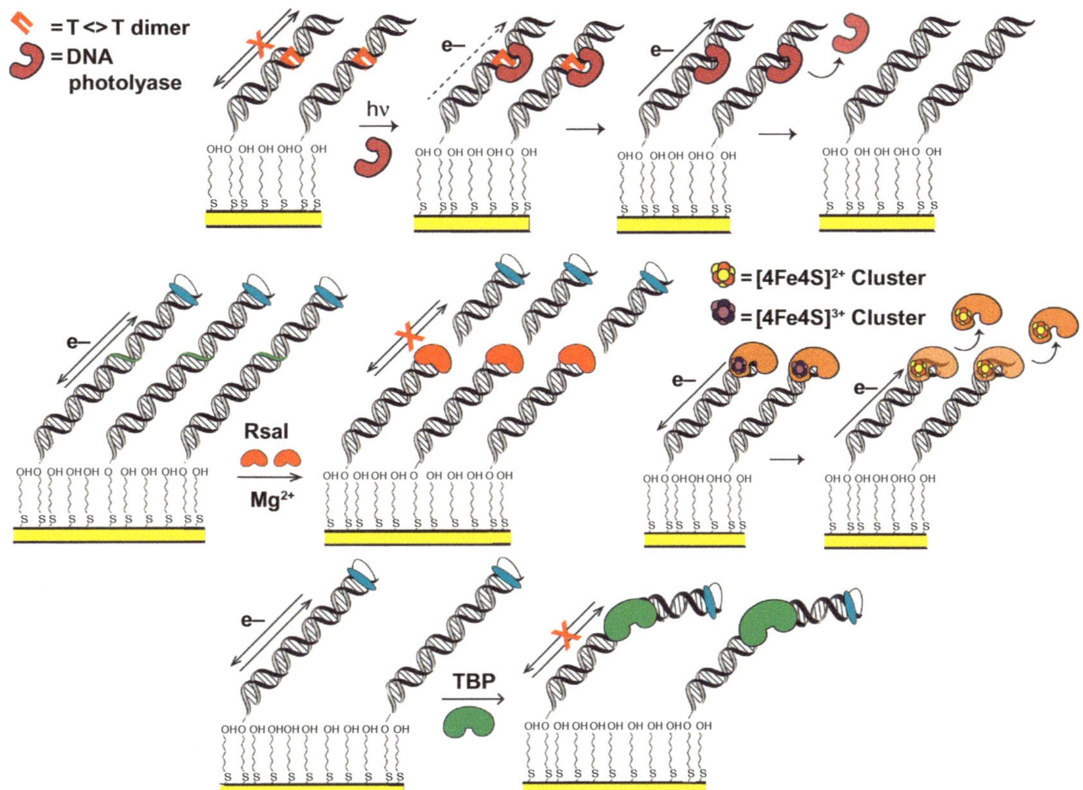

Fig. 1 Electrochemical monitoring of DNA-binding protein activity on DNA-modified electrodes. (*Top*) DNA photolyase binds and repairs a thymine–thymine dimer on a DNA-modified electrode, restoring DNA CT and producing a signal from the flavin cofactor, through repaired DNA (DeRosa, Sancar, & Barton, 2005). (*Center left*) *Rsa*I restriction enzyme cuts duplex DNA, removing covalently attached redox probe. Signal disappears after wash of surface, indicating that *Rsa*I binding and cutting of DNA at recognition site occurs (Slinker, Muren, Renfrew, & Barton, 2011). (*Center right*) A bound [4Fe4S] enzyme is oxidized from the resting $[4Fe4S]^{2+}$ state to the tightly bound $[4Fe4S]^{3+}$ state through DNA CT; it can then be reduced from the tightly bound $[4Fe4S]^{3+}$ state to the more weakly binding resting $[4Fe4S]^{2+}$ state through DNA CT, promoting dissociation (Boal et al., 2009). (*Bottom*) TBP binding kinks duplex DNA, attenuating CT, and diminishing signal from a DNA-intercalating, covalently attached redox probe (Gorodetsky, Ebrahim, & Barton, 2008).

90 degree when bound, significantly perturbing the π-stacking interactions of the DNA duplex (Boon et al., 2002; Gorodetsky, Ebrahim, et al., 2008) (Fig. 1). This interaction perturbs DNA CT, and TBP binding is therefore detectable on DNA-modified electrodes (Gorodetsky, Ebrahim, et al., 2008). It was shown that when TBP binds and kinks duplex DNA containing the TATA box recognition sequence, CT attenuation occurs immediately. The DNA substrate in this assay contained the TBP recognition sequence, as well as a covalent, DNA-intercalating Nile Blue redox probe tethered at the distal end of the DNA duplex from the electrode surface. With nanomolar concentrations of TBP bound to the substrate, the DNA is kinked and the Nile Blue redox signal associated with DNA CT between the electrode surface and the redox probe is lost. The signal, moreover, could be easily regenerated upon washing the surface with KCl to remove TBP. This signal attenuation does not, importantly, occur when other proteins, which do not specifically bind the TATA box site or kink the substrate, are incubated on the electrode surface (Gorodetsky, Ebrahim, et al., 2008), nor when the 5′-TATA-3′ recognition sequence is not available.

DNA-intercalating redox probes can also be used to detect restriction enzyme activity upon binding specific recognition sequences on a DNA duplex (Fig. 1) (Boon et al., 2002; Slinker et al., 2011). Restriction enzymes *Rsa*I and *Pvu*II, for example, were each incubated on a DNA-modified electrode surface with a duplex substrate containing the respective restriction enzyme recognition sequence and a distal, covalently bound redox probe. These enzymes were given any necessary catalytic metal ions to perform their native function (Slinker et al., 2011), and they subsequently bound and cut the DNA substrate at the recognition site. This site was engineered in between the intercalated probe and the electrode surface, so the DNA no longer possessed a redox moiety once the restriction enzyme had cut the duplex at the appropriate site. This again did not occur when the substrate DNA lacked a recognition sequence, demonstrating that the observed effect depended on the reaction assayed. The restriction enzyme assay described is, additionally, adaptable to both the single electrode (Boon et al., 2002) and multiplexed chip (Slinker et al., 2011) setup.

In addition to detecting general protein binding and nuclease activity, these platforms have also facilitated the study of DNA-bound redox processes in biology. The repair of DNA by flavoenzyme DNA photolyase (*Escherichia coli*), for example, can be monitored in real time on DNA-modified electrodes (DeRosa et al., 2005). DNA photolyase is an enzyme that repairs cyclobutane thymine dimer (T<>T) lesions which attenuate DNA CT and are a result of photoinduced [2+2] cycloaddition between adjacent thymine bases. Photolyase repairs these lesions using a reductive catalytic cycle (Sancar, 2003), driven by photoexcitation of the flavin cofactor within the enzyme. The T<>T lesion is flipped out of the DNA helix in this reaction, and the flavin cofactor in photolyase initiates a redox reaction to reverse the damage. When the repaired thymine bases are then flipped back in to the DNA duplex, the substrate is able to perform DNA CT, generating a redox signal from the flavoenzyme (Fig. 1). Initially, no signal is observed on DNA-modified electrodes from photolyase in the presence of T<>T damaged duplex DNA. When the surface was irradiated with blue light, however, activating photolyase repair, a reversible redox signal at 40 mV vs NHE appears on the electrode surface (DeRosa et al., 2005). This signal potential is within the expected range for photolyase, and it appears only after the T<>T lesion has been repaired. The redox signal is, moreover, attenuated when an abasic site is present in the duplex sequence between the gold electrode surface and the T<>T site, demonstrating that the electron transfer reaction is DNA mediated. This redox activity is observable using different electrochemical techniques, such as cyclic voltammetry (CV) and square wave voltammetry (SWV), and it is enhanced upon longer exposure times to blue light; more repaired lesions yield a larger electrochemical signal. These electrodes thus allow for observation of the redox activity involved in several different DNA-bound biochemical reactions.

The surprising discovery of a [4Fe4S] cluster in the base excision repair (BER) glycosylase Endonuclease III (*E. coli*) (Cunningham et al., 1989) led to the investigation of several important questions about the role of these cofactors in DNA repair: Are [4Fe4S] clusters present in other DNA repair enzymes? Do they serve a structural or a biochemical purpose? The discovery of this cluster in Endonuclease III, for example, led to the prediction that it was also present in the homologous BER glycosylase MutY (Michaels, Pham, Nghiem, Cruz, & Miller, 1990), which, similar to Endonuclease III, catalyzes the removal of oxidative damage products from genomic DNA (Kim & Wilson, 2012). The [4Fe4S] cluster would eventually be shown to exist in several BER enzymes, including Endonuclease III, MutY (Guan et al., 1998), and uracil DNA glycosylase (UDG) in *Anisocentropus fulgidus* (Hinks et al., 2002). Several bioinformatics, structural, and spectroscopic studies contributed to these discoveries (Fu, O'Handley, Cunningham, & Johnson, 1992; Guan et al., 1998; Thayer, Ahern, Xing,

Cunningham, & Tainer, 1995). The question of what role the [4Fe4S] clusters played, however, was less straightforward. These clusters are often associated with biological redox chemistry (Rees & Howard, 2003), yet early studies were unable to demonstrate a redox role for these clusters. The DNA-modified electrode platforms developed in our laboratory for protein detection proved instrumental in deciphering and demonstrating the redox chemistry performed by these clusters during DNA repair and represented a completely new tool in characterizing the redox chemistry of these DNA-binding proteins (Fig. 1) (Boal et al., 2009, 2005; Grodick et al., 2014; Mui et al., 2011).

2. DNA-MODIFIED ELECTRODES FOR ELECTROCHEMISTRY

2.1 Designs and Optimization of DNA Electrochemistry Substrates/Monolayers

The adaptability of the platform to various DNA substrates is in part why several different enzymatic reactions can be studied using DNA electrochemistry. Optimal DNA substrates depend on the protein size and enzymatic function and must be suited for appending onto the DNA electrode surface. An alkanethiol moiety is generally tethered to one end of one strand comprising the final duplex oligonucleotide. This can be readily performed using standard phosphoramidite chemistry, or a thiol-modified oligonucleotide can be ordered from a company such as Integrated DNA Technologies (IDT). This moiety is instrumental in attaching the DNA to the electrode, as a covalent Au-thiol bond will form and give rise to a self-assembling DNA monolayer on the working electrode surface (Kelley et al., 1998) (Fig. 2). Pyrene linkers for DNA modification of graphite electrodes can be appended to the end of a DNA substrate in a similar manner (Gorodetsky, Boal, & Barton, 2006; Gorodetsky, Dietrich, et al., 2008; Gorodetsky, Ebrahim, et al., 2008). Special DNA modifications are commercially available as phosphoramidites from companies such as Glen Research and can be easily integrated into an oligonucleotide sequence on programmable devices such as the Applied Biosystems 3400 DNA Synthesizer (Boal et al., 2009, 2005; Grodick et al., 2014; Mui et al., 2011; Pheeney, Arnold, Grodick, & Barton, 2013; Slinker, Muren, Gorodetsky, & Barton, 2010; Slinker et al., 2011).

In addition to ensuring that a DNA substrate is modified for attachment to an electrode surface, the oligonucleotide sequence and design is important

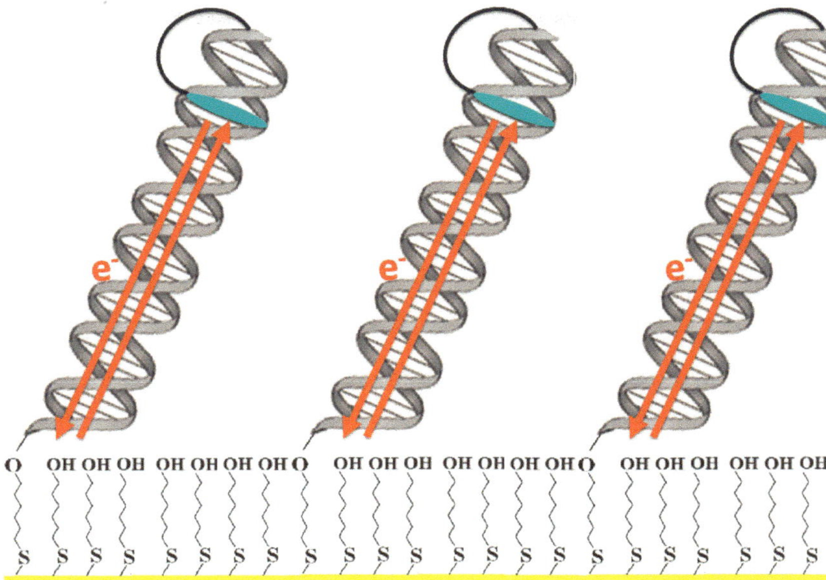

Fig. 2 Electrochemical monitoring of DNA-mediated charge transport processes. In a typical setup, alkanethiol-modified DNA is annealed to its complement and allowed to form a self-assembled monolayer on a gold electrode. Gaps in the Au surface are filled in with 6-mercapto-1-hexanol, passivating the surface, and electrochemistry is carried out in a buffered solution. Redox-active probes, such as the intercalator Nile Blue, can be covalently tethered to one end of the DNA, or simply bound noncovalently. The DNA duplex then serves as a bridge for electron transfer between the probe and the gold electrode. Notably, charge transport through the DNA is very rapid, and electron transfer rates in this system are limited by tunneling through the alkanethiol linker.

for monitoring the desired redox reaction. The most important component of a DNA substrate is the presence of a stable duplex segment, generally 15–40 base pairs in length, though DNA CT through up to 100 base pairs (34 nm) has been observed electrochemically (Slinker et al., 2011). The duplex should contain at least 50% GC pairs, which is easily verified using the OligoAnalyzer tool on the IDT website. This GC content will prevent melting of a duplex on the electrode surface. A 5′- or 3′-ssDNA overhang, generally 3–15 nucleotides in length (Grodick et al., 2014; Mui et al., 2011), can also be engineered onto the end of the oligonucleotide extending into the electrolyte solution.

DNA monolayers on electrodes can additionally be formed with high or low duplex DNA substrate density, optimized for the size, and binding properties of the enzyme assayed (Pheeney et al., 2013). Some examples of the different densities of monolayers that can be formed on Au electrodes are shown in Fig. 3. Larger proteins, for example, may require low-density monolayers to access the substrate. They may also require a longer duplex sequence or single-stranded DNA (ssDNA) overhang segment to accommodate a larger DNA footprint. Finally, oligonucleotide sequences with mismatches, apurinic sites, or even oxidative lesions such as 8-oxo-guanine, the target lesion of MutY, can be engineered into a substrate duplex. This incorporation is achieved readily with phosphoramidite chemistry; many of these special sequences can alternatively be ordered from IDT.

With respect to DNA-modified gold electrodes, we have worked with several platforms over time, each of which has its own particular uses. The overall strategy in forming DNA monolayers is the same in all cases, but each platform has distinct requirements in its preparation. The platform developed by our laboratory is unique because we modify our electrode surfaces with duplex DNA substrates, as opposed to ssDNA substrates. ssDNA adheres to and passivates the gold electrode, making the surface very heterogeneous, and precluding observation of a DNA CT-mediated redox signal (Pheeney et al., 2013). Later, we describe the procedures for DNA film preparation on three devices: the 16-electrode multiplexed chip, the standard gold rod electrode, and a gold on mica atomic force microscopy (AFM) surface adapted to fit a custom electrochemical cell (Fig. 4).

2.2 Preparing a Self-Assembled Monolayer for DNA Electrochemistry

2.2.1 Sixteen-Electrode Chip Setup (Pheeney et al., 2013)

Notes:

This procedure takes ~2 days, with an overnight incubation step.

Incubate DNA monolayer 21–24 h for best results.

Much of the material used for this setup is custom-made, but the monolayer formation protocol is adaptable to different Au electrode surfaces, for example, the single Au on mica surface and rod electrode setup.

Thiol-modified ssDNA substrates should be re-reduced with dithiothreitol (DTT)/Cleland's Reagent and repurified after 2–3 weeks of storage at −20°C in the reduced form.

The monolayer needs to be incubated in a moist environment; a pipette box with water in the bottom works well. The porous surface, raised from the water at the bottom, facilitates incubation of the electrode on a raised platform.

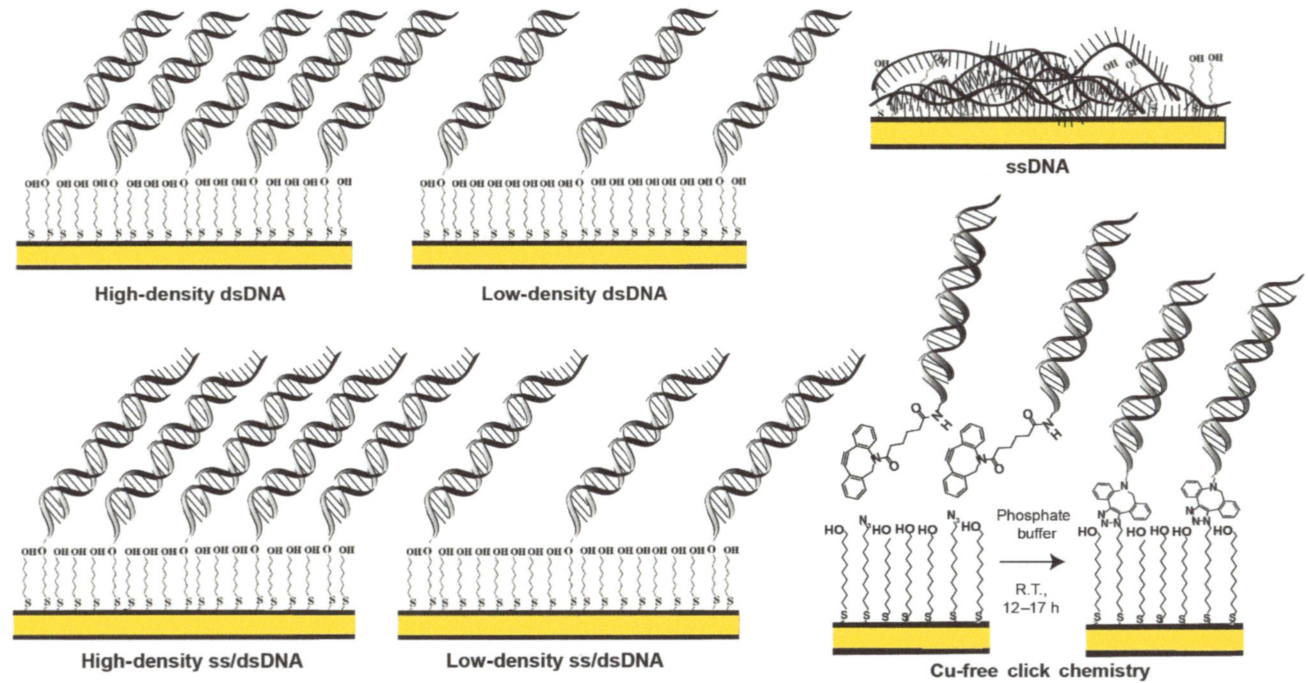

Fig. 3 Different DNA monolayer morphologies formed on DNA-modified Au electrodes. When duplex DNA is incubated with Mg^{2+} on an Au surface (*yellow*), the substrate forms a high-density monolayer of duplex DNA (*top left*). When incubated on Au in the absence of Mg^{2+} a low-density duplex DNA monolayer results. DNA containing a single-stranded overhang segment at the interface of DNA monolayer and electrolyte can also be used to form high-density or low-density monolayers for assaying proteins with a preferred primed end substrate (*bottom left*). When single-stranded DNA is incubated on the Au electrode, the substrate adheres to the surface and passivates the Au, precluding observation of a redox signal (*top right*). Finally, Cu-free click chemistry can be used to form a DNA monolayer on an Au electrode surface (*bottom right*). Azide-terminated alkanethiol-modified Au electrode is incubated in 1:1 mix of mercaptoundecanol and 1-azidoundecane-11-thiol in ethanol for about 4 h. 50 μ*M* DBCO-modified dsDNA in DNA phosphate buffer is incubated with modified Au electrodes for 12–17 h to let the cyclooctyne-based copper-free click reaction proceed. DBCO-modified DNA clicks only to the azide terminal groups, so that the binding density depends on the initial azide content. These monolayers all serve as useful conditions or controls when characterizing redox activity of a DNA-binding enzyme.

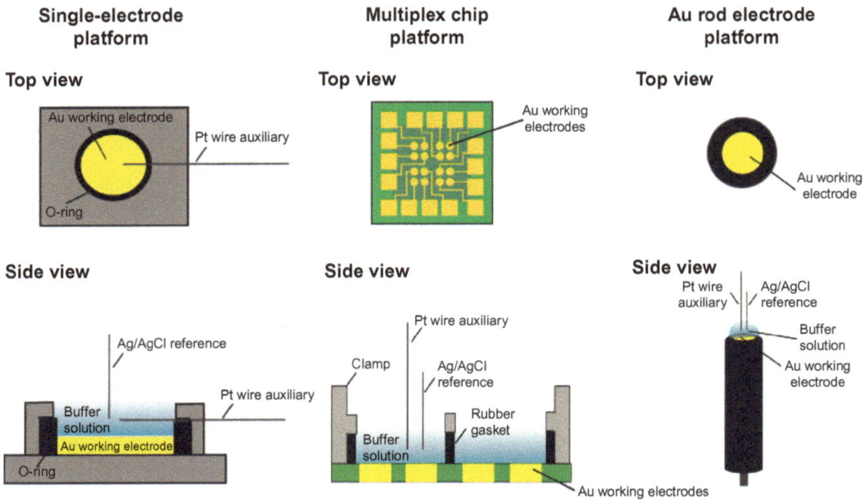

Fig. 4 Different platforms for DNA electrochemistry. Single Au electrodes can be set up on either an Au on mica surface (*left*) or using a rod electrode (*right*). A multiplex platform (*center*) (Pheeney et al., 2013; Slinker et al., 2010) with 16-electrodes separated into four quadrants can also be used to assay multiple DNA substrates on a single surface, with replicates for each condition. Platforms are shown from the *top* (*above*) and from the side (*below*) with components of the setup.

Solutions and Reagents:

1. Sixteen-electrode multiplex chip
2. Buna-N rubber gaskets, plastic clamps for setup
3. Isopropanol, Acetone, MQ water
4. 1 *M* $MgCl_2$
5. 6-Mercapto-1-hexanol (stored under Argon, 100 m*M* stock)
6. Purified, annealed thiol-modified dsDNA substrate/thiol-modified ssDNA for control

Instruments and Supplies:

1. Sonication bath (Branson Ultrasonic)
2. UV ozone cleaner
3. Small screwdriver for chip assembly
4. Chip Incubation Box
5. Argon Gun
6. Ag/AgCl Gel Tip Reference Electrode
7. Platinum Wire

Buffer Conditions:

Thiol-modified dsDNA: 5 m*M* sodium phosphate, pH 7.0, 50 m*M* NaCl

100 m*M* 6-Mercapto-1-hexanol: 5 m*M* phosphate, pH 7.0, 50 m*M* NaCl, 5% glycerol

Electrochemistry buffer (chip washing): 5 m*M* sodium phosphate, pH 7.0, 50 m*M* NaCl, 5% glycerol

TBP buffer (chip washing): 5 m*M* sodium phosphate, pH 7.0, 50 m*M* NaCl, 5% glycerol, 4 m*M* $MgCl_2$, 4 m*M* spermidine

Procedure:

1. Retrieve annealed, thiol-modified dsDNA stock and make the desired stock for the monolayer (dilute 50% to 25 μ*M* with DNA storage buffer, 5 m*M* sodium phosphate, pH 7.0, 50 m*M* NaCl). Final volumes are approximately 20–25 μL in each multiplex electrode quadrant. The prepared substrate stock should be prepared with 10%–20% more volume of DNA solution than will be used in electrochemistry experiment.
   - **(a)** High-density duplex: 25 μ*M* dsDNA, 0.1 *M* $MgCl_2$. For $MgCl_2$ addition, use the 1 *M* $MgCl_2$ stock (American Bioanalytical) and add directly to the 25 μ*M* dsDNA stock to a final concentration of 0.1 *M*.
   - **(b)** Low-density duplex: 25 μ*M* dsDNA
   - **(c)** ssDNA control: Dilute thiol-modified ssDNA stock (approximately 150–800 μ*M* stock) fourfold into 5 m*M* P_i, pH 7.0, 50 m*M* NaCl. Add 1 *M* $MgCl_2$ stock to a final concentration of 0.1 *M*.
2. Allow DNA stocks to thaw from storage at −20°C vortex, centrifuge, and prepare high-density or low-density duplex DNA.
3. Pour deionized water into a sonication bath (Branson Ultrasonic size and model is sufficient) before cleaning chip, clamp, gasket.
4. Place chip in a beaker alone with tweezers, and place the clamp and gasket into a separate beaker with tweezers.
5. Wash chip in one beaker and clamp/gasket in a separate beaker in the sonication bath with the following four wash cycles:
   - **(a)** Chip: 3 washes of 10–20 mL acetone, 1 wash of 10–20 mL 100% isopropanol
   - **(b)** Clamp and gasket: 1 wash of 40–60 mL 50% isopropanol in MQ water, 3 washes of 40–60 mL MQ water
6. Dry the chip thoroughly with an argon gun and place in a UV ozone cleaner. Set the ozone cleaner to 10–20 min cleaning time, depending

on how long the chip has been stored in the hood under argon. Longer ozone cleaning times may be necessary for chips that have been stored outside a clean room environment for longer periods of time.

7. Dry the clamp and gasket thoroughly with the argon gun. Place them on clean surface, such as a clean room wipe or paper towel.
8. When the ozone cleaning cycle has finished, retrieve the chip with tweezers and set it on the center of the platform setup.
9. Align gasket first, then clamp on top of the chip. When the alignment is satisfactory, use the small screwdriver to fasten the setup in place. Tighten the screws thoroughly to avoid leakage of the DNA substrates between quadrants.
10. Deposit 20–25 μL of each dsDNA substrate for monolayers from the prepared stock into the four quadrants. Avoid mixing the stock solutions or mixed monolayers will result.
11. When all monolayers are deposited, cover the top of the clamp with Parafilm and place the chip setup in the incubator box. Incubate the monolayers for 21–24 h.
12. After the monolayers have incubated, wash the electrodes with 20–25 μL volume per quadrant of DNA electrochemistry buffer (5 m*M* sodium phosphate, pH 7.0, 50 m*M* NaCl, 5% glycerol), five cycles through all four quadrants on the chip.
13. Passivate the electrode surface with 1 m*M* 6-mercapto-1-hexanol, a 100-fold dilution of the 100 m*M* stock in DNA electrochemistry buffer. Wash the electrode quadrants in the same manner as performed with DNA electrochemistry buffer, rinsing each quadrant three times with the passivation agent.
14. Incubate the backfilled surface in the humid box for 45 min.
15. After 45 min have passed, wash each quadrant ten more times with DNA electrochemistry buffer, in 20–25 μL volumes per quadrant, to remove mercaptohexanol.
16. Optional: Wash all quadrants twice with TBP buffer (5 m*M* sodium phosphate, pH 7.0, 5% glycerol, 4 m*M* $MgCl_2$, 4 m*M* spermidine), in the same volumes as previous washes. This washing can aid in monolayer formation and produce better CV scans. Add ~150–300 μL of TBP buffer to the top of the solution. Assemble a circuit with a gel tip reference electrode, with a Pt wire fastened securely to the reference setup and proper alligator clip connections (white = reference, red = counter, green = ground) Scan on CHI software to ensure that a monolayer has formed.

CHI Software Parameters for Monolayer Scans:

(a) CV = technique

(b) 100 mV/s scan rate

(c) High voltage = 0.1 V, low voltage = −0.4 V

(d) Sweep Segments: 6

(e) Sensitivity: 1×10^{-7} to 1×10^{-8} for initial buffer scans generally works well.

17. When the presence of a monolayer (observe capacitance of ~40 nA on a 2 mm^2 electrode) has been verified, wash electrodes with protein storage buffer at least five times, through all four quadrants. Repeat the preparation and setup for scanning in TBP buffer, and scan a background of the protein buffer.

2.2.2 Single Au Rod Electrode Setup

Notes:

Total experiment time and monolayer incubation time of 21–24 h for optimal results are the same for this platform as for the multiplex chip. Thiol-modified DNA which has been stored at −20°C for 2–3 weeks or longer should again be re-reduced and purified again before deposition onto this electrode platform.

A humid incubation environment is also necessary for this platform. The volume of DNA incubated on the rod electrodes, ~10–15 μL of dsDNA substrate, is prone to evaporation and should be monitored during the incubation period to prevent evaporation of the DNA solution droplet. (The same incubation chamber appropriate for the chip can be used for this setup.)

Solutions and Reagents:

The same DNA substrates, 6-mercapto-1-hexanol passivating agent stock, buffers, and 1 *M* $MgCl_2$ stock used for the chip setup can be used here.

Instruments and Supplies:

1. Au working rod electrode (1.6 mm diameter model manufactured by Pine Research Instruments or Bioanalytical Systems is a typical example for this platform)
2. Buehler Diamond polish (0.05 μm alumina)
3. Polish Pads
4. Ag/AgCl Gel Tip Reference Electrode
5. Platinum Wire

Buffer Conditions:

The same buffers (DNA storage buffer for substrate storage, DNA Electrochemistry Phosphate Buffer, and TBP buffer) as those used in the chip platform are used in this setup.

Procedure:

1. Thaw annealed, thiol-modified dsDNA stocks and prepare 25 μ*M* high-density or low-density DNA substrate solutions as described in the Procedure for the chip setup.
2. Deposit a small scoop of 0.05 μ*M* alumina polish onto the polishing pad. Mix with water to make a slurry of moderate thickness on the pad surface.
3. Wipe the Au rod electrode with a Kim Wipe and then press the Au surface of the electrode into the slurry on the polishing pad. To ensure thorough and even polishing of the surface, make figure-eight motions with the rod electrode surface, pressing into the slurry each time.
4. Rinse the polished electrode with deionized water until the slurry has been completely removed from the surface. Blot any excess water on the surface by touching a Kim Wipe to the edges of the electrode. Take care to avoid direct contact of the Kim Wipe with the polished surface, instead using the wipe to absorb the excess water.
5. Deposit the dry, polished electrode surface into a humid, secure position where the monolayer can incubate after dsDNA deposition.
6. Deposit high-density or low-density DNA onto the Au surface in a 10–15 μL volume, so that a small droplet forms over the surface area of the Au working electrode. Take care to avoid touching the polished surface with a pipet tip, close the incubation chamber, and allow the monolayer to form for 21–24 h.
7. Wash the surface of the rod electrode with the same reagents as described for the chip: five washes of DNA Electrochemistry Phosphate Buffer, three washes of 1 m*M* 6-mercapto-1-hexanol in DNA Electrochemistry Phosphate Buffer, incubate 45 min, wash 10 times with DNA Electrochemistry Phosphate Buffer. Optional washing with TBP buffer can also be performed here. All washes should be with a 10–15 μL volume of buffer/passivating agent. When pipetting droplets on and off of the electrode surface, take care again to avoid touching the electrode surface, where the new monolayer of DNA has formed.
8. Check for capacitance on the electrode after the final wash. This is performed in the same manner as for the chip; the reference electrode must touch the droplet but not the surface of the DNA electrode.

2.2.3 Single Au Electrode Setup: Au on Mica Working Electrode (Boal et al., 2009, 2005; Mui et al., 2011).

Notes:

Total experiment time and monolayer incubation time of 21–24 h for optimal results are the same for this platform as for the multiplex chip. Thiol-modified DNA which has been stored at −20°C for 2–3 weeks or longer should again be re-reduced and repurified before deposition onto this electrode platform.

A humid incubation environment is also necessary for this platform. The volume of DNA incubated on the rod electrodes, ~10–15 μL of dsDNA substrate, is prone to evaporation and should be monitored during the incubation period to prevent evaporation of the DNA solution droplet. (The same incubation chamber appropriate for the chip can be used for this setup.)

Solutions and Reagents:

The same DNA substrates, 6-mercapto-1-hexanol passivating agent stock, buffers, and 1 *M* $MgCl_2$ stock used for the chip setup can be used here.

Instruments and Supplies:

1. Au on mica surface (Molecular Imaging)
2. Platinum wire
3. Silver paint
4. Rubber O-ring/Metal apparatus to fasten Au on mica surface
5. Ag/AgCl Gel Tip Reference Electrode

Buffer Conditions:

The same buffers (DNA storage buffer for substrate storage, DNA Electrochemistry Phosphate Buffer, and TBP buffer) as those used in the chip platform are used in this setup.

Procedure:

1. Attach Au on Mica onto the metal apparatus which will connect the Au surface in the three-electrode cell to the potentiostat for measurements. Typically, silver paint is an effective bonding agent to connect Au on mica to the apparatus.
2. Assemble the working electrode surface, fixing its area with the O-ring and top section of the apparatus to hold the electrode surface in a constant position.
3. Insert a platinum wire into the electrode solution area, using rubber to prevent leakage through the opening for the platinum counter electrode in the electrochemical cell.

4. Pipet approximately 40–50 μL of 25 μ*M* thiol-modified dsDNA (high density or low density) onto the Au surface. Allow the monolayer to form, incubating in a humid environment for 21–24 h.
5. Wash the surface of the rod electrode with the same reagents as described for the chip: five washes of DNA Electrochemistry Phosphate Buffer, three washes of 1 m*M* 6-mercapto-1-hexanol in DNA Electrochemistry Phosphate Buffer, incubate 45 min, wash 10 times with DNA Electrochemistry Phosphate Buffer. Optional washing with TBP buffer can also be performed here. All washes should be with a 40–50 μL volume of buffer/passivating agent. When pipetting droplets on and off of the electrode surface, take care again to avoid touching the electrode surface, where the new monolayer of DNA has formed.
6. Check for capacitance on the electrode after the final wash. This is performed in the same manner as for the chip; the reference electrode must touch the droplet but not the surface of the DNA electrode.

2.3 DNA-Modified Au Electrodes Using Copper-Free Click Chemistry

Conventional DNA-modified surfaces are prepared through self-assembly of thiolated DNA duplexes on gold electrodes followed by backfilling with an alkanethiol to passivate any remaining exposed surface. By including or excluding 100 m*M* $MgCl_2$ during the incubation, one can form both high-density (30–50 pmol/cm^2) (Kelley, Barton, Jackson, & Hill, 1997) and low-density (15–20 pmol/cm^2) (Boon et al., 2002) monolayers on Au. While straightforward to fabricate, these films pose challenges for control over the spacing of the DNA molecules (Murphy, Cheng, Yu, & Bizzotto, 2009; Sam, Boon, Barton, Hill, & Spain, 2001). Close-packed DNA films limit the accessibility to individual helices during the event of the detection of very large proteins that target-specific sequences of DNA, or hybridization/dehybridization events (Peterson, Heaton, & Georgiadis, 2001). Although adjusting the ionic strength of the deposition solution with Mg^{2+}, some control over the surface density is possible (~15–50 pmol/cm^2), close packing still occurs among many helices (Furst, Hill, & Barton, 2013). In such films, the DNA helices cluster into exceedingly large domains of very high density within a sea of passivating thiol. The extensive clustering of helices can be somewhat problematic because it leads to variability across the electrode surface, with regions of close-packed helices in which access to specific base sequences may be inhibited.

The structural similarity of the components of a mixed monolayer-forming solution is a major determining factor for the degree of homogeneity within the resulting self-assembled monolayer (SAM) (Love, Estroff, Kriebel, Nuzzo, & Whitesides, 2005; Ulman, 1996). Thus an alternative approach to a low-density DNA film is to prepare a homogeneous mixed SAM without DNA, followed by DNA conjugation to the functionalized mixed monolayer (Fig. 3). Previous work showed the preliminary formation of a mixed alkanethiol monolayer on gold containing azide-terminated thiols, followed by copper-catalyzed click chemistry to tether single-stranded oligonucleotides to gold surfaces (Devaraj et al., 2005). While copper-catalyzed click chemistry shows high efficiency with mild reaction conditions, conventional copper (I) catalysts can damage DNA and are difficult to remove after the reaction has occurred.

We have developed a catalyst-free method of DNA conjugation to a mixed monolayer that capitalizes on ring strain to drive the [3+2] cycloaddition (Agard, Prescher, & Bertozzi, 2004; Baskin & Bertozzi, 2007). We first form a mixed azide-terminated monolayer, then add cyclooctyne-labeled DNA that spontaneously couples only to the azide via azide-alkyne cycloaddition. The resulting DNA-modified surfaces obtain a low density, more evenly spaced monolayer, while maintaining surface passivation against the redox reporter. Both electrochemical and imaging methods used to characterize these monolayers have been reported (Furst et al., 2013; Furst, Muren, Hill, & Barton, 2014; Muren & Barton, 2013). This approach offers several advantages over conventional preparations of DNA monolayers: (i) it allows for precise control over the total amount of DNA by simply changing the fraction of thiol-azide present in the preliminary monolayer; (ii) the preliminary self-assembly step results in a passivated surface before the addition of DNA, minimizing undesirable direct interactions between the gold surface and DNA helices; and (iii) because the underlying azide conjugation sites are more evenly distributed in the preliminary monolayer, DNA helices are less prone to cluster into large, high-density domains.

This platform facilitates DNA-mediated CT and is thus extremely sensitive to perturbations in the DNA, providing exquisite electrochemical discrimination between well-matched and -mismatched DNA duplexes. Additionally, this platform provides greater sensitivity to protein binding events than conventional high-density films due to the larger number of accessible surface-exposed binding sites. In particular, low-density films allow for the detection of as little as 4 n*M* TBP and 5 n*M* human methyltransferase DNMT1 (Furst et al., 2013, 2014; Muren & Barton, 2013).

The enhanced detection with copper-free click chemistry adds another sensitive detection tool to the toolbox of electrochemical DNA detection strategies.

Here, we briefly describe the synthesis of azide-terminated alkanethiol linker, the preparation of cyclooctyne-modified DNA from a commercially available source, and the conditions for the copper-free click reaction for DNA-modified electrodes.

2.3.1 Dibenzo-Bicyclooctyne-Modified DNA

From the variety of cyclooctyne-based copper-free click reagents, we use a soluble dibenzo-bicyclooctyne (DBCO)-sulfo-NHS ester sodium salt for conjugation reactions with amino-modified oligonucleotides.

Solutions and Reagents:

1. DBCO-sulfo-NHS Ester (Glen Research)
2. Primary amine modified 5′ DNA samples (IDT)
3. GE Healthcare illustra NAP-5 column
4. Sodium bicarbonate conjugation buffer (pH 9)

Instruments and Supplies:

1. HPLC
2. UV–vis
3. Thermo cycler

Buffer Conditions:

DNA phosphate buffer (5 m*M* sodium phosphate, 50 m*M* NaCl, pH 7).

Procedure:

1. Dissolve DBCO-sulfo-NHS Ester at a concentration of 5.2 mg per 60 μL (~0.17 *M* solution) in water.
2. Use this stock solution to conjugate with amino-modified oligos in sodium bicarbonate conjugation buffer (pH 9).
3. For a 0.2 μmol synthesis of a 5′ end amino-modified oligo: dissolve oligo in 500 μL of conjugation buffer. Add 6 μL of DBCO-sulfo-NHS Ester solution.
4. Vortex mixture and incubate at room temperature overnight.
5. Desalt conjugated oligo on a GE Nap 5 column to remove salts and organics. Nap 5 column protocol is followed from the supplier instruction.
6. Purify DBCO-modified DNA and its complementary strand using reverse-phase HPLC with a polymeric PLRP-S column (Agilent) and characterized by mass spectrometry.

7. To prepare duplexes, the DBCO-modified DNA and its complementary strand stocks were desalted, resuspended in DNA phosphate buffer, and quantified by UV–vis absorption at 260 nm. Equimolar amounts (50 μ*M*) of complementary strands were combined and thermally annealed.

2.3.2 Copper-Free Click Reaction for DNA-Modified Electrodes

Proper cleaning of the gold surface is necessary to obtain high-quality thiol-gold-based SAM. For this purpose, rational methods for preparing highly reproducible gold surfaces, include the oxidative and reductive pretreatments (Campuzano, Pedrero, Montemayor, Fatas, & Pingarron, 2006; Kondo et al., 2007). Briefly, gold substrates could be oxidized to a positive charge state via conventional methods, such as ultraviolet/ozone, oxygen plasma, electrochemical oxidation, and piranha solution oxidation. The freshly prepared oxidized gold surfaces can be chemically reduced to zero state (metallic gold) after they were immersed in ethanol. The synthesis of azide-terminated thiol linker, 1-azidoundecane-11-thiol, is reported and adapted from a previously published procedure (Shon, Kelly, Halas, & Lee, 1999).

Solvents and Reagents:

1. Gold electrode (Au) for voltammetry 1.6 mm diameter (Bioanalytical Systems)
2. 0.05 μm alumina polish powder (Buehler)
3. Piranha solution (1:3 H_2O_2/H_2SO_4)
4. SAM deposition solutions: dissolve the desired ratio of mercaptoundecanol (Sigma) and 1-azidoundecane-11-thiol in ethanol. The total thiol concentration is always 1 m*M*.
5. DBCO-modified double-stranded DNA

Buffer Conditions:

DNA phosphate buffer (5 m*M* sodium phosphate, 50 m*M* NaCl, pH 7.0).

Procedure:

1. Au rod electrode was polished with alumina polish powder for 1 min, rinse with deionized water.
2. The rod electrodes were immersed in piranha solution for 15 min, rinse with deionized water.
3. Immerse in ethanol and sonicate for 10 min, rinse with deionized water.

4. Cycle Au rod in 50 m*M* H_2SO_4 between 1.4 and 0 V vs Ag/AgCl to obtain an Au (111) single-crystal electrode. Rinse Au rod again with deionized water followed by ethanol (Kondo et al., 2007).
5. Immerse the cleaned gold substrates in the SAM deposition solution for 4 h. After deposition, SAM is rinsed in ethanol and water in order to remove excess adsorbate and dried with N_2 to remove residual solvent.
6. Rinse the Au rod with DNA phosphate buffer. Annealed DBCO-modified dsDNA are conjugated with 1-azidoundecane-11-thiol in phosphate buffer at room temperature for 12–17 h.

Note:

1. For step 4, sharp anodic and cathodic peaks were observed at +1.30 and +0.91 V, respectively in CV. The former peak can be assigned to the oxide formation and the latter to the reduction of oxide.
2. For the 16-electrode multiplex chip setup with the copper-free click chemistry, follow Section 2.2.1 procedures 1–9 for chip cleaning and preparation. For SAM preparation and click DNA coupling, follow Section 2.3.2 procedures 5–6.
3. Another copper-free click reaction using a cyclooctyne moiety (OCT) tethered 5′ DNA, a mixed monolayer of mercaptoethanol (MCE) as the passivating agent and 6-azido-1-hexanethiol was also reported (Furst et al., 2013). The azide-terminated SAM was formed by soaking the electrodes in an ethanol solution containing 1 m*M* MCE and 0.25 m*M* 6-azido-1-hexanethiol for 24 h to form a monolayer composed of 20% azide, followed by an OCT-labeled duplexes, OCT–DNA, coupling to the film via azide-alkyne cycloaddition.

2.4 Characterization of DNA Self-Assembled Monolayers

As described earlier, DNA self-assembled monolayers (SAMs) can be formed on gold electrodes by spontaneous assembly of thiolated DNA or by clicking alkyne-modified DNA onto a preformed azide/thiol monolayer. Regardless of how the monolayer was formed, it is important to characterize the DNA surface coverage and monolayer morphology before proceeding with experiments, as both of these parameters can affect the ability of redox probes or proteins to undertake DNA-mediated processes. If the surface coverage is too sparse, probes and proteins may preferentially interact with the surface, while too much crowding provides steric hindrance that can block efficient protein binding.

We have developed several methods to assess these factors, including visualization of DNA-modified surfaces with AFM, quantification of ^{32}P-labeled DNA, and quantification by $[Ru(NH_3)_6]^{3+}$ groove binding (Furst et al., 2013; Sam et al., 2001). AFM is useful in providing a means of visualizing the overall arrangement of DNA on the electrode surface. By repetitive scanning at a high applied voltage, holes can be generated in the surface, permitting the absolute film height to be measured (Furst et al., 2013). Surface area and height can further be used to estimate surface coverage by DNA, although this estimate should be verified by either ^{32}P-labeling or $[Ru(NH_3)_6]^{3+}$ quantification. The primary limitations of AFM in surface characterization are that it does require access to an instrument and that such manipulations of the surface preclude further experiments with the particular film being examined. Nonetheless, AFM is an indispensable technique in characterizing the morphology of novel surfaces or monolayers.

2.4.1 AFM Imaging of DNA Films

Solvents and Reagents:

1. Gold metal (Kurt J. Lesker Industries)
2. Gold AFM surface (Novascan)
3. Ethanol (200 proof)
4. Hexanethiol

Instruments and Supplies:

1. Silicon AFM tips (Nanosensors Advanced TEC, force constant 0.2 N)
2. Metal evaporator
3. Multimode Scanning Probe Microscope (Digital Instruments)

Buffer Conditions:

DNA phosphate buffer (5 m*M* sodium phosphate, 50 m*M* NaCl, pH 7.0).

Procedure:

1. Prepare DNA SAMs on gold electrodes or on Novascan AFM surfaces
2. Deposit 10 nm gold onto the silicon AFM tips using a metal evaporator
3. Soak the AFM tips in 10 m*M* hexanethiol in ethanol for 1 h, and rinse thoroughly with ethanol prior to use
4. Mount surfaces containing DNA films on scanning probe microscope
5. Scan surface in contact mode
6. To measure monolayer height, apply 10 V to the AFM tip and repetitively scan a 1 μm square to remove the film in this region; after hole generation, measure the height profile by scanning in contact mode

2.4.2 ^{32}P Labeling of DNA

^{32}P labeling allows direct quantification of the DNA at the surface, and provides a 1:1 ratio of signal to DNA. ^{32}P is easily appended to the 5′ end of DNA using commercially available T4 polynucleotide kinase and γ-^{32}P ATP, and monolayers can be formed according to standard procedures. However, the safety concerns, limited half-life of the probe (14 days), and difficulty in measuring radioactivity on an electrode make this technique less appealing.

Solvents and Reagents:

1. T4 polynucleotide kinase (New England Biolabs)
2. T4 buffer (New England Biolabs)
3. 10 pmol ssDNA with free 5′ ends
4. γ-^{32}P ATP (Perkin Elmer; 3000–6000 Ci/mmol)
5. MQ water
6. Ethylenediaminetetraacetic acid (EDTA)

Instruments and Supplies:

1. Benchtop incubators
2. MicroBioSpin6 columns (BioRad)
3. Tabletop centrifuge
4. 1.5 mL Eppendorf tubes

Procedure:

1. Prepare reactions mixes (50 μL) in Eppendorf tubes by adding DNA, 5 μL 10 × concentrated T4 buffer, and MQ water; keep on ice
2. Thaw ^{32}P-labeled ATP, and add 40 μCi to each reaction tube (All steps involving radioactivity should be carried out behind a Lucite shield!)
3. Add 1.0 μL T4 polynucleotide kinase (5 units) to each tube, and start reactions by sealing the tube and incubating at 37°C for 30 min
4. Stop reactions by adding EDTA to a final concentration of 10 m*M*, and heat inactivate the kinase by incubation at 85°C for 10 min
5. Isolate DNA by adding quenched reactions to a MicroBioSpin6 column and spinning for 4 min at 1000 × *g*

2.4.3 DNA Quantification Using $[Ru(NH_3)_6]^{3+}$

Due to the experimental ease relative to AFM and ^{32}P-labeling, we generally favor the use of $[Ru(NH_3)_6]^{3+}$ for DNA quantification. This method is quite simple, involving only the addition of $[Ru(NH_3)_6]^{3+}$ to a surface and scanning (Furst et al., 2013). However, unlike ^{32}P, the signal to DNA stoichiometry is not 1:1, as $[Ru(NH_3)_6]^{3+}$ binds electrostatically to the DNA backbone in a ratio of 1 molecule per 3 DNA phosphates. Further drawbacks to this strategy are that it can be easy to underestimate the amount of DNA if saturation is not achieved, and surface accessibility can be an issue. To ensure accurate quantification with $[Ru(NH_3)_6]^{3+}$, increasing concentrations should be added until signal saturation is achieved, with care being taken to use a total monovalent ionic strength of no greater than 5 m*M* in the buffer to ensure access of $[Ru(NH_3)_6]^{3+}$ to the DNA. The background signal can be determined by comparison with an alkane-thiol only SAM, and at this point, the amount of $[Ru(NH_3)_6]^{3+}$ bound to DNA can be determined from either CV or chronocoulomtery. DNA surface coverage is then calculated from the following equation:

$$\Gamma = (Q/nFA)^{*}(\#\mathrm{nt}/\mathrm{Ru}) \quad (1)$$

Γ is DNA surface coverage in mol/cm^2, Q is total measured charge in coulombs from the $Ru^{3+/2+}$ reduction, n is the number of electrons transferred per reduction (1 in the case of $[Ru(NH_3)_6]^{3+}$), F is Faraday's constant (96,485 C/mol), A is electrode area in cm^2, and #nt/Ru is the maximum number of $[Ru(NH_3)_6]^{3+}$ molecules bound per nucleotide (Kissinger & Heineman, 1996). For the sake of comparison, values are typically reported in pmol/cm^2 (Furst et al., 2013). Lastly, it should be noted that, due to the difficulty in washing such small molecules off of the surface, $[Ru(NH_3)_6]^{3+}$ quantification should be the final step if further experiments with redox probes or proteins are planned.

Solvents and Reagents:

1. $[Ru(NH_3)_6]Cl_3$ (Sigma-Aldrich)

Instruments and Supplies:

1. Potentiostat
2. Ag/AgCl reference electrode (Bioanalytical Systems)
3. Pt wire counter electrode (Kurt J. Lesker Industries)
4. DNA monolayers on gold electrode

Buffers Conditions:

DNA phosphate buffer (5 m*M* sodium phosphate, pH 7.0).

Procedure:

1. Prepare DNA SAM in parallel with a monolayer without DNA
2. Add a small quantity (~1 μ*M*) $[Ru(NH_3)_6]Cl_3$ in 5 m*M* DNA phosphate buffer and scan at a low scan rate (20 mV/s is ideal) by CV (or apply a negative potential and use chronocoulometry); the main reductive peak will be near 0 mV vs NHE (~200 mV vs Ag/AgCl)

3. Titrate increasingly high concentrations of $[Ru(NH_3)_6]Cl_3$, scanning each time as in step 2
4. Repeat with the DNA-free surface
5. Quantify peak area and subtract DNA-free charge from that obtained with DNA to determine surface coverage

3. DETECTION OF REDOX-SILENT PROTEINS

Electrochemical assays that rely on the sensitivity of DNA-mediated charge transport (DNA CT) chemistry show particular promise for rapid biosensing. As DNA CT is mediated through the base pair π-stack formed by the double helix, this chemistry has unmatched structural sensitivity to perturbations of the π-stack. The nonredox-active DNA-binding proteins that we detect structurally distort the DNA. With a DNA-modified electrode, when a potential is applied to the electrode, DNA CT facilitates reduction of a redox probe, producing an electrochemical signal. DNA with a structural distortion to the π-stack shows an attenuated signal, relative to unperturbed DNA, thereby allowing for sensitive detection of the structural distortion. As most DNA-binding proteins bind specific DNA sequences, this property may be exploited to specifically detect a protein of interest. Electrodes can easily be modified with customized DNA-containing binding sites aimed at the specific detection of target proteins (Figs. 1 and 6). Thus DNA may be utilized in these electrochemical sensors of protein-DNA interactions as both the recognition element and the transducer.

In order to measure the activities of nonredox-active DNA-binding proteins by DNA CT, a redox-active probe moiety is incorporated at or near the end of the DNA that is distal from the surface. For this purpose, noncovalent (Boon & Barton, 2003; Boon et al., 2003) and covalent (Buzzeo & Barton, 2008; Gorodetsky & Barton, 2007) redox probes have been employed as well as DNA-binding proteins that are redox active (Section 4). In the DNA-modified electrode, CT is mediated from the electrode surface to the redox probe via the intervening path of well-stacked DNA bases. Importantly, experiments with this platform are all performed in aqueous, buffered solution such that the DNA maintains a native, CT-active conformation.

With the electrochemical monitoring of the DNA-mediated CT, we are able to detect the activity of a sequence-specific restriction enzyme. The efficient cleavage by the restriction enzyme attenuates the DNA CT signals detected by a covalently attached Nile Blue redox probe at the 3′ end of the DNA probe (Section 3.1). The TBP severely kinks the DNA by 80 degree (Fig. 1). The attenuation in DNA CT caused by these structural perturbations can be detected by a covalently tethered methylene blue (MB) redox probe in buffer (Section 3.2). Proteins that bind but do not distort the DNA or proteins do not bind DNA, such as bovine serum albumin (BSA), do not cause this signal attenuation. Further work with the MB probe showed that its DNA-mediated signal may be amplified in an electrocatalytic cycle with ferricyanide (Kelley et al., 1997) and used to sensitively detect all base mismatches (Boon et al., 2002) and a variety of DNA lesions (Kelley, Boon, Barton, Jackson, & Hill, 1999) by an attenuation of DNA CT to the MB redox probe. We can therefore sensitively methyltransferase activity with the MB/ferricyanide electrocatalysis electrochemistry (Section 3.3). Lastly, incorporating the highly sensitivity of the electrocatalysis system, we have designed and fabricated a two-electrode electrochemical platform to detect methyltransferase activity from crude cell lysate (Section 3.4).

3.1 Detection of Restriction Enzyme *Alu*I

We can demonstrate detection of DNA-binding proteins by measuring the sequence-specific activity of the *Alu*I restriction endonuclease, which cleaves at the restriction site 5′-AGCT-3′, leaving blunt ends between the G and C bases. Covalent tethering of the redox probe Nile Blue on the DNA is the probe to monitor the restriction enzyme binding. Here we use the 16-electrode multiplex chip (Section 2.2.1, Fig. 4). The chip was prepared with 17-mer Nile Blue-modified DNA, where half of the electrodes were assembled with a sequence containing the *Alu*I recognition site and the other half with a sequence lacking this site. The *Alu*I restriction enzyme was titrated onto the chip, and the integrated CV peak areas were recorded at each concentration (Slinker et al., 2010). The threshold of *Alu*I restriction activity for the sequence containing the restriction site was 400 units/mL, corresponding to a concentration of approximately 10 n*M*. As the total sample volume was 250 μL, this corresponds to 2.5 pmol of enzyme per chip, or 160 fmol of enzyme per electrode. At concentrations greater than 1600 units/mL, the charge at the electrodes lacking the restriction site decreases due to nonspecific restriction activity, also known as star activity. In this case, the DNA without the consensus restriction site contains a pseudo-site differing by only one base (5′-ATCT-3′). Thus, as expected at higher enzyme concentrations, restriction cleavage at this pseudo-site is apparent.

Several important implications arise from these observations. Cleavage by the *Alu*I restriction endonuclease requires that the DNA on these chips is in its native conformation and accessible to the protein; one can therefore consider the DNA electrode surface equivalent to that in solution. Moreover, the observation of sequence-specific cleavage indicates that protein detection with DNA-mediated electrochemistry is highly selective. Also, by extension, incorporation of multiple DNA sequences with different protein binding characteristics on a single chip indicates that multiplex chips can serve as a robust platform to simultaneously monitor reactions on different oligonucleotides. Finally, this assay requires only microliter volumes of low protein concentrations, making it competitive with alternative detection methods.

Solutions and Reagents:

1. *Alu*I restriction enzyme (New England Biolabs), stored at −20°C until use.
2. Nile Blue-modified DNA with the *Alu*I restriction site (5′-AGCT-3′)
3. Nile Blue-modified DNA with the pseudo-site (5′-AGAT-3′)
4. Nile Blue perchlorate (laser grade, Acros)

Instruments and Equipment:

1. Slide-A-Lyzer mini dialysis kit (Pierce)
2. CH760B Electrochemical Analyzer and a 16-channel multiplexer module (CH Instruments)
3. Ag/AgCl reference electrode
4. Pt wire auxiliary electrode
5. Sixteen-electrode multiplex chip

Buffer Conditions:

DNA phosphate buffer (5 m*M* sodium phosphate, 50 m*M* NaCl, pH 7.0)

Testing Phosphate buffer (DNA phosphate buffer supplemented with 4 m*M* $MgCl_2$, 4 m*M* spermidine, 50 μ*M* EDTA and 10% glycerol, pH 7.0)

Tris buffer (50 m*M* Tris–HCl, 10 m*M* EDTA, and 10 m*M* $MgCl_2$, pH 7.8)

Procedure:

1. DNA-modified 16-electrode multiplex chip setup, see Section 2.2.1.
2. Prior to use, the *Alu*I restriction enzyme aliquots were exchanged into Tris buffer using a Pierce Slide-A-Lyzer mini dialysis kit at 4°C with overnight stirring.

3. CV experiments were performed by a CH760B Electrochemical Analyzer and a 16-channel multiplexer module. Chips were tested with a common Pt auxiliary electrode and a common Ag/AgCl reference electrode. Electrochemistry was recorded at ambient temperature in either testing phosphate buffer or Tris buffer.
4. Dialyzed *Alu*I in Tris buffer was titrated onto the chip with test concentrations ranging from 0 to 50 n*M* (0–2000 units/mL). The reaction was allowed to equilibrate at each point of the titration for approximately 30 min before scanning the chip. The integrated CV peak areas were recorded at each concentration.

Note:

1. For the preparation of DNA-modified multiplex chips to measure restriction activity, $MgCl_2$ was excluded from the DNA assembly solution in order to produce a lower density monolayer and grant greater access to the restriction enzyme.
2. For the electrochemical test, reference and counter electrodes can be patterned on the chip surface, though including other metals for a stable reference would increase the complexity of chip fabrication.

3.2 Detection of TBP Binding Activity

The transcriptional activator TBP has been easily detected on DNA-modified electrodes, given the large perturbation in DNA stacking associated with the binding of TBP. TBP binds to a TATA sequence in DNA and kinks the helix 80 degree at that location, leading to a significant DNA-mediated signal attenuation. In the presence of TBP, which binds to the specific TBP binding site (5′-TATAAAG-3′) and kinks the DNA, the charge accumulation is significantly attenuated (Furst et al., 2013). Protein binding, in kinking the DNA, acts essentially as a switch, turning off DNA CT. BSA, which does not bind to DNA, shows no signal change.

Solvents and Reagents:

1. MB-modified DNA with the TBP binding sites (5′-TATAAAG-3′)
2. Modified MB dye for coupling was synthesized as described previously (Pheeney & Barton, 2012)
3. TBP (ProteinOne), stored at −80°C until use
4. BSA (New England Biolabs), stored at −20°C until use
5. Mercaptohexanol (Sigma-Aldrich)

Instruments and Supplies:

1. Sixteen-electrode multiplex chip
2. CH760B Electrochemical Analyzer and a 16-channel multiplexer module (CH Instruments)
3. Ag/AgCl reference electrode
4. Pt wire counter electrode

Buffer Conditions:

Tris buffer (10 m*M* Tris, 100 m*M* KCl, 2.5 m*M* $MgCl_2$, 1 m*M* $CaCl_2$, pH 7.6)

DNA phosphate buffer (5 m*M* sodium phosphate, 50 m*M* NaCl, pH 7.0)

TBP binding buffer (5 m*M* sodium phosphate, 50 m*M* NaCl, 4 m*M* $MgCl_2$, 4 m*M* spermidine, 50 μ*M* EDTA, 10% glycerol, pH 7.0)

Procedure:

1. For the 16-electrode multiplex chip cleaning and preparation for the TBP binding test, see Section 2.2.1.
2. For all electrochemistry, CV scans were performed at a 100 mV/s scan rate over the potential window of 0 mV to −500 mV. SWV was performed at 15 Hz over the same potential range. Signal size was measured as the CV cathodic peak area or the SWV peak area.
3. For all protein binding experiments, after backfilling with mercaptohexanol, electrodes were backfilled with 3 μ*M* BSA in phosphate buffer for 45 min at room temperature. After thorough rinsing by buffer exchange, background scans were performed in the TBP buffer TBP. After removing blank TBP buffer from the common well over the electrodes, a solution of the target protein in binding buffer was then added (200 μL total volume).

Note:

1. In this electrochemical protein detection scheme, the protein binding buffer is also the electrochemical running buffer.

3.3 Methyltransferase Detection With Electrocatalysis

The redox-active intercalator MB binds to DNA and becomes electrochemically active on the DNA electrode as long as the individual duplexes that make up the film are completely Watson–Crick base paired. However, the presence of a single-base mismatch or other base-stacking perturbation between the electrode and the site of intercalation greatly attenuates the electrochemical response (Boon et al., 2002; Kelley, Boon, et al., 1999; Kelley, Jackson, Hill, & Barton, 1999). The sensitivity of DNA CT to perturbations in base pair stacking has been used as a platform for the development of electrochemical sensors for mutational analysis (Boon, Ceres, Drummond, Hill, & Barton, 2000), as well as protein/DNA interactions (Boon et al., 2002).

An electrochemical analysis strategy was developed that has improved sensitivity through the combination of electrocatalysis using MB and ferricyanide $[Fe(CN)_6]^{3-}$ for signal amplification (Fig. 5). When a negative potential is applied to the DNA-modified electrode, the DNA-bound MB is reduced to leucomethylene blue (LB) via DNA CT and enters the solution. LB has a lower binding affinity to DNA than MB. In solution, ferricyanide $[Fe(CN)_6]^{3-}$ is further reduced to ferrocyanide $[Fe(CN)_6]^{4-}$ facilitated by the electrocatalytic reduction by MB. A key element is that electrostatic repulsion between the negatively charged DNA films prevents ferricyanide from penetrating and undergoing reduction without mediation by DNA. Instead the free-floating ferricyanide in solution receives electrons from MB through DNA CT. The auxiliary electrode inserted in solution measures the reduction signal of ferricyanide and shows a reading in current change. The positive oxidation potential reoxidizes the $[Fe(CN)_6]^{4-}$ and LB is reoxidized to MB. The two-step electrocatalytically amplification has been used for methyltransferase DNMT1 activity detection (Muren &

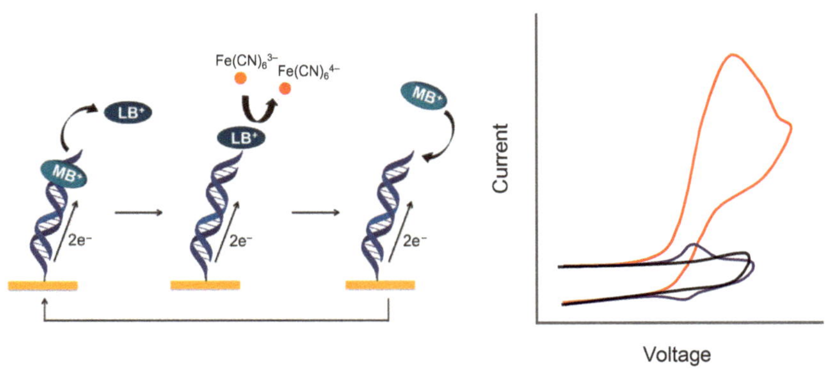

Fig. 5 Electrocatalytic cycle between free methylene blue (MB) and ferricyanide on a DNA-modified electrode. MB in its oxidized form is intercalated into the DNA base stack. Upon reduction of MB to leucomethylene blue (LB) via DNA-mediated CT, the affinity of the LB for DNA is lowered, and LB is no longer intercalated. The reduced LB is capable of reducing ferricyanide that is freely diffusing in solution. The LB is then reoxidized to MB and can reintercalate into the DNA. The ferricyanide acts as a diffusing electron sink in solution for the redox probe MB. Electrostatic repulsion prevents ferricyanide from penetrating the negatively charged DNA film. A cyclic voltammetry at a DNA-modified electrode of ferricyanide (*black*), MB (*blue*), and ferricyanide and MB (*red*).

Barton, 2013). Once a DNA array is established on the Au electrodes platform, electrocatalytic detection is then performed with the three-electrode electrochemical cell. Importantly, whether the direct or catalytic reduction of MB is monitored, reduction of MB has been shown to take place via CT through the DNA base stack. DNA CT electrochemistry therefore provides an exquisitely sensitive means to monitor nucleic acid structure and stacking. Even small perturbations in base pair stacking, as is associated with some base lesions, diminish the efficiency of MB reduction (Boon et al., 2000; Kelley, Boon, et al. 1999).

DNA methylation is the most prominent form of epigenetic gene regulation and is a critical long-term gene silencing mechanism in mammals (Miranda & Jones, 2007). This covalent addition of a methyl group to the carbon-5 position of cytosine at predominantly 5′-CG-3′ sites is catalyzed by DNA methyltransferases, which use the cofactor *S*-adenosyl-L-methionine (SAM) as a methyl donor (Flynn & Reich, 1998). However, aberrant DNA methylation has been associated with multiple disease states (Baylin & Herman, 2000; Chen, Akbarian, Tudor, & Jaenisch, 2001; Esteller, 2002). DNMT1 transmits methylation patterns across cell divisions by completing methylation on newly replicated strands at 5′-CG-3′ sites that carry methylation on the template strand alone (Jeltsch, 2002). Thus DNMT1 is characterized as a maintenance methyltransferase and displays a significant preference for hemimethylated DNA substrates (Jeltsch, 2002). These inherently different activities contribute to the complex roles of methyltransferases that are now being elucidated in a growing number of cancers. We have developed an electrochemical platform that combines the ferricyanide/MB electrocatalysis signal-on detection of human DNMT1 activity (Muren & Barton, 2013). Due to the highly sensitivity of the redox probe, 4 n*M* DNMT1 can be detected with the DNA-modified electrodes.

Solvents and Reagents:

1. Methylene blue (Sigma-Aldrich)
2. Potassium ferricyanide $K_3Fe(CN)_6$ (Sigma-Aldrich)
3. Human DNMT1 (BPS Bioscience)
4. BSA (New England Biolabs, used as received)
5. SAM (New England Biolabs, used as received)
6. Restriction endonucleases *BssH*II (New England Biolabs, used as received)
7. Protease from *Streptomyces griseus* dry powder (Sigma-Aldrich), stored as a 250 μ*M* solution in protease buffer at −20°C

Instruments and Supplies:

1. Sixteen-electrode multiplex chip
2. CH760B Electrochemical Analyzer and a 16-channel multiplexer module (CH Instruments)
3. Ag/AgCl reference electrode
4. Pt wire auxiliary electrode
5. DNA-modified multiplex chip
6. Size exclusion spin column (10 kDa cutoff, Amicon)
7. Incubator

Buffer Conditions:

DNA phosphate buffer (5 m*M* sodium phosphate, 50 m*M* NaCl, pH 7)
Scanning buffer (5 m*M* sodium phosphate, 50 m*M* NaCl, 4 m*M* $MgCl_2$, 4 m*M* spermidine, 50 μ*M* EDTA, 10% glycerol, pH 7)
DNMT1 activity buffer (50 m*M* Tris–HCl, 1 m*M* EDTA, 5% glycerol, pH 7.8)
Protease buffer (5 m*M* sodium phosphate, 40% glycerol, pH 7)
Methylation/restriction (M/R) buffer (10 m*M* Tris–HCl, 50 m*M* NaCl, 10 m*M* $MgCl_2$, pH 7.9)

Procedure:

1. For the 16-electrode multiplex chip cleaning and preparation test, see Section 2.2.1.
2. Rinse the chip with phosphate buffer. Scan DNA phosphate buffer first to ensure there is no extra signal/contamination anywhere. CV scan is from 0.4 to −0.4 V at 0.1 V/s scan rate.
3. Before the protein treatment, check the DNA-modified multiplex chip with MB and ferricyanide in the scanning buffer for surface passivation and the electrocatalysis signal from DNA monolayer. Replace solution three to five times with MB in scan buffer or ferricyanide in scan buffer at desired concentration.
4. DNMT1 with 100 μg/mL of BSA and 160 μ*M* SAM were applied to individual chip quadrants, and chips were incubated at 37°C for 2 h in a humidified container. Then chips were rinsed thoroughly with DNMT1 activity buffer and then protease buffer.
5. Chips were then treated with 1 μ*M* protease in DNA phosphate buffer for 1 h at 37°C. Then chips were rinsed thoroughly with protease buffer and then M/R buffer.
6. Chips were treated with 1500 units/mL of *Bss*HII in M/R buffer at 37°C for 1 h. Then chips were rinsed thoroughly with scanning buffer and

scanned with 200 μL of the MB and ferricyanide mixture in scanning buffer in a common well.

Note:

1. DNMT1 shows strong preferential activity at hemimethylated 5′-^{m}CG-3′ sites, DNA substrates with a hemimethylated *BssH*II restriction site (5′-G^{m}CGCGC-3′) were utilized.
2. *BssH*II requires full methylation of either 5′-CG-3′ site within its recognition sequence to prevent DNA restriction.
3. Buffer exchange of DNMT1 and *BssH*II prior to electrochemistry experiments is necessary to remove DTT, which disrupts DNA-modified electrodes upon heating. Buffer exchange by size exclusion spin column was performed on DNMT1 and *BssH*II. The exchange was performed according to manufacturer instructions at 4°C. DNMT1 was exchanged into DNMT1 activity buffer and *BssH*II was exchanged into M/R buffer.
4. A protease treatment step was introduced to remove bound DNMT1 following DNMT1 treatment, prior to *BssH*II treatment.
5. Including the methyltransferase and restriction enzyme incubations, the total assay time for DNMT1 is about 5 h.
6. Signal size was measured as the CV cathodic peak area. The reported variation in the data represents the standard deviation across all electrodes measured for a given condition.

3.4 Direct Detection of Methyltransferase From Colorectal Cancer Cell Lysate With Two-Electrode Platform

Integrating the high sensitivity of the electrocatalysis system, we have developed a two-electrode platform with the click coupling of low-density DNA monolayers (Section 2.3) for direct detection in crude cancer cell lysates. As opposed to conventional electrochemical readout from the primary DNA-modified electrode, a secondary electrode coupled with ferricyanide/MB electrocatalytic signal amplification, enables more sensitive detection with spatial resolution on the DNA array electrode surface (Fig. 6). Using this two-electrode platform, arrays have been formed that facilitate differentiation between well-matched and -mismatched sequences, detection of transcription factors, and sequence-selective DNA hybridization, all with the incorporation of internal controls (Furst et al., 2013, 2014). For effective clinical detection, the two-electrode platform was multiplexed to contain two complementary arrays, each with 15 electrodes. With the sensitivity and selectivity obtained from the multiplexed, two working electrode array, an electrochemical signal-on assay for activity of the DNMT1 was incorporated.

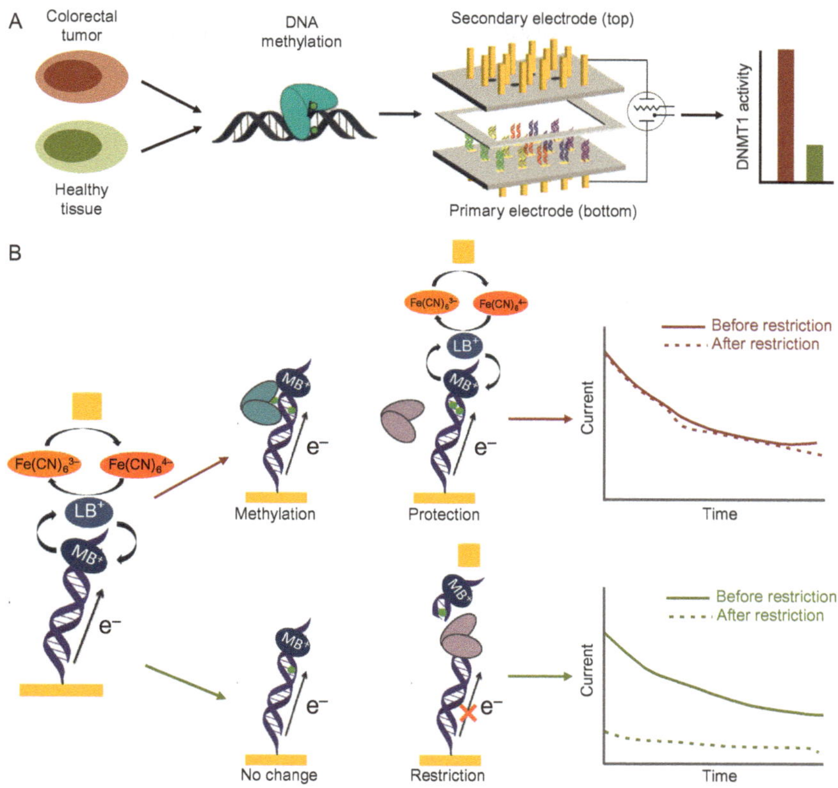

Fig. 6 Overview of electrochemical DNMT1 analysis from tumors with two-electrode platform (*top*). Tumor and healthy tissues are lysed, and nuclear lysate is used to detect DNMT1 methyltransferase activity. The lysate is applied to a multiplexed, two working electrode platform that enables the conversion of methylation events into an electrochemical signal. The electrochemical detection platform contains two electrode arrays, each with 15 electrodes (1 mm diameter each) in a 5 × 3 array. Multiple DNAs are patterned covalently to the substrate electrode by an electrochemically activated click reaction initiated with the patterning electrode array. Once a DNA array is established on the substrate electrode platform, electrocatalytic detection is then performed from the top patterning/detection electrode. Generally, we find hyperactivity of DNMT1 in tumor samples as compared to the healthy adjacent tissue. Signal-on electrochemical assay for DNMT1 detection (*bottom*). *Left*: The *bottom* (primary) electrode modified with a dilute DNA monolayer is responsible for generating electrochemical signals through DNA-mediated (CT) amplified by electrocatalysis. Methylene blue (MB), a DNA-intercalating redox probe, is reduced by DNA CT and enters solution as leucomethylene blue (LB), where it can interact with an electron sink, ferricyanide. Upon interaction with LB, ferricyanide is reduced to ferrocyanide, reoxidizing the LB to MB in the process. Current is generated and detected at the secondary electrode from the reoxidation of ferrocyanide. The current generated is proportional to the amount of ferrocyanide oxidized. To detect DNMT1, crude lysate is added to the electrode. If DNMT1 (*blue*) is

(Continued)

A two-electrode detection system enables the determination of more specific spatial information on a single substrate electrode surface and leads to high sensitivity since the ferricyanide is only reduced at the secondary electrode, optimizing charge transport through the DNA. Our arrays are formed through selective electrochemical patterning of multiple DNA sequences onto a single-electrode surface containing a preformed mixed monolayer. Electrochemical readout is then accomplished via amperometric detection at a spatially isolated probe electrode controlled by a bipotentiostat. Because multiple DNA sequences are patterned onto a single substrate, different sequences can be examined under identical experimental conditions. With our assay, we now have the ability to incorporate both redundancy and internal controls onto the same electrode surface.

Extending from the signal-on DNMT1 assay, we incorporate the two-electrode electrochemical platform enabling label-free measurements from crude cultured colorectal cancer cell lysates (HCT116) and biopsied tumor tissues (Furst et al., 2014). The multiplexed detection system involving patterning and detection from a secondary electrode array combines low-density DNA monolayer patterning and electrocatalytically amplified DNA CT chemistry to measure selectively and sensitively DNMT1 activity within these complex and congested cellular samples. Based on differences in DNMT1 activity measured with this assay, we distinguish colorectal tumor tissue from healthy adjacent tissue. No difficult or time-consuming purification steps are necessary. For each electrode, only ~4000 cultured cells or ~500 μg tissue sample are required. Importantly, because of the multiplexed nature of this platform, we are able to assay for substrate specificity while simultaneously measuring normal tissue and tumor tissue lysates. Therefore, with our platform, healthy tissue is easily distinguished from tumor tissue using very small amounts of sample. More generally, this work represents an important step in new electrochemical biosensing technologies.

Fig. 6—Cont'd capable of methylating DNA (*red arrow*), the DNA on the electrode becomes fully methylated. If the protein is not active, the DNA remains hemimethylated or unmethylated (*green arrow*). A methylation-specific restriction enzyme *BssH*II (*purple*) is then added that cuts the unmethylated or hemimethylated DNA (*green arrow*), significantly attenuating the electrochemical signal, while leaving the fully methylated DNA (*red arrow*) untouched. Constant potential amperometry (*right*) is used to measure the percent change before and after restriction enzyme treatment. If the restriction enzyme does not affect the DNA (*top*), the signals overlay. If, however, the restriction enzyme cuts the DNA, the signal is significantly attenuated (*bottom*).

Solution and Reagents:

1. 12-Azidododecane-1-thiol (Sigma-Aldrich)
2. 11-Mercaptoundecylphosphoric acid (Sigma-Aldrich)
3. HCT116 cells, either parent or DNMT1 −/− (received from Vogelstein Lab)
4. McCoy's 5A media, with 10% FBS, 100 units/mL penicillin, and 100 μg/mL streptomycin
5. Trypsin

Instruments and Supplies:

1. Au rod electrodes (1 mm in diameter)
2. 0.05 μm polish
3. 1.5 mm deep Teflon spacer
4. Tissue culture flasks (Corning Costar)
5. Cell culture incubator, 37°C under a humidified atmosphere containing 5% CO_2
6. Centrifuge
7. −80°C freezer
8. Bipotentiostat (Bioanalytical Systems)
9. Ag/AgCl reference electrode
10. Pt wire auxiliary electrode
11. DNA-modified multiplex chip
12. Size exclusion spin column (10 kDa cutoff, Amicon)
13. Incubator

Buffer Conditions:

1 × Phosphate buffer for cell culture

DNA phosphate buffer (5 m*M* sodium phosphate, 50 m*M* NaCl, pH 7)

Tris buffer (10 m*M* Tris, 100 m*M* KCl, 2.5 m*M* $MgCl_2$, 1 m*M* $CaCl_2$, pH 7.6)

DNMT1 activity buffer (50 m*M* Tris–HCl, 1 m*M* EDTA, 5% glycerol, pH 7.8)

Nuclear protein extraction kit (Pierce from Thermo Scientific)

Bicinchoninic assay (BCA, Pierce)

Procedure:

1. The multiplexed setup consisted of two complementary arrays containing 15 × 1-mm-diameter gold rod electrodes embedded in Teflon. Gold surfaces were polished with 0.05-μm polish before monolayer assembly.
2. Thiol SAMs were formed on one of the plates by incubating with 1 *M* 12-azidododecane-1-thiol and 1 *M* 11-mercaptoundecylphosphoric

acid in ethanol for 18–24 h, followed by rinsing with ethanol and phosphate buffer.

3. The water-soluble $[Cu(phendione)_2]^{2+}$ (phendione =1,10-phenanthroline-5,6-dione) was synthesized by mixing two equivalents of phendione with copper sulfate in water. Covalent attachment of DNA to mixed monolayers containing 50% azide head group and 50% phosphate head group through electrochemically activated click chemistry was accomplished by applying a sufficiently negative potential to the secondary electrode. Specifically, a constant potential of −350 mV was applied to a secondary electrode for 25 min, allowing for precise attachment of the appropriate DNA to a primary electrode. 40 μL of 100 μ*M* catalyst and 80 μL of 50 μ*M* DNA in Tris buffer were added to the platform for covalent attachment.
4. All electrochemistry was performed as constant potential amperometry for 90 s with an applied potential of 320 mV to the patterning/detecting electrode array and −400 mV to the substrate electrode array All scans were performed in Tris buffer with 4 μ*M* MB and 300 μ*M* ferricyanide.
5. To incubate electrodes with desired proteins or cell lysate, a 1.5-mm deep Teflon spacer was clipped to the primary electrode surface. Each electrode is isolated in an individual well that holds 4 μL of solution.
6. HCT116 cells were grown in McCoy's 5A media in tissue culture flasks in a cell culture incubator.
7. Approximately 6 million cells were harvested from adherent cell culture by trypsinization, followed by washing with cold PBS and pelleting by centrifugation at 500 × *g* for 5 min.
8. A nuclear protein extraction kit was used for cell lysis, with buffer then exchanged by size exclusion spin column into DNMT1 activity buffer.
9. Cell lysate was immediately aliquoted and stored at −80°C until use.
10. BCA was used to quantify the total amount of protein in the lysate. The total protein concentration at which the lysate was frozen was 35,000–50,000 μg/mL.
11. Cell lysate was combined with SAM to a final SAM concentration of 160 μ*M*; or the lysate was diluted in DNMT1 activity buffer to the desired total protein concentration and then combined with SAM to a final SAM concentration of 160 μ*M*. Incubate the electrodes at 37°C for 2 h in a humidified chamber.
12. Then the electrodes were treated with protease and restriction enzyme *BssH*II as described in last section.

Note:

1. The two-electrode array contains two sets of fifteen gold electrodes, each embedded in a Teflon plate. Each electrode has a 1 mm diameter (Fig. 6). The two complementary Teflon arrays are assembled with a 150-μm spacer between them, which were previously determined to be the optimal distance such that signals are not diffusion-limited. The electrodes of the primary (bottom) array are modified with DNA of the desired sequences such that DNA-mediated charge transport is detectable. The electrodes of the secondary (top) array are bare for electrochemical detection.

4. REDOX-ACTIVE ENZYMES IN DNA REPAIR MONITORING A REDOX-ACTIVE PROTEIN

Studies of DNA repair enzymes containing [4Fe4S] clusters repeatedly demonstrated that the proteins were isolated in an EPR-silent, $[4Fe4S]^{2+}$ oxidation state (Boal et al., 2005; Cunningham et al., 1989; Hinks et al., 2002) and resistant to a change in cluster redox state even upon addition of powerful chemical oxidants and reductants to the protein solution. This lack of redox activity was observed in several spectroscopic and biophysical studies, leading to the early conclusion that the [4Fe4S] cluster played a structural rather than functional role in Endonuclease III (Cunningham et al., 1989; Fu et al., 1992; Thayer et al., 1995). In the case of MutY, a BER glycosylase with significant homology to Endonuclease III (Michaels et al., 1990); however, the [4Fe4S] cluster was demonstrated to be nonessential for structural integrity of the protein (Markkanen, Dorn, & Hübscher, 2013; Porello, Cannon, & David, 1998). A substrate-sensing role was thus proposed for the cluster in light of this discovery, but a chemical role for the cofactor in these BER enzymes continued to elude observation.

More structural and biochemical studies of both DNA-dissociated and DNA-bound forms of these proteins continued to make progress towards demonstrating a role for the [4Fe4S] cluster. DNA-bound, high-resolution X-ray crystal structures of both Endonuclease III and MutY (Fromme, Banerjee, Huang, & Verdine, 2004; Fromme & Verdine, 2003), as well as DNA-free structures, were determined, and it was clear that the protein conformation in DNA-bound structures of these proteins was not radically different from conformation in the DNA-dissociated structures (Fromme et al., 2004; Fromme & Verdine, 2003; Guan et al., 1998; Thayer et al., 1995).

The [4Fe4S] cluster was, additionally, relatively close to the bound DNA substrate (approximately 20–30 Å from the DNA) in both the MutY and *Endo*III structures. The short distance from [4Fe4S] cluster to DNA is especially striking when considering that labile Fe^{2+} ions from such a cofactor can react with hydroxyl radicals and other reactive oxygen species, which are a natural consequence of aerobic respiration in cells, to damage DNA bases (Imlay, 2013). The design in Nature of placing a potentially harmful metal cofactor in a position so close to bound DNA suggested that the [4Fe4S] cluster plays a more significant role in these enzymes.

To probe directly any redox chemistry associated with the cluster, DNA-mediated electrochemistry studies were carried out (Boal et al., 2005; Gorodetsky et al., 2006). This platform is unique in that it facilitates the study of *DNA-bound* electron transfer activity of proteins, such as the vast family of redox-active, DNA-binding enzymes associated with genomic repair (Boal et al., 2009, 2005; DeRosa et al., 2005; Grodick et al., 2014; Mui et al., 2011). When BER glycosylases MutY (*E. coli*), Endonuclease III (*E. coli*), and UDG (*A. fulgidus*) were initially assayed for DNA-bound redox activity, all three displayed reversible redox signals in the physiologically relevant potential range, corresponding to cycling between the $[4Fe4S]^{2+}$ and $[4Fe4S]^{3+}$ oxidation states (Boal et al., 2005). They additionally all displayed similar midpoint potentials, near ~85 mV vs NHE. Thus these enzymes appeared to be activated for redox activity at physiological potentials when bound to the DNA polyanion. A subsequent study of Endonuclease III on highly oriented pyrolytic graphite (HOPG) electrodes, comparing redox potentials in the presence and absence of DNA (Gorodetsky et al., 2006), provided direct evidence supporting this observation. The HOPG electrode study showed that while the DNA-dissociated potential of the [4Fe4S] cluster in Endonuclease III is outside the physiologically relevant potential range at approximately ~280 mV vs NHE, the DNA-bound potential is near 80 mV vs NHE (Gorodetsky et al., 2006). This 200 mV negative shift in potential corresponds thermodynamically with a 1000-fold increase in DNA-binding affinity for the oxidized, $[4Fe4S]^{3+}$ Endonuclease III relative to the reduced $[4Fe4S]^{2+}$ Endonuclease III. Thus binding of DNA shifts the potential of the [4Fe4S] cluster in these enzymes into the physiologically relevant range, promoting oxidation to the $[4Fe4S]^{3+}$ state and reversible biological redox activity.

DNA electrochemistry allows for observation of redox activity under physiologically relevant conditions and is adaptable to important control experiments for characterization of the redox signal. Since DNA CT is sensitive to base-pair mismatches and apurinic sites (Arnold et al., 2016; Grodick, Muren, & Barton, 2015), a duplex substrate on the DNA-modified electrode containing a mismatch or apurinic site attenuates a DNA-mediated redox signal. This control allows for confirmation that the protein redox activity is mediated by DNA CT. Additionally, charge transfer-deficient protein mutants, such as Endonuclease III Y82A, are easily assayed on this platform (Boal et al., 2009). A perturbed CT pathway through the protein also attenuates the DNA-bound signal and helps to characterize these mutants before they are used in other in vitro experiments, such as activity assays or genetics studies. Finally, proteins that bind nucleotide triphosphates, such as ATP, in their active form can be tested in the presence and absence of their cofactors (Grodick et al., 2014; Mui et al., 2011). This assay allows for comparison of redox activity in the active and inactive forms of the enzyme.

With physiologically relevant assay conditions and adaptability to different DNA and small-molecule substrates, DNA-mediated electrochemistry can be used to monitor several different kinds of enzymatic activity. Helicase activity from the [4Fe4S] enzymes XPD (*Sulfolobus acidocaldarius*) in the nucleotide excision repair (NER) pathway (Mui et al., 2011), as well as DinG (*E. coli*) in the R-loop maturation pathway (Grodick et al., 2014), has been measured on DNA-modified electrodes. XPD is an ATP-dependent, 5′- to 3′-helicase that is part of the TFIIH machinery, which is important in transcription and repair (Fan et al., 2008; Liu et al., 2008). Mutations in the [4Fe4S] domain of XPD furthermore lead to genetic disorders such as trichothiodystrophy (TTD), Cockayne syndrome (CS), and xeroderma pigmentosum (XP) (Fan et al., 2008; Liu et al., 2008; Mui et al., 2011). When assayed on a DNA-modified Au electrode, XPD was shown to have a similar DNA-bound redox potential to the BER glycosylase enzymes, near 80 mV vs NHE. The redox signal for XPD increased in current upon addition of ATP, but did not change in the presence of a non-hydrolyzable ATP analogue (Mui et al., 2011). The positive correlation between ATP hydrolysis by XPD and the DNA-bound redox signal suggests that the active form of XPD is better coupled to the DNA duplex for DNA-mediated CT activity (Mui et al., 2011). A similar midpoint potential and ATP dependence was observed in the DinG signal on these electrodes (Grodick et al., 2014). Electrochemical experiments measuring the DNA-bound redox activity of glycosylase and helicase enzymes thus shaped the foundation for further biophysical and genetics studies (Boal et al., 2009; Grodick et al., 2014; Sontz, Mui, Fuss, Tainer, & Barton, 2012), which together shaped the foundation for a model in which DNA CT mediates the first step in searching the genome for damage.

A multiplexed electrode setup (Slinker et al., 2010, 2011) was adapted for DNA-bound, redox-active protein electrochemistry (Pheeney et al., 2013). Previously, electrochemistry on DNA-processing enzymes was performed on single surfaces; an AFM surface served as the Au working electrode (Boal et al., 2009, 2005; DeRosa et al., 2005; Mui et al., 2011; Romano, Sontz, & Barton, 2011). A different electrode surface was prepared to obtain each replicate of a single experiment, or to compare of CT activity between different substrates, such as well-matched versus mismatched DNA. The multiplexed setup, however (Fig. 4), has 16-Au electrodes at the center of a silicon chip, which can be patterned and fabricated using standard photolithography and metal evaporation techniques (Pheeney et al., 2013; Slinker et al., 2010, 2011). These electrodes are uniform in area and physically divided into four quadrants, so redox activity on as many as four different DNA substrates can be compared in parallel on a single surface. Though the single electrode platform with an Au AFM surface is robust and straightforward, this multiplexed system facilitates more efficient and ultimately more thorough characterization of DNA-bound redox enzyme activity.

4.1 Conditions for Protein Electrochemistry

When running a protein electrochemistry experiment, it is important to ensure that the buffer conditions, potential range, and protein concentrations used are not harmful to the DNA substrates or the protein in solution. It is also important that they facilitate protein binding to DNA, so a signal can be observed. The buffer pH should be near physiological pH; the general range depends on specific storage conditions for a protein and falls between 7.0 and 8.0. Low pH conditions are not recommended, as they can promote glycosidic bond cleavage and depurination of DNA bases on the electrode surface (An et al., 2014). Phosphate is a salt with good buffer capacity, and it should be used as the buffer salt when possible. Some enzymes, such as helicases (Grodick et al., 2014; Mui et al., 2011) and polymerases (O'Brien et al., 2017), however, have phosphate binding sites for native nucleotide triphosphate substrates. These enzymes may have compromised ability to bind DNA or their NTP substrates in electrochemical assays, and in this case a buffer such as Tris or HEPES can be substituted. Tris buffer has a relatively large temperature dependence on pH, however (New England BioLabs,

Inc., 2017) and should be used in electrochemical experiments only in a temperature-controlled environment.

Protein concentration is important when performing DNA-mediated electrochemistry. The general range of concentrations, though optimal conditions depend on the specific protein assayed, is approximately 5–50 μ*M*. Larger proteins often produce better signals with lower protein concentrations on the electrode, as this allows unhindered individual molecules to diffuse to the DNA substrate, and bind the substrate in a redox-active manner (Grodick et al., 2014; Mui et al., 2011). Though optimal concentration for each protein is determined empirically, the important consistency in these experiments is measurement of concentration for molecules of protein loaded with the redox cofactor. Proteins with a [4Fe4S] cluster have an absorbance at 410 nm in the UV–visible spectrum, for example (Cunningham et al., 1989), and the concentration of cluster-loaded protein is most important because those macromolecules, unlike the apoprotein, will be capable of producing a redox signal. For CV experiments, the redox potential scanning range is typically 0.1 to −0.4 V vs Ag/AgCl (0.3 to −0.2 V vs NHE) (Boal et al., 2009; Grodick et al., 2014; Mui et al., 2011; Pheeney et al., 2013). A range of 0.4 to −0.3 V vs NHE is acceptable for a signal with a different potential or widely split reductive and oxidative peaks, however. Bulk electrolysis potentials are generally higher or lower than these limits by only 0.1–0.2 V, to ensure full reduction or oxidation of a sample. These mild potential values are important to use on this platform, as they will not strip the DNA monolayer and preclude the measure of DNA-mediated redox activity.

4.2 *Endo*III and MutY: [4Fe4S] Proteins in BER

We have made extensive use of electrochemistry on DNA-modified gold electrodes as well as HOPG and pyrolytic graphite edge (PGE) in the presence and absence of DNA to investigate the role of the [4Fe4S] cluster in BER proteins (Bartels et al., 2017; Boal et al., 2005; Gorodetsky et al., 2006). Among these proteins, the first to be well characterized were the *E. coli* base excision (BER) glycosylases endonuclease III (*Endo*III) and MutY (Fig. 7) (Cunningham et al., 1989; Fromme & Verdine, 2003; Porello et al., 1998; Thayer et al., 1995). *Endo*III targets oxidized pyrimidines, excising the base as well as nicking the phosphate backbone. MutY is much more specific in its actions, excising adenine mispaired with 8-oxo guanine; unlike *Endo*III, MutY lacks AP lyase activity and cannot

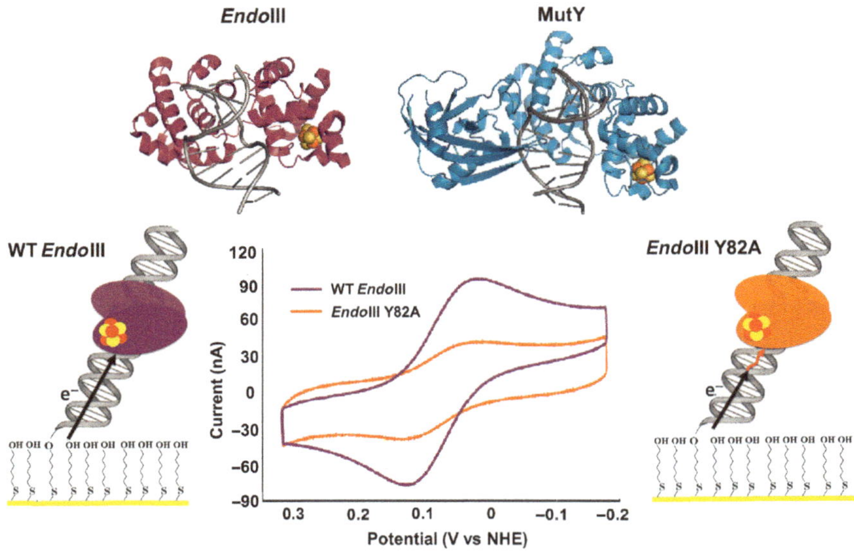

Fig. 7 Electrochemistry of *Endo*III and MutY on DNA-modified gold electrodes. *Endo*III and MutY are BER glycosylases that target sites of oxidative damage in DNA; *Endo*III (*top left*) excises oxidized pyrimidines, while MutY (*top right*) removes adenine mispaired with 8oxoG. When incubated on a DNA-modified electrode, both proteins (*Endo*III depicted) display reversible single-electron redox peaks by CV, a process that can be disrupted by mutating critical amino acid residues in the CT pathway as illustrated by *Endo*III Y82A (*bottom*). Structures are adapted from PDB structures IP59 (*Endo*III) and 1RRQ (MutY); both are from *Geobacillus stearothermophilus*, but each shows high homology to the *E. coli* proteins used in electrochemistry. *The CV is adapted from Pheeney, C. G., Arnold, A. R., Grodick, M. A., & Barton, J. K. (2013). Multiplexed electrochemistry of DNA-bound metalloproteins.* Journal of the American Chemical Society, 135, *11869–11878.*

nick the phosphate backbone. Both of these enzymes were first studied on DNA-modified gold using a 15-mer duplex containing a 5′ thiol modification on one strand (Boal et al., 2005). Using CV, reversible signals occurred at a midpoint potential of 60 mV vs NHE for *Endo*III and 90 mV vs NHE for MutY (Fig. 7). These potentials are at the lower range of HiPIP [4Fe4S] proteins, which utilize the $[4Fe4S]^{3+/2+}$ couple, and EPR experiments in this and other studies independently confirmed that this was the couple observed electrochemically (Boal et al., 2005; Yavin et al., 2005).

Importantly, the total peak charge was significantly attenuated on films consisting of DNA with an abasic site positioned near the monolayer surface, indicating that the signal was DNA-mediated (Boal et al., 2005). This platform may also be used to elucidate the CT pathway between the DNA and the cluster, which is usually positioned ~15 Å from the DNA (Fromme et al., 2004; Fromme & Verdine, 2003) and thus requires the assistance of aromatic amino acid residues to shuttle charge. With this purpose, we have investigated a range of CT-deficient *Endo*III mutants, many of which have disease-relevance in the human homologues (Romano et al., 2011). Among these, *Endo*III Y82A exhibited a particularly striking signal attenuation relative to WT, implicating this residue in the CT pathway (Boal et al., 2009).

Work carried out primarily with *Endo*III has further revealed the importance of monolayer morphology to protein signals (Pheeney et al., 2013). On high-density films, close packing can render much of the DNA inaccessible, but the enhanced rigidity can make DNA-mediated signals easier to achieve. In contrast, low density films offer more accessible DNA, but the increased flexibility of the DNA can cause much of the signal to become surface-mediated, making the ability of a protein to signal through the DNA difficult to determine. Which monolayer is most appropriate depends largely on the protein itself. *Endo*III shows smaller, but DNA-mediated, signals on high-density films, and larger, DNA-bound but surface-mediated signals on low density films; in contrast, bulkier proteins like MUTYH show the opposite effect (Bartels, et al., unpublished manuscript). For this reason, we find it is important to characterize any new protein on both types of DNA film, checking for mismatch or abasic site discrimination in each case as well.

Because *Endo*III can be prepared at relatively high concentrations and in large quantities (Pheeney et al., 2013), it is possible to study this protein on graphite electrodes in the absence of DNA. We have used both HOPG and PGE for this purpose, with HOPG serving to make direct comparisons between DNA-free and DNA-bound proteins and PGE allowing us to determine the effects of nearby amino acids and non-DNA molecules on cluster potential. Using HOPG, SWV revealed a large, reversible DNA-mediated signal at 20 mV vs NHE, while a smaller, irreversible signal was present around 250 mV vs NHE, and the DNA-free $[4Fe4S]^{2+/+}$ couple was also observed near −300 mV vs NHE; these signals were also visible by CV, but the DNA-free signal was small with widely split peaks, impeding quantification (Gorodetsky et al., 2006). On a DNA-modified electrode, the protein is brought to the surface, but this does not occur without DNA modifications. Whether one is using HOPG or PGE, we have found stock concentrations of 50–75 μ*M* protein to be ideal; lower concentrations can be observed, but small and shallow peaks make the signal difficult to quantify.

To compare *Endo*III and MutY in the absence of DNA, and to determine the effect of point mutations on cluster potential, we turned to thin film voltammetry on PGE with carbon nanotubes (CNTs), which provided large and readily quantifiable signals in the absence of DNA (Bartels et al., 2017). DNA tends to passivate the electrode surface in this system, however, so, while it is possible to see an effect of DNA-binding, the signals are far from ideal and noticeable only by SQWV. Limitations aside, using this system, we were able to show that DNA-binding has a large effect on the potential in both *Endo*III and MutY, while poly-L-glutamate and several *Endo*III point mutations (E200K, Y205H, and K208E) containing an altered charge distribution in the Fe–S domain do not appreciably alter the potential.

Overall, we have used DNA-modified gold, HOPG, and PGE electrodes to study BER, complemented by spectroscopic techniques, to study long-range electron transfer in [4Fe4S] BER glycosylases. Each of the electrochemical platforms described is useful for answering particular questions about these enzymes, and together they tell a more complete story than any single method alone. DNA-modified gold is well-characterized and easy to use, facilitating extensive characterization of DNA-bound proteins. This includes measurement of the DNA-bound potential, abasic site, or mismatch discrimination experiments to determine if the signal is DNA-mediated, and experiments with mutant proteins to identify the intraprotein CT pathway. The larger potential window provided by graphite electrodes allows both relevant redox couples of the [4Fe4S] proteins to be observed, and make direct comparisons of DNA-free and DNA-bound proteins possible. Among graphite electrodes, HOPG is ideal for comparing potentials on and off of DNA, while PGE is superior for examining differences between DNA-free proteins. In summary, the combination of these three platforms can provide a wealth of information, and their use in our lab has greatly improved the understanding of previously puzzling [4Fe4S] clusters in BER.

4.3 XPD: [4Fe4S] Proteins in NER

Some time after the discovery and characterization of DNA processing [4Fe4S] enzymes in BER, [4Fe4S] clusters were reported in proteins operating in several other repair pathways, including NER. In particular, our laboratory has worked extensively with the specialized helicases XPD (Fig. 8) and DinG, both superfamily 2 helicases (Grodick et al., 2014; Mui et al., 2011). XPD (Rad3 in yeast) is a component of the archaeal and eukaryotic transcription factor IIH (TFIIH) complex, responsible for unwinding DNA surrounding bulky lesions such as thymine dimers to facilitate their removal

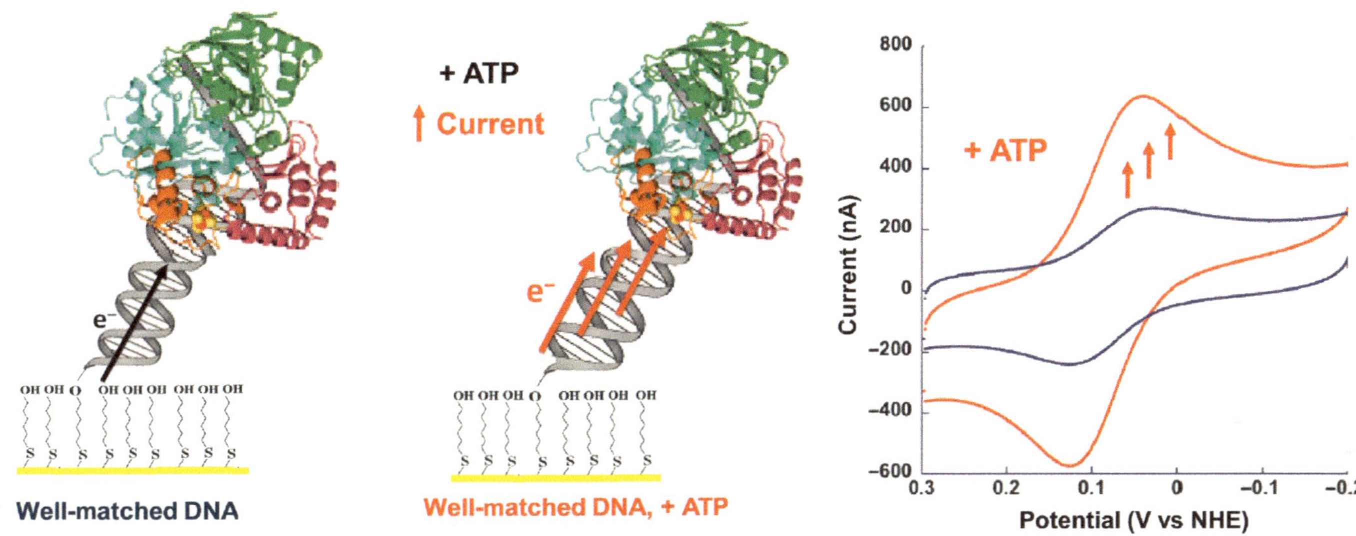

Fig. 8 Electrochemistry of the NER helicase XPD on a substrate containing a 9-mer 5′ single-stranded overhang. On this substrate, electrochemical experiments with XPD yielded a signal similar in potential and general form to those from BER proteins (*left*, *blue* CV at *right*). The addition of ATP, known to stimulate helicase activity in XPD, resulted in a substantial increase in current as a result of enhanced electronic coupling between the [4Fe4S] cluster and the DNA base stack (*middle*, *red* CV at *right*). *All images in this figure were adapted from Mui, T. P., Fuss, J. O., Ishida, J. P., Tainer, J. A., & Barton, J. K. (2011) ATP-stimulated, DNA-mediated redox signaling by XPD, a DNA repair and transcription helicase.* Journal of the American Chemical Society, 133, *16378–16381.*

by endonucleases during NER (Fan et al., 2008). Although structurally homologous to XPD, DinG is a bacterial helicase specialized for unwinding RNA–DNA hybrids at sites of replication fork/transcription collisions (Ren, Duan, & Ding, 2009).

Electrochemistry with helicases on DNA-modified gold electrodes required some modifications from previous work. Unlike *Endo*III and MutY, which can bind nonspecifically to dsDNA, XPD, as a $5' \rightarrow 3'$ helicase, preferentially binds duplexes containing a $3'$ single-stranded overhang. To meet these criteria, a DNA substrate consisting of 20 base pairs of duplexed DNA and a 9-mer single-stranded overhang was prepared (Fig. 8). With respect to substrates of this type, we have had success with single-stranded overhangs ranging from 9–20 nucleotides in length (Mui et al., 2011); however, since ssDNA can be problematic on gold electrodes due to its tendency to adhere to the surface, overhangs substantially longer than 20 nucleotides are not recommended. Finally, as with *Endo*III and MutY, the DNA monolayers were all formed in the absence of Mg^{2+} to improve substrate accessibility.

Electrochemistry with low density monolayers of the 20:29-mer substrate resulted in a substantial signal from XPD centered around 80 mV vs NHE, similar in both appearance and potential to *Endo*III and MutY (Mui et al., 2011) (Fig. 8). As long as the duplexed region of the DNA extends below the bound protein, mismatch discrimination experiments can be performed, and the incorporation of a CA mismatch into the base of this substrate did lead to substantial charge attenuation. Overall, these experiments revealed that, despite the single-stranded overhang, the surface-bound DNA was recognizable to a specialized helicase, and mismatch discrimination further indicated that this DNA was accessible in an upright conformation even to a protein as large as XPD (*S. acidocaldarius* XPD is ~64 kDa).

In further contrast with *Endo*III and MutY, XPD activity requires ATP hydrolysis to unwind DNA; this afforded an opportunity to observe enzymatic activity on an electrode. Addition of ATP, but not the poorly hydrolyzed analogue ATP γ-S, led to a sharp increase in current (Fig. 8), indicative of improved coupling to the electrode upon helicase activity (Mui et al., 2011). It appears that XPD is able to signal its activity. Likely, this arises as a result of a conformational change associated with activity leading to more effective electronic coupling on the electrode. As mentioned previously, the electrochemical signal from a diffusing [4Fe4S] protein grows in overtime and eventually stabilizes before decreasing, which can complicate observation of signal enhancement by accessory factors such as ATP. To do so, ATP was added to the surface only after the signal had stabilized, and the change was recorded as the increase in current over time (current after ATP addition increased much more sharply than in either normal signal growth or following ATP γ-S addition).

In summary, DNA-modified gold electrodes can be prepared with diverse substrates, including those with single-stranded overhangs, and this platform can also be used to study the effects of cofactor binding and enzymatic activity on [4Fe4S] cluster coupling. Together, these features allow one to study a wide range of proteins under similarly diverse conditions, as exemplified by XPD.

4.4 Eukaryotic DNA Primase

The redox activity of human DNA primase on DNA illustrates the range of protein detection possible on the DNA-modified electrode platform (O'Brien et al., 2017). Eukaryotic primases are heterodimeric enzymes that contain a [4Fe4S] cluster cofactor in the large, regulatory subunit (Kuchta & Stengel, 2010). DNA primase, as well as the [4Fe4S] cluster domain of DNA primase, each have the intrinsic ability to bind with modest affinity a primed DNA substrate (Sauguet, Klinge, Perera, Maman, & Pellegrini, 2010; Vaithiyalingam, Warren, Eichman, & Chazin, 2010). It was initially observed on DNA-modified electrodes that the [4Fe4S] domain, as isolated in the $[4Fe4S]^{2+}$ redox state, does not bind DNA tightly enough to couple the redox cofactor into the DNA duplex for CT. In light of the previously demonstrated disparity in DNA binding between the oxidized $[4Fe4S]^{3+}$ state and the reduced, $[4Fe4S]^{2+}$ state (Gorodetsky et al., 2006) of these enzymes, bulk electrolysis was performed in anaerobic conditions to monitor and compare the redox activity of an oxidized vs a reduced sample of the DNA primase [4Fe4S] domain. Upon performing bulk electrolysis to produce electrochemically reduced or oxidized samples, CV scans demonstrated that the reduced $[4Fe4S]^{2+}$ protein, similar to the purified sample before electrolysis, was electrochemically inactive on DNA. The oxidized sample, however, displayed a large reductive peak that disappears after a single scan in CV to negative potentials but is regenerated upon iterative electrochemical oxidations. Electrochemical oxidation or reduction of a protein sample before CV scanning, as was previously demonstrated with *Endo*III (Boal et al., 2005), facilitates detection of redox activity in identical enzymes under different cluster oxidation states. In the case of primase, the significant difference between the redox activity of the $[4Fe4S]^{3+}$ and $[4Fe4S]^{2+}$ states first observed on DNA-modified electrodes provided significant insight into the relationship between a change in [4Fe4S] cluster oxidation state and the primer synthesis/redox signaling activity of the enzyme.

Using bulk electrolysis to oxidize and reduce samples of DNA-binding [4Fe4S] proteins, before electrochemically monitoring the redox activity with CV or SWV scanning, thus provides a method of directly comparing DNA-mediated redox activity of the protein in the oxidized, $[4Fe4S]^{3+}$ state, and the reduced, $[4Fe4S]^{2+}$ state. This method of electrochemical oxidation is advantageous, as it produces a sample of protein in a specific redox state without the damaging effects of chemical oxidants or reductants. Although electrochemical oxidation and reduction avoids potentially damaging the cluster, which for example can degrade to the $[3Fe4S]^{+}$ species upon oxidation by $Co(phen)_3^{3+}$ (phen = 1,10-phenanthroline) (Boal et al., 2005), there are important limits to this method of cluster oxidation/reduction. The sample cannot be stoichiometrically oxidized/reduced on a feasible timescale for a protein experiment on a DNA electrode. Generally, yields of 60%–80% oxidized protein result upon bulk electrolysis for ~5–10 min. The other important consideration in these electrochemical experiments is that the sample must be electrochemically converted on a DNA-modified electrode in an anaerobic atmosphere, with deoxygenated reagents, to avoid atmospheric oxidation of the cluster (Imlay, 2013) and to fully control the redox state of the sample assayed.

5. GRAPHITE ELECTRODES FOR DIRECT ELECTROCHEMISTRY IN THE PRESENCE AND ABSENCE OF DNA

DNA-modified gold is an invaluable tool in studying redox-active proteins as well as in detection, as described in subsequent sections, but the limited potential window available on gold monolayers is a disadvantage when attempting to compare potentials of redox-active [4Fe4S] proteins on and off DNA. On DNA-modified gold electrodes, the potentials of DNA-bound [4Fe4S] proteins all ranged between 65 and 95 mV vs NHE, but little to no signal occurred on films lacking DNA (Boal et al., 2005) and in-solution DNA-free proteins were largely inert to electron transfer (Cunningham et al., 1989; Porello et al., 1998). Taken together, these observations implied that DNA-binding shifted the potential to some extent, but the degree of this change could not readily be investigated because the potential window on gold SAMs is limited by desorption of alkanethiols from gold at reducing potentials (≤ -0.4 V vs NHE) (Imabayashi et al., 1997; Walczak et al., 1991) and gold oxidation at higher potentials. Although gold oxidation occurs at ~1.5 V for bare gold in concentrated acid, this potential decreases by hundreds of mV with increasing pH and is further lowered in the presence of alkanethiol monolayers (Benck, Pinaud, Gorlin, & Jaramillo, 2014; Esplandiú, Hagenström, & Kolb, 2001). Finally, the window on gold was too narrow to observe the effect of DNA binding on the lower potential $[4Fe4S]^{2+/+}$ couple (potentials for this couple typically range from −300 to −700 mV vs NHE).

Pyrolytic graphite electrodes, with an available potential window of 2 V, offered a solution to these problems (Fig. 9) (Gorodetsky et al., 2006).

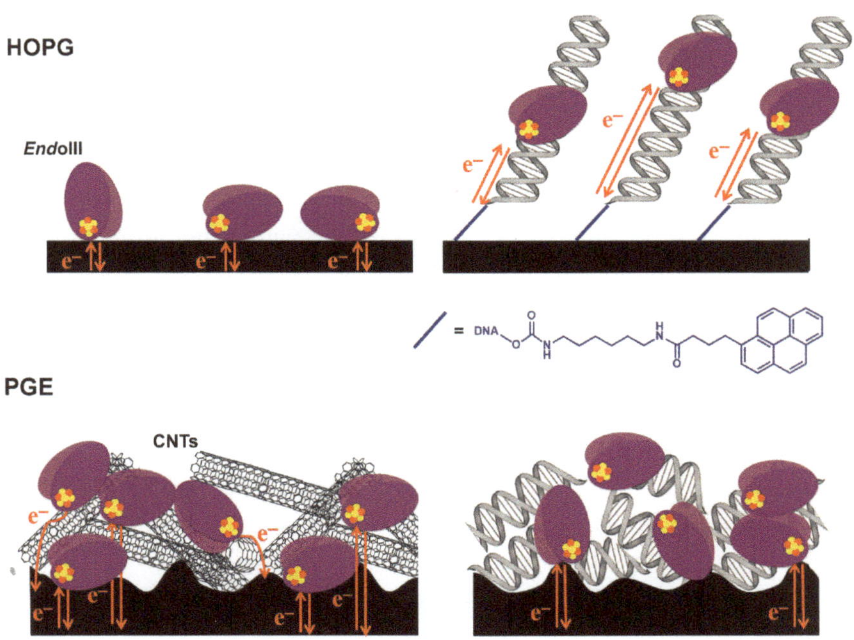

Fig. 9 Graphite platforms for protein electrochemistry. Two general platforms are commonly used for protein electrochemistry: HOPG (*top*) and PGE (*bottom*). HOPG consists of a pristinely flat, strongly hydrophobic surface, while PGE is rough and often contains surface oxides that lower the hydrophobicity. Proteins can adsorb directly to HOPG, although this interaction is weak, but the ability of pyrene-modified DNA to form a noncovalent bond with the surface allows a direct comparison of DNA-free and DNA-bound proteins. In contrast, PGE provides ample surface area for binding DNA-free proteins, and signals can be further enhanced by the addition of carbon nanotubes (CNTs); DNA can be incorporated into a film on PGE, but in this environment, it tends to passivate the surface and the random orientation prevents the observation of DNA-mediated signals.

Carbon electrodes are commonly used in the study of redox-active proteins, and several forms of graphite surface have been well characterized (Armstrong, Bond, Hill, Oliver, & Psalti, 1989; Banks & Compton, 2006; Blanford & Armstrong, 2006), including HOPG and PGE. Both of these electrodes are formed from stacks of conductive graphite sheets, but they differ in the nature of the exposed electroactive surface (Banks & Compton, 2006): HOPG exposes the largely flat basal plane of the uppermost sheet, and electron transfer is through the stacked sheets, while the PGE surface consists of the perpendicular edge plane with electron transfer occurring through individual sheets. These properties lend advantages and disadvantages to each surface, and the choice is largely dictated by the nature of the experimental questions at hand; indeed, we have successfully used both in our investigations of DNA-binding proteins containing [4Fe4S] clusters.

Redox-active proteins can be observed directly on HOPG, although the scarcity of suitable binding sites generally results in small, highly split redox peaks in electrochemistry (Armstrong et al., 1989; Gorodetsky et al., 2006). However, because the HOPG surface is composed of a layer of sp^2-hybridized carbon, noncovalent stacking interactions between the surface- and pyrene-modified molecules are favored and it is possible to form monolayers with pyrene-modified molecules. In our laboratory, we have established a procedure for preparing pyrene-modified DNA; characterization by AFM, $[Ru(NH_3)_6]^{3+}$, and ^{32}P labeling indicates that these monolayers are similar to those formed on gold electrodes (Gorodetsky & Barton, 2006). Pyrene modification of DNA is relatively simple, involving a series of couplings that can be undertaken under ambient conditions, requiring only the exclusion of water. Once prepared, pyrenated DNA monolayers are relatively straightforward to form on HOPG, and the surface can be backfilled with octane; which blocks the exposed surface analogously to alkanethiols on gold.

5.1 Pyrene Modification of DNA

Solvent and Reagents:

1. Dry acetonitrile
2. Dry methanol
3. Dry dioxane
4. Dry dichloromethane
5. Diisopropylethylamine (DIEA)

6. 1,1′-Carbonyldiimidazole (CDI)
7. Hexanediamine
8. 1-Pyrenebutyric acid *N*-hydroxysuccinimide ester
9. Ammonium hydroxide
10. HPLC-grade acetonitrile
11. 50 m*M* ammonium acetate (filtered)
12. Ethanol
13. 3 *M* NaCl

Instruments and Supplies:

1. Cylindrical glass cell with frit at base
2. Rubber septum
3. Parafilm
4. Shaker
5. 3 mL syringe
6. Needles
7. Aspirator
8. Tabletop centrifuge
9. Benchtop incubator
10. HPLC
11. PLRPS column for HPLC
12. UV–visible spectrophotometer

Buffer Conditions:

DNA phosphate buffer (5 m*M* sodium phosphate, pH 7.0, 50 m*M* NaCl).

Procedure:

1. Add freshly synthesized ssDNA bound to solid CPG beads to a glass cell connected to an aspirator; seal the exposed end with a septum secured by Parafilm. If the DNA was prepared directly on a DNA synthesizer, the terminal 5′-DMT group should be cleaved; alternatively, if the DNA is not attached to a solid support, one end must be blocked with a phosphate group to achieve selective functionalization.
2. Wash the beads successively with 3 mL dry acetonitrile (4 ×) and 3 mL dioxane (3 ×).
3. Add 1 mL CDI in dioxane and shake at RT for 3 h.
4. After 3 h, wash 3 × with 3 mL dioxane.
5. Add 1 mL hexanediamine in dioxane and shake for 30 min at RT.
6. Wash successively with 3 mL dioxane (3 ×), 3 mL dichloromethane (3 ×), 3 mL acetonitrile (3 ×), and 3 mL methanol (3 ×).

7. Add 1 mL 1-pyrenebutyric acid *N*-hydroxysuccinimide ester (dissolved in 90% dichloromethane/10% DIEA) and shake overnight at RT.
8. Wash successively with 3 mL dichloromethane (3 ×), 3 mL acetonitrile (3 ×), and 3 mL methanol.
9. Cleave DNA from CPG beads with 800 μL fresh NH_4OH (8 h, 60°C).
10. Isolate cleaved DNA by centrifugation in spin columns; discard the beads and dry the flow-through on a speed vacuum overnight.
11. Dissolve DNA in 600 μL phosphate buffer (5 m*M* sodium phosphate, pH 7.0, 50 m*M* NaCl) and purify by HPLC (acetonitrile/50 m*M* ammonium acetate in a gradient ranging from 95% to 85% ammonium acetate over 35 min, followed by a return to 95% over 5 min), making sure to collect the peak with absorbance at 260, 280, and 345 nm.
12. Freeze the sample in liquid nitrogen and dry overnight by lyophilization.
13. Desalt DNA: add 100 μL water, 50 μL 3 *M* NaCl, and 1 mL 100% ethanol; freeze in liquid nitrogen, spin down (12,000 rpm, 25 min), and dry overnight on a speed vacuum.
14. Quantify DNA by UV–vis and anneal with complement in a 1:1 ratio in phosphate buffer; annealing should be carried out by 5′-incubation at 95°C followed by slow cooling to RT over 2–3 h.

5.2 DNA Monolayer Assembly on HOPG

Although very useful for DNA-mediated electrochemistry, the highly hydrophobic HOPG surface is not ideal for direct electrochemistry without DNA, and PGE is instead favored for this purpose. PGE has a rough surface that provides abundant, readily accessible electroactive sites which improve the coupling of redox-active species to the electrode. To minimize protein diffusion, it is common practice to immobilize redox-active enzymes in a thin film when working with PGE. Proteins immobilized in such films have been shown to be catalytically active even after several days, indicating that adsorption to the electrode does not alter their native conformations (Baffert et al., 2012). Although the PGE surface can be very effective in facilitating direct electrochemistry, the incorporation of conductive single-walled CNTs into the film can greatly enhance signal size by providing additional three-dimensional area for protein binding (Yin, Lu, Wu, & Cai, 2005). To anchor the CNTs in place and prevent film dispersal, the entire protein-CNT film can be secured by a capping layer such as Nafion (Yin et al., 2005). Indeed, we have achieved large signals from DNA-free [4Fe4S] proteins using several (3–6) CNT-protein layers capped with 5% aqueous Nafion (Bartels et al., 2017). It should be noted that, despite the advantages of PGE in direct protein electrochemistry, this surface is not ideal for studying DNA-bound proteins. In particular, homogenous DNA films cannot be prepared on this surface, necessitating the addition of DNA along with the protein. In this state, the orientation of the DNA prevents experiments with mismatch or abasic site discrimination, which rely on upright DNA, and it causes severe surface passivation; furthermore, CNTs were found to hinder DNA binding by *Endo*III, and thus had to be excluded from these thin films (Bartels et al., 2017). Nonetheless, very small signals can still be obtained with DNA present, but they are of much lower quality than those obtained on gold or HOPG.

Solvents and Reagents:

1. Pyrene-modified DNA (preannealed)
2. Octane
3. Glycerol
4. Ethanol
5. MQ Water
6. Protein storage buffer

Instruments and Supplies:

1. HOPG electrode, either a commercial rod electrode (Pine Research Instrumentation provides a high-quality product) or surfaces (SPI Supplies sells these in several different grades)
2. 5 μm Silica polish (if using a rod electrode)
3. 3 *M* Double-sided Scotch tape (if using surfaces)
4. Electrochemical cell (if desired; inverted drop cells can be purchased from Pine Research Instrumentation or custom made)
5. Sonicator (if using a rod electrode)
6. 50 mL Falcon tubes

Buffer Conditions:

DNA phosphate buffer (5 m*M* sodium phosphate, pH 7.0, 50 m*M* NaCl).

Procedure:

1. a. If using a commercially prepared rod electrode, clean the surface with 0.05 μm silica polish. The polish can then be rinsed off by sonication for 0.5–1 min first in ethanol and then in water.

   b. Alternatively, if mounting a square of HOPG in a drop cell, clean HOPG can be exposed by pressing 3 *M* Scotch tape on the square and rapidly pulling back. This process will ideally remove a single layer, but it often takes several attempts before a truly pristine surface is obtained. Large defects are visible as pits on the surface and should be kept to a minimum; as with rod electrodes, a cursory buffer scan should be used to verify the absence of electroactive impurities.

2. To ensure the absence of surface defects and electroactive impurities, a CV scan of the surface should be taken in the protein storage buffer. Defects representing surface oxides on exposed edge plane give a broad, reversible electrochemical signal around 200 mV vs NHE. These defects can enhance direct electrochemistry of adsorbed proteins, but too many of them will inhibit the attachment of pyrenated DNA.
3. Once the surface is sufficiently clean, add 25 μ*M* annealed DNA (pyrenated strand + complement) in phosphate buffer to the surface and incubate overnight; if high-density surfaces are desired, 100 m*M* $MgCl_2$ may be added along with the DNA in this step.
4. Rinse two to three times with phosphate buffer to remove unbound DNA, taking care not to scratch the surface. This step should be carried out very gently if pipetting to avoid removing excess DNA; if this is problematic, the electrode can instead be immersed in a Falcon tube containing 5–10 mL phosphate buffer.
5. If DNA-free protein adsorption is a problem, the remaining bare electrode surface can be blocked by backfilling with octane. To backfill, add phosphate buffer containing 20% glycerol and 10% octane by volume and incubate 15–30 min at RT; rinse in phosphate buffer containing 20% glycerol as in step 4. This process should be repeated two to three times.
6. After backfilling, the electrode should be rinsed and scanned in protein storage buffer. Once the background scans have been taken, protein solution can be added to the surface and scanned. High concentrations are ideal if possible; in the case of *Endo*III, 50 μ*M* protein worked well, but even stronger signals were obtained using 200 μ*M* protein.

5.3 Protein Thin-Film Voltammetry With CNTs

Solutions and Reagents:

1. Ethanol
2. MQ water
3. Single-walled CNTs
4. Nafion (diluted to 5% in water, purchased as 10% solution from Sigma-Aldrich)

Instruments and Supplies:

1. PGE electrode (we use the relevant product from Pine Research Instrumentation, but others are available)
2. 400 grit sandpaper or diamond polish
3. Sonicator
4. Electrochemical cell (if desired)
5. Potentiostat
6. Reference electrode (usually Ag/AgCl)
7. Counter electrode (Pt wire)

Buffer Conditions:

Protein storage buffer

Procedures:

1. To roughen the surface for protein adsorption, the electrode should first be abraded with 400 grit sandpaper. This is done by applying water to a sheet of sandpaper and gently "polishing" the electrode. Sandpaper provides deep ridges for protein adsorption, and worked well for *Endo*III and MutY. If sandpaper treatment is ineffective, diamond polish can be used as an alternative; diamond polish generates pitting across the surface, which may be more suitable for some proteins.
2. To remove electroactive impurities, sonicate the electrode in ethanol (30–60 s) followed by water (30–60 s). At this point, a drop of protein storage buffer can be applied and the background scanned to ensure the absence of impurities. It should be noted that the rough edge plane is oxide rich, and surface species will generate a broad peak around 200 mV vs NHE; this should not interfere with protein voltammetry and can even assist in adsorption.
3. To prepare thin films, first dry on a layer of CNTs, and then apply several layers of protein solution as follows:
   - **(a)** Suspend single-walled CNTs in water (0.25 mg/mL) by vigorous sonication and add a droplet to the electrode surface, and dry in air or under an argon stream.
   - **(b)** Apply a droplet of protein solution (ideally 50–75 μ*M*) to the surface, and dry as with the CNTs. Repeat several times until the surface appears coated with protein.
   - **(c)** Secure the film with 5% Nafion in water; this should be applied to the electrode and dried as with the other layers.
4. Once the thin film is dried and secured, apply a 30–50 μL droplet to the vertical electrode, drop reference and auxiliary electrodes into the

bulk solution, and scan. Alternatively, the electrode can be inverted and placed in an electrochemical cell containing a bulk buffer solution.

In summary, graphite electrodes have the advantage of providing a wide accessible potential window that covers the entire range of potentials accessed by [4Fe4S] proteins, allowing the observation of redox events not possible on gold. Specifically, particular types of graphite electrode are better for addressing distinct aspects of protein electrochemistry. Because HOPG can facilitate both direct and DNA-mediated electrochemistry, this electrode is the best choice for making direct comparisons of DNA-free and DNA-bound potentials. However, the hydrophobicity of HOPG makes direct electrochemistry challenging, and if one is interested in studying redox-active proteins in the absence of DNA, thin-film voltammetry on PGE is the best option.

ACKNOWLEDGMENTS

We are grateful to the NIH for their support of this research and to our coworkers and collaborators for their insights, efforts, and hard work.

REFERENCES

Agard, N. J., Prescher, J. A., & Bertozzi, C. R. (2004). A strain-promoted [3+2] azide-alkyne cycloaddition for covalent modification of blomolecules in living systems. *Journal of the American Chemical Society*, *126*, 15046–15047.

An, R., Jia, Y., Wan, B., Zhang, Y., Dong, P., Li, J., et al. (2014). Non-enzymatic depurination of nucleic acids: Factors and mechanisms. *PloS One*, *9*, e115950.

Armstrong, F. A., Bond, A. M., Hill, H. A. O., Oliver, N., & Psalti, I. S. M. (1989). Electrochemistry of cytochrome c, plastocyanin, and ferredoxin at edge- and basal-plane graphite electrodes interpreted via a model based on electron transfer at electroactive sites of microscopic dimensions in size. *Journal of the American Chemical Society*, *111*, 9185–9189.

Arnold, A. R., Grodick, M. A., & Barton, J. K. (2016). DNA charge transport: From chemical principles to the cell. *Cell Chemical Biology*, *23*, 183–197.

Baffert, C., Sybirna, K., Ezanno, P., Lautier, T., Hajj, V., Meynial-Salles, I., et al. (2012). Covalent attachment of FeFe hydrogenases to carbon electrodes for direct electron transfer. *Analytical Chemistry*, *84*, 7999–8005.

Banks, C. E., & Compton, R. G. (2006). New electrodes for old: From carbon nanotubes to edge plane pyrolytic graphite. *Analyst*, *131*, 15–21.

Bartels, P.L.; O'Brien, E.; Barton, J.K. in collaboration with McDonnell, K.; Chemler, J.A.; et al. unpublished manuscript.

Bartels, P. L., Zhou, A., Arnold, A. R., Nuñez, N. N., Crespilho, F. N., David, S. S., et al. (2017). Electrochemistry of the [4Fe4S] cluster in base excision repair proteins: Tuning the redox potential with DNA. *Langmuir*, *33*, 2523–2530.

Baskin, J. M., & Bertozzi, C. R. (2007). Bioorthogonal click chemistry: Covalent labeling in living systems. *QSAR & Combinatorial Science*, *26*, 1211–1219.

Baylin, S. B., & Herman, J. G. (2000). DNA hypermethylation in tumorigenesis: Epigenetics joins genetics. *Trends in Genetics*, *16*, 168–174.

Benck, J. D., Pinaud, B. A., Gorlin, Y., & Jaramillo, T. F. (2014). Substrate selection for fundamental studies of electrocatalysts and photoelectrodes: Inert potential windows in acidic, neutral, and basic electrolyte. *PloS One*, *9*, e107942.

Blanford, C. F., & Armstrong, F. A. (2006). The pyrolytic graphite surface as an enzyme substrate: Microscopic and spectroscopic studies. *Journal of Solid State Electrochemistry*, *10*, 826–832.

Boal, A. K., Genereux, J. C., Sontz, P. A., Gralnick, J. A., Newman, D. K., & Barton, J. K. (2009). Redox signaling between repair proteins for efficient lesion detection. *Proceedings of the National Academy of Sciences of the United States of America*, *106*, 15237–15242.

Boal, A. K., Yavin, E., Lukianova, O. A., O'Shea, V. L., David, S. S., & Barton, J. K. (2005). DNA-bound redox activity of DNA repair glycosylases containing [4Fe-4S] clusters. *Biochemistry*, *44*, 8397–8407.

Boon, E. M., Barton, J. K., Pradeepkumar, P. I., Isaksson, J., Petit, C., & Chattopadhyaya, J. (2002). An electrochemical probe of DNA stacking in an antisense oligonucleotide containing a C3'-endo-locked sugar. *Angewandte Chemie, International Edition*, *41*, 3402–3405.

Boon, E. M., & Barton, J. K. (2003). DNA electrochemistry as a probe of base pair stacking in A-, B-, and Z-form DNA. *Bioconjugate Chemistry*, *14*, 1140–1147.

Boon, E. M., Ceres, D. M., Drummond, T. G., Hill, M. G., & Barton, J. K. (2000). Mutation detection by electrocatalysis at DNA-modified electrodes. *Nature Biotechnology*, *18*, 1096–1100.

Boon, E. M., Jackson, N. M., Wightman, M. D., Kelley, S. O., Hill, M. G., & Barton, J. K. (2003). Intercalative stacking: A critical feature of DNA charge-transport chemistry. *The Journal of Physical Chemistry B*, *107*, 11805–11812.

Buzzeo, M. C., & Barton, J. K. (2008). Redmond red as a redox probe for the DNA-mediated detection of abasic sites. *Bioconjugate Chemistry*, *19*, 2110–2112.

Campuzano, S., Pedrero, M., Montemayor, C., Fatas, E., & Pingarron, J. M. (2006). Characterization of alkanethiol-self-assembled monolayers-modified gold electrodes by electrochemical impedance spectroscopy. *Journal of Electroanalytical Chemistry*, *586*, 112–121.

Chen, R. Z., Akbarian, S., Tudor, M., & Jaenisch, R. (2001). Deficiency of methyl-CpG binding protein-2 in CNS neurons results in a Rett-like phenotype in mice. *Nature Genetics*, *27*, 327–331.

Cunningham, R. P., Asahara, H., Bank, J.F, Scholes, C. P., Salerno, J. C., Surerus, K., et al. (1989). Endonuclease III is an iron-sulfur protein. *Biochemistry*, *28*, 4450–4455.

DeRosa, M. C., Sancar, A., & Barton, J. K. (2005). Electrically monitoring DNA repair by photolyase. *Proceedings of the National Academy of Sciences of the United States of America*, *102*, 10788–10792.

Devaraj, N. K., Miller, G. P., Ebina, W., Kakaradov, B., Collman, J. P., Kool, E. T., et al. (2005). Chemoselective covalent coupling of oligonucleotide probes to self-assembled monolayers. *Journal of the American Chemical Society*, *127*, 8600–8601.

Esplandiú, M. J., Hagenström, H., & Kolb, D. M. (2001). Functionalized self-assembled Alkanethiol monolayers on au(111) electrodes: 1. Surface structure and electrochemistry. *Langmuir*, *17*, 828–838.

Esteller, M. (2002). CpG island hypermethylation and tumor suppressor genes: A booming present, a brighter future. *Oncogene*, *21*, 5427–5440.

Fan, L., Fuss, J. O., Chemg, Q. J., Arval, A. S., Hammel, M., Roberts, V. A., et al. (2008). XPD helicase structures and activities: Insights into the cancer and aging phenotypes from XPD mutations. *Cell*, *133*, 789–800.

Flynn, J., & Reich, N. (1998). Murine DNA (cytosine-5-)-methyltransferase: Steady-state and substrate trapping analyses of the kinetic mechanism. *Biochemistry*, *37*, 15162–15169.

Fromme, J. C., Banerjee, A., Huang, S. J., & Verdine, G. L. (2004). Structural basis for removal of adenine mispaired with 8-oxoguanine by MutY adenine DNA glycosylase. *Nature*, *427*, 652–656.
Fromme, J. C., & Verdine, G. L. (2003). Structure of a trapped endonuclease III-DNA covalent intermediate. *The EMBO Journal*, *22*, 3461–3471.
Fu, W., O'Handley, S., Cunningham, R. P., & Johnson, M. K. (1992). The role of the iron-sulfur cluster in *Escherichia coli* endonuclease III: A resonance Raman study. *The Journal of Biological Chemistry*, *267*, 16135–16137.
Furst, A. L., Hill, M. G., & Barton, J. K. (2013). DNA-modified electrodes fabricated using copper-free click chemistry for enhanced protein detection. *Langmuir*, *29*, 16141–16149.
Furst, A. L., Muren, N. B., Hill, M. G., & Barton, J. K. (2014). Label-free electrochemical detection of human methyltransferase from tumors. *Proceedings of the National Academy of Sciences of the United States of America*, *111*, 14985–14989.
Gorodetsky, A. A., & Barton, J. K. (2006). Electrochemistry using self-assembled DNA monolayers on highly oriented pyrolytic graphite. *Langmuir*, *22*, 7917–7922.
Gorodetsky, A. A., & Barton, J. K. (2007). DNA-mediated electrochemistry of disulfides on graphite. *Journal of the American Chemical Society*, *129*, 6074–6075.
Gorodetsky, A. A., Boal, A. K., & Barton, J. K. (2006). Direct electrochemistry of endonuclease III in the presence and absence of DNA. *Journal of the American Chemical Society*, *128*, 12082–12083.
Gorodetsky, A. A., Dietrich, L. E., Lee, P. E., Demple, B., Newman, D. K., & Barton, J. K. (2008a). DNA binding shifts the redox potential of transcription factor SoxR. *Proceedings of the National Academy of Sciences of the United States of America*, *105*, 3684–3689.
Gorodetsky, A. A., Ebrahim, A., & Barton, J. K. (2008b). Electrical detection of TATA binding protein at DNA-modified microelectrodes. *Journal of the American Chemical Society*, *130*, 2924–2925.
Grodick, M. A., Muren, N. B., & Barton, J. K. (2015). DNA charge transport within the cell. *Biochemistry*, *54*, 962–973.
Grodick, M. A., Segal, H. M., Zwang, T. J., & Barton, J. K. (2014). DNA-mediated signaling by proteins with 4Fe-4S clusters is necessary for genomic integrity. *Journal of the American Chemical Society*, *136*, 16470–16478.
Guan, Y., Manuel, R. C., Arvai, A. S., Parikh, S. S., Mol, C. D., Miller, J. H., et al. (1998). MutY catalytic core, mutant, and bound adenine structures define specificity for DNA repair enzyme superfamily. *Nature Structural Biology*, *5*, 1058–1064.
Hinks, J. A., Evans, M. C. W., de Miguel, Y., Sartori, A. A., Jiricny, J., & Pearl, L. H. (2002). An iron-sulfur cluster in the family 4 uracil-DNA glycosylases. *The Journal of Biological Chemistry*, *277*, 16936–16940.
Imabayashi, S., Iida, M., Hobara, D., Feng, Z. Q., Niki, K., & Kakiuchi, T. (1997). Reductive desorption of carboxylic-acid terminated alkanethiol monolayers from Au(111) surfaces. *Journal of Electroanalytical Chemistry*, *428*, 33–38.
Imlay, J. A. (2013). The molecular mechanisms and physiological consequences of oxidative stress: Lessons from a model bacterium. *Nature Reviews. Microbiology*, *11*, 443–454.
Jeltsch, A. (2002). Beyond Watson and Crick: DNA methylation and molecular enzymology of DNA methyltransferases. *Chembiochem*, *3*, 274–293.
Kelley, S. O., Barton, J. K., Jackson, N. M., & Hill, M. G. (1997). Electrochemistry of methylene blue bound to a DNA-modified electrode. *Bioconjugate Chemistry*, *8*, 31–37.
Kelley, S. O., Barton, J. K., Jackson, N. M., McPherson, L. D., Potter, A. B., Spain, E. M., et al. (1998). Orienting DNA helices on gold using applied electric fields. *Langmuir*, *14*, 6781–6784.
Kelley, S. O., Boon, E. M., Barton, J. K., Jackson, N. M., & Hill, M. G. (1999a). Single-base mismatch detection based on charge transduction through DNA. *Nucleic Acids Research*, *27*, 4830–4837.

Kelley, S. O., Jackson, N. M., Hill, M. G., & Barton, J. K. (1999b). Long-range electron transfer through DNA films. *Angewandte Chemie, International Edition, 38*, 941–945.

Kim, Y.-J., & Wilson, D. M., III. (2012). Overview of base excision repair biochemistry. *Current Molecular Pharmacology, 5*, 3–13.

Kissinger, P. T., & Heineman, W. R. (Eds.), (1996). *Laboratory techniques in electroanalytical chemistry* (2nd ed.). New York, NY: Marcel Dekker.

Kondo, T., Morita, J., Hanaoka, K., Takakusagi, S., Tamura, K., Takahasi, M., et al. (2007). Structure of Au(111) and Au(100) single-crystal electrode surfaces at various potentials in sulfuric acid solution determined by in situ surface X-ray scattering. *Journal of Physical Chemistry C, 111*, 13197–13204.

Kornberg, R. D. (2007). The molecular basis of eukaryotic transcription. *Proceedings of the National Academy of Sciences of the United States of America, 104*, 12955–12961.

Kuchta, R. D., & Stengel, G. (2010). Mechanism and evolution of DNA primases. *Biochimica et Biophysica Acta, 1804*, 1180–1189.

Liu, H., Rudolf, J., Johnson, K. A., McMahon, S. A., Oke, M., Carter, L., et al. (2008). Structure of the DNA repair helicase XPD. *Cell, 133*, 801–812.

Love, J. C., Estroff, L. A., Kriebel, J. K., Nuzzo, R. G., & Whitesides, G. M. (2005). Self-assembled monolayers of thiolates on metals as a form of nanotechnology. *Chemical Reviews, 105*, 1103–1169.

Markkanen, E., Dorn, J., & Hübscher, U. (2013). MUTYH DNA glycosylase: The rationale for removing undamaged bases from the DNA. *Frontiers in Genetics, 4*, 1–20.

Michaels, M. L., Pham, L., Nghiem, Y., Cruz, C., & Miller, J. H. (1990). MutY, an adenine glycosylase active on G-A mispairs, has homology to endonuclease III. *Nucleic Acids Research, 18*, 3841–3844.

Miranda, T. B., & Jones, P. A. (2007). DNA methylation: The nuts and bolts of repression. *Journal of Cellular Physiology, 213*, 384–390.

Mui, T. P., Fuss, J. O., Ishida, J. P., Tainer, J. A., & Barton, J. K. (2011). ATP-stimulated, DNA-mediated redox signaling by XPD, a DNA repair and transcription helicase. *Journal of the American Chemical Society, 133*, 16378–16381.

Muren, N. B., & Barton, J. K. (2013). Electrochemical assay for the signal-on detection of human DNA methyltransferase activity. *Journal of the American Chemical Society, 135*, 16632–16640.

Murphy, J. N., Cheng, A. K. H., Yu, H. Z., & Bizzotto, D. (2009). On the nature of DNA self-assembled monolayers on au: Measuring surface heterogeneity with electrochemical in situ fluorescence microscopy. *Journal of the American Chemical Society, 131*, 4042–4050.

New England BioLabs, Inc. (2017). *pH vs. temperature for Tris buffer.* Retrieved from https://www.neb.com/tools-and-resources/usage-guidelines/ph-vs-temperature-for-tris-buffer.

O'Brien, E., Holt, M. E., Thompson, M. K., Salay, L. E., Ehlinger, A. C., Chazin, W. J., et al. (2017). The [4Fe4S] cluster of human DNA primase functions as a redox switch using DNA charge transport. *Science, 355*, 813, eaag1789.

O'Neill, M. A., Becker, H.-C., Wan, C., Barton, J. K., & Zewail, A. H. (2003). Ultrafast dynamics in DNA-mediated electron transfer: Base gating and the role of temperature. *Angewandte Chemie, International Edition, 42*, 5896–5900.

Peterson, A. W., Heaton, R. J., & Georgiadis, R. M. (2001). The effect of surface probe density on DNA hybridization. *Nucleic Acids Research, 29*, 5163–5168.

Pheeney, C. G., Arnold, A. R., Grodick, M. A., & Barton, J. K. (2013). Multiplexed electrochemistry of DNA-bound metalloproteins. *Journal of the American Chemical Society, 135*, 11869–11878.

Pheeney, C. G., & Barton, J. K. (2012). DNA electrochemistry with tethered methylene blue. *Langmuir, 28*, 7063–7070.

Porello, S. L., Cannon, M. J., & David, S. S. (1998). A substrate recognition role for the $[4Fe\text{-}4S]^{2+}$ cluster of the DNA repair glycosylase MutY. *Biochemistry*, *37*, 6465–6475.
Rees, D. C., & Howard, J. B. (2003). The interface between the biological and inorganic worlds: Iron-sulfur metalloclusters. *Science*, *300*, 929–931.
Ren, B., Duan, X., & Ding, H. (2009). Redox control of the DNA damage-inducible protein DinG helicase activity via its iron-sulfur cluster. *The Journal of Biological Chemistry*, *284*, 4829–4835.
Romano, C. A., Sontz, P. A., & Barton, J. K. (2011). Mutants of the base excision repair glycosylase, endonuclease III: DNA charge transport as a first step in lesion detection. *Biochemistry*, *50*, 6133–6145.
Sam, M., Boon, E. M., Barton, J. K., Hill, M. G., & Spain, E. M. (2001). Morphology of 15-mer duplexes tethered to Au(111) probed using scanning probe microscopy. *Langmuir*, *17*, 5727–5730.
Sancar, A. (2003). Structure and function of DNA photolyase and cryptochrome blue-light photoreceptors. *Chemical Reviews*, *103*, 2203–2237.
Sauguet, L., Klinge, S., Perera, R. L., Maman, J. D., & Pellegrini, L. (2010). Shared active site architecture between the large subunit of eukaryotic primase and DNA photolyase. *PloS One*, *5*, e10083.
Shon, Y. S., Kelly, K. F., Halas, N. J., & Lee, T. R. (1999). Fullerene-terminated alkanethiolate SAMs on gold generated from unsymmetrical disulfides. *Langmuir*, *15*, 5329–5332.
Slinker, J. D., Muren, N. B., Gorodetsky, A. A., & Barton, J. K. (2010). Multiplexed DNA-modified electrodes. *Journal of the American Chemical Society*, *132*, 2769–2774.
Slinker, J. D., Muren, N. B., Renfrew, S. E., & Barton, J. K. (2011). DNA charge transport over 34 nm. *Nature Chemistry*, *3*, 228–233.
Sontz, P. A., Mui, T. P., Fuss, J. O., Tainer, J. A., & Barton, J. K. (2012). DNA charge transport as a first step in coordinating the detection of lesions by repair proteins. *Proceedings of the National Academy of Sciences of the United States of America*, *109*, 1856–1861.
Thayer, M. M., Ahern, H., Xing, D., Cunningham, R. P., & Tainer, J. A. (1995). Novel DNA binding motifs in the DNA repair enzyme endonuclease III crystal structure. *The EMBO Journal*, *14*, 4108–4120.
Ulman, A. (1996). Formation and structure of self-assembled monolayers. *Chemical Reviews*, *96*, 1533–1554.
Vaithiyalingam, S., Warren, E. M., Eichman, B. F., & Chazin, W. J. (2010). Insights into eukaryotic priming from the structure and functional interactions of the 4Fe-4S cluster domain of human DNA primase. *Proceedings of the National Academy of Sciences of the United States of America*, *107*, 13684–13689.
Walczak, M., Popenoe, D. D., Deinhammer, R. S., Lamp, B. D., Chung, C., & Porterm, M. D. (1991). Reductive desorption of alkanethiolate monolayers at gold: A measure of surface coverage. *Langmuir*, 7, 2687–2693.
Yavin, E., Boal, A. K., Stemp, E. D. A., Boon, E. M., Livingston, A. L., O'Shea, V. L., et al. (2005). Protein-DNA charge transport: Redox activation of a DNA repair protein by a guanine radical. *Proceedings of the National Academy of Sciences of the United States of America*, *102*, 3546–3551.
Yin, Y., Lu, Y., Wu, P., & Cai, C. (2005). Direct electrochemistry of redox proteins and enzymes promoted by carbon nanotubes. *Sensors*, *5*, 220–234.

CHAPTER SIXTEEN

Preparation of Stable Nitrogen Mustard DNA Interstrand Cross-Link Analogs for Biochemical and Cell Biological Studies

Alejandra Castaño*, Upasana Roy*, Orlando D. Schärer*,†,‡,1
*Stony Brook University, Stony Brook, NY, United States
†Center for Genomic Integrity, Institute for Basic Science, Ulsan, Korea
‡Ulsan National Institute of Science and Technology, Ulsan, Korea
[1]Corresponding author: e-mail address: orlando.scharer@ibs.re.kr

Contents

Abstract

Nitrogen mustards (NMs) react with two bases on opposite strands of a DNA duplex to form a covalent linkage, yielding adducts called DNA interstrand cross-links (ICLs). This prevents helix unwinding, blocking essential processes such as replication and transcription. Accumulation of ICLs causes cell death in rapidly dividing cells, especially cancer cells, making ICL-forming agents like NMs valuable in chemotherapy. However, the repair of ICLs can contribute to chemoresistance through a number of pathways that remain poorly understood. One of the impediments in studying NM ICL repair mechanisms has been the difficulty of generating site-specific and stable NM ICLs. Here, we describe two methods to synthesize stable NM ICL analogs that make it possible to study DNA ICL repair. As a proof of principle of the suitability of these NM ICLs for

Methods in Enzymology, Volume 591
ISSN 0076-6879
http://dx.doi.org/10.1016/bs.mie.2017.03.007

biochemical and cell biological studies, we use them in primer extension assays with Klenow polymerase. We show that the NM ICL analogs block the polymerase activity and remain intact under our experimental conditions.

1. INTRODUCTION

Bis-(2-chloroethyl)-amine derivatives or nitrogen mustards (NMs) are bifunctional alkylating agents widely used in the clinic to treat a variety of cancers (Chabner et al., 2005). The bifunctional nature of NMs allows for the formation of DNA intra- and interstrand cross-links (ICLs). It has been shown that although DNA ICLs make up only 1%–5% of the total adducts, they are responsible for the cytotoxicity as they provide a complete block to essential processes such as DNA replication and transcription (Noll, Mason, & Miller, 2006; Schärer, 2005). The ability of ICLs to induce apoptosis particularly in replicating cells provides some degree of selective cytotoxicity toward rapidly dividing cancer cells in a therapeutic setting (Deans & West, 2011). However, cellular resistance is often observed in patients treated with NMs, in large part due to ICL repair pathways that remove NM ICLs from the genome. Several ICL repair pathways exist, and the best understood of these is coupled to replication and has been biochemically reconstituted in *Xenopus* egg extracts (Knipscheer, Raschle, Schärer, & Walter, 2012; Räschle et al., 2008). Replication-coupled ICL repair makes use of genes involved in a number of pathways, including Fanconi anemia (FA), translesion DNA synthesis (TLS), homologous recombination (HR), and nucleotide excision repair (NER). The factors involved in this pathway recognize ICLs at stalled replication forks, remove the ICL, and eventually reestablish the replication fork. In another, less understood replication-coupled ICL repair pathway, the replication fork initially bypasses the intact ICL in a process known as replication traverse, presumably followed by ICL removal at a later stage (Huang et al., 2013). In addition, ICL repair is also known to occur outside of replication in an NER- and TLS-dependent manner, but the mechanistic details remain poorly understood as well (Clauson, Schärer, & Niedernhofer, 2013).

A major limitation in studying repair of NM ICLs has been the difficulty in generating substrates suitable for biochemical and cell biological studies. Two main challenges exist: first, treatment of a DNA duplex with NMs yields the desired ICLs only as a small fraction of products, with monoadducts and intrastrand cross-links making up the majority of adducts. This makes the isolation of sufficient amounts of ICLs difficult (Millard, Raucher, & Hopkins, 1990; Povirk & Shuker, 1994). Second, the alkylation of guanosine at N7 yields a positive charge on the purine ring (**1**, Fig. 1A), rendering the base prone to spontaneous depurination, leading to loss of the ICL, formation of abasic sites and eventually strand breaks (Fig. 1B; Gates, 2009).

To overcome these limitations, we have developed two different methods to generate stable analogs of NM ICLs that closely mimic the native NM ICL substrates (Fig. 1A). Our first approach is based on the previously synthesized 7-deazaguanine phosphoramidite (7CdG, **2**, Fig. 1A) precursors bearing masked diols that are easily incorporated into DNA oligomers by solid DNA synthesis (Angelov, Guainazzi, & Schärer, 2009; Guainazzi, Campbell, Angelov, Simmerling, & Schärer, 2010; Mukherjee, Guainazzi, & Schärer, 2014). After incorporation into 5′-GNC-3′ sequences, two complementary oligonucleotides containing 7CdG-residues

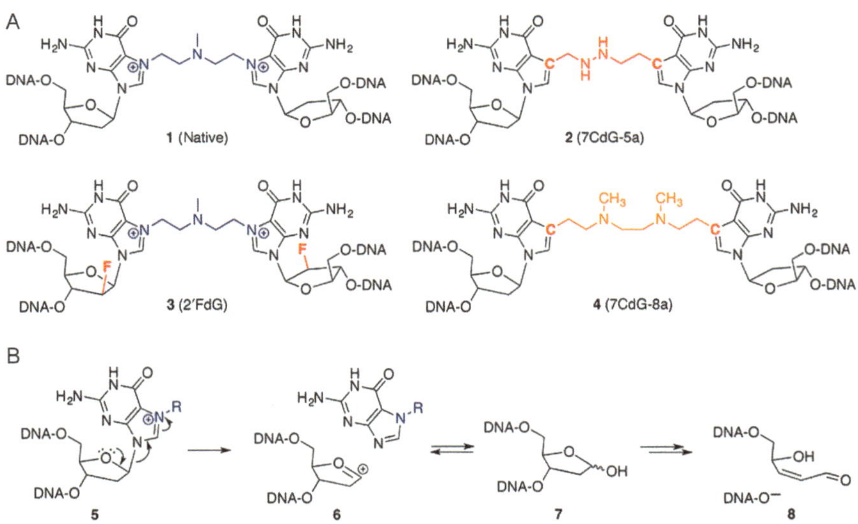

Fig. 1 Native and stable NM ICLs. (A) Nitrogen mustard (NM) interstrand cross-links (ICLs) formed by reaction of DNA with NM (**1**) and stable analogs in which the glycosidic bond is stabilized by using 7-deazaguanine bases (**2**, 7CdG-5a), to remove the positive charge in the purine ring or 2′-fluorodeoxyribose sugars (**3**, 2′FdG), to prevent depurination by destabilizing the positive charge formed during glycosidic bond cleavage. The approach to generate 7CdG ICL additionally allows for the synthesis of structurally diverse ICLs, such as **4** (7CdG-8a). (B) The positive charge in the native NM ICL destabilizes the glycosidic bond (**5**) and leads to depurination (**6**) resulting in the formation of an abasic site (**7**) and strand cleavage (**8**).

with alkyldiol side chains at the 7 position are annealed. The diols are oxidized to aldehydes with sodium periodate and coupled with hydrazine resulting in a high yield of the NM ICL mimic **2** (Figs. 1A and 2B). By using alkylaldehyde chains and diamines of different lengths, structurally diverse NM-like ICLs can be generated, for example, the 5-atom ICL (7CdG-5a, **2**), a close structural mimic of the native NM ICL, or the 8-atom ICL (7CdG-8a, **4**), that has a longer linkage and no distortion in the DNA (Fig. 1A).

In our second approach, NM ICLs are rendered resistant to glycosidic bond cleavage by incorporation of 2′-deoxy-2′-β-fluoroarabino guanosine (2′FdG, **3**, Fig. 1A) in the 5′-GNC-3′ sequences via solid-phase DNA synthesis. The electronegative fluorine atom at the 2′ carbon of the sugar base is strongly electron withdrawing and dramatically destabilizes the transient positive charge formed during the glycosidic bond cleavage reaction (**6**, Fig. 1B), thus de facto eliminating spontaneous depurination (Lee, Bowman, Ueno, Wang, & Verdine, 2008; Watts, Katolik, Viladoms, & Damha, 2009).

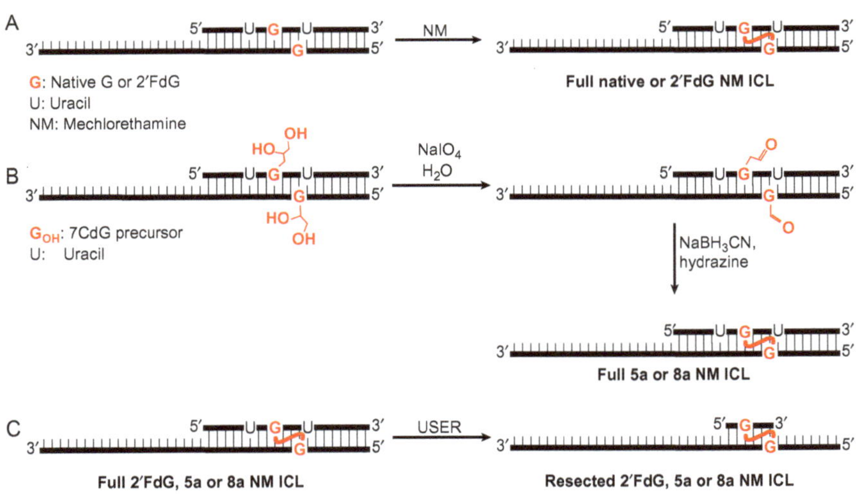

Fig. 2 Scheme for the formation of NM ICLs. (A) Native or 2′FdG NM ICLs are formed upon reaction of a duplex DNA containing a site-specific 5′-GNC-3′ sequence with excess mechlorethamine. (B) 5a or 8a 7CdG NM ICLs are formed by oxidation of the diol-bearing 7-deazaguanosines (incorporated within a 5′-GNC-3′ sequence in a duplex) with sodium periodate to the aldehydes followed by reductive amination using hydrazine or DMEDA in the presence of sodium cyanoborohydride. (C) Resected NM ICLs are produced by cleavage of uracil residues flanking the ICL with USER mix (UDG/EndoVIII).

The generation of various synthetic stable NM ICL analogs and the relative advantages and disadvantages of our approaches are described. We used these ICLs in primer extension assays with Klenow polymerase to assess the stability of the ICLs. Additionally, we describe a method to generate substrates in which the duplex around the ICL can be resected by incorporation of uracil residues in the oligonucleotide followed by enzymatic cleavage with UDG/EndoVIII (USER, Fig. 2C). These methods allow the synthesis of a panel of NM-like ICLs for the study of ICL repair mechanisms.

2. MATERIALS

2.1 Reagents and Buffers

- 1 m*M* DNA oligonucleotides in 1 × TE: 5′-CCCTCTUCTG* TCCUTCTTTC-3′ (20mer) and 5′-GAAAGAAGG*ACAGAAGA GGGTACCATCATAGAGTCAGTG-3′ (39mer, where G* represents dG, 2′FdG, or 7CdG) (Notes 2 and 4).
- Use ultrapure water 18 *M*Ω cm for the preparation of solutions and in all reactions. All the reagents are of analytical grade purity.
- 1 × TE (10 m*M* Tris–HCl, pH 7.4, 1 m*M* EDTA).
- 40 m*M* sodium cacodylate (pH 7): mix 0.86 g sodium cacodylate trihydrate (Sigma-Aldrich) in 70 mL water. Adjust pH to 7 by addition of 0.2 *M* HCl and bring final volume to 100 mL with Milli-Q water.
- 5 m*M* mechlorethamine hydrochloride 98% (5 m*M* NM): use fresh solution prior reaction with DNA by mixing 1 mg of mechlorethamine (Sigma-Aldrich) in 1 mL 40 m*M* sodium cacodylate pH 7 (Note 1).
- 50 m*M* sodium periodate solution: dissolve 0.1 g sodium metaperiodate in 10 mL ultrapure water. Store in the dark at 4°C.
- 0.5 *M* sodium cyanoborohydride solution: dissolve 31.0 mg sodium cyanoborohydride in 1 mL ultrapure water. Store in the dark at 4°C.
- 5 m*M* hydrazine solution: add 2.5 μL of 64%–65% hydrazine monohydrate solution (Sigma-Aldrich) to 10 mL ultrapure water. Store in the dark at 4°C.
- 5 m*M* DMEDA solution: add 5.4 μL dimethylethylenediamine (Sigma-Aldrich) to 10 mL ultrapure water. Store in the dark at 4°C.
- 1 *M* sodium phosphate buffer (pH 5.4).
- 5 m*M* sodium borate buffer (pH 8) for electroelution: mix 1.95 g sodium tetraborate in 800 mL water. Adjust pH to 8 with boric acid and bring up volume to 1 L with Milli-Q water.

- 9:1 MALDI-TOF MS matrix solution: mix 1 mL of 50 mg/mL 3-hydroxypicolinic acid (Protea Bioscience) in 50% acetonitrile/Milli-Q water and 1 mL of 50 mg/mL ammonium citrate (Sigma-Aldrich) in Milli-Q water. Store at 4°C in the dark.
- 50% (v/v) acetonitrile:water.
- 5 × Tris–borate–EDTA buffer (TBE): dissolve 54 g Tris base, 27.5 g boric acid, 20 mL 0.5 *M* EDTA in 800 mL ultrapure water. Make up volume to 1 L with ultrapure water (pH 8.0).
- 0.5 × TBE denaturing polyacrylamide gel electrophoresis (DPAGE) running buffer: dilute 100 mL 5 × TBE stock solution with 900 mL H_2O.
- 20% denaturing polyacrylamide gel electrophoresis (20% DPAGE) casting solution: stir vigorously 210.2 g ultrapure urea, 250 mL acrylamide/Bis 19:1, 40% (w/v) solution, 50 mL 5 × TBE in 400 mL ultrapure water. Once all components have dissolved bring volume up to 500 mL with ultrapure water.
- 0% denaturing polyacrylamide gel electrophoresis (0% DPAGE) casting solution: mix 210.2 g urea, 50 mL 5 × TBE in 400 mL ultrapure water. Once all components have dissolved bring volume up to 500 mL with ultrapure water.
- TEMED.
- 10% ammonium persulfate (APS) solution (mix 1 g APS in 10 mL H_2O, store at 4°C).
- 80% formamide/orange G loading buffer: mix 800 μL formamide, 0.5 mg orange G, and 200 μL H_2O). Loading buffer can be stored at room temperature.
- 80% formamide/xylene cyanol/bromophenol blue tracking buffer: mix 800 μL formamide, 0.5 mg xylene cyanol, 0.5 mg bromophenol blue, and 200 μL H_2O. Tracking buffer can be stored at room temperature.
- USER enzyme mix (NEB, M5505S).
- 1 × SYBR Gold solution (Life Technologies): dissolve 50 μL 10,000 × SYBR Gold into 500 mL 1 × TBE in amber plastic bottle. Store at 4°C.

2.2 Equipment and Consumables

- NanoDrop or UV spectrophotometer
- ZipTip c18 pipette tips (Millipore)
- MALDI plate (MTP Anchorchip, Bruker Daltonics)
- AutoFlex II MALDI-TOF mass spectrometer (Bruker Daltonics)

- FlexAnalysis 3.0 software
- Thermomixer compact (Eppendorf)
- Heat Block
- Elutrap™ device (Schleicher & Schuell)
- Bio-Trap membranes (BT1 glycerinized and BT2 dry, Whatman)
- Sub-Cell GT Horizontal Electrophoresis system, 15 × 25 cm tray (BioRad) to hold the Elutrap device
- Bench-top centrifuge (Sorvall Legend Mach 1.6R with swing bucket rotor)
- 0.5 mL and 1.5 mL microcentrifuge tubes
- 15 mL Falcon tubes
- Amicon Ultra-0.5 3 kDa MWCO (centrifugal filter device, Millipore)
- Amicon Ultra-4 3 kDa MWCO (centrifugal filter device, Millipore)
- BioRad PowerPac (400 W) power supply
- Electrophoresis tank (model V15.17 Whatman)
- V-series electrophoresis glass sandwich plates (Apogee): long (19.7 cm wide × 19.1 cm long) and short (19.7 cm wide × 16.0 cm long)
- 1.5 mm semi-prep comb (with a single and a long well), 1.5 mm spacers, 0.75 mm spacers, and analytical combs
- TLC glass plates with fluorescence indicator for UV shadowing (EMD/Millipore)
- Scalpel and tweezers
- Handheld UV lamp (254 nm)
- Typhoon 9400 fluoroimager (GE Healthcare)
- ImageQuant Software to analyze fluorescent images

3. METHODS

3.1 Preparation of NM ICL Analogs

3.1.1 Native and 2′FdG-Containing NM ICL Analogs

1. In a 1.5-mL microcentrifuge tube, mix 100 μL 1 m*M* 20mer 5′-CCCTCTUCTG*TCCUTCTTTC-3′ and 100 μL 1 m*M* 39mer 5′-GAAAGAAGG*ACAGAAGAGGGTACCATCATAGAGTCAGTG-3′ (where G* represents dG or 2′FdG) in 200 μL of 40 m*M* sodium cacodylate pH 7 (250 μ*M* final duplex concentration) (Note 2).
2. Heat this 250 μ*M* DNA solution at 95°C for 5 min in a preheated heat block. Switch off heating block and let it cool slowly to allow annealing of the oligos.

3. NM ICL reaction: add 3 equiv. of freshly made 5 m*M* NM solution (pH 7) to the annealed DNA solution (Note 1). Incubate the reaction mixture at 37°C for 3 h in a thermomixer in the dark.
4. The resulting NM ICLs (~7% yield for 2′FdG and ~2% for canonical dG) can be purified by DPAGE as described in Section 3.3 (Note 3).

3.1.2 7CdG-Containing NM ICL Analogs

1. In a 1.5-mL microcentrifuge tube, mix 25 nmol each of the 7CdG modified 20mer (5′-CCCTCTUCTG*TCCUTCTTTC-3′) and 39mer (5′-GAAAGAAGG*ACAGAAGAGGGTACCATCATAGAGTCAG TG-3′) where G* denotes the modified 7CdG, in 125 μL of 100 m*M* NaCl. Heat this mixture to 95°C for 5 min and cool slowly to allow the oligos to anneal (Note 4).
2. Add 10 μL of 50 m*M* sodium periodate solution and 15 μL 1 *M* sodium phosphate buffer (pH 5.4). Incubate this mixture at 4°C overnight in the dark and allow oxidation to occur.
3. Transfer the mixture to a Millipore Amicon column (3 K MWCO) and add 0.1 *M* sodium phosphate buffer (pH 5.4) up to a volume of 500 μL. Centrifuge at 11,000 rpm for 30 min.
4. Discard flow through and repeat twice.
5. Collect the final solution by inverting the Amicon column into a new collection tube. Centrifuge at 2000 rpm for 2 min.
6. Transfer oxidized oligos from the collection tube to a new 1.5-mL microcentrifuge tube.
7. Add 10 μL 5 m*M* aqueous solution of the amine (hydrazine or DMEDA) and 10 μL 0.5 *M* sodium cyanoborohydride. Incubate overnight in the dark at room temperature to allow the cross-linking reaction to take place.
8. The ICL formation can be analyzed by loading 5 pmol of the reaction on a 15% denaturing PAGE gel. The DNA can be visualized by SYBR Gold staining.
9. The ICL can be isolated and purified by denaturing PAGE (Section 3.3).

3.2 Resection of NM ICLs

1. In a 1.5-mL microcentrifuge tube, dilute 500 pmol of purified ICL in 80 μL ultrapure water. Add 10 μL 0.1 *M* Tris–Cl (pH 7.6) and 10 μL USER enzyme mix. Incubate at 37°C for 6 h.
2. The completion of digestion can be checked by loading an aliquot (~5 pmol) on an analytical denaturing PAGE and visualizing

the DNA by SYBR Gold staining (Section 3.3). Once the digestion is complete, the resected ICL is ready to be purified by denaturing PAGE.

3.3 Purification and Characterization of NM ICLs

1. Assemble the DPAGE sandwich using the long and short glass plates with 2 × 1.5 mm spacers and rest horizontally on a solid support (Note 5).
2. In a 100-mL beaker, prepare 60 mL 15% DPAGE by mixing 45 mL 20% DPAGE gel casting solution, 14.4 mL 0% DPAGE gel casting solution, 0.6 mL 10% APS, and 20 μL TEMED.
3. Using a serological pipette, cast the gel solution into glass sandwich without introducing bubbles. Place the 1.5 mm semi-prep comb into the sandwich preventing the formation of bubbles and allow the gel solution to polymerize for at least 40 min.
4. Suspend the NM ICL to be purified in an equal volume of 80% formamide/orange G buffer. Denature the NM ICL sample by heating the solution to 95°C for 5 min and place sample vial on ice immediately for 5 min or until ready to load.
5. Fix the polymerized 15% DPAGE sandwich gel vertically in electrophoresis tank (longer plate facing outward) and fill the buffer reservoirs with 0.5 × TBE running buffer to completely cover the wells. Carefully remove the comb and rinse the formed wells with running buffer. Attach the temperature probe to the outer gel glass and connect the tank to the PowerPac, setting it to 20 W and 50°C. Prerun the gel for 40 min or until the temperature reaches 50°C. Rinse the wells with 0.5 × TBE to remove excess urea.
6. Load the denatured DNA solution into the wide well and load tracking dye buffer to the single well.
7. Run the PowerPac at 20 W and 50°C until the bromophenol blue dye reaches the bottom of the glass plate and orange G dye exits the gel (Note 6).
8. Disconnect the power supply and the temperature probe, and dislodge the glass sandwich. With a plastic wedge, carefully separate the plates without breaking the gel.
9. Place the gel on top of a TLC plate covered with saran wrap.
10. In a dark room, hold the UV lamp directly above the gel (254 nm wavelength) and slice the resolved bands: NM ICL 59mer (top band), uncross-linked 39mer (middle band), and uncross-linked 20mer (bottom band) with a clean scalpel (Fig. 3A and C) (Note 7). Turn off UV lamp.

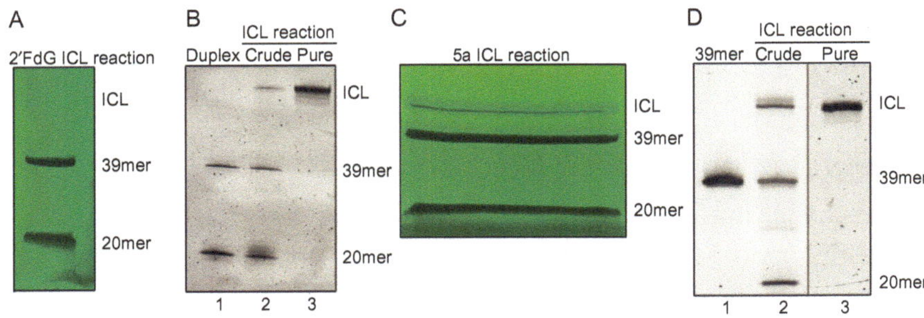

Fig. 3 Purification of stable NM ICL analogs. (A) Separation of a 25-nmol 2′FdG NM ICL reaction by 15% DPAGE and visualization by UV shadowing. (B) 3 pmol of the annealed duplex (lane 1), crude reaction mixture (lane 2), and the purified 2′FdG ICL (lane 3) were resolved by 15% denaturing PAGE and visualized by SYBR Gold staining. (C) Separation of 30 nmol of a 5a 7CdG NM ICL formation reaction by 10% denaturing PAGE and visualization by UV shadowing. (D) 5 pmol of the 39mer (lane 1), crude ICL reaction (lane 2), and the purified 5a ICL (lane 3) were resolved by 10% denaturing PAGE and visualized by SYBR Gold staining. The positions of the ICL, 39mer and 20mer single-stranded oligonucleotides, are indicated.

11. Cut the gel bands further into ~1 cm gel pieces and save in labeled Falcon tubes until next step.
12. DNA isolation from the gel is done via electroelution utilizing an Elutrap system. Rinse two BT1 glycerinized membranes with ultrapure water and mount tightly at the ends of the Elutrap device. Mount two dry BT2 membranes, ~0.8 cm (collection chamber) and ~1.6 cm (gel piece chamber), away from the BT1 membrane at the positive end of the trap.
13. Load the purified NM ICL gel pieces into the gel piece chamber and place the Elutrap device horizontally into the Sub-Cell GT electrophoresis tray. Fill all the reservoirs with 5 m*M* sodium borate buffer (pH 8) covering the gel pieces, the collection chamber and the outer electrophoresis tank.
14. Cover the electrophoresis tank and connect it to the power supply. Run at 200 V for 15 min. Pipette out the buffer in the collection chamber into a clean-labeled vial and place on ice.
15. Measure DNA content via NanoDrop and record the absorbance. Calculate the ng/μL NM ICL recovered.
16. Refill the collecting chamber with 5 m*M* sodium borate buffer and repeat steps 14–15 at least two more times until no DNA is detected in the collected fraction.

17. Pool all the saved fractions into a Amicon Ultra-3K 4 mL filter device and concentrate the DNA using the bench-top centrifuge at 3500 rpm, 4°C for 45 min. Buffer exchange with 1 × TE at least twice.
18. Pipette the NM ICL solution out of the Amicon column and save into a new labeled vial.
19. Take 2 μL of NM ICL solution and dilute with 48 μL water. Briefly vortex and measure UV absorbance via NanoDrop. Calculate the ng/μL DNA in both diluted sample and stock solution.
20. MALDI-TOF MS is used to verify the mass of the cross-link. Spot 1 μL of MALDI matrix onto a MALDI plate and let dry at room temperature.
21. ZipTip the diluted NM ICL sample (at least 10 pmol) following the manufacturer's protocol (Millipore).
22. Load 1 μL zip-tipped NM ICL on spot containing MALDI matrix and allow to dry at room temperature.
23. Mount the MALDI plate to MALDI carrier and insert it into the AutoFlex II MALDI-TOF mass spectrometer (previously calibrated with known high-molecular-weight oligo calibrants).
24. Measure the mass/charge ratio (m/z) in a linear negative mode using ion source acceleration voltage of 20.00 kV at a frequency of 50 Hz across a m/z of 7000–20,000 Da. To achieve a high signal-to-noise ratio, represent each spectrum integrating at least 600 individual laser shots.
25. The collected spectra are analyzed and visualized using the FlexAnalysis 3.0 software (Table 1).
26. To verify the purity of the NM ICL and to quantify any degraded products arising during purification steps, run a 15% analytical DPAGE gel. Use 0.75 mm spacers and a 14-well comb in the DPAGE sandwich. Always run the appropriate controls (e.g., the 20mer and 39mer used for original reaction) along with the NM ICL. Load 2 pmol in 10 μL 80% formamide/orange G loading dye (Note 6). Run the gel as described earlier.

Table 1 MALDI-TOF MS of Synthesized NM ICLs

Name	Calculated Mass $[M-H]^-$ (Da)	Experimental Mass $[M-H]^-$ (Da)	Error (%)
Full 5a deaza NM ICL	18179.9	18136.4	0.2
Full 8a deaza NM ICL	18194	18226.9	0.2
Full FANA NM ICL	18200.95	18251	0.3

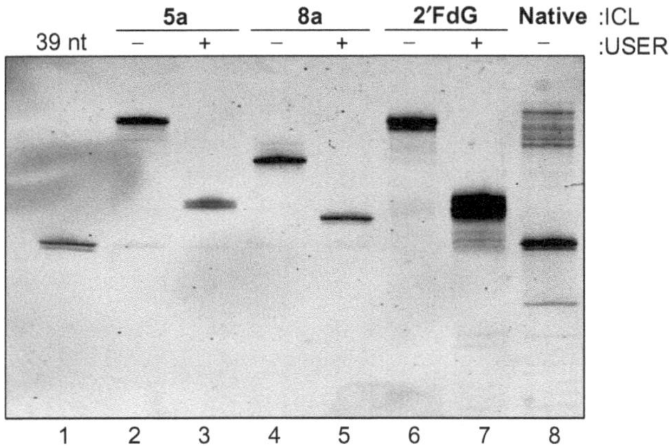

Fig. 4 Analysis of purified NM ICLs by 15% PAGE. 3 pmol 39mer (lane 1), NM ICL analogs (lanes 2–7), and native NM ICL (lane 8) were resolved by 15% denaturing PAGE and DNA visualized by SYBR Gold staining. The NM ICL analogs were either untreated (lanes 2, 4, 6) or treated with 0.1 U/μL USER mix for 6 h at 37°C to generate the resected forms of the ICL (lanes 3, 5, 7). Note that the native NM ICL decomposed during purification, isolation, and analysis, while the modified NM ICLs show no significant decomposition.

27. To visualize the gel via fluorescence detection, suspend the gel in a solution containing 1 × SYBR Gold for 30 min in the dark.
28. Place gel onto Typhoon fluoroimager plate and scan gel using the appropriate filter (Figs. 3B and D, and 4).
29. Analyze bands with ImageQuant software and calculate percentage of degradation (if any).

3.4 Primer Extension Assays With NM ICL Analogs

1. Dilute the purified NM ICL analogs and FAM-labeled primer P15 (5′-FAM-CACTGACTCTATGATG) to 1 μ*M* in 1 × TE.
2. Mix 7.5 μL 1 μ*M* ICL, 2.5 μL 1 μ*M* primer P15, 5 μL 10 × annealing buffer (100 m*M* Tris–Cl, pH 8, 500 m*M* NaCl), and 35 μL of ultrapure water. Incubate overnight at room temperature to allow annealing (Note 8).
3. For primer extension assays, mix 1 μL of the annealed mixture, 1 μL NEB2 buffer, 1 μL 1 m*M* dNTPs, and 6 μL ultrapure water.
4. Incubate at 37°C and add 1 μL 10 n*M* Klenow (*exo*-).
5. Incubate for 5 min at 37°C and then add 10 μL formamide loading buffer.
6. Heat to 95°C for 5 min and then snap chill on ice.
7. Load on a 10% DPAGE gel and visualize bands by scanning the gel using Typhoon fluoroimager (Fig. 5).

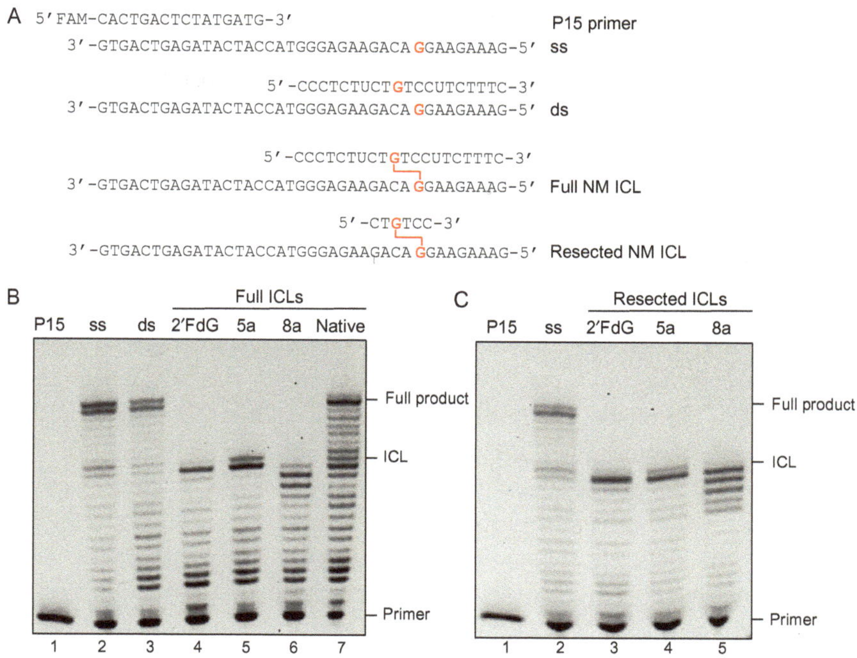

Fig. 5 NM ICL analogs block the polymerase reaction by Klenow (*exo-*) fragment. (A) Substrates used for primer extension assay. The modified base and cross-link are highlighted in *red*. (B) Primer extension assay of full length NM ICL templates with Klenow using noncross-linked controls (lanes 2 and 3), and full NM ICL-containing templates (lanes 4–7). Note that decomposition of the ICL in the native NM ICL allows for bypass, while the stable analogs block the polymerase. (C) Primer extension with resected NM ICLs. Primer (lane 1), noncross-linked control (lane 2), and resected NM ICLs (lanes 3–5). All substrates were annealed to the FAM-labeled primer P15 and incubated with 1 n*M* Klenow for 5 min at 37°C. Products were resolved by 10% denaturing PAGE.

4. NOTES

1. *Caution*: Mechlorethamine is a carcinogenic/mutagenic agent! Please refer to its MSDS for proper handling and disposal.
2. Native and 2′FdG-containing 20mer and 39mer DNA oligos (Integrated DNA Technologies) are ordered HPLC purified and diluted to 1 m*M* stock solutions in 1 × TE.
3. DNA mixture can be buffer exchanged with 1 × TE utilizing an Amicon Ultra-0.5 column (3 K MWCO) at 12 K rpm and stored at −20°C if the newly formed NM ICL is not purified immediately.

4. The 7CdG phosphoramidite building block needs to be synthesized as described or can be requested from the authors (Angelov et al., 2009; Guainazzi et al., 2010; Mukherjee et al., 2014). It is incorporated into the 20mer and 39mer oligos using Expedite DNA synthesizer. The final 7CdG containing oligos used for NM ICL reactions are deprotected by treatment with concentrated NH_4OH solution at 50°C for 12h and purified by Agilent BondElut C18 columns.
5. The sizes of the spacers and combs can vary depending on the amount of DNA to be purified.
6. For analytical gels, (0.75mm spacers) 30mL DPAGE solution is sufficient. Running times can be adjusted depending the DNA sizes to be resolved. Here, we are resolving a 20mer, 39mer, and 59mer, thus the bromophenol blue migrates to about 42bp in a 15% DPAGE. We use orange G in the loading buffer to visualize the loading. Orange G migrates faster than bromophenol blue thus it completely exits the gel.
7. Make sure to wear appropriate personal protection equipment (UV-resistant goggles, long sleeve lab coat, and nitrile gloves) when shining UV light to the gel. Clean the scalpel every time you slice the different types of gel bands to avoid contamination.
8. Heating of ICLs to 95°C should be avoided to preserve the stability of the ICL. All annealing reactions should be done overnight at room temperature.

5. SUMMARY AND CONCLUSIONS

This chapter outlines methods to synthesize stable NM ICL analogs for biochemical and cell biological studies. Formation of native NM ICLs by reaction of duplex oligonucleotide with NMs typically gives rise to a mixture of products with a very low yield of the desired ICL. Additionally, NM ICLs are hydrolytically unstable, as depurination of the positively charged base in a native NM ICL leads to loss of the ICL and strand breaks (Fig. 1B), thereby severely limiting their use for functional studies. We compared the stability of our modified analogs to that of native NM ICLs (Figs. 4 and 5) and showed that the modified ICLs are stable under the experimental conditions used, whereas the native NM ICLs are not. We describe two approaches to generate analogs that mitigate the limitations of native NM ICLs. In one, the glycosidic bond is stabilized by introduction of a fluorine substituent at the 2′ position (2′FdG) (**3**, Fig. 1A), which eliminates depurination, yielding a stable ICL during purification and polymerase assays (Figs. 4 and 5). Although the yield of ICL formation is low, the 2′FdG NM ICL can be synthesized from commercially available DNA oligonucleotides and NMs without the need to conduct organic synthesis. Importantly, it contains a positively charged purine ring and only differs from the native ICL by addition of a fluorine atom at the 2′ position in the β-orientation, which has been shown to only minimally affect DNA structure (Martin-Pintado et al., 2012). The 7CdG NM ICL requires the synthesis of a specific phosphoramidite precursor (Angelov et al., 2009; Guainazzi et al., 2010), but can be used in any sequence context to generate a site-specific, stable NM ICL analog in a high-yielding cross-linking reaction. This approach can furthermore be modified to produce structural variants such as NM ICLs that contain cross-links of various lengths, allowing for the synthesis of ICLs of different structures for functional studies (Angelov et al., 2009; Mukherjee et al., 2014; Roy, Mukherjee, Sharma, Frank, & Schärer, 2016). 7CdG NM analogs have also been successfully incorporated into substrates for biochemical and cell-based DNA repair studies (Ho, Guainazzi, Derkunt, Enoiu, & Schärer, 2011; Hodskinson et al., 2014; Pizzolato, Mukherjee, Schärer, & Jiricny, 2015; Räschle et al., 2008; Roy et al., 2016). A possible drawback of the 7CdG NM ICLs is that there are three substitutions in the cross-link compared to the native NM ICL, and the cross-linked purines are not positively charged. Although our modeling studies have shown that this induces only minor effects on the DNA structure (Guainazzi et al., 2010), the main stalling point of Klenow at the 2′FdG ICL and 7CdG ICL differs by one nucleotide (compare Fig. 5B, lanes 4 and 5), indicating that the charge may influence interaction with proteins. Thus, we anticipate that the two types of modified NM ICLs will each be useful for specific applications.

In summary, we describe the generation of two types of stable NM ICL analogs that have been and will continue to be of use to the scientific community to study biological pathways triggered by ICLs.

REFERENCES

Angelov, T., Guainazzi, A., & Schärer, O. D. (2009). Generation of DNA interstrand cross-links by post-synthetic reductive amination. *Organic Letters*, *11*(3), 661–664. http://dx.doi.org/10.1021/ol802719a.

Chabner, B. A., Amrein, P. C., Druker, B. J., Michaelson, M. D., Mitsiades, C. S., Goss, P. E., et al. (2005). Antineoplastic agents. In L. L. Brunton, J. S. Lazo, & K. L. Parker (Eds.), *Goodman & Gilman's the pharmacological basis of therapeutics* (pp. 1315–1403). New York: McGraw-Hill.

Clauson, C., Schärer, O. D., & Niedernhofer, L. (2013). Advances in understanding the complex mechanisms of DNA interstrand cross-link repair. *Cold Spring Harbor Perspectives in Medicine*, *3*(10), a012732.

Deans, A. J., & West, S. C. (2011). DNA interstrand crosslink repair and cancer. *Nature Reviews. Cancer*, *11*(7), 467–480. http://dx.doi.org/10.1038/nrc3088.

Gates, K. S. (2009). An overview of chemical processes that damage cellular DNA: Spontaneous hydrolysis, alkylation, and reactions with radicals. *Chemical Research in Toxicology*, *22*(11), 1747–1760. http://dx.doi.org/10.1021/tx900242k.

Guainazzi, A., Campbell, A. J., Angelov, T., Simmerling, C., & Schärer, O. D. (2010). Synthesis and molecular modeling of a nitrogen mustard DNA interstrand crosslink. *Chemistry*, *16*(40), 12100–12103. http://dx.doi.org/10.1002/chem.201002041.

Ho, T. V., Guainazzi, A., Derkunt, S. B., Enoiu, M., & Schärer, O. D. (2011). Structure-dependent bypass of DNA interstrand crosslinks by translesion synthesis polymerases. *Nucleic Acids Research*, *39*(17), 7455–7464. http://dx.doi.org/10.1093/nar/gkr448.

Hodskinson, M. R., Silhan, J., Crossan, G. P., Garaycoechea, J. I., Mukherjee, S., Johnson, C. M., et al. (2014). Mouse SLX4 is a tumor suppressor that stimulates the activity of the nuclease XPF-ERCC1 in DNA crosslink repair. *Molecular Cell*, *54*(3), 472–484. http://dx.doi.org/10.1016/j.molcel.2014.03.014.

Huang, J., Liu, S., Bellani, M. A., Thazhathveetil, A. K., Ling, C., de Winter, J. P., et al. (2013). The DNA translocase FANCM/MHF promotes replication traverse of DNA interstrand crosslinks. *Molecular Cell*, *52*(3), 434–446. http://dx.doi.org/10.1016/j.molcel.2013.09.021.

Knipscheer, P., Raschle, M., Schärer, O. D., & Walter, J. C. (2012). Replication-coupled DNA interstrand cross-link repair in Xenopus egg extracts. *Methods in Molecular Biology*, *920*, 221–243. http://dx.doi.org/10.1007/978-1-61779-998-3_16.

Lee, S., Bowman, B. R., Ueno, Y., Wang, S., & Verdine, G. L. (2008). Synthesis and structure of duplex DNA containing the genotoxic nucleobase lesion N7-methylguanine. *Journal of the American Chemical Society*, *130*(35), 11570–11571. http://dx.doi.org/10.1021/ja8025328.

Martin-Pintado, N., Yahyaee-Anzahaee, M., Campos-Olivas, R., Noronha, A. M., Wilds, C. J., Damha, M. J., et al. (2012). The solution structure of double helical arabino nucleic acids (ANA and 2′F-ANA): Effect of arabinoses in duplex-hairpin interconversion. *Nucleic Acids Research*, *40*(18), 9329–9339. http://dx.doi.org/10.1093/nar/gks672.

Millard, J. T., Raucher, S., & Hopkins, P. B. (1990). Mechlorethamine cross-links deoxyguanosine residues at 5′-GNC sequences in duplex DNA fragments. *Journal of the American Chemical Society*, *112*(6), 2459–2460. http://dx.doi.org/10.1021/ja00162a079.

Mukherjee, S., Guainazzi, A., & Schärer, O. D. (2014). Synthesis of structurally diverse major groove DNA interstrand crosslinks using three different aldehyde precursors. *Nucleic Acids Research*, *42*(11), 7429–7435. http://dx.doi.org/10.1093/nar/gku328.

Noll, D. M., Mason, T. M., & Miller, P. S. (2006). Formation and repair of interstrand cross-links in DNA. *Chemical Reviews*, *106*(2), 277–301. http://dx.doi.org/10.1021/cr040478b.

Pizzolato, J., Mukherjee, S., Schärer, O. D., & Jiricny, J. (2015). FANCD2-associated nuclease 1, but not exonuclease 1 or flap endonuclease 1, is able to unhook DNA interstrand cross-links in vitro. *The Journal of Biological Chemistry*, *290*(37), 22602–22611. http://dx.doi.org/10.1074/jbc.M115.663666.

Povirk, L. F., & Shuker, D. E. (1994). DNA damage and mutagenesis induced by nitrogen mustards. *Mutation Research*, *318*(3), 205–226.

Räschle, M., Knipscheer, P., Enoiu, M., Angelov, T., Sun, J., Griffith, J. D., et al. (2008). Mechanism of replication-coupled DNA interstrand crosslink repair. *Cell*, *134*(6), 969–980. http://dx.doi.org/10.1016/j.cell.2008.08.030.

Roy, U., Mukherjee, S., Sharma, A., Frank, E. G., & Schärer, O. D. (2016). The structure and duplex context of DNA interstrand crosslinks affects the activity of DNA polymerase eta. *Nucleic Acids Research*, *44*(15), 7281–7291. http://dx.doi.org/10.1093/nar/gkw485.

Schärer, O. D. (2005). DNA interstrand crosslinks: Natural and drug-induced DNA adducts that induce unique cellular responses. *Chembiochem*, *6*(1), 27–32. http://dx.doi.org/10.1002/cbic.200400287.

Watts, J. K., Katolik, A., Viladoms, J., & Damha, M. J. (2009). Studies on the hydrolytic stability of 2′-fluoroarabinonucleic acid (2′F-ANA). *Organic & Biomolecular Chemistry*, 7(9), 1904–1910. http://dx.doi.org/10.1039/b900443b.

AUTHOR INDEX

Note: Page numbers followed by "*f*" indicate figures, "*t*" indicate tables, and "*np*" indicate footnotes.

A

B

D

E

H

I

J

M

P

Q

R

S

T

U

V

Z

SUBJECT INDEX

Note: Page numbers followed by "*f*" indicate figures, and "*t*" indicate tables.

L

M

N

O

FSC
www.fsc.org
MIX
Papier | Fördert gute Waldnutzung
FSC® C083411

Kolibri 360 GmbH
Ferdinand-Jühlke-Straße 7
99095 Erfurt, Deutschland
produktsicherheit@kolibri360.de